Advances in Experimental Medicine and Biology

Volume 1497

Advances in Experimental Medicine and Biology provides a platform for scientific contributions in the main disciplines of the biomedicine and the life sciences. This series publishes thematic volumes on contemporary research in the areas of microbiology, immunology, neurosciences, biochemistry, biomedical engineering, genetics, physiology, and cancer research. Covering emerging topics and techniques in basic and clinical science, it brings together clinicians and researchers from various fields.

Advances in Experimental Medicine and Biology has been publishing exceptional works in the field for over 40 years, and is indexed in SCOPUS, Medline (PubMed), EMBASE, Reaxys, EMBiology, the Chemical Abstracts Service (CAS), and Pathway Studio.

2023 CiteScore: 5.9

Avia Rosenhouse-Dantsker
Editor

Decoding Ion Channels Structure and Function

Computational and Experimental Studies

Springer

Editor
Avia Rosenhouse-Dantsker
Department of Chemistry
University of Illinois Chicago
Chicago, IL, USA

ISSN 0065-2598 ISSN 2214-8019 (electronic)
Advances in Experimental Medicine and Biology
ISBN 978-3-032-07522-2 ISBN 978-3-032-07523-9 (eBook)
https://doi.org/10.1007/978-3-032-07523-9

This Springer imprint is published by the registered company Springer Nature Switzerland AG
The registered company address is: Gewerbestrasse 11, 6330 Cham, Switzerland

In Memoriam: Sergei Y. Noskov

Over the span of his much too brief scientific career, Sergei Noskov displayed extraordinary intellectual energy and versatility. His background was fundamental physical chemistry. During his PhD thesis at the Institute of Solution Chemistry of the Russian Academy of Sciences in Ivanovo, he worked on statistical mechanical questions about liquid mixtures under the joint mentorship of Professor Arkadiy Kolker and Professor Mikhail Kiselev, including a collaboration with Professor Bernd Michael Rode at the Institute of Theoretical Chemistry at the University of Innsbruck in Austria. Even at this early stage, Sergei demonstrated his remarkable ability to seamlessly navigate collaborations with multiple researchers.

After his successful PhD, Sergei left to join the lab of Professor Carmay Lim at Taipei University in Taiwan as a postdoctoral researcher, where he worked on protein-protein interactions and on the effects of mutations within the core domain of the DNA-binding tumor suppressor p53 protein. After a few years, he joined my lab at Cornell Weill Medical College in New York City as a postdoctoral researcher. He first worked on Brownian Dynamics simulations of ion conduction through the large pore formed by the alpha-hemolysin toxin in collaboration with Wonpil Im, and on the molecular basis of ion selectivity in the KcsA potassium channel. For a long time, the classical explanation for the high selectivity of potassium channels was that the protein ligands in the narrow pore had the perfect geometry to favorably coordinate K^+ ions but were too distant for the smaller Na^+ ions. Inspired from host-guest chemistry, where ion-binding specificity to fairly rigid chelator molecules is observed, this perspective is referred to as the "snug fit" mechanism. Suddenly, the availability of a high-resolution crystallographic structure of a potassium channel allowed us to put this idea to the test. Rapidly, Sergei's molecular dynamics simulations of the KcsA potassium channel showed that the fluctuations of the carbonyl ligands lining the selectivity filter were, in fact, larger than the size difference between K^+ and Na^+, undermining the "snug fit" explanation based on a selectivity filter with a precise geometry. I still have most of our email exchanges from that exciting time when we were writing our first paper on the subject. Soon after, I moved my lab to the University of Chicago, and Sergei stayed with me to continue testing our ideas about ion selectivity in a variety of systems. Our work generated a fair amount of controversy, but Sergei always remained calm, open minded, and thoughtful throughout this tumultuous period.

In 2007, Sergei took an independent position as assistant professor at the Institute for Biocomplexity and Informatics in the Department of Biological Sciences at the University of Calgary in Canada. There, he found a stimulating environment and excellent colleagues in computational chemistry, such as Peter Tieleman and Dennis Salahub, as well as many exciting new colleagues in biology and physiology. He also acclimated very well to the Canadian culture, being already a huge hockey fan. Rapidly, his productivity exploded in many directions in the following years, extending from computational biology to the innovative application of machine learning in metabolomics data analysis.

Being in an independent position illuminated a striking aspect of Sergei's character, which was his great collegiality and ability to collaborate with other scientists on a very broad range of subjects. A complete list of all collaborators would be too long, but in about a decade he worked and published on a whole host of subjects, including potassium channels, multiscale modeling of cardiac arrhythmias, drug design, DNA diffusion in nanopores, and force field developments. In doing so, Sergei collaborated with an impressive number of prominent scientists (e.g., Peter Larsson, Colleen Clancy, Louis De Felice, Sergey Bezrukov, Lei Shi, Diomedes Logothetis, Avia Rosenhouse-Dantsker, Tatiana Rostovtseva, Ira Kurz, Serdar Durdagi, Vladimir Yarov-Yarovoy, Harel Weinstein, Siewert-Jan Marrink, and many more). These books in memory of Sergei Noskov are a testimony to the large footprint of his work in computational biophysics. I am not sure I know anyone in computational biophysics who has had as much reach and contact within such a diverse community of scientists in such a short time. Some of his most persistent efforts were dedicated to understanding the hERG potassium channel, its impact on cardiology, and the mechanism of drug binding and blockade.

Sergei's unwavering passion for science and dedication to nurturing future generations of scientists exemplified his commitment to advancing research and prioritizing collective progress over personal accolades. While some young scientists are careful to mark their territory and avoid working with former mentors, Sergei was not the least encumbered by this. He published an article with his thesis advisor, Professor Arkadiy Kolker, in 2016; one with his first postdoctoral advisor, Carmay Lim, in 2020; and even one with me in 2021. May his legacy of excellence, intellectual honesty, collegiality, and enthusiasm continue to inspire future generations of scientists, as we honor the lasting impact he had on countless members of the community.

University of Chicago, Chicago, IL, USA Benoit Roux

Preface

Unraveling the intricate mechanisms underlying ion channel structure and function requires a multifaceted approach. In this regard, the synergistic integration of experimental and computational approaches is exceptionally potent. However, it wasn't until the past decade that computational tools garnered sufficient power to play a prominent role alongside experimental techniques. Accordingly, this volume highlights contemporary experimental and computational methods in the study of ion channels, providing examples that illustrate how these approaches advance our understanding of structure and function across a diverse range of channel types.

Specifically, the volume includes two parts. The first part presents computational and experimental approaches that are used in ion channel research and lead to profound insights into complex phenomena in this field. Among the computational approaches discussed are a theoretical framework aimed at clarifying the role of architectural and structural forces in ion selectivity as well as advances in molecular dynamics simulations that seek to elucidate ion channel gating. Among the experimental approaches are structural techniques, such as crystallography and cryo-electron microscopy, fluorescence-based techniques, including single-molecule total internal reflection fluorescence (TIRF) microscopy and fluorescence spectroscopy, and mutagenesis-centered integrative approaches. The second part explores the structure and function of various channel types, including the voltage-gated hERG (Kv11.1) potassium channel, two-pore domain potassium channels, the ryanodine receptor intracellular calcium channel, transient receptor potential vanilloid (TRPV) channels, pentameric ligand-gated ion channels, and aquaporin ion channels.

In summary, this collection of chapters highlights cutting-edge approaches for studying ion channel structure and function, presenting methodological advances with broader applications in the field as well as targeted insights into different ion channels. The book is dedicated to the memory of Prof. Sergei Y. Noskov who focused his research on computational and theoretical techniques and actively engaged in numerous collaborations, integrating computational and experimental approaches to study ion channels. Accordingly, this volume emphasizes the application of this interdisciplinary strategy within the ion channel field.

This book is one of two volumes dedicated to the memory of Prof. Noskov. The companion volume (AEMB 1496) is titled *Unlocking Protein*

Structure-Function Relationships: Computational and Experimental Approaches and Applications. Together, the two volumes (1496, 1497) offer a comprehensive collection of chapters on computational and experimental methods for studying the structure-function relationship of proteins with a diverse array of applications, including those from the ion channel field.

The editor is deeply thankful to all the authors who brought this project to life through their enthusiasm, excitement, and dedication. The editor would also like to thank the anonymous book proposal reviewers for their invaluable suggestions at the very beginning of this project. Last but not least, the editor is grateful to her collaborators and colleagues for providing a nurturing environment that supported the completion of this collection.

Chicago, IL, USA Avia Rosenhouse-Dantsker

Contents

Part I

Advances in Computational and Experimental Approaches for Ion Channel Research

1 The Role of Architectural Forces in Ion Selectivity

Benoît Roux

Abstract

A theoretical framework is presented to clarify the role of architectural and structural forces in ion selectivity. It expresses the relative free energy of bound ions in terms of a reduced subsystem corresponding to the local degrees of freedom coupled to the rest of the protein. The latter is separated into a first contribution that includes all the forces keeping the ion and the coordinating ligands confined to a microscopic sub-volume but do not prevent the ligands from adapting to a smaller ion, while the second contribution regroups the remaining forces that serve to dictate the precise geometry of the coordinating ligands best adapted to a given ion. The theoretical framework makes it possible to delineate two important limiting cases. In the limit where the geometric forces are dominant (rigid binding site), selectivity is controlled via the cavity size according to the familiar "snug-fit" mechanism of host-guest chemistry. In the limit where the geometric forces are negligible, the ion and ligands behave as a dynamical "confined droplet" that is free and adapt to the ion's size. In this case, selectivity is controlled by the number and the chemical type of ion-coordinating ligands.

Keywords

Binding site · Free energy · Solvation · Structural rigidity · Structural flexibility · Ion coordination

1.1 Introduction

Understanding the microscopic mechanisms by which small ions such as Na^+ and K^+ associate with high specificity to protein-binding sites has fascinated scientists for decades. Because these ions are strongly bound to water molecules in bulk solutions, protein-binding sites must provide coordinating groups capable of compensating for the loss of hydration. Selectivity arises when this energetic compensation is more favorable for one ion type than another, relative to their difference in hydration free energy. Simple monovalent cations like K^+ and Na^+ play critical roles in physiological processes. The remarkable discrimination some proteins achieve between these two ions, differing only slightly in atomic radii (about 0.38 Å), remains a profound puzzle.

Several mechanisms have been proposed to explain ion selectivity. Mullins suggested that

B. Roux (✉)
Department of Biochemistry and Molecular Biology, University of Chicago, Chicago, IL, USA
e-mail: roux@uchicago.edu

A. Rosenhouse-Dantsker (ed.), *Decoding Ion Channels Structure and Function*, Advances in Experimental Medicine and Biology 1497, https://doi.org/10.1007/978-3-032-07523-9_1

selectivity is predominantly controlled by the number of ligands coordinating the ion (Mullins 1959). Eisenman emphasized the importance of ion-ligand interactions characterized by the ligand dipole properties via the concept of “field strength” (Eisenman 1962; Szabo et al. 1973). Another compelling explanation is the “snug-fit” mechanism (Armstrong and Bezanilla 1973), which posits that protein-binding sites are structurally optimized to form a cavity of precise size, favoring one ion while excluding others. This concept aligns with host-guest chemistry principles (Andreoli et al. 1967; Mueller and Rudin 1967; Pedersen and Frensdor 1972; Michaux and Reisse 1982; Gokel et al. 1983; Dietrich 1985; Schneider et al. 2008). The determination of the structure of the KcsA channel to atomic resolution by X-ray crystallography (Doyle et al. 1998; Zhou et al. 2001) reinforced the snug-fit argument (Hille et al. 1999; Yellen 2002; Berg et al. 2002; Berg 2003; Garofoli and Jordan 2003; Beckstein et al. 2003; MacKinnon 2004; Ban et al. 2004; Gouaux and Mackinnon 2005). The availability of high-resolution X-ray structure of membrane proteins with highly selective ion-binding site offers a great opportunity to advance our understanding of ion selectivity in those systems using detailed computational methods. Despite their approximation nature, the atomic models used in molecular dynamics (MD) simulations appear to be sufficiently accurate to gain mechanistic insights about ion selectivity.

The goal with the present monograph is to review the theoretical and computational work carried out in collaboration with Sergei Y. Noskov over several years to address the fundamental aspects of ion-binding selectivity (Bernèche and Roux 2001; Noskov et al. 2004, 2006; Noskov and Roux 2007, 2008; Yu et al. 2009, 2010; Roux et al. 2011). A statistical mechanical framework shedding light on the existence of two distinct limiting mechanisms giving rise to ion selectivity is presented (Yu et al. 2010), and illustrated with application to the ion-selective binding sites of the KcsA channel (Doyle et al. 1998; Zhou et al. 2001) and the LeuT transporter (Yamashita et al. 2005).

1.2 Theoretical Analysis and Results

1.2.1 Statistical Mechanical Reduction

We consider an ion-binding site in a protein that is in equilibrium with bulk solvent. Selectivity is governed by the relative free energy of ions i and j

$$\Delta\Delta G_{ij} = \Delta G_{ij}^{\text{site}} - \Delta G_{ij}^{\text{bulk}}, \qquad (1.1)$$

where $\Delta G_{ij}^{\text{bulk}} = [G_i^{\text{bulk}} - G_j^{\text{bulk}}]$ is the free energy difference between ions i and j in the bulk solvent, and $\Delta G_{ij}^{\text{site}} = [G_i^{\text{site}} - G_j^{\text{site}}]$ is the free energy difference between ions i and j in the binding site. The relative free energy of ions i and j in the binding site, which controls the binding selectivity, can be written in terms of configurational integrals,

$$e^{-\beta\Delta G_{ij}^{\text{site}}} = \frac{\int_{\text{site}} d\mathbf{R}\; e^{-\beta U_i(\mathbf{R})}}{\int_{\text{site}} d\mathbf{R}\; e^{-\beta U_j(\mathbf{R})}}, \qquad (1.2)$$

where $\mathbf{R}$ represents all degrees of freedom in the system, $\beta = 1/k_{\text{B}}T$, and U_i and U_j represent the total potential energy with ion i or j, respectively.

While Eq. (1.2) can be employed to carry out free energy computations from all-atom MD simulations, identifying direct causal relationships in such complex systems remains challenging. For instance, strong ion-ligand interactions are balanced by numerous long-range interactions, which arise from the structural packing of atoms that confer stability to the three-dimensional protein configuration. Those long-range interactions may, or may not, be sensitive to the type of ion bound in the site. To make progress, we seek to reduce the complexity of Eq. (1.2) by integrating out all the degrees of freedom that are not immediately relevant to local interactions. To this end, we distinguish a reduced subsystem comprising only the bound ion and the n most important ligands that are directly participating in the binding site. By ligands we mean to designate the molecular groups coordinating the ion directly, together with the nearest covalently at-

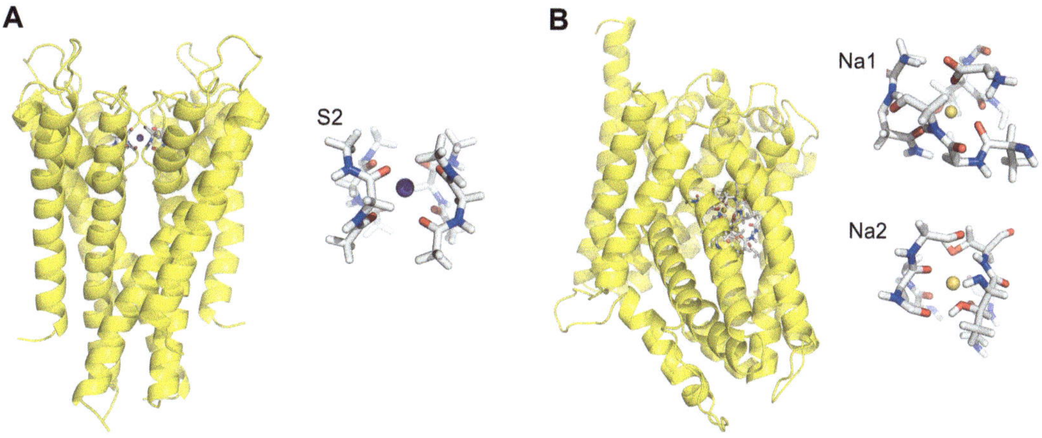

Fig. 1.1 Two membrane proteins with selective ion-binding sites are shown, the KcsA K^+ channel in A, and the leucine transporter LeuT in B. In A, the full KcsA structure (PDB code: 1K4C) is shown in yellow ribbon, with the central K^+-selective binding site S2 formed by the backbone carbonyl of Val76 and Gly77 chosen to construct the reduced model consisting of four diglycine molecules. In B, the full LeuT structure (PDB code: 2A69) is shown in yellow ribbon, with the atoms incorporated into the reduced models of the two Na^+-selective binding sites Na1 and Na2. Reproduced from reference (Yu et al. 2010)

tached atoms. Reduced subsystems for the KcsA channel and the LeuT transporter are illustrated in Fig. 1.1. We represent the coordinates of the local "inner" subsystem by **X**, and the remaining coordinates of the "outer" subsystem by **Y**. Equation (1.2) is reexpressed as

$$e^{-\beta\Delta G_{ij}^{\text{site}}} = \frac{\int_{\text{site}} d\mathbf{X}\, d\mathbf{Y}\; e^{-\beta U_i(\mathbf{X},\mathbf{Y})}}{\int_{\text{site}} d\mathbf{X}\, d\mathbf{Y}\; e^{-\beta U_j(\mathbf{X},\mathbf{Y})}} = \frac{\int_{\text{site}} d\mathbf{X}\; e^{-\beta[U_i^{\text{il}}(\mathbf{X})+U^{\text{ll}}(\mathbf{X})+\Delta W^{\text{site}}(\mathbf{X})]}}{\int_{\text{site}} d\mathbf{X}\; e^{-\beta[U_j^{\text{il}}(\mathbf{X})+U^{\text{ll}}(\mathbf{X})+\Delta W^{\text{site}}(\mathbf{X})]}}, \tag{1.3}$$

where U^{il} and U^{ll} represent the ion-ligand and ligand-ligand interactions in the inner region, respectively, and $\Delta W^{\text{site}}(\mathbf{X})$ is an effective potential that incorporates all the influence of the atoms in the outer region of the system (protein, membrane, solvent). The statistical mechanical reduction is achieved by introducing a fixed assignment of the participating molecular moieties. Effectively, no spatial restriction is actually imposed on the ion or ligands in the reduced model, and in that sense, the separation between the "inner" and "outer" regions is flexible.

By definition, the effective potential of mean force ΔW^{site} in Eq. (1.3) is exact and rigorously incorporates all influence of the three-dimensional protein structure on the ion and ligands. In other words, ΔW^{site} is formally associated with all so-called architectural forces arising from the protein. It can be safely assumed that, for a fixed configuration **X** for the ion and its coordinating ligand in the reduced subsystem, $\Delta W^{\text{site}}(\mathbf{X})$ does not depend explicitly on the ion type. This is a reasonable assumption because the bound ion in the fixed configuration **X** is surrounded by the coordinating ligands and is not in direct contact with the remaining atoms of the outer region. Because the outer atoms are some distance away from the bound ion, their interactions with the bound ion is mainly electrostatic. Those long-range interactions are the same as long as an ion of the same charge is bound in the reduced subsystem. For example, MD simulations indicate that ΔW^{site} for Na^+ and K^+ bound to the site S2 in the KcsA channel differ by less than 0.2 kcal/mol.

At a conceptual level, it is useful to separate the architectural forces acting on the reduced subsystem into two distinct contributions, which we will call "confinement" and "geomet-

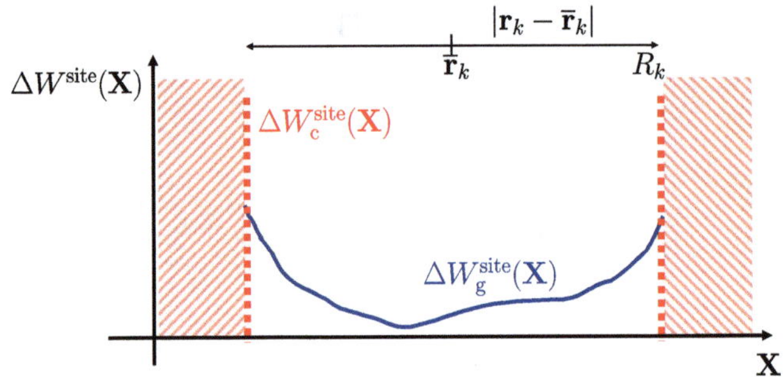

Fig. 1.2 Schematic representation of the character of the architectural forces from the protein structure, including the "confinement" contribution $\Delta W_{\mathrm{c}}^{\mathrm{site}}$ (red-shaded regions) imposing an upper bound on the fluctuations of the bound ion and its coordinating ligands, and the weaker "geometric" contribution $\Delta W_{\mathrm{g}}^{\mathrm{site}}$ (blue line) yielding a precise position to the bound ion and its coordinating ligands. The total potential of mean force (PMF) is $\Delta W^{\mathrm{site}} = \Delta W_{\mathrm{c}}^{\mathrm{site}} + \Delta W_{\mathrm{g}}^{\mathrm{site}}$

ric," $\Delta W^{\mathrm{site}} = \Delta W_{\mathrm{c}}^{\mathrm{site}} + \Delta W_{\mathrm{g}}^{\mathrm{site}}$. The first contribution, $\Delta W_{\mathrm{c}}^{\mathrm{site}}$, accounts for the generic effect of the protein structure surrounding the reduced subsystem imposing an upper bound on the fluctuations of the bound ion and its coordinating ligands. Without the surrounding protein structure, the atoms of the reduced subsystem would flee away and the binding site would not maintain its integrity. By virtue of the surrounding protein structure, the instantaneous atomic fluctuations of the atoms within the reduced subsystem are bounded and can never exceed some maximum value. This is the generic effect of architectural confinement.

Different models of the confinement can be constructed. For simplicity, for each atom k in the reduced subsystem, we define the architectural confinement as the smallest spherical volume encompassing all the dynamic excursions of that atom, independently of the type of ion that is bound in the site. Each spherical volume is parametrically defined by the position of the center $\overline{\mathbf{r}_k}$ and radius R_k; those parameters can be extracted from MD simulations of the full system. Accordingly, the architectural confinement contribution is defined as

$$e^{-\beta \Delta W_{\mathrm{c}}^{\mathrm{site}}(\mathbf{X})} = \Pi_k \, H\left(|\mathbf{r}_k - \overline{\mathbf{r}_k}| - R_k\right) \equiv H_{\mathrm{c}}(\mathbf{X}), \qquad (1.4)$$

where the H's are Heaviside step functions. This construction defines a minimal default model with probability distribution

$$\rho_0(\mathbf{X}) = H_{\mathrm{c}}(\mathbf{X}) \, e^{-\beta[U_i^{\mathrm{il}}(\mathbf{X}) + U^{\mathrm{ll}}(\mathbf{X})]} \qquad (1.5)$$

that incorporates the generic effect of architectural confinement by the surrounding protein structure in an idealized fashion: Each atom k of the reduced subsystem can fluctuate and make arbitrary excursions, as long as it remains confined a distance R_k from its mean position $\overline{\mathbf{r}_k}$. By construction, the confinement function H_{c} has been chosen to match the upper bound of all atomic fluctuations, whether an ion of type i or j is bound (H_{c} is independent of ion type). This is illustrated in Fig. 1.2. This model provides a simplified idealization of architectural confinement, upon which additional factors can be added. We will refer to the situation embodied by Eq. (1.5) as the generic *confined droplet* model.

The second contribution, $\Delta W_{\mathrm{g}}^{\mathrm{site}}$, accounts for the remaining architectural forces from the surrounding protein that enforce a precise geometry to the atoms of the inner region (blue line in Fig. 1.2). This contribution controls the magnitude of the fluctuations of each atom k within the generic confinement region. To help determine an optimal model for $\Delta W_{\mathrm{g}}^{\mathrm{site}}$, we adopt a cross-entropy method used previously by Hummer et al. (1996) in studies of the hydrophobic solvation and write

$$\eta = -\int d\mathbf{X} \, \rho(\mathbf{X}) \, \ln\left[\rho(\mathbf{X})/\rho_0(\mathbf{X})\right]. \qquad (1.6)$$

We seek to maximize the relative entropy η in Eq. (1.6) under the constraint that the probability distribution $\rho(\mathbf{X})$ is normalized and that the quadratic fluctuations of the atoms in the reduced subsystem must match the value extracted from all-atom MD simulations. The constrained optimization problem is solved by introducing the Lagrange multiplier λ_g to satisfy the condition on the normalization and the overall fluctuations. The resulting normalized distribution is

$$\rho(\mathbf{X};\lambda_g) = \frac{H_c^{site}(\mathbf{X})\, e^{-\beta[U_i^{il}(\mathbf{X})+U^{ll}(\mathbf{X})+\Delta W_g^{site}(\mathbf{X};\lambda_g)]}}{\int d\mathbf{X}\, H_c^{site}(\mathbf{X})\, e^{-\beta[U_i^{il}(\mathbf{X})+U^{ll}(\mathbf{X})+\Delta W_g^{site}(\mathbf{X};\lambda_g)]}}, \tag{1.7}$$

where

$$\Delta W_g^{site}(\mathbf{X};\lambda_g) = \lambda_g \sum_k (\mathbf{r}_k - \overline{\mathbf{r}_k})^2. \tag{1.8}$$

In principle, it would be possible to build models that satisfy higher order moments of the atomic fluctuations following the same procedure. The Lagrange multiplier λ_g has the dimension of a harmonic spring constant (kcal/mol/Å^2), and its optimal value should be determined self-consistently by performing averages with the distribution $\rho(\mathbf{X};\lambda_g)$ to match the reference values extracted from all-atom MD with ions i and j. Two extreme qualitative behaviors are expected: (1) If the binding site is very stiff, the root-mean-square (RMS) fluctuations of the reduced subsystem are small and λ_g is expected to be very large. (2) If the binding site is flexible, the RMS fluctuations of the reduced subsystem are large and λ_g is expected to be small. When $\lambda_g = 0$, the distribution $\rho(\mathbf{X};\lambda_g)$ returns to the generic confined droplet model $\rho_0(\mathbf{X})$ of Eq. (1.5).

1.2.2 Relative Free Energies Ions in a Binding Site

The relative binding free energy of ions i and j in the binding site is

$$e^{-\beta\Delta G_{ij}^{site}} = \frac{\int_{site} d\mathbf{X}\, H_c(\mathbf{X})\, e^{-\beta[U_i^{il}(\mathbf{X})+U^{ll}(\mathbf{X})+\Delta W_g^{site}(\mathbf{X};\lambda_g)]}}{\int_{site} d\mathbf{X}\, H_c(\mathbf{X})\, e^{-\beta[U_j^{il}(\mathbf{X})+U^{ll}(\mathbf{X})+\Delta W_g^{site}(\mathbf{X};\lambda_g)]}}. \tag{1.9}$$

The difference in solvation free energy in the bulk region plays also an important role, but it only adds a constant offset that is the same for all protein-binding sites. In Eq. (1.9), H_c given by Eq. (1.4) is a product of spherical Heaviside step functions chosen to match the upper bound of all atomic fluctuations regardless of the bound ion type, and $\Delta W_g^{site}(\mathbf{X};\lambda_g)$, given by Eq. (1.8), is a harmonic potential with force constant λ_g adjusted to optimally match the atomic RMS fluctuations from the all-atom simulations. By construction, the only terms that depend on ion type i or j in Eq. (1.9) are the ion-ligand interaction energies $U_i^{il}(\mathbf{X})$ and $U_j^{il}(\mathbf{X})$ from the reduced subsystem.

It is useful to derive some simplified expressions for ΔG_{ij}^{site}. To this end, let us define the 1-ion self free energy G_i^{site},

$$e^{-\beta G_i^{site}} = \frac{\int_{site} d\mathbf{X}\, H_c^{site}(\mathbf{X})\, e^{-\beta W_i(\mathbf{X})}}{\int_{site} d\mathbf{X}\, H_c^{site}(\mathbf{X})}, \tag{1.10}$$

where $W_i = U_i^{il} + U^{ll} + \Delta W_g^{site}(\lambda_g)$. The self free energy may be expressed as

$$e^{-\beta G_i^{site}} = \int dw\, \mathcal{P}_i(w)\, e^{-\beta w}, \tag{1.11}$$

where the density of state $\mathcal{P}_i(w)$ is defined as

$$\mathcal{P}_i(w) = \frac{\int_{site} d\mathbf{X}\, \delta[W_i(\mathbf{X})-w]\, H_c^{site}(\mathbf{X})\, e^{-\beta W_i(\mathbf{X})}}{\int_{site} d\mathbf{X}\, H_c^{site}(\mathbf{X})}. \tag{1.12}$$

The self free energy G_i^{site} can be developed as cumulant expansion, yielding

$$G_i^{site} \approx \langle w \rangle_{(i)} - \frac{\beta}{2}\left\langle (w - \langle w \rangle_i)^2 \right\rangle_{(i)}. \tag{1.13}$$

To lowest order, the quadratic fluctuations are not expected to depend strongly upon the ion type. In this case, the relative free energy of ions i and j

in the binding site is dominated by

$$\begin{aligned}\Delta G_{ij}^{\text{site}} &= G_i^{\text{site}} - G_j^{\text{site}} \\ &\approx \langle w\rangle_{(i)} - \langle w\rangle_{(j)} \\ &\approx \left[\langle U_i^{\text{il}} + U^{\text{ll}}\rangle_{(i)} - \langle U_j^{\text{il}} + U^{\text{ll}}\rangle_{(j)}\right] \\ &\quad + \left[\langle \Delta W_{\text{g}}^{\text{site}}\rangle_{(i)} - \langle \Delta W_{\text{g}}^{\text{site}}\rangle_{(j)}\right].\end{aligned} \tag{1.14}$$

The first and second terms in Eq. (1.14), $\langle U_i^{\text{il}}\rangle_{(i)} - \langle U_j^{\text{il}}\rangle_{(j)}$ and $\langle U^{\text{ll}}\rangle_{(i)} - \langle U^{\text{ll}}\rangle_{(j)}$, correspond to the difference in the average ion-ligand and ligand-ligand interactions, respectively. The last term, $\langle \Delta W_{\text{g}}^{\text{site}}\rangle_{(i)} - \langle \Delta W_{\text{g}}^{\text{site}}\rangle_{(j)}$, corresponds to the difference in the geometric architectural contribution to the potential of mean force (PMF). The latter depends implicitly on λ_{g}, and it is identically equal to zero if $\lambda_{\text{g}} = 0$.

1.2.3 Two Distinct Physical Limits

The formal separation of ΔW^{site} makes it possible to delineate two important limiting cases. The first corresponds to the situation where the geometric structural forces associated with ΔW_{g} are very strong. As a consequence, the configuration of the binding site is very stiff, its geometry is strictly enforced, and the average ion-ligand interaction energy dominates the relative free energy difference,

$$\lim_{\lambda_{\text{g}}\to\infty} \Delta G_{ij}^{\text{site}} \approx \langle U_i^{\text{il}}\rangle_{(i)} - \langle U_j^{\text{il}}\rangle_{(j)}, \tag{1.15}$$

where the brackets with subscripts represent thermal averages for ions i and j. This expression is consistent with the concept of a fixed cavity size as in host-guest chemistry (Gouaux and Mackinnon 2005), traditionally called the *snug-fit* mechanism of ion selectivity in the ion channel literature (Armstrong and Bezanilla 1973). In this case, the selectivity is predominantly controlled by the ion-ligand interactions relative to the hydration free energy. The surrounding protein structure is very stiff and precisely dictates the configurational geometry of the ligands that is best adapted to an ion of a given size.

A second limiting case corresponds to the situation where the geometric structural forces are much weaker and essentially negligible. This is the generic confined droplet model of Eq. (1.5) described above. The ion and the ligands are allowed to fluctuate freely and adapt to coordinate a smaller ion but are otherwise restricted to remain within a small spatial region strictly enforced by the function $H_{\text{c}}^{\text{site}}$. The relative free energy is dominated by Noskov et al.

$$\lim_{\lambda_{\text{g}}\to 0} \Delta G_{ij}^{\text{site}} \approx \langle U_i^{\text{il}} + U^{\text{ll}}\rangle_{(i)} - \langle U_j^{\text{il}} + U^{\text{ll}}\rangle_{(j)}. \tag{1.16}$$

This situation corresponds to the confined droplet limit, representing a flexible binding site (Noskov et al. 2004, 2006). In the confined droplet, the probability of the configurations is dictated by the strong ion-ligand interactions causing an induced-fit coordination of the ion (i.e., even an ion of the "wrong" size ends up being well-coordinated). Structurally, the ion-ligand system is allowed to fluctuate and displays some liquid-like features such as allowing a finite width for first peak in the ion-ligand radial distribution function (Noskov et al. 2004).

Thermodynamically, the mean ion-ligand interaction energy does not give rise to selectivity in the confined droplet limit, in contrast with the host-guest mechanism with fixed cavity radius as described by Eq. (1.15). In the confined droplet, both the favorable ion-ligand and unfavorable ligand-ligand interactions contribute and compete, in a nontrivial manner, to control selectivity (Noskov et al. 2004). The lack of selectivity from the mean ion-ligand interaction energy has been documented very clearly by Asthagiri et al. (2006). There are some similarities with Eisenman's classic work, in which structurally featureless binding sites are selective by virtue of the chemical type of ion-coordinating ligands (Eisenman 1962). This view is also closely related to the model proposed by Nonner and coworkers to explain the selectivity of Ca^{2+} channels in which the ion and the negatively charged carboxylate groups of the protein are allowed to dynamically fluctuate within a confined sub-volume (Nonner et al. 2000; Boda et al. 2008).

Spatial confinement is an essential component of the context that gives rise to selectivity in the limit where $\Delta W_g \rightarrow 0$ in Eq. (1.16). Factors, through mutations for example, site. Interestingly, a search through the structural database reveals that a common feature of ion-binding sites in proteins is the existence of a shell of non-polar hydrophobic groups surrounding the ion-coordinating ligands (Yamashita et al. 1990). Presumably, this serves to provide a region of low dielectric serving to protect the confinement and enhance the local electrostatic interactions. While confinement is ultimately ensured by the 3D fold of the protein, with all its *nonlocal* long-range complexities, it is important to realize that the character of ΔW_c^{site} does not need to be complex for supporting selectivity. Nontrivial free energy patterns can emerge from the *local* interactions between the confined ion and the ligand, even in the context of a relatively generic and featureless ΔW_c^{site}. Therefore, in the confined droplet limit, the protein structure plays a critical role to support selectivity, albeit, not by enforcing a specific cavity size with a ligand geometry to sub-Ångström precision (Noskov et al. 2004).

1.2.4 Illustrative Systems: Ion Binding to KcsA and LeuT

To illustrate the concepts presented above, we first consider the binding site S2 of the KcsA channel, which is the most selective when the narrow pore is in its conductive conformation (Bernèche and Roux 2001; Noskov et al. 2004). The conductive state observed in crystal structure obtained at high K^+ concentration is the conformation that is relevant for ion selectivity and that we consider here (Valiyaveetil et al. 2006). To further appreciate the difference between the snug-fit and confined droplet limits, it is helpful to compare the above results with those obtained on LeuT, a bacterial homologue of Na^+/Cl^--dependent neurotransmitter transporters. LeuT possesses two ion-binding sites, called Na1 and Na2, both of which are highly selective for Na^+ over K^+ (Yamashita et al. 2005). Free energy MD simulations with explicit membrane and solvent show that they are highly selective for Na^+ over K^+ in good accord with experiments (Noskov and Roux 2008). In LeuT, the site Na1 comprises six ligands, including a negatively charged carboxylate from the bound leucine substrate, while the site Na2 comprises five neutral ligands.

Following the formal developments of Eqs. (1.4), (1.7), and (1.8), we construct a reduced computational model of the site S2 of KcsA. The reduced model comprises four diglycine peptides corresponding to the backbone of Val76 and Gly77 of the four subunits in the KcsA channel, for a total of eight ion-coordinating carbonyl ligands. This generic confined droplet model does not prevent collapse of the ion-coordinating ligand onto Na^+. The ΔW_g is designed to enforce the optimal coordinating geometry for K^+. Analysis indicates that the value of λ_g that matches the atomic fluctuations from the all-atom MD is on the order of 0.5–1.0 kcal/mol/Å^2.

The calculated free energy $\Delta\Delta G_{Na,K}$ for the KcsA- and LeuT-binding sites is shown in Fig. 1.3 as a function of the strengths of the geometric forces. In the confined droplet limit, the S2 site in KcsA is selective for K^+ over Na^+ by about 6 kcal/mol. As the geometric structural potential is increased via the Lagrange multiplier λ_g, the fluctuations progressively decrease and selectivity increases. When the architecture is enforced by a force constant λ_g of more than 10–100 kcal/mol/Å^2, the ion-ligand interactions start to make an increasingly important contribution to $\Delta\Delta G_{Na,K}$ as the ligand-ligand interactions decrease progressively down to zero. This is the snug-fit limit. Selectivity start to be strictly dictated by the cavity size only when the structural distortion of the coordinating ligand upon by the binding of Na^+ is causing a sufficiently large energy penalty, i.e., when $\lambda_g \geq 100$ kcal/-mol/Å^2. At such value of λ_g, the RMS fluctuations of the ligands become smaller than about 0.2-0.3 Å, corresponding to the size difference between Na^+ and K^+. At the opposite limit, when

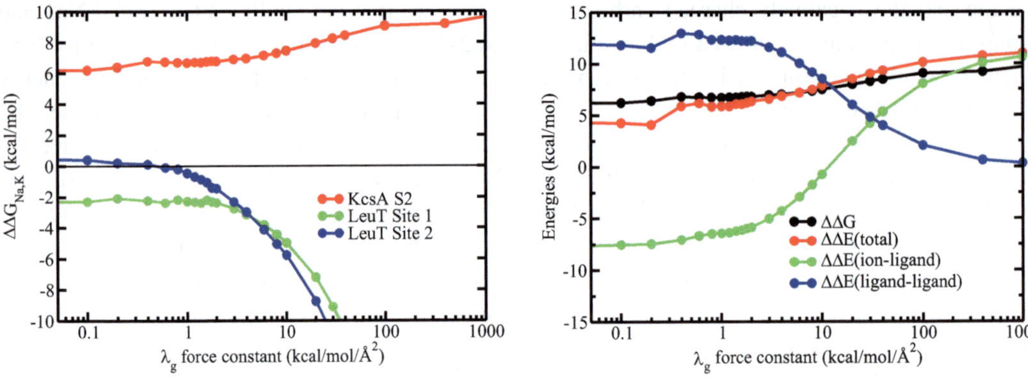

Fig. 1.3 Ion selectivity in the binding sites of KcsA and LeuT. (Left) Ion selectivity $\Delta\Delta G_{\mathrm{Na,K}}$ as a function of the geometric force constant λ_{g} calculated from free energy perturbation (FEP) MD simulations of the reduced models of KcsA- and LeuT-binding sites. (Right) Decomposition of the various energetic terms to the selectivity in the reduced model of the S2-binding site of KcsA. The ion selectivity, total energy, ion-ligand interaction energy U^{il}, and ligand-ligand interaction energy U^{ll} in the reduced model consisting of four diglycine molecules as a function of the strength of the geometric structural forces. The calculations were carried out using the nonpolarizable CHARMM potential function (MacKerell et al. 1998). Reproduced from reference Yu et al. (2010)

$\lambda_{\mathrm{g}} \leq 10$ kcal/mol/Å^2, both the ion-ligand and ligand-ligand interactions contribute significantly to $\Delta\Delta G_{\mathrm{Na,K}}$. This is the confined droplet regime. It is particularly noteworthy that the selectivity for K^+ over Na^+ is robust down to $\lambda_{\mathrm{g}} = 0$ kcal/mol/Å^2, where the atomic fluctuations are more than twice as large as the size difference between Na^+ and K^+. This implies that no strain energy contribution from the geometric structural potential is needed to maintain selectivity in the confined droplet limit.

In LeuT, while the reduced model of the Na1 site retains a selectivity comparable to that from all-atom simulations even in the confined droplet limit, the reduced model of the site Na2 does not display any selectivity in the absence of additional structural forces to stabilize an optimal cavity size for Na^+. The mechanism giving rise to Na^+ selectivity in these two sites in LeuT corresponds to two different limits. In Na1, there is selectivity for Na^+ over K^+ in the confined droplet limit by virtue of the strong electrostatic field arising from the negatively charged carboxylate group (Eisenman 1962). The $\Delta\Delta G_{\mathrm{Na,K}}$, which is on the order of about -2 kcal/mol, reflects the local thermodynamics determined by the composition of the confined droplet (number and type of ligands). In contrast, the selectivity of the site Na2 arises solely by virtue of geometric structural forces, because the coordinating ligands forming the Na2-binding site are taken from backbone and side-chain atoms of nearest neighboring residues along the polypeptide chain. For instance, the Na^+ ion is coordinated by the carbonyl oxygen of Gly20 and Val23, and by the carbonyl oxygen of Ala351, as well the side-chain atoms Oγ of Thr354 and the Oδ of Ser355. Effectively, the protein exploits the local degrees of freedom from the covalent structure (bonds, angles) to establish a binding cavity best adapted to the size of Na^+ that is robust to thermal fluctuations. This situation is reminiscent of K^+-selective valinomycin, a small cyclic ionophore rendered fairly rigid by virtue of its covalent and hydrogen bonding structure (Åqvist et al. 1992; Noskov et al. 2006; Varma and Rempe 2008).

1.2.5 Selectivity of Simplified Binding Site Models

The present analysis shows that, in the confined droplet limit, which is representative of flexible protein-binding site, both the number and chemical type of the ligands coordinating the ion play a crucial role in ion selectivity. To further explore how flexible protein-binding sites achieve ion selectivity, we can represent binding sites as clusters of ligands confined within a small volume, mimicking the spatial restrictions imposed by the protein environment (Roux 2010). A total of 1,077 reduced binding site models were constructed, each comprising four to nine oxygen-coordinating ligands, such as carbonyls, hydroxyls, water molecules, or carboxylates. These models allowed ligands to interact freely with ions within a confined spherical volume, simulating the flexibility of real protein-binding sites. The simulations revealed that both the number and type of ligands significantly influence ion selectivity. For example, binding sites with carbonyl ligands tended to favor K^+ over Na^+, whereas sites with carboxylate ligands favored Na^+. These observations are consistent with Eisenman's concept of "field strength" (Eisenman 1962; Szabo et al. 1973). The results demonstrated a wide range of selectivity profiles, with some models strongly favoring one ion over another and others showing only marginal selectivity. The study found that ion selectivity arises from a complex interplay between ion-ligand attraction and ligand-ligand repulsion. Smaller ions like Na^+ generate stronger ligand-ligand repulsion due to closer spacing of the electronegative atoms coordinating the ion. This balance between attractive and repulsive interactions depends on both the ion size and the chemical properties of the ligands. Interestingly, the study finds correlations between selectivity trends for monovalent ions (Na^+/K^+) and divalent ions (Ca^{2+}/Ba^{2+}), suggesting common underlying principles that govern size-based selectivity, even without rigid geometrical constraints (Nonner et al. 2000; Boda et al. 2008).

1.2.6 Selectivity and Coordination Number

As shown in Fig. 1.3, the mechanisms able to support ion selectivity are greatly affected by the number and by the type of ion-coordinating ligands forming a binding site. Particular binding sites require additional forces from the protein to favor an optimal ligand configuration to be selective (the site Na2 in LeuT), whereas others remain robustly selective up to the confined droplet limit (the sites Nal in LeuT and S2 in KcsA). For the latter, geometric structural forces are utilized by the protein to enhance, rather than oppose, the trend arising spontaneously in the confined droplet limit.

A number of studies have suggested that ion selectivity is predominantly controlled by the number of ligands coordinating the ion and that their chemical type plays no role (Varma and Rempe 2007, 2008; Thomas et al. 2007; Bostick and Brooks 2007). This seductive idea, based on the equivalence of ion-coordinating ligands, goes back to classic work by Mullins (1959). In the "topological control" hypothesis presented by Bostick and Brooks (2007), this was put on a more formal footing by suggesting that differences in the probabilities $P_i(n)$ and $P_j(n)$ of the spontaneous occurrence of a n-coordinated state for ions i and j in the bulk phase can be translated into an intrinsic "topological" free energy $\Delta J_{ij}(n) = k_B T \ln[P_i(n)/P_j(n)]$, which directly reflects the relative free energy $\Delta\Delta G_{ij}$ governing ion selectivity in Eq. (1.1). The general idea is depicted schematically in Fig. 1.4.

To examine the fundamental basis of the topological control hypothesis, we reexpress the relative free energy $\Delta G_{ij}^{\text{bulk}}$ of ions i and j in the bulk solvent in terms of constrained n-coordinated states (Yu et al. 2009). This is achieved by introducing a Kroenecker discrete delta function $\delta_{nn'}$, equal to 1 only when the instantaneous coordination number of the ion n' is equal to n (Deng and Roux 2008). By definition, $P_i(n)$ is equal to $\langle \delta_{nn'} \rangle_{(i)}$. It follows that

$$
\begin{aligned}
e^{-\beta \Delta G_{ij}^{\text{bulk}}} &= \frac{\int_{\text{bulk}} d\mathbf{R}\, e^{-\beta U_i}}{\int_{\text{bulk}} d\mathbf{R}\, e^{-\beta U_j}} \\
&= \frac{\int_{\text{bulk}} d\mathbf{R}\, \delta_{nn'}\, e^{-\beta U_i}}{\int_{\text{bulk}} d\mathbf{R}\, \delta_{nn'}\, e^{-\beta U_j}} \times \frac{\langle \delta_{nn'} \rangle_{(j)}}{\langle \delta_{nn'} \rangle_{(i)}} \\
&= e^{-\beta[\Delta G_{ij}^{\text{bulk}}(n) - \Delta J_{ij}(n)]}, \qquad (1.17)
\end{aligned}
$$

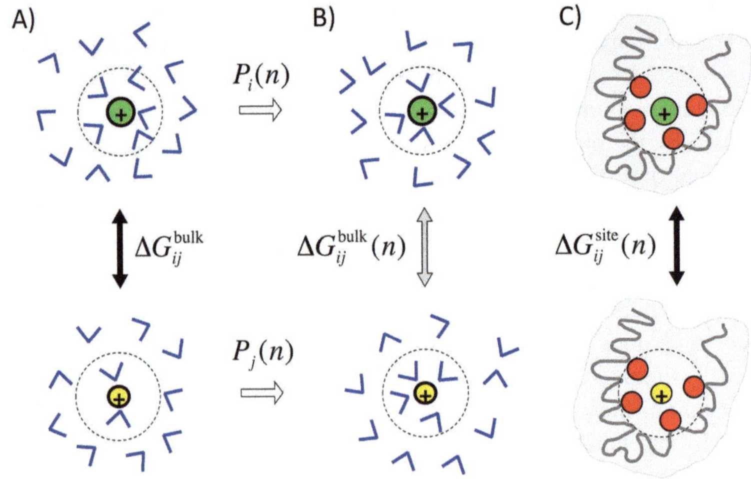

Fig. 1.4 Schematic representation of the role of coordination states in bulk solvent in ion selectivity. In (A), the ions i and j are free in solution; the free energy difference is $\Delta G_{ij}^{\text{bulk}}$. In (B), the ions i and j are immersed in the bulk phase but constrained to be coordinated by n solvent molecules; the free energy difference is $\Delta G_{ij}^{\text{bulk}}(n)$. The probabilities for ions i and j to spontaneously realize a n-coordinated state in solution are $P_i(n)$ and $P_j(n)$, respectively. In (C), the ions i and j are in the protein-binding site surrounded by n ligands; the free energy difference is $\Delta G_{ij}^{\text{site}}(n)$. Ion-binding sites in proteins are typically embedded in a nonpolar low dielectric shell (colored gray) (Yamashita et al. 1990). The notation $\Delta G_{ij}^{\text{site}}(n)$ is meant to emphasize that the binding site is effectively constrained to provide n coordinating ligands. Taken from ref Yu et al. (2009) with permission

where $\Delta G_{ij}^{\text{bulk}}(n)$ is the free energy difference between ions i and j in the bulk for a solvation shell constrained to have a coordination number equal to n. Accordingly, the total free energy difference from Eq. (1.1) may be expressed as

$$\Delta\Delta G_{ij} = \Delta G_{ij}^{\text{site}}(n) - \Delta G_{ij}^{\text{bulk}}(n) + \Delta J_{ij}(n). \tag{1.18}$$

The step-by-step procedure leading to Eq. (1.18) is illustrated schematically in Fig. 1.4. According to Eq. (1.18), the quantity $\Delta J_{ij}(n)$ controls ion selectivity only if the free energies $\Delta G_{ij}^{\text{site}}(n)$ and $\Delta G_{ij}^{\text{bulk}}(n)$ are approximately equal. The implication is that the relative free energies of ions i and j in an n-coordinated state in two different environments (bulk water or protein-binding site) are equal. As discussed above, the n ligands in the binding site are held in place via a range of architectural forces, including covalent bonding, hydrogen bonding, and van der Waals packing (Fig. 1.4c). In contrast, the solvent molecules in the n-coordinated state are freely fluctuating within the confined sub-volume of the coordination shell and surrounded by a high dielectric liquid (Fig. 1.4b).

Table 1.1 Relative free energy of Na^+/K^+ in 8-coordinated state (kcal/mol)[1]

$\Delta\Delta G$	$\Delta G^{\text{site}}(n)$	ΔG^{bulk}	$\Delta G^{\text{bulk}}(n)$	$\Delta J(n)$
5.93	− 12.37	− 18.30	− 16.96	1.61

[1] The simulations were performed with the CHARMM program (Brooks et al. 2009) and the PARAM27 force field (MacKerell et al. 1998). The simulation methodology has been described elsewhere (Bernèche and Roux 2001; Noskov et al. 2004; Noskov and Roux 2007; Yu et al. 2009).

To ascertain the validity of the topological control hypothesis of Eq. (1.18), we now examine the equivalence between $\Delta G_{ij}^{\text{site}}(n)$ and $\Delta G_{ij}^{\text{bulk}}(n)$ by characterizing the free energy contributions corresponding to Fig. 1.4 for ions binding to the site S2 of the KcsA channel. The results are given in Table 1.1. The topological constraint $\Delta J_{\text{Na,K}}$ displays a slight preference in favor of K^+, but differs quantitatively from $\Delta\Delta G_{\text{Na,K}}$. The probabilities of the 8-coordinated states of K^+ and Na^+ in the KcsA channel calculated from a trajectory of 2 ns are shown in Fig. 1.5. The constraint for a 8-coordinated state yields a topological free energy $\Delta J_{\text{Na,K}}$ of about 1-2 kcal/mol. This pref-

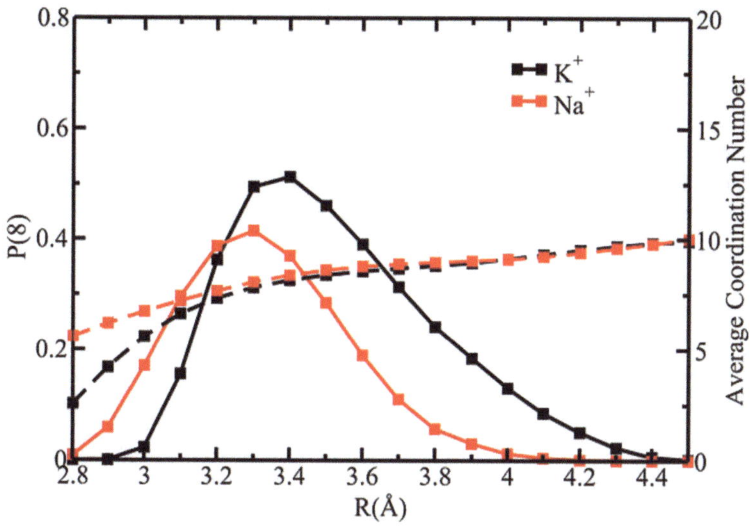

Fig. 1.5 Probability to find exactly eight oxygen ligands (backbone carbonyls, water, hydroxyl, or carboxylate) around K^+ or Na^+ in the binding site S2 of the KcsA channel (solid lines). At the larger distances, the average number of ligands is 10 when the eight carbonyls and the two nearest water molecules are included (dashed lines). The probabilities of the 8-coordinated states of K^+ and Na^+ in the KcsA channel were calculated from a trajectory of 2 ns. Taken from reference Yu et al. (2009) with permission

erence for K^+ over Na^+ reflects the phenomenon of over-coordination (Varma and Rempe 2007). However, based on this comparison, the free energies $\Delta G^{\text{site}}_{\text{Na,K}}(8)$ and $\Delta G^{\text{bulk}}_{\text{Na,K}}(8)$ are clearly not equivalent. The topological free energy $\Delta J_{\text{Na,K}}$ for $n = 8$ is roughly equal to 1-2 kcal/mol according to the present calculations, which is too small to account for the free energy for the site S2 estimated from MD calculations (Bernèche and Roux 2001; Noskov et al. 2004).

While the topological reduction in terms of coordination number of Eq. (1.18) assigns the ligands on the basis of a fixed spatial boundary and a distance R criterion, one may note that the statistical mechanical reduction of Eq. (1.3) is achieved by first assigning the molecular participants in the system. However, no spatial restriction is actually imposed on the ion or ligands in the reduced model. Quasichemical theory (QCT) (Pratt and Laviolette 1998) and the small-system grand canonical ensemble (SGCE) (Gerardo Soto-Campos et al. 1998) are also statistical mechanical reductions relying on a fixed spatial separation, where the inner region is referred to the "sub-volume" (Gerardo Soto-Campos et al. 1998) or the "observation volume" (Merchant and Asthagiri 2009). QCT provides an effective framework for studying ions in bulk solvents (Rempe et al. 2000; Rempe and Pratt 2001; Rempe et al. 2004; Asthagiri et al. 2004; Roux and Yu 2010). However, a straightforward application of QCT to protein-binding sites poses challenges due to the structural constraints imposed by the protein's covalently bonded ligands. Within the context of a fixed boundary, it is critical to properly account for the fluctuations of the ligands in and out of the observation volume to achieve a valid statistical mechanical reduction. When one is considering the simple situation of an ion i solvated in a bulk liquid, the solvation free energy in the reduced subsystem can be expressed in terms of constrained configurational integrals with fixed number of ligands n, weighted by the probability $P_i(n)$ of that configuration. While this formal route is strictly valid for an ion solvated in a bulk liquid where the solvent molecules are chemically identical and can freely exchange their position, it cannot be readily transposed to an ion-binding site in a protein. Keeping track of the number of protein ligands within the observation volume provides an incomplete characterization because the protein ligands cannot freely exchange their position along the covalently linked chain like the solvent molecules in a bulk liquid. To achieve a valid statistical mechanical reduction, each ligand must be treated as a distinct entity when reporting the coordinated states of an ion bound in a protein site, and one must specify the complete coordinated state comprising the ligands $\mathbf{L}_n \equiv \{L_1, L_2, \ldots L_n\}$ within the observation volume. These considerations lead to the following expression for the relative free energy of ions i and j in a binding site (Roux 2012)

$$
\begin{aligned}
e^{-\beta\Delta G_{ij}^{\text{site}}}
&= \frac{\int_{\text{site}} d\mathbf{R}\, e^{-\beta U_i}}{\int_{\text{site}} d\mathbf{R}\, e^{-\beta U_j}} \\
&= \frac{P_j(\mathbf{L}_n; R)}{P_i(\mathbf{L}_n; R)} \\
&\times \frac{\int d\mathbf{X}_n\, H_{\text{in}}(\mathbf{X}; R)\, e^{-\beta[U_i^{\text{il}}(\mathbf{X})+U^{\text{ll}}(\mathbf{X})+\Delta W^{\text{site}}(\mathbf{X}; R)]}}{\int d\mathbf{X}\, H_{\text{in}}(\mathbf{X}; R)\, e^{-\beta[U_j^{\text{il}}(\mathbf{X})+U^{\text{ll}}(\mathbf{X})+\Delta W^{\text{site}}(\mathbf{X}; R)]}} \\
&= \frac{P_j(\mathbf{L}_n; R)}{P_i(\mathbf{L}_n; R)} \times e^{-\beta\Delta G_{ij}^{\text{site}}(\mathbf{L}_n; R)}, \qquad (1.19)
\end{aligned}
$$

where $[U^{\text{il}} + U^{\text{ll}}]$ is the total potential energy of the ion and the n ligands, and ΔW^{site} is an effective potential that incorporates the influence of the remaining degrees of freedom. The probabilities $P_i(\mathbf{L}_n; R)$ and $P_j(\mathbf{L}_n; R)$ are the probability of finding a specific set of n ligands $\mathbf{L}_n$ within the observation volume with ion i or j. All microscopic information about the relative flexibility of the protein-binding site is implicitly incorporated into the probabilities $P_j(\mathbf{L}_n; R)$ and $P_i(\mathbf{L}_n; R)$. While the restricted free energy $\Delta G_{ij}^{\text{site}}(\mathbf{L}_n; R)$ provides useful information, the treatment remains incomplete without a meaningful analysis of the binding site flexibility via $P_j(\mathbf{L}_n; R)$ and $P_i(\mathbf{L}_n; R)$. Ignoring the contribution from the probabilities $P_j(\mathbf{L}_n; R)$ and $P_i(\mathbf{L}_n; R)$ is equivalent to postulating that the protein-binding site is very rigid, regardless of the type of bound ion. In the treatment of ion solvation in bulk liquids, closed-form expressions for P_i and P_j can be written form in terms of simple properties such as the solvent density and the excess chemical potential of the solvent molecules (Pratt and Laviolette 1998; Deng and Roux 2008; Roux and Yu 2010). Such simplifications are possible because solvent molecules can be assumed to be identical and able to exchange with one another. However, no such simplifications are not possible in the case of a protein-binding site because the coordinating ligands are not strictly identical. A number of studies have ignored these formal issues and have employed a formal reduction strictly designed to treat ion solvation in a bulk liquid to analyze the mechanism of ion selectivity in proteins-binding sites (Varma and Rempe 2007, 2008; Bostick and Brooks 2009). Because of the difficulties discussed above, conclusions drawn from a straight transfer of QCT to protein-binding sites are of unclear significance.

1.3 Conclusion

The fundamental principles governing equilibrium binding selectivity are relatively simple and straightforward. Selectivity arises from the relative free energy of ions in the binding site compared to that in the bulk solution. However, despite this apparent simplicity, dissecting the various factors at play within a specific protein-binding site to draw definitive conclusions about the microscopic mechanisms behind ion selectivity remains challenging. To enable a rational analysis of the different factors at play, we developed a statistical mechanical framework to express the relative free energy of bound ions in terms of local interactions and the influence of the surrounding protein architecture (Yu et al. 2010). The influence of the surrounding protein architecture is represented as an effective potential of mean force (PMF) acting on the local subsystem comprising the ion and its most relevant ligands. Two significant limiting cases were identified. In the first limit, where geometric structural forces dominate, selectivity is governed by ion-ligand interactions, which define the precise cavity size, consistent with the well-known "snug fit" mechanism in host-guest chemistry. In the second limit, where geometric structural forces are negligible, ligands are free to fluctuate and adapt locally to a smaller ion. In this "confined droplet" regime, selectivity emerges from the local thermodynamics of ion-ligand and ligand-ligand interactions, determined by the number and chemical nature of coordinating ligands. This mechanism balances these interactions and demonstrates robust selectivity under well-defined conditions. The proposition that significant insights into the selectivity of a biological binding site can emerge from a dynamic, liquid-like environment without imposing rigid geometric constraints on ion-coordinating ligands remains somewhat provocative. Notably, the fluctuations of the carbonyl ligands lining the selectivity filter of the K^+ channel are *larger* than the size difference between K^+ and Na^+ (approximately 0.38 Å) (Bernèche and Roux 2000; Bernèche and Roux 2001; Noskov et al. 2004). This analysis was carried out using a nonpolarizable potential function with fixed partial charges (MacKerell et al. 1998). While nonadditive interactions caused by induced polarization are expected to affect trends quantitatively (Roux 1993), work with polarizable force field and quantum mechanical models indicate that the general principles derived from these studies remain valid (Harder et al. 2006; Bogdan et al. 2013; Lev et al. 2013; Li et al. 2015; Ngo et al. 2021).

Seeking a purely structural explanation for ion selectivity is intuitive. However, it may be overly simplistic to expect a single explanation to account for ion selectivity across all binding sites. For many years, the predominant explanation relied on the concept of geometrical "misfit," where selectivity arises from an ion's inability to fit into a nearly rigid binding site. This idea, borrowed directly from host-guest chemistry (Andreoli et al. 1967; Mueller and Rudin 1967; Pedersen and Frensdor 1972; Michaux and Reisse 1982; Gokel et al. 1983; Dietrich 1985; Schneider et al. 2008), was adapted early on to describe ion channels (Bezanilla and Armstrong 1972). The crystal structure of the KcsA channel (Doyle et al. 1998) further reinforced this argument (Hille et al. 1999; Yellen 2002; Berg et al. 2002; Berg 2003; Garofoli and Jordan 2003; Beckstein et al. 2003; MacKinnon 2004; Ban et al. 2004; Gouaux and Mackinnon 2005). While a rigid binding site can indeed achieve selectivity, it is not the only mechanism –a selective binding site is not necessarily rigid. As demonstrated in this analysis, ion selectivity can also emerge within a confined droplet environment, where it is significantly influenced by the number, physical properties, and chemical nature of the coordinating ligands. The importance of the number of coordinating ligands was noted in the classic work of Mullins (1959). Similarly, the significance of their chemical and physical character is reminiscent of the mechanism proposed by Eisenman to describe equilibrium selectivity in glass electrodes (Eisenman 1962; Szabo et al. 1973). He highlighted the role of "field strength," which correlates with the dipole moment of the coordinating ligands. High-field sites favor cations with smaller ionic radii, while low-field sites are selective for larger ions. Remarkably, Eisenman's model, grounded in simple principles, successfully predicted the experimentally observed selectivity sequences of glass electrodes by varying the dipole moment of coordinating ligands (Eisenman 1962; Szabo et al. 1973). This central idea underscores that the intrinsic physical and chemical properties of the coordinating ligands significantly impact ion selectivity (Eisenman 1962; Szabo et al. 1973). Furthermore, as shown here with variations in the geometric rigidity parameter λ_g, the relative importance of architectural rigidity in the binding site—as suggested in later work by Eisenman and colleagues (Åqvist et al. 1992)—also plays a critical role.

The snug-fit and confined droplet limits represent two conceptually extreme physical models of ion selectivity. Real systems are expected to predominantly exhibit the characteristics of one of these idealized extremes. Some binding sites achieve robust selectivity in the confined droplet limit, while others require additional protein forces to stabilize an optimal ligand configuration for selectivity. In the case of the S2 site in KcsA, analysis indicates that selectivity is achieved across a wide range of structural flexibilities. Conversely, the analysis of the LeuT transporter reveals two distinct mechanisms. The Na1 site is selective primarily due to the (high-field) negatively charged ligand provided by the leucine substrate. Meanwhile, the Na2 site achieves selectivity through the architectural stiffness of the covalent structure that connects the coordinating ligands from adjacent residues. The proximity of the ion-coordinating ligands along the amino acid sequence shows how the the primary structure is leveraged to produce local structural stiffness of the binding site. Obviously, equilibrium free energy differences cannot provide information about the mechanism of ion conduction through a channel, which is inherently a nonequilibrium process. Nonetheless, it is likely that the flexibility of the pore is essential for maintaining low free energy barriers and enabling rapid ion conduction (Bernèche and Roux 2001, 2003; Roux et al. 2004). This suggests that binding sites in K^+ channels should be expected to match more closely the confined droplet limit in terms of their selectivity. In contrast, membrane transporters like LeuT operate at much slower rates and can function with relatively rigid binding sites. These examples highlight that understanding ion selectivity across diverse biomolecular systems requires evaluating the interplay of ligand number, ligand type, and the architectural forces that enforce their spatial configuration. Such assessments are crucial for capturing the distinct mechanisms underlying selectivity in different biological contexts.

Acknowledgments This work was supported by the National Institute of Health via grant R35-GM152124.

References

Åqvist, JJ, Alvarez O, Eisenman G (1992) Ion-selective properties of a small ionophore in methanol studied by free energy perturbation simulations. J Phys Chem 96:10019–10025

Andreoli TE, Tieffenberg M, Tosteson DC (1967) The effect of valinomycin on the ionic permeability of thin lipid membranes. J Gen Physiol 50(11):2527–2545

Armstrong C, Bezanilla F (1973) Currents related to movement of the gating particles of the sodium channels. Nature 242:459–461

Asthagiri D, Pratt LR, Paulaitis ME, Rempe SB (2004) Hydration structure and free energy of biomolecularly specific aqueous dications, including zn2+ and first transition row metals. J Am Chem Soc 126:1285–1289

Asthagiri D, Pratt L, Paulaitis M (2006) Role of fluctuations in a snug-fit mechanism of KcsA channel selectivity. J Chem Phys 125:24701

Ban F, Kusalik P, Weaver DF (2004) Density functional theory investigations on the chemical basis of the selectivity filter in the K+ channel protein. J Am Chem Soc 126:4711–4716

Beckstein O, Biggin P, Bond P, Bright JN, Domene C, Grottesi A, Holyoake J, Sansom MS (2003) Ion channel gating: insights via molecular simulations. FEBS Lett 555:85–90

Armstrong CM (2003) Voltage-gated K channels. Sci STKE 2003:re10

Berg JM, Tymoczko JL, Stryer L (2002) Biochemistry, 5th edn. W.H. Freeman and Co., New York

Bernèche S, Roux B (2000) Molecular dynamics of the KcsA K(+) channel in a bilayer membrane. Biophys J 78:2900–2917

Bernèche S, Roux B (2001) Energetics of ion conduction through the k+ channel. Nature 414:73–77

Bernèche S, Roux B (2003) A microscopic view of ion conduction through the KcsA K+ channel. Proc Natl Acad Sci 100:8644–8648

Bezanilla F, Armstrong CM (1972) Negative conductance caused by entry of sodium and cesium ions into the potassium channels of squid axons. J Gen Physiol 60:588–608

Boda D, Nonner W, Henderson D, Eisenberg B, Gillespie D (2008) Volume exclusion in calcium selective channels. Biophys J 94:3486–3496

Bogdan L, Roux B, Noskov S (2013) Relative free energies for hydration of monovalent ions from QM and QM/MM simulations. J Chem Theor Comp 9:4165–4175

Bostick DL, Brooks CL (2007) Selectivity in K^+ channels is due to topological control of the permeant ion's coordinated state. Proc Natl Acad Sci USA 104:9260–9265

Bostick DL, Brooks CL (2009) Statistical determinants of selective ionic complexation: ions in solvent, transport proteins, and other "hosts". Biophys J 96:4470–4492

Brooks BR, Brooks III CL, Mackerell Jr AD, Nilsson L, Petrella RJ, Roux B, Won Y, Archontis G, Bartels C, Boresch S, Caflisch A, Caves L, Cui Q, Dinner AR, Feig M, Fischer S, Gao J, Hodoscek M, Im W, Kuczera K, Lazaridis T, Ma J, Ovchinnikov V, Paci E, Pastor RW, Post CB, Pu JZ, Schaefer M, Tidor B, Venable RM, Woodcock HL, Wu X, Yang W, York DM, Karplus M (2009) CHARMM: the biomolecular simulation program. J Comp Chem 30(10, Sp. Iss. SI):1545–1614

Deng YQ, Roux B (2008) Computation of binding free energy with molecular dynamics and grand canonical Monte Carlo simulations. J Chem Phys 128:115103

Dietrich B (1985) Coordination chemistry of alkali and alkaline-earth cations with macrocyclic ligands. B. Dietrich, J. Chem. Ed. 62 (1985) 954–964.

Doyle D, Cabral J, Pfuetzner R, Kuo A, Gulbis J, Cohen S, Chait B, MacKinnon R (1998) The structure of the potassium channel: molecular basis of K+ conduction and selectivity. Science 280:69–77

Eisenman G (1962) Cation selective electrodes and their mode of operation. Biophys J 2(supp 2):259–323
Garofoli S, Jordan PC (2003) Modeling permeation energetics in the KcsA potassium channel. Biophys J 84:2814–2830
Gerardo Soto-Campos G, Corti DS, Reiss H (1998) A small system grand ensemble method for the study of hard-particle systems. J Chem Phys 108:2563–2570
Gokel G, Goli D, Mlinganti C, Echegoyen L (1983) Clarification of the hole-size cation-diameter relationship in crown ethers and a new method for determining calcium cation homogeneous binding constant. J Am Chem Soc 105(23):6786–6788
Gouaux E, Mackinnon R (2005) Principles of selective ion transport in channels and pumps. Science 310:1461–1465
Harder E, Anisimov VM, Vorobyov IV, Lopes PEM, Noskov SY, MacKerell AD, Roux B (2006) Atomic level anisotropy in the electrostatic modeling of lone pairs for a polarizable force field based on the classical Drude oscillator. J Chem Theor Comp 2(6):1587–1597
Hille B, Armstrong CM, MacKinnon R (1999) Ion channels: from idea to reality. Nat Med 5:1105–119
Hummer G, Pratt LR, Garcia AE (1996) Free energy of Ionic Hydration. J Phys Chem 100:1206–1215
Lev B, Roux B, Noskov SY (2013) Relative Free Energies for Hydration of Monovalent Ions from QM and QM/MM Simulations. J Chem Theo Comp 9(9):4165–4175
Li H, Ngo VA, Da Silva MC, Callahan KM, Salahub DR, Roux B, Noskov SY (2015) Representation of Ion-Protein Interactions using the Drude Polarizable Force-Field. J Phys Chem B 119(29):9401–9416
MacKerell, AJ, Bashford D, Bellot M, Dunbrack R, Evanseck J, Field M, Fischer S, Gao J, Guo H, Ha, DJ-MS, Kuchnir L, Kuczera K, Lau F, Mattos C, Michnick S, Ngo T, Nguyen D, Prodhom B, Reiher III W, Roux B, Schlenkrich M, Smith J, Stote R, Straub J, Watanabe M, Wiorkiewicz-Kuczera J, Karplus M (1998) All-atom empirical potential for molecular modeling and dynamics studies of proteins. J Phys Chem B 102:3586–3616
MacKinnon R (2004) Potassium channels and the atomic basis of selective ion conduction (Nobel Lecture). Angew Chem Int Ed Engl 43:4265–4277
Merchant S, Asthagiri D (2009) Thermodynamically dominant hydration structures of aqueous ions. J. Chem. Phys. 130, 195102
Michaux G, Reisse J (1982) Solution thermodynamics studies: 6. Enthalpy-entropy compensation for the complexation reaction of some crown ethers with alkaline cations. Quantitative interpretation of the complexing properties of 18 Crown-6. J Am Chem Soc 104(25):6895–6899
Mueller P, Rudin DO (1967) Development of k^+-na^+ discrimination in experimental bimolecular lipid membranes by macrocyclic antibiotics. Biochem Biophys Res Commun 26(4):398–404
Mullins L (1959) The penetration of some cations into muscle. J Gen Physiol 42:817–829
Ngo V, Li H, MacKerell AD, Allen TW, Roux B, Noskov S (2021) Polarization effects in water-mediated selective cation transport across a narrow transmembrane channel. J Chem Theory Comput 17(3):1726–1741
Nonner W, Catacuzzeno L, Eisenberg B (2000) Binding and selectivity in L-type calcium channels: a mean spherical approximation. Biophys J 79:1976–1992
Noskov S, Roux B (2007) Importance of hydration and dynamics on the selectivity of the KcsA and NaK channels. J Gen Physiol 129:135–143
Noskov SY, Roux B (2008) Control of ion selectivity in LeuT: two Na+ binding sites with two different mechanisms. J Mol Biol 377(3):804–818
Noskov S, Berneche S, Roux B (2004) Control of ion selectivity in k(+) channels by dynamic and electrostatic properties of carbonyl ligands. Nature 431:830–834
Noskov S, Berneche S, Roux B (2006) Ion selectivity in potassium channels. Biophys Chem 124:279–291
Pedersen C, Frensdor H (1972) Macrocyclic polyethers and their complexes. Angew Chem Intl Ed 11(1):16–25
Pratt L, Laviolette R (1998) Quasi-chemical theories of associated liquids. Mol Phys 94:909–915
Rempe S, Pratt L (2001) The hydration number of Na+ in liquid water. Fluid Phase Equil. 183(Sp. Iss. SI):121–132
Rempe S, Pratt L, Hummer G, Kress J, Martin R, Redondo A (2000) The hydration number of Li+ in liquid water. J Am Chem Soc 122(5):966–967
Rempe S, Asthagiri D, Pratt L (2004) Inner shell definition and absolute hydration free energy of K+(aq) on the basis of quasi-chemical theory and ab initio molecular dynamics. Phys. Chem. Chem. Phys. 6(8):1966–1969
Roux B (1993) Nonadditivity in cation-peptide interactions: a molecular dynamics and Ab initio study of Na^+ in the gramicidin channel. Chem. Phys. Lett. 212:231–240
Roux B (2010) Exploring the ion selectivity properties of a large number of simplified binding site models. Biophys J 98:2877–2885
Roux B (2012) Ion binding sites and their representations by reduced models. J Phys Chem B 116(23):6966–6979
Roux B, Yu H (2010) Assessing the accuracy of approximate treatments of ion hydration based on primitive quasichemical theory. J Chem Phys 132:234101
Roux B, Allen TW, Bernèche S, Im W (2004) Theoretical and computational models of biological ion channels. Quat Rev Biophys 37:15–103
Roux B, Berneche S, Egwolf B, Lev B, Noskov SY, Rowley CN, Yu H (2011) Ion selectivity in channels and transporters. J Gen Physiol 137:415–426
Schneider R, Ader C, Lange A, Giller K, Hornig S, Pongs O, Becker S, Baldus M (2008) Solid-state NMR spectroscopy applied to a chimeric potassium channel in lipid bilayers. J Am Chem Soc 130:7427–7435
Szabo G, Eisenman G, Laprade R, Ciani SM, Krasne S (1973) Experimentally observed effects of carriers on the electrical properties of bilayer membranes–equilibrium domain. In: With a contribution on the molecular basis of ion selectivity. *Membranes*. pp 179–328

Thomas M, Jayatilaka D, Corry B (2007) The predominant role of coordination number in potassium channel selectivity. Biophys J 93:2635–2643
Valiyaveetil F, Leonetti M, Muir T, Mackinnon R (2006) Ion selectivity in a semisynthetic K+ channel locked in the conductive conformation. Science 314:1004–1007
Varma S, Rempe S (2007) Tuning ion coordination architectures to enable selective partitioning. Biophys J 93:1093–1099
Varma S, Rempe SB (2008) Structural transitions in ion coordination driven by changes in competition for ligand binding. J Am Chem Soc 130:15405–15419
Yamashita M, Wesson L, Eisenman G, Eisenberg D (1990) Where metal ions bind in proteins. Proc Natl Acad Sci USA 87:5648–5652
Yamashita A, Singh S, Kawate T, Jin Y, Gouaux E (2005) Crystal structure of a bacterial homologue of Na^+/Cl^--dependent neurotransmitter transporters. Nature 437:215–223
Yellen G (2002) The voltage-gated potassium channels and their relatives. Nature 419:35–42
Yu H, Noskov SY, Roux B (2009) Hydration number, topological control, and ion selectivity. J Phys Chem B 113:8725–8730
Yu H, Noskov SY, Roux B (2010) Two mechanisms of ion selectivity in protein binding sites. Proc Natl Acad Sci USA 107(47):20329–20334
Zhou M, Morais-Cabral, JH, Mann S, MacKinnon R (2001) Potassium channel receptor site for the inactivation gate and quaternary amine inhibitors. Nature 411:657–661

Understanding Ion Channel Gating Mechanisms by Molecular Dynamics Simulations

2

Meng Cui, Yongcheng Lu, and Diomedes E. Logothetis

Abstract

Drug discovery of membrane proteins, such as ion channels, presents an unprecedented opportunity through recent breakthroughs in the determination of atomic resolution 3D cryo-electron microscopy structures of these proteins. Structure-based drug discovery utilizing in silico methods enables the study of the dynamic connectivity of stable conformations induced by the drug to achieve its effect. With the ever-expanding computational power, simulations of this type reveal protein dynamics in the nano-, micro-, and even millisecond time scales. In the course of hundreds of nanoseconds to a few microseconds, such simulations can capture ion permeation as a result of the gating of ion channel proteins. We call these MD simulations Gating (GMD) as we can see gating and ion permeation during the simulations. Coupled with these technological breakthroughs, our understanding of key common steps regulating the activation of ion channels has matured, allowing us to understand in depth the mechanisms involved in activating an ion channel to perform its function. The insights we have been able to obtain from these GMD simulations have provided an unprecedented understanding of the biological mechanisms of activation of ion channels by PIP_2 and other gating molecules involved and have prepared the road for structure-based drug discovery for this important class of membrane proteins. In this review, we summarize some of these discoveries using GMD from our group as well as others in the field.

M. Cui (✉)
Department of Pharmaceutical Sciences, School of Pharmacy and Pharmaceutical Sciences, Bouvé College of Health Sciences, Northeastern University, Boston, MA, USA

Center for Drug Discovery, Northeastern University, Boston, MA, USA
e-mail: m.cui@northeastern.edu

Y. Lu
Department of Pharmaceutical Sciences, School of Pharmacy and Pharmaceutical Sciences, Bouvé College of Health Sciences, Northeastern University, Boston, MA, USA

D. E. Logothetis
Department of Pharmaceutical Sciences, School of Pharmacy and Pharmaceutical Sciences, Bouvé College of Health Sciences, Northeastern University, Boston, MA, USA

Center for Drug Discovery, Northeastern University, Boston, MA, USA

Affiliate of Chemistry and Chemical Biology, Northeastern University, Boston, MA, USA

Affiliate of Bioengineering, Northeastern University, Boston, MA, USA

Affiliate of Roux Institute of Northeastern University, Portland, ME, USA
e-mail: d.logothetis@northeastern.edu

A. Rosenhouse-Dantsker (ed.), *Decoding Ion Channels Structure and Function*, Advances in Experimental Medicine and Biology 1497, https://doi.org/10.1007/978-3-032-07523-9_2

Keywords

GIRK channel · cholesterol · MD simulations · Protein-ligand interactions · Ion channel activation · Conformational changes · Interaction network analysis · Electrostatic potential

Abbreviations

CHS	Cholesteryl hemisuccinate
cryo-EM	cryo-electron microscopy
CTD	Cytosolic domain
GIRK	G-protein inwardly rectifying K^+
GMD	Gating molecular dynamics
HB	Hydrogen bond
HBC	Helix bundle crossing
HP	Hydrophobic
IC	Ion channel
Kir	Inwardly rectifying potassium
PCA	Principal Component Analysis
PCE	patch-clamp electrophysiology
PIP_2	Phosphatidylinositol 4,5-bisphosphate
SB	Salt bridge
SF	Selectivity filter
TMD	Transmembrane domain

2.1 Introduction

Ion channels (ICs) are membrane proteins, present in every cell of the body, that allow a rapid and large ion movement across cellular membranes. Ion movement is fundamental to many life processes, such as the electrical communication in excitable cells (e.g., neurons, muscle, endocrine cells) that underlie complex processes, including our ability to rapidly respond to environmental stimuli, driving our heartbeat and our pancreatic release of insulin. Ion movement is also fundamental to the function of non-excitable cells like those of epithelia (e.g., in the lung and gut), moving ions and water through osmosis between the lumen of a cavity and the blood perfusing the organ. Examples of epithelial transport through ion channels are NaCl in the lung that hydrates the mucous carpet that clears bacteria and particles we inhale, or HCl in the stomach that acidifies the lumen and turns food into chyme, beginning the digestion process. Defects in these processes result in disease.

Drug discovery targeting ion channels requires detailed structural information, which can be obtained through X-ray crystallography or cryo-electron microscopy (cryo-EM). Recent breakthroughs in resolving ion channel structures at atomic resolution using cryo-EM have opened new opportunities to utilize in silico approaches for studying dynamic conformational changes induced by drugs to achieve their effects. With the ever-expanding computational power, particularly through GPU acceleration, molecular dynamics simulations of ion channels can now reach nano-, micro-, and even millisecond time scales, enabling the capture of ion permeation events resulting from channel gating. These technological breakthroughs enable us to uncover the key steps involved in ion channel activation and gain deeper insights into the mechanisms by which ion channels perform their functions.

Ion movement can be measured as current with unprecedented precision and sensitivity (down to the single-molecule level). Experimental measurements of ion flow through ion channel proteins through patch-clamp electrophysiology (PCE) are accomplished in the millisecond to second time scales. To bridge the gap between molecular dynamics (MD) and PCE measurements to more accurately couple theory and experiment, longer simulations would be needed to make this possible. In some cases, the key to success in gating ion channels in silico and achieving ion permeation in the simulations is the inclusion of the phospholipid phosphatidylinositol bisphosphate (PIP_2), a cofactor required for the final step of stabilizing (opening) or destabilizing (closing) the gating transition. PIP_2 regulates the constriction points (gates) along the permeation pathway that ions use to flow into and out of the cell across the plasma membrane. PIP_2 has been called the "Master" regulator of membrane proteins, and although it is a minor component of the inner leaflet of eukaryotic plasma membranes, comprising only 1% of their total phospholipid pool, it is a critical signaling molecule. The two phosphates at positions 4′ and 5′ of the inositol ring confer a negative charge to the phospholipid headgroup at the lipid-to-cytosol interface that interacts with positively charged amino acids of channel proteins to allosterically control the conformation of channel gates. Previous studies have revealed that multiple mechanisms by which channel activity is either stimulated or inhibited involve allosteric strengthening or weakening of channel-PIP_2 interactions, respectively. Although many ion channels exhibit stimulatory effects in response to PIP_2, such as voltage-gated potassium (Kv) channels, inwardly rectifying potassium (Kir) channels, and calcium-activated potassium (KCa) channels, some channels are inhibited by PIP_2 (Hille et al. 2015; Logothetis et al. 2015). Examples include Nav1.4 (Gada et al. 2023), Kv1.3 (Matsushita et al. 2009), and the cyclic nucleotide-gated channel SthK (Thon et al. 2024). In certain cases, PIP_2 regulation is complex or even contradictory, as observed in channels like Kv1.2 (Rodriguez-Menchaca et al. 2012), TRPC5 (Ningoo et al. 2021), and TRPV4 (Harraz et al. 2018; Takahashi et al. 2014). For example, PIP_2 depletion leads to bidirectional regulation of the voltage-dependent Kv1.2 channel. Specifically, PIP_2 depletion causes a leftward shift in the voltage dependence of activation, increasing current initially, but simultaneously decreases the open probability, resulting in an overall reduction in current amplitude (Rodriguez-Menchaca et al. 2012). PIP_2 also plays a dual role in regulating TRPC5 channel activity, both facilitating activation and contributing to desensitization. Upon activation of phospholipase C (PLC), PIP_2 is hydrolyzed into diacylglycerol (DAG) and inositol trisphosphate (IP_3). DAG can activate TRPC5 channels, while IP_3 triggers intracellular calcium release, which also modulates channel activity. Concurrently, the depletion of PIP_2 and the activation of protein kinase C (PKC) by DAG contribute to TRPC5 channel desensitization (Ningoo et al. 2021). Furthermore, not only the physiological activators of ion channels but also small-molecule drug activators or inhibitors open or close the gates by strengthening or weakening channel-PIP_2 interactions, respectively.

We have been able to utilize static stable conformations of ion channels depicted by atomic resolution structures determined by X-ray crystallography or cryo-EM and model them in a lipid environment, docking PIP_2 in a putative binding site, and within microsecond-long Gating Molecular Dynamics (GMD) simulations, we have captured channel gating and permeation of ions with only one of multiple putative PIP_2 binding sites. The first example of in silico gating and permeation was achieved six years ago with a class of K^+ channels (G-protein inwardly rectifying K^+ or GIRK), which we have been studying for over 35 years (Logothetis et al. 1987; Cui et al. 2021). This achievement demonstrated the power of this approach (Li et al. 2019). The trajectories of the simulations contain a wealth of mechanistic information revealing how gating occurs and what are the precise changes in channel-PIP_2 interactions that mediate gating. More recently, we have shown that small-molecule activators of GIRK channels can cause ion permeation in GMD sub-microsecond-long simulations (Li et al. 2019; Cui et al. 2022, 2024; Xu et al. 2020). Within these trajectories we have discerned (a) the molecular mechanism by which a urea-based scaffold of small-molecule activators is able to open the channel gates by enhancing channel-PIP_2 interactions, (b) how a small chemical modification switches the effect of the small molecule from being an activator to being an inhibitor through weakening channel-PIP_2 interactions, and (c) how another subtle chemical modification makes the small molecule specific among different GIRK isoforms distinguishing those expressed in neurons from those in heart cells. Notably, this computational approach yields a dynamic "movie," rather than static pictures, of how a small molecule specifically stimulates or inhibits channel gating by adjusting channel-PIP_2 interactions within the time scale of hundreds of nanoseconds. The insights obtained from these GMD simulations have provided an unprecedented understanding of the biological mechanisms of activation of ion channels by PIP_2 and other gating molecules and have set the stage for structure-based drug discovery for this important class of membrane proteins. In this review, we summarize these discoveries using GMD from our group as well as others in the field.

2.2 Inwardly Rectifying Potassium (Kir) Channels Gating by PIP_2

Our first MD simulation studies on Kir channels were conducted in 2012. They were based on a high-resolution structure of Kir3.1 chimera (PDB entry: 2QKS) and were carried out in both closed and open conformations of the cytosolic (G-loop) gate (Meng et al. 2012).

The high-resolution structure of the Kir3.1 chimera closely resembles that of KcsA in the selectivity filter region, likely reflecting the conserved structural features required for rapid and selective K^+ conduction. K^+ ions were found not only within the selectivity filter but also at non-filter sites along the extended pore. The presence of multiple ions both inside and outside the filter supports the hypothesis that the high electrical valence of voltage-dependent blockers arises from their coupled movement with permeant ions. The presence of two distinct conformations within a single crystal form suggests that rigid-body motions of the cytoplasmic pore subunits are coupled to the position and conformation of the G-loop, which may serve as a gate within the cytoplasmic pore. PIP_2 may influence channel gating by binding to specific sites on the protein and/or by inserting into the inner membrane leaflet near the channel. Several amino acids on the surface of the cytoplasmic pore are well positioned to interact with PIP_2 and modulate pore function (Nishida et al. 2007).

The simulation results showed that the closed state of the channel with PIP_2 underwent a conformational change to an intermediate state between the closed and open states. The key elements in the G-loop gating transition were identified as a PIP_2-driven movement of the N-terminus and C-linker that removed intermolecular constraints and led to CD-loop stabilization of the G-loop gate in the open state. This was our first MD simulation study on PIP_2-induced channel gating that was consistent with existing experimental data. However, the MD simulations were conducted for only 100 ns, and no helix bundle crossing (HBC) gate opening, or ion permeation events were observed even in the presence of an electric field (corresponding to a 200 mV membrane potential).

2.3 Kir Ion Channel Mutant Facilitates Channel Opening and Ion Permeation

To achieve Kir channel opening and observe ion permeations, we conducted another 100 ns MD simulation study based on a Kir3.1 chimera channel mutant M170P, which corresponds to the S170P mutant in Kir3.1 (Fig. 2.1) (Meng et al. 2016). Previous studies showed that the Kir3.1 (S170P) mutant was a constitutively active channel (Jin et al. 2002). During these 100 ns MD simulations in the presence of an electric field equivalent to a membrane potential of 200 mV, the HBC gate reached 0.6 nm in diameter, allowing three partially hydrated K^+ ions to pass. Both transmembrane helices, TM1 and TM2, cooperatively rotated in a counterclockwise direction (extracellular side view) with the aid of PIP_2. All transmembrane helices were required to adopt a counterclockwise rotation to stabilize the HBC gate in the open state. The overall hydrophobic interactions between TM1 and TM2 from adjacent subunits were weakened during the transition of the HBC gate from the closed to the open state. PIP_2 binding enhanced the bending of TM2 at G169 by 7° (greater in the holo than the apo) to open the HBC gate. Introduction of a proline mutation reduced the energy barrier for opening the HBC gate by approximately 1.4 kcal/mol. This was the first time we observed K^+ ion permeation during such a short period of simulation time (100 ns). However, for the opening of the wild-type Kir channel by PIP_2, longer simulation times were needed.

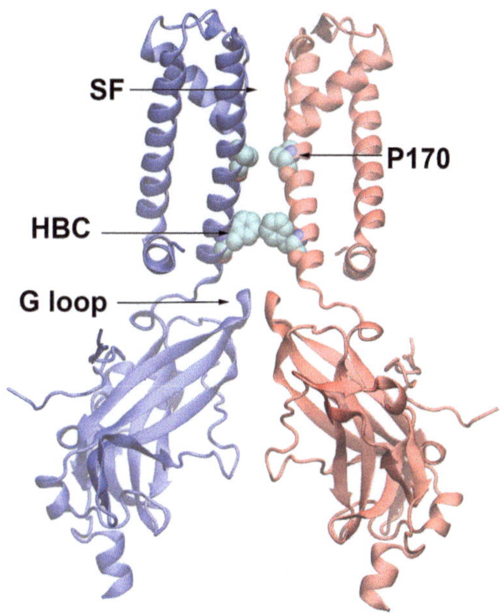

Fig. 2.1 M170P Mutation in the Kir3.1 chimera facilitates the channel gating. The Kir3.1-M170P chimera (based on PDBID: 2QKS) is rendered as new cartoon. Only two diagonally facing subunits of the channel are shown for clarity. The structural elements, SF (Selectivity Filter), HBC (Helix Bundle Crossing) gate, and G-loop gate are labeled on the structure

2.4 Na^+ and Gβγ Regulate GIRK Channel Activity Through PIP_2

With the advances in GPU computers and simulation software, such as Amber, longer timescale MD simulations on Kir channels became possible. In 2019, we conducted 500 ns MD simulations on the Kir3.2 (GIRK2, PDB entry: 4KFM) channel in the presence of different regulators, including PIP_2, Na^+, and the G-protein βγ dimer (Gβγ) (Fig. 2.2) (Li et al. 2019). The crystal structure of the mammalian GIRK2 channel was resolved at 3.5 Å in complex with Gβγ subunits, Na^+, and DiC_8- PIP_2. Four Gβγ subunits bind at the interfaces between the four K^+ channel subunits, inducing a pre-open state characterized by a conformation intermediate between the closed and open states. The structure also provides insight into how PIP_2 and Na^+ contribute to the multi-ligand regulation of GIRK channels (Whorton and MacKinnon 2011). During simulations of GIRK2-Apo in the absence of any regulators, no K^+ ion permeation events were observed. In contrast, the GIRK2-PIP_2, GIRK2-PIP_2-Na^+, and GIRK2-PIP_2-Gβγ systems exhibited 3–4 ions passing through the channel gates. Notably, in the GIRK2-PIP_2-Gβγ-Na^+ system, a substantial increase was observed with 36 ions passing through the channel gates. These simulation results suggested that Na^+ ions exerted control over the cytosolic gate of the channel through an anticlockwise rotation, whereas Gβγ stabilized the transmembrane gate in the open state through a rocking movement of the cytosolic domain. Each of these effects altered the channel's interactions with PIP_2, contributing to the stabilization of the open state of the respective gate. These investigations into GIRK channel dynamics unveiled, for the first time, a comprehensive structural model that aligned with the extensive body of literature on GIRK channel function.

2.5 Ion Permeation Mechanism for GIRK2 Channel

To further understand K^+ ion permeation mechanisms, Li et al. conducted 1 μs all-atom GMD simulations with a 200 mV electric field on three GIRK2 channel systems in complex with PIP_2, Na^+ or Gβγ regulators, including GIRK2-PIP_2-Na^+, GIRK2-PIP_2-Gβγ, and GIRK2-PIP_2-Gβγ-Na^+ (Li et al. 2020). Prior studies indicated that, despite the activation of both the HBC and G-

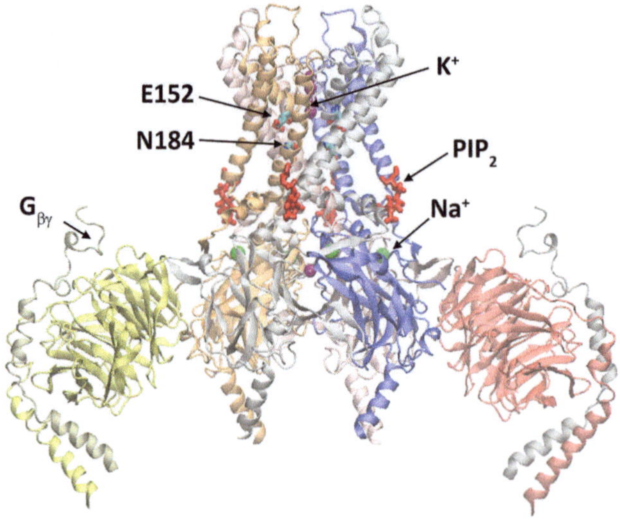

Fig. 2.2 Modulation of GIRK2 channel gating by PIP_2, Na^+, and Gβγ. The GIRK2 channel and Gβγ subunits are rendered as a new cartoon and colored by chain (PDB: 4KFM). PIP_2 is depicted as licorice and colored red. K^+ and Na^+ ions are shown as spheres in purple and green, respectively. Only two Gβγ subunits are shown for clarity

loop gates in the GIRK channel by regulators, the permeation of K^+ ions remained inefficient (Li et al. 2019). The 1 μs all-atom GMD simulation pointed to a third "gate", suggesting that the ion permeation limitation could be attributed to the SF rather than the HBC or the G-loop. Ion permeation through the SF was facilitated by water molecules and aided by a tilt in the SF. Furthermore, these studies have demonstrated that four K^+ ions are required in the central cavity of the channel to facilitate ion permeation through the opened HBC gate. The four negatively charged residues E152 in GIRK2 create an energy barrier for the permeation of K^+ ions through electrostatic interactions, indicating the need for the presence of three K^+ ions to neutralize and reduce the barrier. After the fourth K^+ ion enters the cavity, it can traverse through the HBC gate with a diminished energy barrier. These findings contribute to a deeper understanding of the dynamic ion permeation mechanisms through GIRK2 channels. A subsequent GMD study by Li et al. further confirmed the proposed mechanism using virtual mutations of the residue E152 of the GIRK2 mutant R201A that exhibits low conductivity in the presence of PIP_2. The studies showed that E152 was the key residue facilitating ion permeation through the central cavity and HBC gate. In contrast, mutation of residue N184 (located in the TM pore cavity), which was previously reported to significantly enhance K^+ selectivity over Na^+ and improve inward rectification of GIRK2 channels when replaced with negatively charged residues (N184D or N184E) (Bichet et al. 2006), did not produce notable effects when substituted with other residues, such as leucine or glycine (Li et al. 2024).

2.6 GIRK Channel Activity Is Regulated by Small Molecules

ML297 was discovered as a GIRK1/2 channel activator (Kaufmann et al. 2013). Interestingly, minor structural changes led to a channel inhibitor, GAT1587 (Wen et al. 2014). To elucidate the molecular mechanism underlying the action of these small molecules on channel activity, we conducted sub-microsecond GMD on the models of GIRK1/2-PIP_2-ML297 (Fig. 2.3) and GIRK1/4-PIP_2-GAT1587 systems (Gazgalis et al. 2022). The simulations enabled the monitoring of the dynamics of two channel gates,

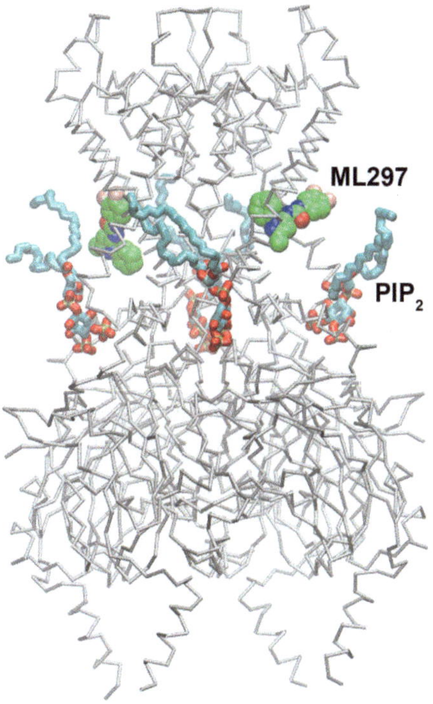

Fig. 2.3 Modulation of GIRK1/2 channel gating by small molecules. The GIRK1/2 channel is rendered as a Cα trace colored gray. PIP_2 is depicted as licorice and colored by atom types. ML297 (activator) is shown as van der Waals (vdW) representations, colored by atom type

the HBC and G-loop, as well as their control by PIP_2, and captured the channel gating transitions and K^+ ion permeations. A comparison of the two systems pinpointed three hydrophobic residues (F87, Y91, and W95) in the TM1 of GIRK1, which form a hydrophobic wire in the ML297 system but not in the GAT1587 system. This interaction prompted the bending of TM2, aligning two acidic GIRK1 residues (E141 and D173) for K^+ ion coordination in the permeation pathway, thereby facilitating K^+ ion conduction. The movement of the Slide Helix, propelled by ML297/GAT1587-induced TM1 movement, implicated conformational changes in the CD-loop interactions, regulating the gating of the G-loop gate. Accordingly, both molecules influence the HBC gate by either strengthening or weakening channel-PIP_2 interactions.

2.7 Cholesteryl Hemisuccinate (CHS) Activates GIRK2 Channel

Cholesterol, an essential lipid, that plays critical roles in maintaining the integrity and fluidity of cell membranes, can activate GIRK channels (Rosenhouse-Dantsker et al. 2010; Corradi et al. 2022). A cryo-EM structure of GIRK2 in complex with CHS (cholesteryl hemisuccinate, a cholesterol analog) and diC$_8$-PIP_2 (a short acyl chain soluble analog of PIP_2) was solved in 2021 (PDBID: 6XEV) (Mathiharan et al. 2021), providing the opportunity to understand the molecular mechanism by which cholesterol activates GIRK channels using GMD (Fig. 2.4). Based on the cryo-EM structure, we conducted microsecond-long GMD simulations on three systems: GIRK2-APO, GIRK2/PIP_2, and GIRK2/PIP_2/CHS. These simulations were followed by systematic analyses of the GMD trajectories, including Principal Component Analysis (PCA), conformation changes, salt bridge, hydrogen bond, hydrophobic residue pair correlation network, binding sites of CHS and PIP_2, and electrostatic potential of the channel, to gain molecular insights into how PIP_2 and CHS modulate GIRK2 channel gating (Cui et al. 2024).

The GMD simulation results suggested that both the HBC and G-loop gates were closed in the GIRK2-APO state, and there was no K^+ ion permeation during the 1 μs simulations. In the presence of PIP_2, both gates were near their open states (around 6 Å, based on the average minimum distances between two diagonal gate residue groups F192 (HBC gate) or M313-M319 (G-loop gate), yielding two K^+ ion permeation events. In the presence of PIP_2 and CHS, both gates transitioned to an open conformation, with the minimum distances remaining near those observed in the initial cryo-EM structure. 43 K^+ ions were observed passing through both gates of the GIRK2 channel. These results indicate that CHS facilitates GIRK2 channel opening in the presence of PIP_2.

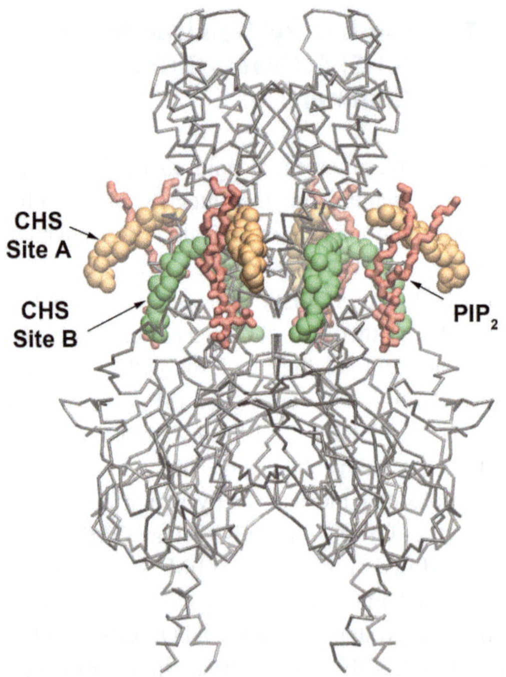

Fig. 2.4 Modulation of GIRK2 channel gating by CHS. The GIRK2 channel is rendered as a Cα trace colored gray. PIP_2 is depicted as licorice and colored red. CHS are shown as van der Waals (vdW) representations, colored by orange and green for site A and site B binding sites, respectively

To study K^+ ion permeation, we calculated dwell times of K^+ ions in the GIRK2 channel. The results suggested that K^+ ions spent most of their time in the SF region, with 67.6% observed for the GIRK2/PIP_2 system and 54.3% for the GIRK2/PIP_2/CHS system. In contrast, K^+ ions spent less time in the central cavity and G-loop regions: 16.0% and 3.3%, respectively, for the GIRK2/PIP_2 system; and 10.0% and 3.4%, respectively, for the GIRK2/PIP_2/CHS system. The observation that fewer K^+ ions were found in the central cavity (SF-HBC) of the GIRK2-PIP_2-CHS system compared to the GIRK2-PIP_2 system was consistent with increased ion passage and HBC gate opening in the GIRK2-PIP_2-CHS system.

Additionally, the percentages of K^+ ion distribution within the channel for the three systems were also calculated. During the simulations, in the central cavity region between the SF and the HBC gate of the channel, the percentages of K^+ ion occupation were as follows for the GIRK2-APO state: 0% (1 ion), 16.2% (2 ions), 42.9% (3 ions), 40.9% (4 ions), and 0% (5 ions). In contrast, for the GIRK2/PIP_2 system, the percentages were 0% (1 ion), 15.0% (2 ions), 16.0% (3 ions), 47.6% (4 ions), and 21.5% (5 ions). For the GIRK2/PIP_2/CHS system, the percentages were 0.3% (1 ion), 14.89% (2 ions), 51.6% (3 ions), 31.6% (4 ions), and 1.7% (5 ions). It is worth noting that ions spent more time near residues E152 and N184 than any other residues in the central cavity region of the channel. This observation is consistent with the previously published results by Li et al. (2020, 2024). The residue E152 in the central cavity right below the SF in GIRK2 creates an electrostatic energy barrier for K^+ ion permeation, which requires the presence of three K^+ ions to neutralize and reduce the barrier. After the fourth K^+ ion enters the cavity, it can traverse through the HBC gate with a diminished energy barrier. To better understand this mechanism, we calculated electrostatic potential maps inside the GIRK2 channel to illustrate it. As shown in Fig. 2.5, the internal electrostatic potential barrier was lowered by adding four PIP_2 molecules (compare Fig. 2.5a and b; Fig. 2.5g). However, single K^+ ions still could not pass across the energy barrier created by the four negatively charged E152 residues, even in the presence of PIP_2 and an electric field. This energy barrier was eliminated once additional K^+ ions entered the cavity and neutralized residue E152. At least two K^+ ions were required to lower the energy barrier, allowing further K^+ permeation (Fig. 2.5c, d, e, and g).

2.8 Long-Timescale MD Simulations for Kir Channel Gating and Modulation

Recently, Jogini et al. conducted long-timescale MD simulations (2–200 μs, under applied voltages of 155–375 mV) on the Kir2.2 channel (PDB ID 3SPI) to further elucidate channel gating and modulation by PIP_2 and spermine (Jogini et al. 2023). The X-ray crystal structure of the Kir2.2 channel was resolved in complex

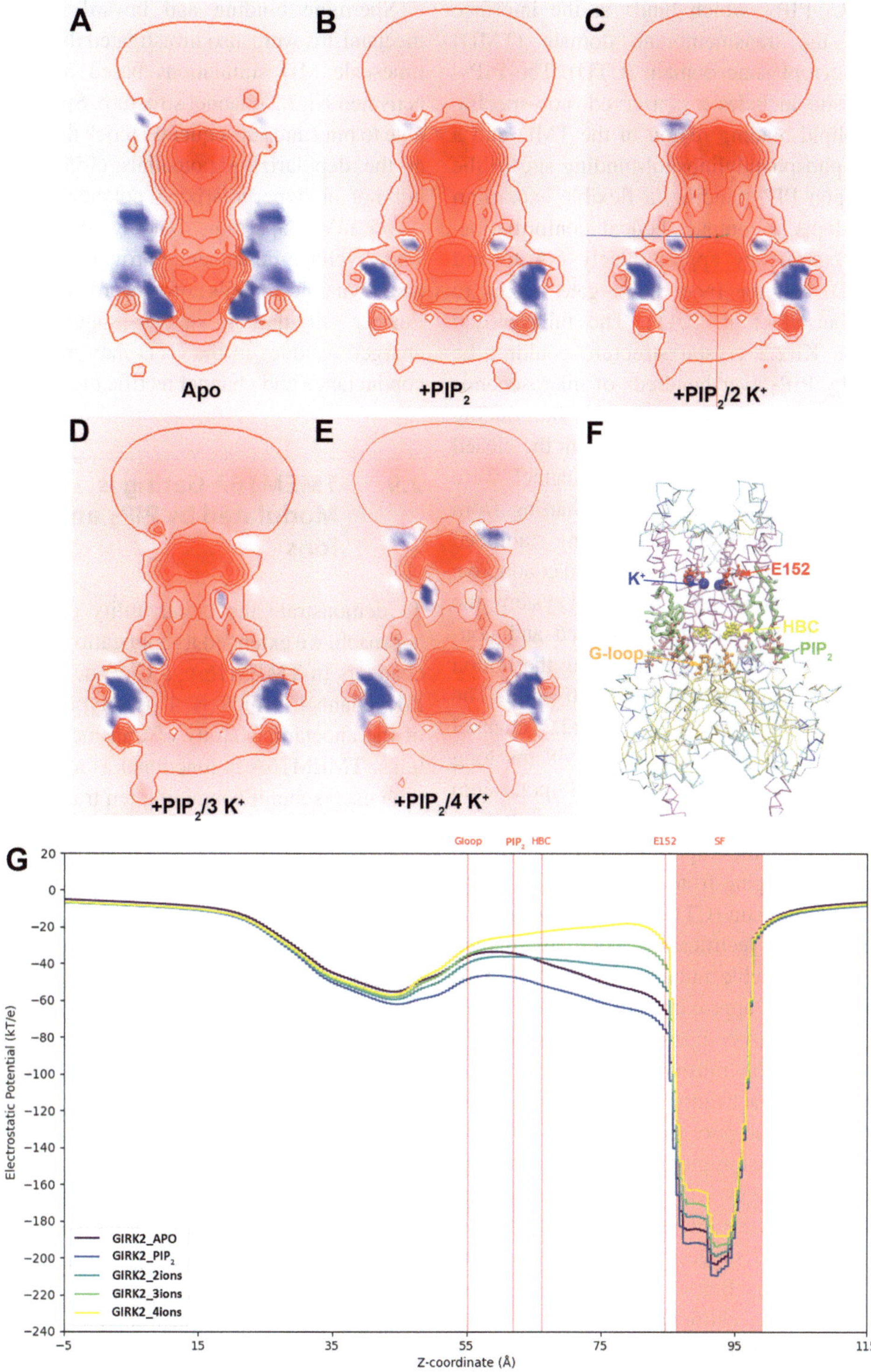

Fig. 2.5 Electrostatic potential maps of GIRK2. (**a**) GIRK2-APO; (**b**) GIRK2-PIP$_2$; (**c**) GIRK2-PIP$_2$/2 K$^+$; (**d**) GIRK2-PIP$_2$/3 K$^+$; (**e**) GIRK2-PIP$_2$/4 K$^+$. The negative and positive electrostatic potential regions are colored by red and blue, respectively. The values of the electrostatic potential contours are −10, −20, −30, −40, −50, −60, −70, −80, and − 100 kT/e, from outside to inside. The Grasp program was used for electrostatic potential calculations and figure generation. (**f**) 3D structure of GIRK2 in complex with PIP$_2$. (**g**) Electrostatic potential plots along the K$^+$ ion permeation pathway

with DiC_8-PIP_2, which binds at the interface between the transmembrane domain (TMD) and the cytoplasmic domain (CTD). The PIP_2-binding site includes a conserved, non-specific phospholipid-binding region in the TMD and a specific phosphatidylinositol-binding site in the CTD. Upon PIP_2 binding, a flexible expansion linker adopts a compact helical conformation, the CTD shifts by approximately 6 Å toward the TMD, and the inner helix gate begins to open (Hansen et al. 2011). The fully closed wild-type Kir2.2 crystal structure couldn't be opened by PIP_2 over hundreds of microseconds of MD simulations. However, after introducing the G178R mutation, transitions from the closed state to the open state of the channel were observed during the simulations. Starting from an open state structure, PIP_2 binding stabilized the wild-type channel in the open and conductive state. In contrast, in a PIP_2-depleted system, the open pore transitioned into a closed and non-conducting conformation, supporting the notion that PIP_2 is critical for channel activation. The estimated conductance of the Kir2.2 channel was 12.9 ± 2.9 pS based on two 48 ms long simulations where 929 and 1461 potassium ions permeated, respectively. This result was consistent with the experimentally determined conductance ranging from 2–30 pS. Long-range cytoplasmic domain (CTD) electrostatics on ion permeation and rectification were investigated using a virtual double mutation of E225A/E300A, which yielded approximately sixfold lower permeation rates compared to the wild-type channel. When the transmembrane (TM) helices collapsed at the activation gate, the Slide Helix shifted toward the pore axis. Without PIP_2, the arginine/lysine-rich region in TM1/TM2 instead formed salt bridges with Slide Helix residues D69 and D76. These interactions became tighter as the pore closed, suggesting a role in stabilizing the closed state. Based on this observation, the authors ran additional simulations using a D69A/D76A double mutant to test the role of these residues in Kir channel deactivation. These simulations resulted in a stable, open, and conducting channel, even in the absence of PIP_2.

Spermine binding and inward rectification mechanisms were also investigated through long-timescale MD simulations based on the fully activated Kir2.2 channel structure. Spermine was able to bind and subsequently block the open pore at the depolarizing potentials (155–310 mV), but not at hyperpolarizing potentials (−310 – −155 mV). Spermine binding and unbinding were voltage-dependent, providing structural and dynamic details of inward rectification. Further simulation studies suggested some charged residues in the CTD may modulate ion conductance and channel rectification.

2.9 TMEM16A Gating Is Modulated by PIP_2 and Ca^{2+} Ions

To demonstrate the broad utility of the GMD approach, we extended its application beyond Kir channels to TMEM16A, a Ca^{2+}-activated chloride channel, which is a prototypical member of the anoctamin family of transmembrane proteins. TMEM16A is functional as a homodimer, with each subunit containing ten transmembrane helices (α1–10) and a putative Cl^- permeation pathway. Several functional studies have demonstrated that TMEM16A gating is dependent on the presence of PIP_2 and Ca^{2+} ions. However, the molecular mechanism by which PIP_2 activates TMEM16A channel has not been clear. We thus conducted GMD simulations on TMEM16A in the presence of PIP_2 and/or Ca^{2+} ions. Consistent with the literature (Yu et al. 2019), we identified four potential PIP_2 binding sites on the TMEM16A channels, but only one of them exhibited Cl^- ion permeation during 2 μs GMD simulations (Fig. 2.6). The interactions of PIP_2 with the α4 and α6 helices of TMEM16A were characterized through both its phosphate headgroups and acyl chains, revealing the mechanism that underpins channel gating. Overall, this study demonstrated that GMD simulations, coupled with functional validation, could serve as an effective strategy to dissect molecular mechanisms that underlie PIP_2-enabled ion channel

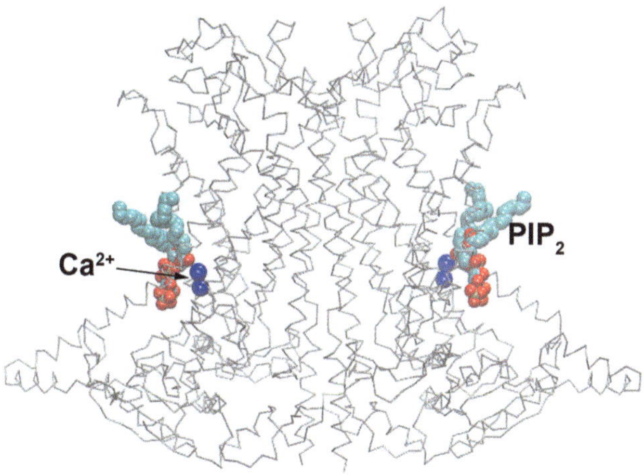

Fig. 2.6 Modulation of TMEM16A channel gating by PIP_2 and Ca^{2+}. TMEM16A channel is rendered as Cα trace colored gray (PDB: 5OYB). PIP_2 is depicted as van der Waals (vdW) representation and colored by atom types. Ca^{2+} ions are shown as spheres in blue

gating. Hence, the GMD approach can provide a robust platform to unravel further insights into the structural dynamics that govern the activity of TMEM16A channels. This approach has potential implications for dissecting mechanisms underlying modulators of the channels and predicting structure-activity relationships for drug development (Xu et al. 2025).

2.10 Conclusions and Outlook

In this review, we summarized examples of applications of GMD simulations to study ion channel conformational changes, gating, and activation, providing a better understanding of channel function at the molecular level. To observe channel gating, MD simulations on the timescale of hundreds of nanoseconds to microseconds are required. The initial conformations of ion channels can significantly impact the observation of channel gating. For example, Kir2.2, when starting from a fully closed conformation, could not open even during hundreds of microseconds of MD simulations. Ion channel activity can be modulated by mutations, small molecules, ions, and proteins, which can be observed during MD simulations on the hundreds-of-nanoseconds timescale. The lipid molecule PIP_2 plays a critical role in the gating of many ion channels, including facilitating HBC gate opening in Kir channels. With its five negative charges, PIP_2 also alters the internal electrostatic potential and lowers the energy barrier for ion permeation. An external electric field mimicking the membrane potential is also required during GMD simulations to observe ion permeation, as it provides the driving force for ions to overcome the internal energy barrier of the channels. GMD trajectories contain a wealth of information that can be extracted and analyzed to gain molecular-level insights into the channel gating mechanism. The current limitation of GMD is its timescale, which is restricted to the hundreds-of-nanoseconds to microsecond range. This limitation could be overcome with future advancements in powerful GPU supercomputers and improved parallelization efficiency of software.

Since channel-PIP_2 interactions are critical determinants of allosteric channel gating future detailed studies of high-resolution structures of different channel isoforms (e.g., for GIRKs: GIRK4, GIRK1/4, GIRK1/2, GIRK2/3, GIRK1/3; for IRKs: IRK1, IRK3, IRK4, IRK5, IRK6; for TMEM16s: TMEM16B, TMEM16A/B, TMEM16F, etc.) will allow a detailed understanding of how strategic changes in structure result in channel-PIP_2 interaction changes to accordingly control gating. In addition, the allosteric nature of the control that channel-PIP_2 interactions exert on channel gating makes them prime targets of small molecules that could stabilize open or closed channel conformations. Structure-based drug design that is now based not on still experimental atomic resolution structures but rather on the dynamic behavior of channels captured by GMD simulations offers a powerful and exciting direction in drug design for ion channel proteins.

Acknowledgments The computations were supported by the ITS (Information Technology Services) Research Computing at Northeastern University and the Argonne Leadership Computing Facility (ALCF) at Argonne National Laboratory. The work was supported by NIH R21DC020136 grant award (M.C.), and NIH grants R01HL059949 and R01NS131467 (D.E.L.).

Data Availability Statement All data are contained within the manuscript.

Conflict of Interest The authors declare that they have no conflicts of interest with the contents of this article.

References

Bichet D, Grabe M, Jan YN, Jan LY (2006) Electrostatic interactions in the channel cavity as an important determinant of potassium channel selectivity. Proc Natl Acad Sci USA 103:14355–14360. https://doi.org/10.1073/pnas.0606660103

Corradi V, Bukiya AN, Miranda WE, Cui M, Plant LD, Logothetis DE et al (2022) A molecular switch controls the impact of cholesterol on a Kir channel. Proc Natl Acad Sci U S A 119:e2109431119. https://doi.org/10.1073/pnas.2109431119

Cui M, Cantwell L, Zorn A, Logothetis DE (2021) Kir Channel molecular physiology, pharmacology, and therapeutic implications. Handb Exp Pharmacol 267:277–356. https://doi.org/10.1007/164_2021_501

Cui M, Xu K, Gada KD, Shalomov B, Ban M, Eptamini-taki GC et al (2022) A novel small-molecule selective activator of homomeric GIRK4 channels. J Biol Chem 298:102009. https://doi.org/10.1016/j.jbc.2022.102009

Cui M, Lu Y, Ma X, Logothetis DE (2024) Molecular mechanism of GIRK2 channel gating modulated by cholesteryl hemisuccinate. Front Physiol 15:1486362. https://doi.org/10.3389/fphys.2024.1486362

Gada KD, Kamuene JM, Chandrashekar A, Kissell RC, Yauch AK, Plant LD (2023) PI(4,5)P2 regulates the gating of NaV1.4 channels. J Gen Physiol 155. https://doi.org/10.1085/jgp.202213255

Gazgalis D, Cantwell L, Xu Y, Thakur GA, Cui M, Guarnieri F et al (2022) Use of a molecular switch probe to activate or inhibit GIRK1 heteromers in silico reveals a novel gating mechanism. Int J Mol Sci 23. https://doi.org/10.3390/ijms231810820

Hansen SB, Tao X, MacKinnon R (2011) Structural basis of PIP2 activation of the classical inward rectifier K+ channel Kir2.2. Nature 477:495–498. https://doi.org/10.1038/nature10370

Harraz OF, Longden TA, Hill-Eubanks D, Nelson MT (2018) PIP(2) depletion promotes TRPV4 channel activity in mouse brain capillary endothelial cells. Elife 7. https://doi.org/10.7554/eLife.38689

Hille B, Dickson EJ, Kruse M, Vivas O, Suh BC (2015) Phosphoinositides regulate ion channels. Biochim Biophys Acta 1851:844–856. https://doi.org/10.1016/j.bbalip.2014.09.010

Jin T, Peng L, Mirshahi T, Rohacs T, Chan KW, Sanchez R et al (2002) The (beta)gamma subunits of G proteins gate a K(+) channel by pivoted bending of a transmembrane segment. Mol Cell 10:469–481. https://doi.org/10.1016/s1097-2765(02)00659-7

Jogini V, Jensen MO, Shaw DE (2023) Gating and modulation of an inward-rectifier potassium channel. J Gen Physiol 155. https://doi.org/10.1085/jgp.202213085

Kaufmann K, Romaine I, Days E, Pascual C, Malik A, Yang L et al (2013) ML297 (VU0456810), the first potent and selective activator of the GIRK potassium channel, displays antiepileptic properties in mice. ACS Chem Neurosci 4:1278–1286. https://doi.org/10.1021/cn400062a

Li D, Jin T, Gazgalis D, Cui M, Logothetis DE (2019) On the mechanism of GIRK2 channel gating by phosphatidylinositol bisphosphate, sodium, and the Gbetagamma dimer. J Biol Chem 294:18934–18948. https://doi.org/10.1074/jbc.RA119.010047

Li DL, Hu L, Wang L, Chen CL (2020) Permeation mechanisms through the selectivity filter and the open helix bundle crossing gate of GIRK2. Comput Struct Biotechnol J 18:3950–3958. https://doi.org/10.1016/j.csbj.2020.11.039

Li D, Shi D, Wang L (2024) Structural insights in the permeation mechanism of an activated GIRK2 channel. Biochim Biophys Acta Biomembr 1866:184231. https://doi.org/10.1016/j.bbamem.2023.184231

Logothetis DE, Kurachi Y, Galper J, Neer EJ, Clapham DE (1987) The beta gamma subunits of GTP-binding proteins activate the muscarinic K+ channel in heart. Nature 325:321–326. https://doi.org/10.1038/325321a0

Logothetis DE, Petrou VI, Zhang M, Mahajan R, Meng XY, Adney SK et al (2015) Phosphoinositide control of membrane protein function: a frontier led by studies on ion channels. Annu Rev Physiol 77:81–104. https://doi.org/10.1146/annurev-physiol-021113-170358

Mathiharan YK, Glaaser IW, Zhao Y, Robertson MJ, Skiniotis G, Slesinger PA (2021) Structural insights into GIRK2 channel modulation by cholesterol and PIP(2). Cell Rep 36:109619 10.1016/j.celrep.2021.109619

Matsushita Y, Ohya S, Suzuki Y, Itoda H, Kimura T, Yamamura H et al (2009) Inhibition of Kv1.3 potassium current by phosphoinositides and stromal-derived factor-1alpha in Jurkat T cells. Am J Physiol Cell Physiol 296:C1079–C1085. https://doi.org/10.1152/ajpcell.00668.2008

Meng XY, Zhang HX, Logothetis DE, Cui M (2012) The molecular mechanism by which PIP(2) opens the intracellular G-loop gate of a Kir3.1 channel. Biophys J 102:2049–2059. https://doi.org/10.1016/j.bpj.2012.03.050

Meng XY, Liu S, Cui M, Zhou R, Logothetis DE (2016) The molecular mechanism of opening the helix bundle crossing (HBC) Gate of a Kir Channel. Sci Rep 6:29399. https://doi.org/10.1038/srep29399

Ningoo M, Plant LD, Greka A, Logothetis DE (2021) PIP2 regulation of TRPC5 channel activation and desensitization. J Biol Chem 296:100726. https://doi.org/10.1016/j.jbc.2021.100726

Nishida M, Cadene M, Chait BT, MacKinnon R (2007) Crystal structure of a Kir3.1-prokaryotic Kir channel chimera. EMBO J 26:4005–4015. https://doi.org/10.1038/sj.emboj.7601828

Rodriguez-Menchaca AA, Adney SK, Tang QY, Meng XY, Rosenhouse-Dantsker A, Cui M et al (2012) PIP2 controls voltage-sensor movement and pore opening of Kv channels through the S4-S5 linker. Proc Natl Acad Sci U S A 109:E2399–E2408. https://doi.org/10.1073/pnas.1207901109

Rosenhouse-Dantsker A, Leal-Pinto E, Logothetis DE, Levitan I (2010) Comparative analysis of cholesterol sensitivity of Kir channels: role of the CD loop. Channels (Austin) 4:63–66. https://doi.org/10.4161/chan.4.1.10366

Takahashi N, Hamada-Nakahara S, Itoh Y, Takemura K, Shimada A, Ueda Y et al (2014) TRPV4 channel activity is modulated by direct interaction of the ankyrin domain to PI(4,5)P(2). Nat Commun 5:4994. https://doi.org/10.1038/ncomms5994

Thon O, Wang Z, Schmidpeter PAM, Nimigean CM (2024) PIP2 inhibits pore opening of the cyclic nucleotide-gated channel SthK. Nat Commun 15:8230. https://doi.org/10.1038/s41467-024-52469-1

Wen W, Wu W, Weaver CD, Lindsley CW (2014) Discovery of potent and selective GIRK1/2 modulators via 'molecular switches' within a series of 1-(3-cyclopropyl-1-phenyl-1H-pyrazol-5-yl)ureas. Bioorg Med Chem Lett 24:5102–5106. https://doi.org/10.1016/j.bmcl.2014.08.061

Whorton MR, MacKinnon R (2011) Crystal structure of the mammalian GIRK2 K+ channel and gating regulation by G proteins, PIP2, and sodium. Cell 147:199–208. https://doi.org/10.1016/j.cell.2011.07.046

Xu Y, Cantwell L, Molosh AI, Plant LD, Gazgalis D, Fitz SD et al (2020) The small molecule GAT1508 activates brain-specific GIRK1/2 channel heteromers and facilitates conditioned fear extinction in rodents. J Biol Chem 295:3614–3634. https://doi.org/10.1074/jbc.RA119.011527

Xu J, Santa-Cruz A, Chandrashekar A, Kawano T, Zaka M, Cui M, Diomedes DE, Plant LD (2025) Orthosteric interactions with PIP 2 activate TMEM16A channels. bioRxiv, 2025 Feb 27, https://doi.org/10.1101/2025.02.24.639951

Yu K, Jiang T, Cui Y, Tajkhorshid E, Hartzell HC (2019) A network of phosphatidylinositol 4,5-bisphosphate binding sites regulates gating of the Ca(2+)-activated Cl(−) channel ANO1 (TMEM16A). Proc Natl Acad Sci USA 116:19952–19962. https://doi.org/10.1073/pnas.1904012116

Mutagenesis-Centered Integrative Approaches for Identifying Binding Sites in Ion Channels and Uncovering Modulatory Mechanisms

3

Avia Rosenhouse-Dantsker

Abstract

Recent advances in structural techniques have provided unprecedented insights into protein structure and function. Yet, obtaining insights into the structure and function of transmembrane proteins, whether alone or in complex with modulators or other proteins, remains challenging. The aim of this chapter is to demonstrate how mutagenesis studies, in conjunction with sequence, structural, and computational information, can assist in identifying binding regions in ion channels and provide mechanistic insights into their modulation. The examples presented comprise various potassium channels and include the identification of binding sites for ions such as sodium and for lipids, including PI(4,5)P_2 and cholesterol, as well as the characterization of channel-protein interactions involving the βγ subunits of G proteins. Further examples focus on elucidating modulatory mechanisms, such as the reversal of the effect of cholesterol on an ion channel from enhancement to suppression. Additionally, recent advances in double-mutant cycle analysis and deep learning are enabling more detailed mapping of interaction surfaces, quantification of interaction strengths, and more precise identification of key residues. Together, these studies exemplify the utility of mutagenesis as a powerful perturbation approach for probing ion channel function and modulation.

Keywords

Site-directed mutagenesis · Binding site · Double-mutant cycle analysis · Ion channel · Protein-protein interactions · Interaction homology

A. Rosenhouse-Dantsker (✉)
Department of Chemistry, University of Illinois Chicago, Chicago, IL, USA
e-mail: dantsker@uic.edu

3.1 Introduction

In recent years, structural biology has witnessed remarkable advancements, driven by innovations in techniques such as cryo-electron microscopy (cryo-EM) (Shoemaker and Ando 2018; Corum et al. 2024), which have overcome the crystal packing limitations (Janin and Rodier 1995) of X-ray crystallography and the size limitations (Wider and Wüthrich 1999) of nuclear magnetic resonance (NMR). X-ray crystallography requires proteins to be crystallized, a process that can be particularly challenging for large, flexible, or membrane proteins (Carpenter et al. 2008). Moreover, the resulting crystal packing can impose ar-

A. Rosenhouse-Dantsker (ed.), *Decoding Ion Channels Structure and Function*, Advances in Experimental Medicine and Biology 1497, https://doi.org/10.1007/978-3-032-07523-9_3

tificial constraints on the protein's conformation, potentially impacting its native structure. On the other hand, studies of large protein complexes using NMR are limited by the complexity of the signals, making it challenging to interpret such structures (Frueh et al. 2013; Frueh 2014). Cryo-EM enables the determination of structures of large and complex biomolecules in environments that mimic their native conditions, capturing them in multiple conformations and providing insights into their dynamic structures. Yet, despite recent advancements in structural methodologies that are continually expanding the boundaries and facilitating atomic-resolution structural determination for an ever-growing list of biomolecules, elucidating the structure and function of transmembrane proteins—whether alone or in complex with modulators or other proteins—remains a formidable challenge. This is primarily due to their inherent complexities and the hydrophobic characteristics of their transmembrane domain (Birch et al. 2020). Furthermore, despite significant advances in artificial lipid environments, the complexities involved in accurately replicating the native cellular environments and interactions remain (Asuhara and Morigaki 2020; Woubshete et al. 2024). Moreover, atomic-resolution structures represent energetically stable conformations of proteins, and their relevance to the dynamic protein function can be greatly informed by functional studies probing the role of a particular amino acid residue and its interactions with the rest of the protein.

A simple, yet powerful, approach that can contribute to the fundamental understanding of protein function and modulation is site-directed mutagenesis (Bachman 2013; Carrigan et al. 2011). Application of this approach is especially straightforward in studies of ion channel proteins (Yang and Jiang 2013), which can be investigated in the whole cell or individually using various electrophysiological techniques. Mutagenesis can provide invaluable insights into the functional role of specific residues in channel activity, ion selectivity, and gating mechanisms. It can also aid in studies on ion channel modulation, including the determination of binding sites of modulators and the identification of interaction sites between ion channels and other proteins. Thus, mutagenesis can be an effective tool for studying ion channels, especially when integrated with sequence, structural, and computational approaches. The objective of this chapter is to present diverse examples illustrating the fundamental insights into complex interactions and regulatory mechanisms that govern ion channel function, which can be obtained by introducing targeted site-directed mutations and analyzing their impact.

3.2 Identification of the Binding Site of Sodium Ions in Inwardly Rectifying Potassium (Kir3) Channels: Combining Targeted Mutagenesis with Computational Approaches

The identification of the sodium binding site in Kir channels began in the late 1990s by integrating mutagenesis with sequence analysis. Several inwardly rectifying potassium channels, including Kir3.2 and Kir3.4, demonstrate sensitivity to sodium ions. In contrast, Kir3.1, for example, does not exhibit sodium sensitivity. To exploit these differential sodium responses, chimeras of Kir3.2 and Kir3.1 were engineered (Ho and Murrell-Lagnado 1999a, b) as an initial step to pinpoint the sodium binding site within these channels. This approach identified the C-terminus as a potential location for a sodium-sensing region. Considering that sodium ions are positively charged, the proximal C-termini of Kir3.2 and Kir3.4 were compared with the C-terminus of Kir3.1. The objective was to identify residues, such as aspartate or glutamate, which are negatively charged at physiological pH and are present in homotetramers that are Na^+-sensitive (Kir3.2 and Kir3.4) but absent in Na^+-insensitive channels (Kir3.1). This sequence comparison led to two residues, which in Kir3.2 were D228 and E234. The role of these residues was assessed through site-directed mutagenesis: Each of the two residues was mutated to the corresponding residue in Kir3.1: D228N and E234S. Functional studies using two-electrode voltage clamp in *Xenopus* oocytes expressing these mutants indicated that D228 played a critical role in the sodium sensitivity of Kir3.2. Similarly, the corresponding mutation in Kir3.4, D223N, also impacted its sodium sensitivity (Zhang et al. 1999).

Nearly a decade later, in the late 2000s, this aspartate guided the design of a computational study aimed to test the hypothesis that this residue was part of a sodium binding site, and, if so, identify which other residues were also involved (Rosenhouse-Dantsker et al. 2008). Harnessing the crystallographic structure of the cytosolic domain of Kir3.1, available at that time (PDB ID: 1N9P) (Nishida and MacKinnon 2002), enabled the application of computational tools to guide further experiments. At the equivalent position to D228 of Kir3.2 and D223 of Kir3.4 was N217 in Kir3.1. Hence, to utilize this model for this purpose, the asparagine was mutated in silico to an aspartate to mimic a sodium-sensitive channel. Indeed, this mutation has been previously shown to confer sodium sensitivity to a chimera of Kir3.1 that possessed the transmembrane (TM) domain of Kir3.2 (Ho and Murrell-Lagnado 1999b). The mutated cytosolic domain was then solvated using the primary hydration shell method (Beglov and Roux 1995). As sodium ions and water molecules are equivalent in electron density maps (Di Cera et al. 1995), water molecules located within 5 Å from D217 were substituted with sodium ions, one at a time. Of the seven water molecules located within this distance from the aspartate, five drifted away from the cytosolic domain during energy minimization. The remaining two converged to the same location, defining a putative sodium binding site (Rosenhouse-Dantsker et al. 2008) (Fig. 3.1). Accordingly, aside from the side-chain carbonyl of the aspartate, the side-chain of H222, the backbone carbonyls of M223 and L264, and a water molecule were also involved in coordinating the sodium ion.

The role of the histidine side-chain was evaluated through mutagenesis studies designed by comparing the sequences of the above three Kir channels with additional Kir channels. For example, although Kir6.2 possesses an aspartate, it lacks sodium sensitivity. Notably, however, it possesses a methionine instead of a histidine, suggesting that mutating the histidine to a methionine impairs sodium sensitivity. Indeed, the H228M mutation in Kir3.4* (Kir3.4_S143T) (Vivaudou et al. 1997) significantly reduced the sensitivity of this channel to sodium (Rosenhouse-Dantsker et al. 2008). In contrast, mutations of adjacent residues (N226K, I229L) did not impact the sensitivity of the channel to sodium. Kir2.1, which similar to Kir3.1 is insensitive to sodium, possesses an asparagine and a histidine at the equivalent positions. Consistent with the above observations, the N216D mutation conferred sodium sensitivity to Kir2.1, whereas incorporating the H221M mutation in addition to this asparagine-to-aspartate mutation reversed the effect. Hence, the N216D_H221M double mutant of Kir2.1 was insensitive to sodium. The involvement of the aspartate and histidine in the sodium binding site was further validated using various mutants of the heterotetramers Kir3.1/Kir3.2 and Kir4.1/Kir5.1 (Rosenhouse-Dantsker et al. 2008). Aside from pinpointing the sodium binding site, these studies also provided insights into the mechanism underlying the effect of sodium on sodium-sensitive Kir channels, indicating its role in strengthening the interaction with Phosphatidylinositol-4,5-bisphosphate ($PI(4,5)P_2$), a phosphoinositide required for the activation of all Kir channels (Ho and Murrell-Lagnado 1999a, b; Zhang et al. 1999; Rosenhouse-Dantsker et al. 2008).

Several years later, in the early 2010s, a structure of Kir3.2 crystallized with thallium (Whorton and MacKinnon 2011) in lieu of sodium, which, as noted above, is indistinguishable from water in electron density maps. This structure confirmed the location of the sodium binding site described above (Rosenhouse-Dantsker et al. 2008), underscoring the effectiveness of integrating site-directed mutagenesis with sequence and structure-based computational approaches.

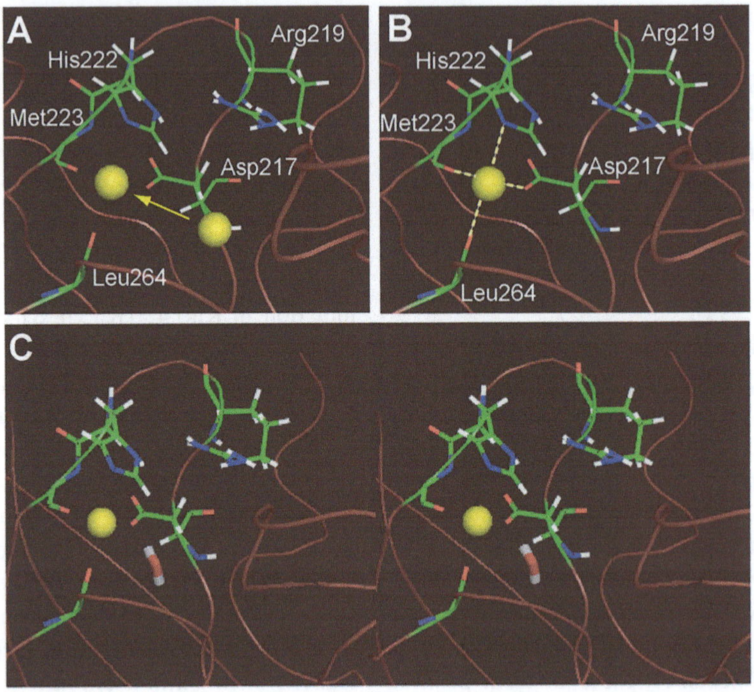

Fig. 3.1 Sodium binding site in Kir3.1_N217D. (**a**) 4.6 Å displacement of a Na^+ ion to its coordination site during minimization of the channel. The model was based on the crystallographic structure of Kir3.1 (PDB ID: 1N9P). (**b**) H-bonding pattern between the Na^+ ion and the residues that surround it in its coordination site in Kir3.1_N217D. (**c**) Stereo view of the sodium coordination site in Kir3.1_N217D that shows a nearby water molecule. Generated using the PyMOL Molecular Graphics System (version 2.0, Schrödinger, LLC). (Adapted from Rosenhouse-Dantsker et al. (2008))

3.3 Identification of the Binding Site of PI(4,5)P_2 in the Voltage-Gated Potassium Channel Kv1.2: Distinguishing Between Closed and Open Conformations Using Site-Directed Mutagenesis

PI(4,5)P_2 is a pivotal lipid regulator in ion channel modulation, and, as noted above, is required for the activation of all Kir channels. Prior to the determination of structures of Kir channels complexed with PI(4,5)P_2, mutagenesis studies identified PI(4,5)P_2-sensitive residues in various Kir channels using electrophysiological techniques, such as two-electrode voltage clamp, patch clamp, and single-channel recordings in *Xenopus* oocytes (Fan and Makielski 1997; Lopes et al. 2002; Schulze et al. 2003; Shyng et al. 2000; Huang et al. 1998; Rosenhouse-Dantsker et al. 2023; Logothetis et al. 2007a, b; Jin et al. 2008; Bukiya and Rosenhouse-Dantsker 2023). Structural analyses of other protein types, such as PH domains, complexed with PI(4,5)P_2 have shown that the negatively charged headgroup of PI(4,5)P_2 (Fig. 3.2) predominantly interacts with positively charged residues (Rosenhouse-Dantsker and Logothetis 2007). Recent structural studies have confirmed that in Kir channels, putative PI(4,5)P_2 binding sites are also primarily composed of positively charged residues and are located at the interface between the cytosolic and TM domains (Hansen et al. 2011; Wang et al. 2014; Niu et al. 2020). Additionally, the binding of PI(4,5)P_2 to the channels may be further strengthened through interactions between the hydrophobic acyl chains of this essential lipid and hydrophobic TM residues (Huang 2007;

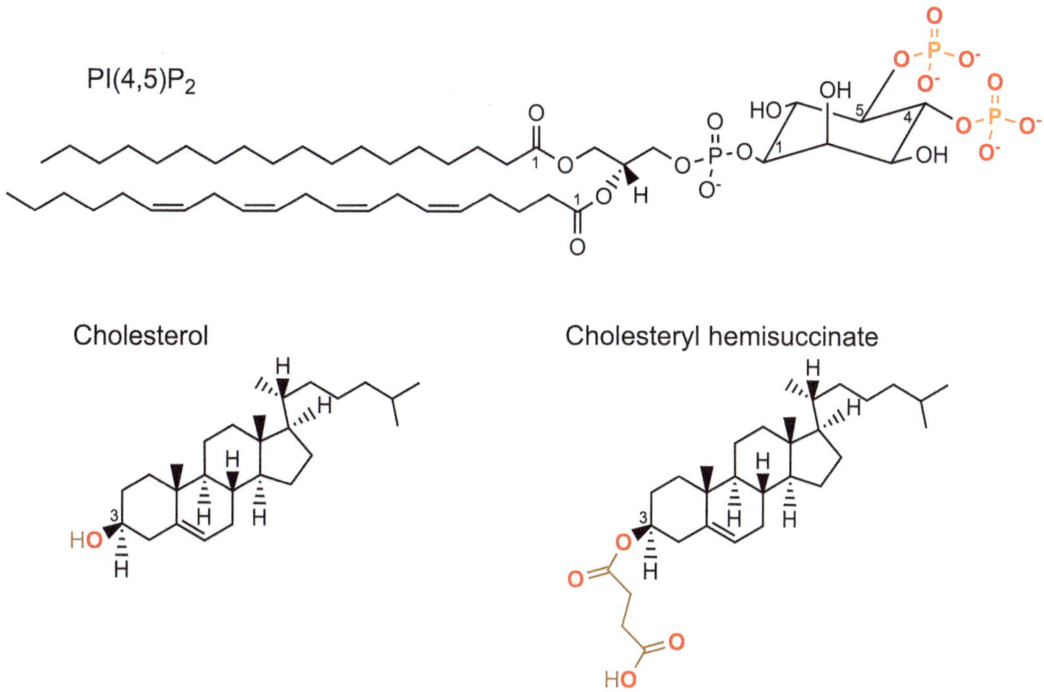

Fig. 3.2 Structures of phosphatidylinositol-4,5-bisphosphate (PI(4,5)P_2), cholesterol, and cholesterol hemisuccinate. Drawn using ChemDraw Version 20.1.0.112 (PerkinElmer Informatics). (Adapter from Bukiya and Rosenhouse-Dantsker (2023))

Liou et al. 1999; Rohács et al. 1999; Zeng et al. 2002; Rusinova et al. 2013).

In a study on the modulation of Kv1.2 by PI(4,5)P_2, mutagenesis has been crucial in elucidating how PI(4,5)P_2 interacts with and affects the channel (Rodriguez-Menchaca et al. 2012). Utilizing targeted site-directed mutagenesis alongside structural information and computational approaches, it was demonstrated that three pivotal positively charged residues—R147 in the N-terminus and K322 and R326 in the S4–S5 linker—concurrently interact with the negatively charged headgroup of PI(4,5)P_2 (Rodriguez-Menchaca et al. 2012) in a model depicting the closed conformation of the channel (Fig. 3.3). This model was constructed through the integration of voltage-clamp fluorometry and various computational techniques, including homology modeling, de novo modeling, and molecular dynamics simulations (Pathak et al. 2007; Khalili-Araghi et al. 2010). A comparison of the orientations of the side-chains of K322 and R326, located at the interface of the cytosolic and TM domain, with those of positively charged residues interacting with PI(4,5)P_2 in the crystal structure of the Kir2.2-PI(4,5)P_2 complex (PDB ID: 3SPI) (Hansen et al. 2011) revealed a similar spatial arrangement. R147, located in the N-terminus, interacted with the inositol ring of PI(4,5)P_2, which was oriented toward the cytosolic domain (Rodriguez-Menchaca et al. 2012).

Notably, PI(4,5)P_2 modulates the Kv1.2 channel through a dual effect, exhibiting distinct kinetics and sensitivities to PI(4,5)P_2 concentrations, and involving different molecular determinants (Rodriguez-Menchaca et al. 2012). First, PI(4,5)P_2 binding constrains the movement of the voltage sensor, reflected as a rightward shift in the activation curve. This shift indicates decreased voltage sensitivity, requiring greater depolarization to open the channel, an effect that favors the channel remaining in the closed state. Indeed, while engagement of all three identified residues

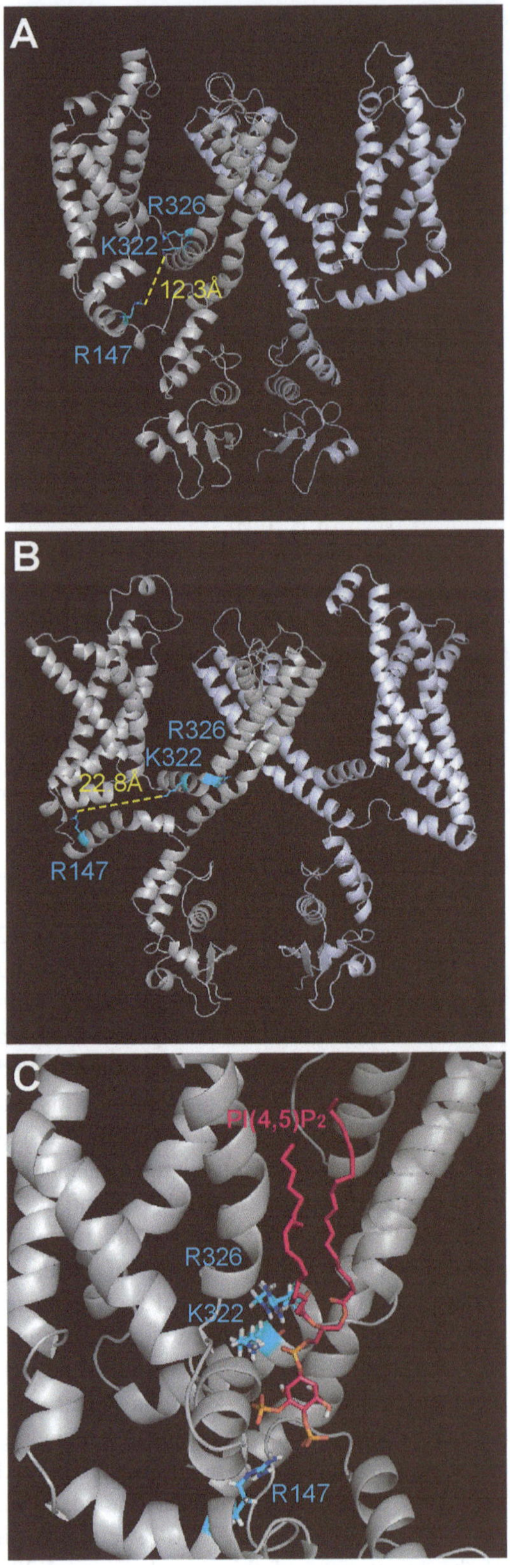

Fig. 3.3 PIP_2 interactions with critical Kv1.2 residues. (**a**) A model of the closed conformation of Kv1.2 showing the distance between the side-chains of R147 and K322. The model was constructed through an integration of voltage-clamp fluorometry and various computational techniques, as described in Khalili-Araghi et al. (2010). (**b**) A model of the open conformation of Kv1.2 showing the distance between the side-chains of R147 and K322. The model was based on the structure of the Kv1.2 and Kv2.1 chimera, PDB ID: 2R9R. (**c**) A model of the closed structure of Kv1.2 illustrating how PIP_2 could bind to the channel, engaging the three critical residues K322, R326, and R147 identified experimentally using site-directed mutagenesis. (Adapted from Rodriguez-Menchaca et al. (2012))

(R147, K322, and R326) with $PI(4,5)P_2$ was observed in the closed conformation of the channel, R147 was too distant from the putative $PI(4,5)P_2$ binding site in an open conformation constructed based on the structure of the Kv1.2 and Kv2.1 chimera, PDB ID: 2R9R (Rodriguez-Menchaca et al. 2012; Long et al. 2007) (Fig. 3.3). Furthermore, the rotation of R326 in the open conformation, impeded its interaction with $PI(4,5)P_2$ in the open conformation. Second, $PI(4,5)P_2$ also stabilizes the conducting state of the channel, increasing its open probability (P_O) (Rodriguez-Menchaca et al. 2012). This comprehensive understanding of $PI(4,5)P_2$'s regulatory role in the modulation of the Kv1.2 channel stems from the combined power of site-directed mutagenesis and molecular modeling.

3.4 Identification of Cholesterol Regulatory and Binding Sites in Inwardly Rectifying Potassium Channels: A Mutagenesis-Centered Approach

Kir2.1 exhibits enhanced cholesterol sensitivity compared to Kir2.3, with an initial study identifying three key residues in the cytosolic CD loop that influence this sensitivity (Epshtein et al. 2009). Notably, cholesterol depletion of CHO cells using methyl-β-cyclodextrin showed that the L222I mutation in Kir2.1 abrogated its sensitivity to cholesterol, whereas the reverse I214L mutation in Kir2.3 enhanced it.

To explore the role of the cytosolic domain in cholesterol sensitivity, a two-pronged mutagenesis strategy was employed, focusing on sequence differences between Kir2.1 and Kir2.3 (Rosenhouse-Dantsker et al. 2011). Initially, single point mutations of Kir2.1 residues proximal to the CD loop were investigated, identifying several key residues. Subsequently, to efficiently screen the remaining differences between Kir2.1 and Kir2.3 in the cytosolic domain, eight segment mutants were designed, each incorporating several point mutations, mimicking chimera construction, albeit for shorter segments rather than swapping domains. Functional testing was conducted by measuring whole-cell basal currents in *Xenopus* oocytes after cholesterol enrichment. This mutagenesis approach revealed that a cytosolic belt of residues surrounding the channel's pore near the TM domain affected cholesterol sensitivity. However, as supported by molecular docking, this effect did not arise from the formation of a direct binding site. Instead, it was proposed that the belt residues constituted a cholesterol regulatory site, impacting channel gating allosterically (Rosenhouse-Dantsker et al. 2011).

As a hydrophobic molecule, cholesterol is expected to preferentially interact with hydrophobic residues. Additionally, the ring structure of cholesterol (Fig. 3.2) is expected to interact with aromatic residues (Ounjian et al. 2019). Both types of residues have been incorporated into several cholesterol-binding motifs, including the one-segment CRAC motif proposed in 1998 (Li and Papadopoulos 1998) and its inverted form, the CARC motif, proposed in 2013 (Fantini and Barrantes 2013), as well as the two-segment CCM cholesterol-binding motif proposed in 2008 (Hanson et al. 2008). With the identification of these cholesterol-binding motifs in various proteins, it is not surprising that cholesterol can bind directly to ion channels to influence their function, while also exerting indirect effects through alterations in membrane properties (Bukiya and Rosenhouse-Dantsker 2023; Rosenhouse-Dantsker et al. 2012).

This notion was exemplified through a study of Kir2.1 designed to determine the binding site of cholesterol in an ion channel. To this end, numerous mutations were made in the inner and outer helices of the TM domain of the channel (Rosenhouse-Dantsker et al. 2013), as well as in the slide and pore helices that flank the TM domain. These helices include multiple hydrophobic and aromatic residues that could potentially interact with cholesterol. To identify which of these residues interact directly with cholesterol, residues surrounding the cholesterol molecule in molecular docking and molecular dynamics simulations were selected for testing, as their position suggested involvement in cholesterol recognition. In addition to targeting residues that interact directly with cholesterol, the study also examined secondary cholesterol-sensitive residues proximal to the primary cholesterol-binding residues. Considering the dense arrangement of residues in the TM domain, mutation of these secondary residues could potentially impact the primary residues, thereby indirectly influencing cholesterol binding (Rosenhouse-Dantsker et al. 2013).

Substituting hydrophobic or aromatic residues in the TM domain and its flanking helices with residues that possess different properties (e.g., polar or charged) could hypothetically have detrimental effects on the structure of this domain, and hence impact channel function due to reasons independent of cholesterol. Therefore, designing conservative or semi-conservative mutations was the preferred course of action. These mild mutations replaced residues with others of similar properties to minimize the impact on the hydrophobicity of the TM domain and enable probing the effects of the mutations on cholesterol sensitivity. The limitation of this approach is that the minimal effects of such mutations may preclude observing changes in cholesterol sensitivity even if the residue in question directly binds to cholesterol. Therefore, to identify the location of the cholesterol-binding site, a large number of residues was screened. This was accomplished predominantly using the two-electrode voltage-clamp technique on *Xenopus* oocytes expressing the mutated channels. The principal putative cholesterol-binding region identified in these experiments was located at the center of the TM domain. The findings from these studies defined non-annular cholesterolophilic surfaces, largely shielded from phospholipids, that could interact with cholesterol to modulate channel function (Rosenhouse-Dantsker et al. 2013).

In Kir3.4*(Kir3.4_S143T) (Vivaudou et al. 1997) and Kir3.2^ (Kir3.2_E152D) (Yi et al. 2001) channels, a similar primary cholesterol-binding region was identified, albeit with a slight shift toward the extracellular domain compared to Kir2.1 (Bukiya and Rosenhouse-Dantsker 2017, 2023; Bukiya et al. 2017). This shift was associated with the opposite effect of cholesterol, as cholesterol enhances the activity of these two Kir3 channels while suppressing Kir2.1. Mutagenesis studies of Kir3.4*, aimed at elucidating cholesterol's differential effects based on sequence differences with Kir2.1 or Kir3.1—both suppressed by cholesterol—pointed to the critical role of the M182 residue (Corradi et al. 2022) (Fig. 3.4). This study, which employed a chimera and single point mutants, demonstrated that the M182I mutation in Kir3.4* converts the channel's response to cholesterol from enhancement to suppression. Consistent with this result, extensive coarse-grained and atomistic simulations indicated that this mutation shifts the cholesterol-binding site toward the cytosolic domain, positioning it similarly to the principal binding site in Kir2.1. Computational analyses further indicated that this mutation modifies the distribution of cholesterol around the channel, underscoring the reciprocal relationship between the channel-protein sequence and cholesterol distribution, which affects the channel's response to this lipid (Corradi et al. 2022).

Separately, research on Kir3.2^ utilized a synergistic computational-experimental approach combining molecular docking analysis and mutagenesis to characterize its putative cholesterol-binding sites (Bukiya et al. 2017). To maintain channel functionality and minimize effects on basal activity, mutations substituted residues with others of a similar type but differing in side-chain size or structure. The residues assessed through mutagenesis were classified into four distinct categories based on their interaction profiles with cholesterol:

(a) Directly interacting residues: Located at the bottom of the binding pocket, mutations of these residues were expected to significantly impact cholesterol sensitivity.
(b) Indirectly influencing residues: Mutations of these secondary residues were expected to affect cholesterol sensitivity through their interactions with directly interacting residues.
(c) Mutation-tolerant residues: Positioned near, but not tightly packed against cholesterol, these residues were predicted to tolerate mild mutations without significantly affecting cholesterol sensitivity.

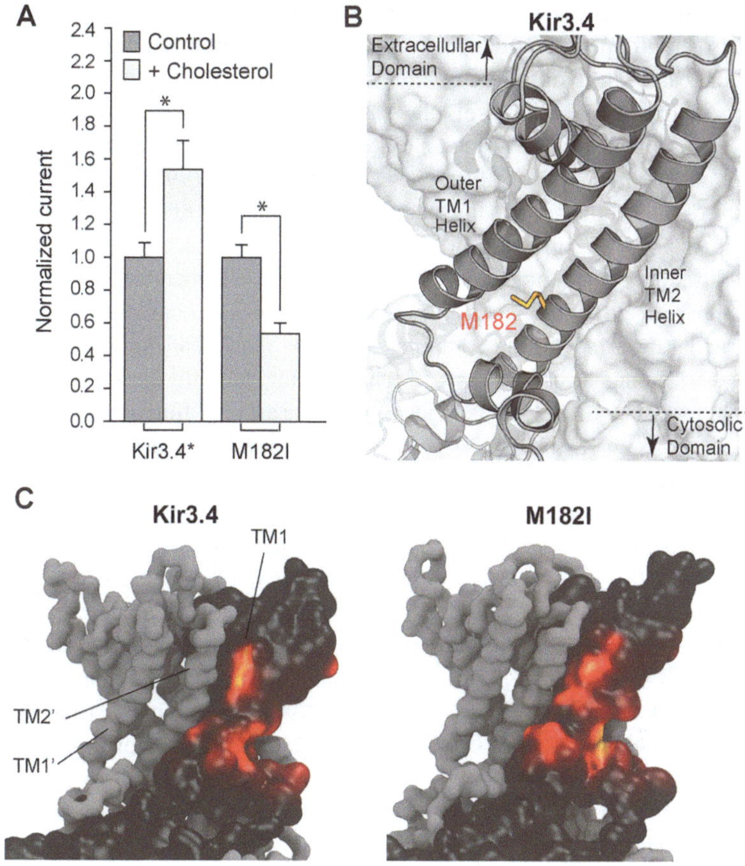

Fig. 3.4 Effect of the M182I mutation on the impact of cholesterol on Kir3.4* and the distribution of cholesterol around the channel. (**a**) Whole-cell basal currents recorded in *Xenopus* oocytes at −80 mV showing the effect of cholesterol enrichment on Kir3.4* and the M182I mutant. The currents were normalized relative to untreated oocytes expressing the same channel (n = 13–21). Significant difference is indicated by an asterisk (*, $p \leq 0.05$). Error bars, S.E.M. (**b**) Model of the transmembrane domain of subunit D in the last structure in the molecular dynamics simulations of Kir3.4, showing the location of the M182 residue in proximity to the cytosolic domain of the channel. For clarity, the backbones of the four subunits of the tetrameric channel are shown as a gray surface. (**c**) Occupancy data of cholesterol showing that the M182I mutation affects the distribution of cholesterol in the inner leaflet. For each construct, occupancy data was mapped on the subunit surface using a color scale ranging from black for no or low occupancy to yellow for high occupancy values. To compare Kir3.4 (left) with its M182I mutant (right), the highest occupancy value was set to be the same. For clarity, the backbones of the remaining subunits of the tetramer are shown as a gray surface. The outer transmembrane helix (TM1) of the channel subunit and the outer and inner transmembrane helices of the neighboring subunit (TM1', TM2') are highlighted. (Adapted from (Corradi et al. 2022))

(d) Non-interacting residues: With their side-chains oriented away from the binding site, mutations of these residues were not expected to affect cholesterol sensitivity.

This strategic approach identified residues critical for cholesterol sensitivity in Kir3.2^, demonstrating a strong correlation between the mutagenesis experiments and computational predictions (Bukiya et al. 2017).

In the absence of a Kir channel structure in complex with cholesterol, a cryo-EM structure of Kir3.2 (PDB ID: 6XEV) has revealed similar binding regions for cholesterol hemisuccinate, albeit with a different orientation (Mathiharan et al. 2021). Cholesterol hemisuccinate, a dicarboxylic acid monoester, is formed by condensing cholesterol's hydroxyl group with succinic acid's carboxyl group, increasing its hydrophilicity (Fig. 3.2). The hydroxyl group in cholesterol plays a key role in its alignment within the channel. Hence, the structural and hydrophobic difference between cholesterol and cholesterol hemisuccinate likely accounts for their differing orientations (Ohvo-Rekilä et al. 2002; Hung et al. 2007; Ermilova and Lyubartsev 2018).

As demonstrated above, integrating computational methods with mutagenesis strategies has been instrumental in elucidating cholesterol regulatory and binding sites in Kir channels. These approaches are also valuable for identifying sequence variations in Kir channels that determine whether cholesterol enhances or suppresses channel function.

3.5 Identification of a Protein-Protein Interaction Surface Between Kir3 Channels and the βγ Subunits of the G Protein via Mutagenesis-Validated Interaction Homology

The types of residues that interact with specific ligands, such as those discussed above, provide insights into the location of binding sites within proteins (Fig. 3.5). Sodium ions are typically coordinated by negatively charged residues (e.g., aspartate) and by residues with side-chain atoms that possess lone pairs of electrons (e.g., histidine with a lone electron pair on a nitrogen atom in the imidazole ring) (Ho and Murrell-Lagnado 1999a, b; Zhang et al. 1999; Rosenhouse-Dantsker et al. 2008). Additionally, water molecules often play a role in sodium ion coordination (Rosenhouse-Dantsker et al. 2008). The negatively charged headgroup of $PI(4,5)P_2$ generally interacts with positively charged residues (e.g., arginine, lysine) (Fan and Makielski 1997; Lopes et al. 2002; Schulze et al. 2003; Shyng et al. 2000; Huang et al. 1998; Rosenhouse-Dantsker et al. 2023; Logothetis et al. 2007a, b; Jin et al. 2008; Bukiya and Rosenhouse-Dantsker 2023; Rosenhouse-Dantsker and Logothetis 2007). Meanwhile, the hydrophobic cholesterol molecule typically binds within a hydrophobic environment, interacting primarily with hydrophobic and aromatic residues (Ounjian et al. 2019; Li and Papadopoulos 1998; Fantini and Barrantes 2013; Hanson et al. 2008).

Protein surfaces are composed of multiple residues and therefore exhibit a greater level of complexity compared to these individual ligands. Different regions of a protein's surface may interact with distinct proteins, facilitating a diverse array of functional interactions. However, there may be some overlap between regions that interact with different proteins, allowing for competing interactions.

An example of such competitive interaction involves the βγ subunits of the G protein (Gβγ). Upon G protein-coupled receptor (GPCR) stimulation, the GDP bound to the Gα subunit is exchanged for GTP (Bourne et al. 1990; Bourne et al. 1991). This exchange triggers the dissociation of Gβγ from Gα, allowing Gβγ to interact with and regulate various downstream effectors, such as ion channels, adenylyl cyclase, and phospholipase C, thereby playing a crucial role in GPCR signaling pathways, facilitating diverse cellular responses (Smrcka 2008; Tuteja 2009; Dupré et al. 2009; Smrcka and Fisher 2019).

For example, the Gβγ subunits bind to Kir3 channels and are involved in their activation (Logothetis et al. 1987; Reuveny et al. 1994; Kofuji et al. 1995; Krapivinsky et al. 1995). Before the crystallographic structure of the pre-open conformation of Kir3.2 in complex with Gβγ was determined in 2013 (PDB ID: 4KFM] (Whorton and MacKinnon 2013), several studies sought to identify key residues for this interaction. These studies employed the chimera strategy, combining segments from a Kir3 channel (Kir3.2 or Kir3.4) with the G protein-insensitive Kir2.1 channel, followed by single point mutations (He et al. 2002; Finley et al. 2004; Ivanina et al. 2003). This approach led to the identification of the N-terminus and the βD-βE and βL-βM sheets of the C-terminus as critical regions. Within these regions, studies of mutant channels expressed in Xenopus oocytes pointed to H69, L273, L344,

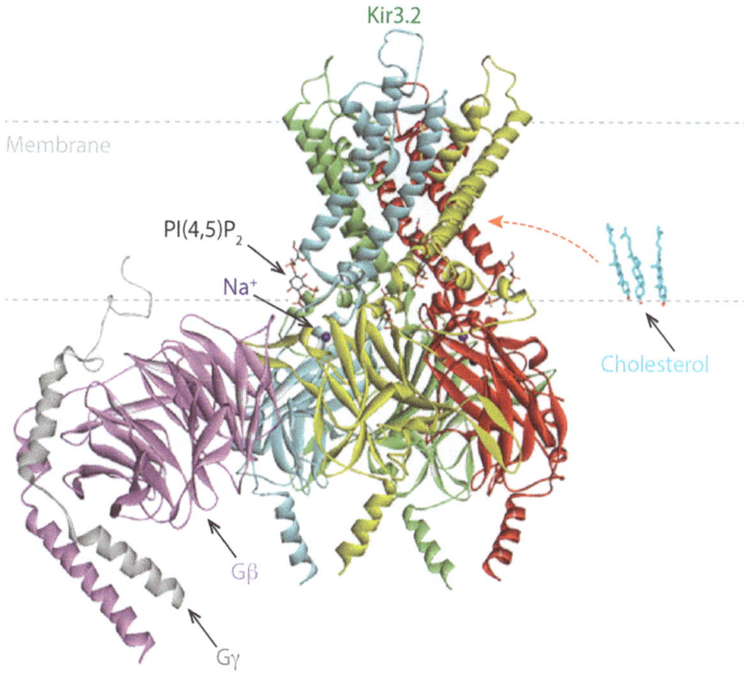

Fig. 3.5 Modulators of Kir3.2. Ribbon representation of Kir3.2 (based on PDB (Horovitz et al. 2019) ID 4KFM) (colored by subunit: yellow, red, green, dark cyan) in complex with PI(4,5)P_2, the βγ subunits of the G protein (purple and gray, respectively), and sodium ions (dark purple). Also shown is the putative primary binding site of one cholesterol molecule (cyan). For clarity, only one Gβγ complex and one putative cholesterol-binding site are shown. Generated using Discovery Studio 3.1 Visualizer v21.1.0.20298 (Dassault Systèmes Biovia Corp). (Adapted from Bukiya and Rosenhouse-Dantsker (2023))

and G347 in Kir3.2 (corresponding to H64, L268, L339, and G342 in Kir3.4; H57, L262, L333, and G336 in Kir3.1). Of these residues, the L344E mutation, located on the outer surface of the channel, had the strongest impact. The interaction of Kir3 channels with Gβγ occurs through the β subunit, where various mutagenesis studies have identified critical residues for interacting with the channels. A notable example is W99.

Comparing the structures of various proteins that interact with Gβγ reveals that W99 is consistently positioned in the binding region, forming interactions with effector residues. Aside from Kir3.2 (Whorton and MacKinnon 2013), this has been observed for a growing number of effectors that interact with Gβγ, including G protein-coupled receptor kinase 2 (GRK2) also known as β-adrenergic receptor kinase (βARK1) (Lodowski et al. 2003), phosducin (Gaudet et al. 1996), PIP_3-dependent Rac exchanger 1 (P-Rex1) (Cash et al. 2019), phospholipase C β3 (*PLC*β3) (Falzone and MacKinnon 2023), and adenylyl cyclase 5 (Yen et al. 2024). Considering that the Gβγ dimer is inactive when bound to the Gα subunit, this pattern suggests that various effectors bind to a surface region on Gβγ that partially overlaps with the binding region utilized by Gα (Lambright et al. 1996; Wall et al. 1998; Ford et al. 1998). As such, competitive binding serves as a dynamic switch between Gα and effector proteins in many Gβγ-mediated signaling pathways. This observation indicates that proteins, regardless of their structural similarity, may interact with a common partner, namely Gβγ, through these shared regions—a phenomenon we term interaction homology (Fig. 3.6).

Patterns observed in complexes of Gβγ with certain proteins can thus be used to predict which residues may mediate interactions with other pro-

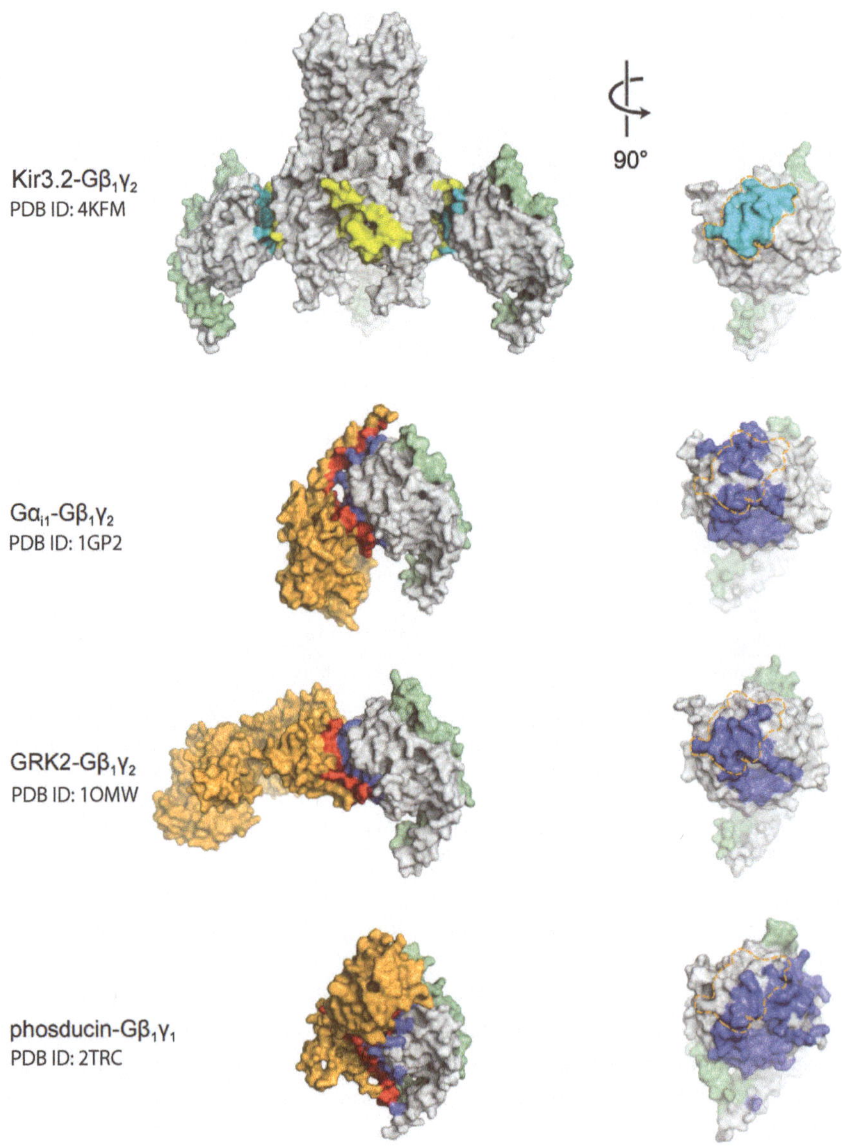

Fig. 3.6 Interaction homology among binding regions for Kir3.2 and other effectors on Gβγ. Partial overlap among the binding regions on the surface of Gβγ where several effectors, including Kir3.2, Gα, GRK2 (βARK1), and phosducin, bind, demonstrating the concept of interaction homology. The panels on the left are the complexes of Gβγ with these effectors, whereas the panels on the right show the Gβ binding regions that interact with each effector. Gβ is colored gray, and Gγ is colored green. Kir3.2 is colored gray, whereas Gα, GRK2 (βARK1), and phosducin are colored orange. The binding region for Kir3.2 on Gβ is colored cyan, and the binding regions on Gα, GRK2 (βARK1), and phosducin are colored purple. The orange dashed line outlines the binding region for Kir3.2 on Gβ, providing a reference on the surface where other effectors bind. The PDB ID of each complex is shown to the left of the corresponding panels. (Adapted from Whorton and MacKinnon (2013))

teins for which the complex structures with Gβγ have not been determined. In this context, mutagenesis approaches can be instrumental in evaluating potential binding regions in the interacting proteins. Based on available complexed structures, mutations can be designed in Gβγ to test predicted binding regions. In the target proteins, residues can be selected for mutagenesis based

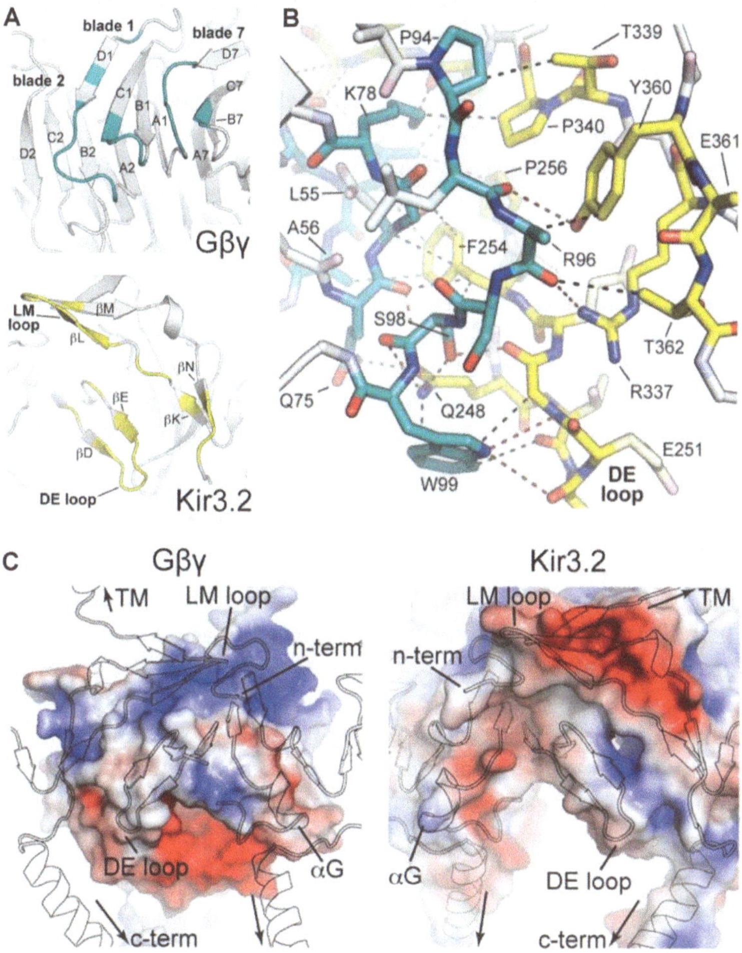

Fig. 3.7 Interaction surface between Kir3.2 and Gβγ. (**a**) A cartoon rendering of the binding interface of Gβ (top) and Kir3.2 (bottom). (**b**) A close-up of the Kir3.2-Gβ interaction surface, focused on the DE loop of Kir3.2. Selected hydrogen bonds and van der Waals interactions are shown as dashed lines as a visual aid. The interactions between W99 of Gβ and E251 and Q248 of Kir3.2 are shown at the bottom of the panel. (**c**) surface representations of Gβγ and Kir3.2, colored according to calculated electrostatic surface potentials (red: -100 mV, blue: +100 mV). The proteins are shown in the same orientation as in panel a, albeit slightly zoomed out. (Adapted from Whorton and MacKinnon (2013))

on the type of interactions observed in known complexes (Fig. 3.7). For example, W99 is within interaction range from E186 of Gα (Wall et al. 1998). Similarly, W99 is within interaction range from E251, E85, and E294 of Kir3.2 (Whorton and MacKinnon 2013), phosducin (Gaudet et al. 1996), and PLCβ3 (Falzone and MacKinnon 2023), respectively. These findings suggest that screening the effect of mutating glutamates in the target protein, potentially guided by a computational model, may help identify the binding region with Gβγ in the absence of a structural model of the complex. In Kir3.2, W99 is also within interaction range from Q248 (Whorton and MacKinnon 2013), whereas in βARK1, N586 is within interaction range from W99 (Lodowski et al. 2003). Considering that asparagine shares structural similarity with glutamine, they may have comparable roles in facilitating interactions with Gβγ. These examples underscore the potential value of mutagenesis-validated interaction homology in revealing how anchor points interact with effector proteins of varying folds, shapes, or sizes.

With recent advances in deep learning (Lee 2023; Cui et al. 2025), computational tools can now be combined with mutagenesis and interaction homology to prioritize mutations for experimental testing and to enhance our ability to map protein-protein interaction surfaces with more precision. These innovations are also rapidly transforming protein-protein interaction network analysis by integrating diverse sequence and structural data, enabling comprehensive characterizations of interactions and functional sites, and thereby paving the way for new avenues in experimental discovery (Kiouri et al. 2025).

3.6 The Double-Mutant Cycle Analysis for Quantifying Coupling: Implementation and Recent Advances

Another application of mutagenesis involves the use of thermodynamic double-mutant cycle analysis to quantify the energetic coupling between residues within proteins and in protein-ligand interactions (Carter et al. 1984; Horovitz and Fersht 1990) and analyze the cooperativity of intra- and inter-molecular interactions. Using this approach, if two mutations independently affect channel gating, their individual free-energy changes would be additive. However, if the two residues interact, the total free-energy change for the double mutant will differ from the sum of the changes in free-energy observed for the single mutants. This difference is termed the coupling energy, reflecting the interaction strength between the residues. A coupling energy exceeding 0.89 kcal/mol typically indicates a significant interaction.

This approach has been utilized to study the coupling between residues for many years. One example from the ion channel field includes the application of scanning mutagenesis and thermodynamic mutant cycle analysis to explore the functional interaction surfaces of an inhibitor and the Shaker voltage-gated potassium channel, identifying energetic interactions and spatial constraints for the channel's pore loops (Ranganathan et al. 1996). In a later study, the allosteric communication trajectory of the Shaker channel was investigated through high-order thermodynamic coupling analysis. To this end, interactions were examined for three types of pairings: two single mutants, one single mutant and one double mutant, and two double mutants (Sadovsky and Yifrach 2007).

Another example is the use of this approach to identify interactions between residues in H_V1, the human voltage-gated proton channel, providing structural information on the closed and open states of the channel (Chamberlin et al. 2014). In a subsequent study, the double-mutant cycle analysis was used to probe functional interactions and significant thermodynamic couplings between a conserved phenylalanine in the S2 helix and arginine residues in the S4 helix of the H_V1 channel during its voltage-dependent activation (Wu et al. 2022). This study underscores the utility of this approach in the absence of a structure for the activated state of the channel.

A recent review of this established method highlighted its growing utility, especially when combined with advancements in native mass spectrometry (MS) and high-throughput approaches (Horovitz et al. 2019). Together, these integrated methodologies enable the detection and quantification of both direct and long-range interactions in proteins. They also facilitate the unraveling of higher-order interactions and provide insights into cooperativity and interaction networks.

A more recent article published in 2024 offered a brief review of double-mutant cycle analysis, tracing its application over the four decades since its inception in 1984 (Lyu et al. 2024). The article demonstrated the expansion of this approach to triple-mutant cycles to characterize complex protein-lipid interactions, involving multiple residues, to gain a detailed understanding of their thermodynamic contributions. Using this approach, single, double, and triple mutations of target residues were introduced to calculate the coupling free energy, the coupling enthalpy, and the coupling entropy among the residues, revealing their energetic contributions to specific interactions (Lyu et al. 2024). These recent advancements underscore a renewed interest in the double-mutant cycle analysis while also marking the expansion of this strategy to multiple-mutant cycle analysis, reflecting its evolving importance in unraveling complex interactions.

3.7 Conclusions and Outlook

Targeted site-directed mutagenesis has proven to be a versatile and effective strategy for dissecting how specific residues contribute to the modulation of ion channel function. Additionally, this perturbation approach has provided unique insights into residues that define binding sites, enabling the mapping of interaction surfaces and regulatory partners. In particular, the integration of experimental mutagenesis with computational modeling and structural analysis has led to a deeper understanding of how different ligands bind to various ion channels, including, for example, the Kir and Kv channels discussed in this chapter. Importantly, the approaches described here offer a framework for probing mechanisms in ion channel modulation and for studying other proteins even when high-resolution structural data are lacking.

The application and further development of interaction homology supported by mutagenesis and computational approaches holds great promise for uncovering new principles of protein-protein interactions. Although site-directed mutagenesis alone cannot differentiate between binding and allosteric sites, its integration with computational methodologies has significantly mitigated this limitation. Moreover, recent advancements in the established technique of double-mutant cycle analysis, particularly its amalgamation with approaches, such as native MS and high-throughput techniques, highlight the potential integration of site-directed mutagenesis with other methods to accelerate discoveries and deepen our understanding of protein function and regulation. Additionally, the evolution of the double-mutant cycle analysis into triple-mutant cycle analysis underscores the potential for ongoing innovation, even within a technique that has been foundational for over four decades.

Collectively, these advances not only revitalize classic methodologies but also set the stage for future breakthroughs in the understanding of protein function and modulation. Looking ahead, the integration of site-directed mutagenesis with deep learning is an emerging direction, poised to transform experimental strategies and reveal new insights into binding sites and interaction networks.

References

Asuhara K, Morigaki K (2020) New lipid membrane technologies for reconstitution, analysis, and utilization of 'living' membrane proteins. Biophys Physicobiol 17:125–129

Bachman J (2013) Site-directed mutagenesis. Methods Enzymol 529:241–248

Beglov D, Roux B (1995) Dominant solvation effects from the primary shell of hydration: approximation for molecular dynamics simulations. Biopolymers 35(2):171–178

Birch J, Cheruvara H, Gamage N, Harrison PJ, Lithgo R, Quigley A (2020) Changes in membrane protein structural biology. Biology (Basel) 9(11):401

Bourne HR, Sanders DA, McCormick F (1990) The GTPase superfamily: a conserved switch for diverse cell functions. Nature 348(6297):125–132

Bourne HR, Sanders DA, McCormick F (1991) The GTPase superfamily: conserved structure and molecular mechanism. Nature 349(6305):117–127

Bukiya AN, Rosenhouse-Dantsker A (2017) Synergistic activation of G protein-gated inwardly rectifying potassium channels by cholesterol and PI(4,5)P_2. Biochim Biophys Acta Biomembr 1859(7):1233–1241

Bukiya AN, Rosenhouse-Dantsker A (2023) From crosstalk to synergism: the combined effect of cholesterol and PI(4,5)P_2 on inwardly rectifying potassium channels. Adv Exp Med Biol 1422:169–191

Bukiya AN, Durdagi S, Noskov S, Rosenhouse-Dantsker A (2017) Cholesterol up-regulates neuronal G protein-gated inwardly rectifying potassium (GIRK) channel activity in the hippocampus. J Biol Chem 292(15):6135–6147

Carpenter EP, Beis K, Cameron AD, Iwata S (2008) Overcoming the challenges of membrane protein crystallography. Curr Opin Struct Biol 18(5):581–586

Carrigan PE, Ballar P, Tuzmen S (2011) Site-directed mutagenesis. Methods Mol Biol 700:107–124

Carter PJ, Winter G, Wilkinson AJ, Fersht AR (1984) The use of double mutants to detect structural changes in

the active site of the tyrosyl-tRNA synthetase (Bacillus stearothermophilus). Cell 38(3):835–840
Cash JN, Urata S, Li S, Ravala SK, Avramova LV, Shost MD, Gutkind JS, Tesmer JJG, Cianfrocco MA (2019) Cryo-electron microscopy structure and analysis of the P-Rex1-Gβγ signaling scaffold. Sci Adv 5(10):eaax8855
Chamberlin A, Qiu F, Rebolledo S, Wang Y, Noskov SY, Larsson HP (2014) Hydrophobic plug functions as a gate in voltage-gated proton channels. Proc Natl Acad Sci USA 111(2):E273–E282
Corradi V, Bukiya AN, Miranda WE, Cui M, Plant LD, Logothetis DE, Tieleman DP, Noskov SY, Rosenhouse-Dantsker A (2022) A molecular switch controls the impact of cholesterol on a Kir channel. Proc Natl Acad Sci U S A 119(13):e2109431119
Corum MR, Venkannagari H, Hryc CF, Baker ML (2024) Predictive modeling and cryo-EM: a synergistic approach to modeling macromolecular structure. Biophys J 123(4):435–450
Cui J, Yang S, Yi L, Xi Q, Yang D, Zuo Y (2025) Recent advances in deep learning for protein-protein interaction: a review. BioData Min 18(1):43
Di Cera E, Guinto ER, Vindigni A, Dang QD, Ayala YM, Wuyi M, Tulinsky A (1995) The Na+ binding site of thrombin. J Biol Chem 270(38):22089–22092
Dupré DJ, Robitaille M, Rebois RV, Hébert TE (2009) The role of Gbetagamma subunits in the organization, assembly, and function of GPCR signaling complexes. Annu Rev Pharmacol Toxicol 49:31–56
Epshtein Y, Chopra AP, Rosenhouse-Dantsker A, Kowalsky GB, Logothetis DE, Levitan I (2009) Identification of a C-terminus domain critical for the sensitivity of Kir2.1 to cholesterol. Proc Natl Acad Sci USA 106(19):8055–8060
Ermilova I, Lyubartsev AP (2018) Cholesterol in phospholipid bilayers: positions and orientations inside membranes with different unsaturation degrees. Soft Matter 15(1):78–93
Falzone ME, MacKinnon R (2023) *Gβγ* activates *PIP2* hydrolysis by recruiting and orienting *PLCβ* on the membrane surface. Proc Natl Acad Sci U S A 120(20):e2301121120
Fan Z, Makielski JC (1997) Anionic phospholipids activate ATP-sensitive potassium channels. J Biol Chem 272(9):5388–5395
Fantini J, Barrantes FJ (2013) How cholesterol interacts with membrane proteins: an exploration of cholesterol-binding sites including CRAC, CARC, and tilted domains. Front Physiol 4:31
Finley M, Arrabit C, Fowler C, Suen KF, Slesinger PA (2004) betaL-betaM loop in the C-terminal domain of G protein-activated inwardly rectifying K(+) channels is important for G(betagamma) subunit activation. J Physiol 555(Pt 3):643–657
Ford CE, Skiba NP, Bae H, Daaka Y, Reuveny E, Shekter LR, Rosal R, Weng G, Yang CS, Iyengar R, Miller RJ, Jan LY, Lefkowitz RJ, Hamm HE (1998) Molecular basis for interactions of G protein betagamma subunits with effectors. Science 280(5367):1271–1274
Frueh DP (2014) Practical aspects of NMR signal assignment in larger and challenging proteins. Prog Nucl Magn Reson Spectrosc 78:47–75
Frueh DP, Goodrich AC, Mishra SH, Nichols SR (2013) NMR methods for structural studies of large monomeric and multimeric proteins. Curr Opin Struct Biol 23(5):734–739
Gaudet R, Bohm A, Sigler PB (1996) Crystal structure at 2.4 angstroms resolution of the complex of transducin betagamma and its regulator, phosducin. Cell 87(3):577–588
Hansen SB, Tao X, MacKinnon R (2011) Structural basis of PIP2 activation of the classical inward rectifier K+ channel Kir2.2. Nature 477(7365):495–498
Hanson MA, Cherezov V, Griffith MT, Roth CB, Jaakola VP, Chien EY, Velasquez J, Kuhn P, Stevens RC (2008) A specific cholesterol binding site is established by the 2.8 A structure of the human beta2-adrenergic receptor. Structure 16(6):897–905
He C, Yan X, Zhang H, Mirshahi T, Jin T, Huang A, Logothetis DE (2002) Identification of critical residues controlling G protein-gated inwardly rectifying K(+) channel activity through interactions with the beta gamma subunits of G proteins. J Biol Chem 277(8):6088–6096
Ho IH, Murrell-Lagnado RD (1999a) Molecular determinants for sodium-dependent activation of G protein-gated K^+ channels. J Biol Chem 274(13):8639–8648
Ho IH, Murrell-Lagnado RD (1999b) Molecular mechanism for sodium-dependent activation of G protein-gated K^+ channels. J Physiol 520 Pt 3(Pt 3):645–651
Horovitz A, Fersht AR (1990) Strategy for analysing the co-operativity of intramolecular interactions in peptides and proteins. J Mol Biol 214(3):613–617
Horovitz A, Fleisher RC, Mondal T (2019) Double-mutant cycles: new directions and applications. Curr Opin Struct Biol 58:10–17
Huang CL, Feng S, Hilgemann DW (1998) Direct activation of inward rectifier potassium channels by PIP2 and its stabilization by Gbetagamma. Nature 391(6669):803–806
Hung WC, Lee MT, Chen FY, Huang HW (2007) The condensing effect of cholesterol in lipid bilayers. Biophys J 92(11):3960–3967
Huang CL (2007) Complex roles of PIP2 in the regulation of ion channels and transporters. Am J Physiol Renal Physiol 293(6):F1761–F1765
Ivanina T, Rishal I, Varon D, Mullner C, Frohnwieser-Steinecke B, Schreibmayer W, Dessauer CW, Dascal N (2003) Mapping the Gbetagamma-binding sites in GIRK1 and GIRK2 subunits of the G protein-activated K+ channel. J Biol Chem 278(31):29174–29183
Janin J, Rodier F (1995) Protein-protein interaction at crystal contacts. Proteins 23(4):580–587
Jin T, Sui JL, Rosenhouse-Dantsker A, Chan KW, Jan LY, Logothetis DE (2008) Stoichiometry of Kir channels with phosphatidylinositol bisphosphate. Channels (Austin) 2(1):19–33
Khalili-Araghi F, Jogini V, Yarov-Yarovoy V, Tajkhorshid E, Roux B, Schulten K (2010) Calculation of the gating

charge for the Kv1.2 voltage-activated potassium channel. Biophys J 98(10):2189–2198

Kiouri DP, Batsis GC, Chasapis CT (2025) Structure-based approaches for protein-protein interaction prediction using machine learning and deep learning. Biomolecules 15(1):141

Kofuji P, Davidson N, Lester HA (1995) Evidence that neuronal G-protein-gated inwardly rectifying K+ channels are activated by G beta gamma subunits and function as heteromultimers. Proc Natl Acad Sci USA 92(14):6542–6546

Krapivinsky G, Krapivinsky L, Wickman K, Clapham DE (1995) G beta gamma binds directly to the G protein-gated K+ channel, IKACh. J Biol Chem 270(49):29059–29062

Lambright DG, Sondek J, Bohm A, Skiba NP, Hamm HE, Sigler PB (1996) The 2.0 A crystal structure of a heterotrimeric G protein. Nature 379(6563):311–319

Lee M (2023) Recent advances in deep learning for protein-protein interaction analysis: a comprehensive review. Molecules 28(13):5169

Li H, Papadopoulos V (1998) Peripheral-type benzodiazepine receptor function in cholesterol transport. Identification of a putative cholesterol recognition/interaction amino acid sequence and consensus pattern. Endocrinology 139(12):4991–4997

Liou HH, Zhou SS, Huang CL (1999) Regulation of ROMK1 channel by protein kinase A via a phosphatidylinositol 4,5-bisphosphate-dependent mechanism. Proc Natl Acad Sci U S A 96(10):5820–5825

Lodowski DT, Pitcher JA, Capel WD, Lefkowitz RJ, Tesmer JJ (2003) Keeping G proteins at bay: a complex between G protein-coupled receptor kinase 2 and Gbetagamma. Science 300(5623):1256–1262

Logothetis DE, Kurachi Y, Galper J, Neer EJ, Clapham DE (1987) The beta gamma subunits of GTP-binding proteins activate the muscarinic K+ channel in heart. Nature 325(6102):321–326

Logothetis DE, Jin T, Lupyan D, Rosenhouse-Dantsker A (2007a) Phosphoinositide-mediated gating of inwardly rectifying K(+) channels. Pflugers Arch 455(1):83–95

Logothetis DE, Lupyan D, Rosenhouse-Dantsker A (2007b) Diverse Kir modulators act in close proximity to residues implicated in phosphoinositide binding. J Physiol 582(Pt 3):953–965

Long SB, Tao X, Campbell EB, MacKinnon R (2007) Atomic structure of a voltage-dependent K+ channel in a lipid membrane-like environment. Nature 450(7168):376–382

Lopes CM, Zhang H, Rohacs T, Jin T, Yang J, Logothetis DE (2002) Alterations in conserved Kir channel-PIP2 interactions underlie channelopathies. Neuron 34(6):933–944

Lyu J, Zhang T, Marty MT, Clemmer D, Russell DH, Laganowsky A (2024) Double and triple thermodynamic mutant cycles reveal the basis for specific MsbA-lipid interactions. elife 22:RP91094

Mathiharan YK, Glaaser IW, Zhao Y, Robertson MJ, Skiniotis G, Slesinger PA (2021) Structural insights into GIRK2 channel modulation by cholesterol and PIP_2. Cell Rep 36(8):109619

Nishida M, MacKinnon R (2002) Structural basis of inward rectification: cytoplasmic pore of the G protein-gated inward rectifier GIRK1 at 1.8 A resolution. Cell 111(7):957–965

Niu Y, Tao X, Touhara KK, MacKinnon R (2020) Cryo-EM analysis of PIP_2 regulation in mammalian GIRK channels. elife 9:e60552

Ohvo-Rekilä H, Ramstedt B, Leppimäki P, Slotte JP (2002) Cholesterol interactions with phospholipids in membranes. Prog Lipid Res 41(1):66–97

Ounjian J, Bukiya AN, Rosenhouse-Dantsker A (2019) Molecular determinants of cholesterol binding to soluble and transmembrane protein domains. Adv Exp Med Biol 1135:47–66

Pathak MM, Yarov-Yarovoy V, Agarwal G, Roux B, Barth P, Kohout S, Tombola F, Isacoff EY (2007) Closing in on the resting state of the shaker K(+) channel. Neuron 56(1):124–140

Ranganathan R, Lewis JH, MacKinnon R (1996) Spatial localization of the K+ channel selectivity filter by mutant cycle-based structure analysis. Neuron 16(1):131–139

Reuveny E, Slesinger PA, Inglese J, Morales JM, Iñiguez-Lluhi JA, Lefkowitz RJ, Bourne HR, Jan YN, Jan LY (1994) Activation of the cloned muscarinic potassium channel by G protein beta gamma subunits. Nature 370(6485):143–146

Rodriguez-Menchaca AA, Adney SK, Tang QY, Meng XY, Rosenhouse-Dantsker A, Cui M, Logothetis DE (2012) PIP2 controls voltage-sensor movement and pore opening of Kv channels through the S4-S5 linker. Proc Natl Acad Sci USA 109(36):E2399–E2408

Rohács T, Chen J, Prestwich GD, Logothetis DE (1999) Distinct specificities of inwardly rectifying K(+) channels for phosphoinositides. J Biol Chem 274(51):36065–36072

Rosenhouse-Dantsker A, Logothetis DE (2007) Molecular characteristics of phosphoinositide binding. Pflugers Arch 455(1):45–53

Rosenhouse-Dantsker A, Sui JL, Zhao Q, Rusinova R, Rodríguez-Menchaca AA, Zhang Z, Logothetis DE (2008) A sodium-mediated structural switch that controls the sensitivity of Kir channels to PtdIns(4,5)P(2). Nat Chem Biol 4(10):624–631

Rosenhouse-Dantsker A, Logothetis DE, Levitan I (2011) Cholesterol sensitivity of KIR2.1 is controlled by a belt of residues around the cytosolic pore. Biophys J 100(2):381–389

Rosenhouse-Dantsker A, Mehta D, Levitan I (2012) Regulation of ion channels by membrane lipids. Compr Physiol 2(1):31–68

Rosenhouse-Dantsker A, Noskov S, Durdagi S, Logothetis DE, Levitan I (2013) Identification of novel cholesterol-binding regions in Kir2 channels. J Biol Chem 288(43):31154–31164

Rosenhouse-Dantsker A, Gazgalis D, Logothetis DE (2023) $PI(4,5)P_2$ and cholesterol: synthesis, regulation, and functions. Adv Exp Med Biol 1422:3–59

Rusinova R, Hobart EA, Koeppe RE 2nd, Andersen OS (2013) Phosphoinositides alter lipid bilayer properties. J Gen Physiol 141(6):673–690

Sadovsky E, Yifrach O (2007) Principles underlying energetic coupling along an allosteric communication trajectory of a voltage-activated K+ channel. Proc Natl Acad Sci USA 104(50):19813–19818

Schulze D, Krauter T, Fritzenschaft H, Soom M, Baukrowitz T (2003) Phosphatidylinositol 4,5-bisphosphate (PIP2) modulation of ATP and pH sensitivity in Kir channels. A tale of an active and a silent PIP2 site in the N terminus. J Biol Chem 278(12):10500–10505

Shoemaker SC, Ando N (2018) X-rays in the cryo-electron microscopy era: structural biology's dynamic future. Biochemistry 57(3):277–285

Shyng SL, Cukras CA, Harwood J, Nichols CG (2000) Structural determinants of PIP(2) regulation of inward rectifier K(ATP) channels. J Gen Physiol 116(5):599–608

Smrcka AV (2008) G protein βγ subunits: central mediators of G protein-coupled receptor signaling. Cell Mol Life Sci 65(14):2191–2214

Smrcka AV, Fisher I (2019) G-protein βγ subunits as multi-functional scaffolds and transducers in G-protein-coupled receptor signaling. Cell Mol Life Sci 76(22):4447–4459

Tuteja N (2009) Signaling through G protein coupled receptors. Plant Signal Behav 4(10):942–947

Vivaudou M, Chan KW, Sui JL, Jan LY, Reuveny E, Logothetis DE (1997) Probing the G-protein regulation of GIRK1 and GIRK4, the two subunits of the KACh channel, using functional homomeric mutants. J Biol Chem 272(50):31553–31560

Wall MA, Posner BA, Sprang SR (1998) Structural basis of activity and subunit recognition in G protein heterotrimers. Structure 6(9):1169–1183

Wang W, Whorton MR, MacKinnon R (2014) Quantitative analysis of mammalian GIRK2 channel regulation by G proteins, the signaling lipid PIP2 and Na+ in a reconstituted system. elife 3:e03671

Whorton MR, MacKinnon R (2011) Crystal structure of the mammalian GIRK2 K+ channel and gating regulation by G proteins, PIP2, and sodium. Cell 147(1):199–208

Whorton MR, MacKinnon R (2013) X-ray structure of the mammalian GIRK2-βγ G-protein complex. Nature 498(7453):190–197

Wider G, Wüthrich K (1999) NMR spectroscopy of large molecules and multimolecular assemblies in solution. Curr Opin Struct Biol 9(5):594–601

Woubshete M, Cioccolo S, Byrne B (2024) Advances in membrane mimetic systems for manipulation and analysis of membrane proteins: detergents, polymers, lipids and scaffolds. ChemPlusChem 89(6):e202300678

Wu X, Zhang L, Hong L (2022) The role of Phe150 in human voltage-gated proton channel. iScience 25(11):105420

Yang W, Jiang LH (2013) Site-directed mutagenesis to study the structure-function relationships of ion channels. Methods Mol Biol 998:257–266

Yen YC, Li Y, Chen CL, Klose T, Watts VJ, Dessauer CW, Tesmer JJG (2024) Structure of adenylyl cyclase 5 in complex with Gβγ offers insights into ADCY5-related dyskinesia. Nat Struct Mol Biol 31(8):1189–1197

Yi BA, Lin YF, Jan YN, Jan LY (2001) Yeast screen for constitutively active mutant G protein-activated potassium channels. Neuron 29(3):657–667

Zeng WZ, Liou HH, Krishna UM, Falck JR, Huang CL (2002) Structural determinants and specificities for ROMK1-phosphoinositide interaction. Am J Physiol Renal Physiol 282(5):F826–F834

Zhang H, He C, Yan X, Mirshahi T, Logothetis DE (1999) Activation of inwardly rectifying K+ channels by distinct PtdIns(4,5)P2 interactions. Nat Cell Biol 1(3):183–188

Advances in Crystallography and Cryo-Electron Microscopy for the Study of Membrane Proteins

4

Abhin Megta, Ankita Punetha, Jyoti Kumari, Hui Wei, Stephannie Rosario-Garrido, James A. Tranos, and Vasileios I. Petrou

Abstract

The field of structural biology has undergone a remarkable transformation over the past decade, fueled by cutting-edge advancements in X-ray crystallography and cryo-electron microscopy (cryo-EM). Central to both approaches is the critical step of sample preparation, which includes heterologous expression of membrane proteins in host systems and extraction from the membrane environment using suitable membrane mimetics. In crystallography, enhanced techniques, such as lipidic cubic phase (LCP) crystallization and serial femtosecond crystallography (SFX), have pushed the boundaries of structure determination to microcrystals and conformationally labile targets. In cryo-EM, technical advancements across many levels have fueled the "resolution revolution," enabling cryo-EM to reach near-atomic resolution and driving single-particle analysis of increasingly small, dynamic, and heterogeneous macromolecular assemblies. Emerging modalities further extend our capability to tackle previously inaccessible questions. Cryo-electron tomography (cryo-ET) is extracting structural insights from native cellular environments, and micro-electron diffraction (MicroED) has opened new frontiers for nanoscale structural studies from submicron crystals at unprecedented resolution. Collectively, these advances have improved our ability to study the structure of challenging targets, with profound implications for structure-based drug discovery, and an evolving paradigm shift toward dynamic visualization of biomolecular processes, signaling the dawn of a new era in structural biology.

Keywords

Membrane proteins · Structural biology · Cryo-EM · Cryo-ET · MicroED · Protein purification

Authors Abhin Megta and Ankita Punetha have equally contributed to this chapter.

A. Megta · A. Punetha · J. Kumari · S. Rosario-Garrido · J. A. Tranos · V. I. Petrou (✉)
Department of Microbiology, Biochemistry, and Molecular Genetics, New Jersey Medical School, Rutgers Biomedical Health Sciences, Newark, NJ, USA

Center for Immunity and Inflammation, New Jersey Medical School, Rutgers Biomedical Health Sciences, Newark, NJ, USA
e-mail: vasileios.petrou@rutgers.edu

H. Wei
Department of Microbiology, Biochemistry, and Molecular Genetics, New Jersey Medical School, Rutgers Biomedical Health Sciences, Newark, NJ, USA

A. Rosenhouse-Dantsker (ed.), *Decoding Ion Channels Structure and Function*, Advances in Experimental Medicine and Biology 1497, https://doi.org/10.1007/978-3-032-07523-9_4

4.1 Introduction

Structural biology aims to determine the three-dimensional (3D) structures of biological macromolecules, to understand their biological functions and mechanisms at the molecular level. Structural knowledge is vital for gaining mechanistic insights into molecular operation, cellular processes, and disease mechanisms (Thompson and Kovall 2019). Historically, determining the structures of membrane proteins has been challenging. However, the last decade has seen substantial technological advancements in structure determination techniques that have revolutionized the field. Membrane proteins (MPs) are essential for numerous cellular functions, and are the targets for over 50% of FDA-approved small-molecule drugs (Santos et al. 2017). Therefore, understanding the structure and function of membrane proteins is crucial. Atomic-level structural details provide insight into how molecules interact with each other, how enzymes catalyze reactions, and how ligands, cofactors, ions, or drugs bind to specific sites on proteins, which is essential for rational drug design.

Macromolecular structures can be obtained experimentally using traditional techniques like X-ray crystallography and Nuclear Magnetic Resonance (NMR) spectroscopy, or relatively newer techniques like cryo-electron microscopy (cryo-EM), which can reach atomic or near-atomic resolution. In addition, machine learning approaches like AlphaFold and RoseTTAFold can predict protein structures with high accuracy in many cases. Although predicted structural models may not always represent macromolecular structure as it exists in the cellular context, or in the presence of ligands or water molecules, they can still be used as search models for structure determination techniques like X-ray crystallography or cryo-EM (Subramaniam and Kleywegt 2022). Structural models, whether derived experimentally or computationally, serve as a basis for structure–function studies. Traditionally, X-ray crystallography has been used to solve structures ranging from small proteins to viruses, while solution NMR has been used for small proteins and studying dynamics or ligand binding. Membrane proteins have been a challenging target, since in many cases they may be difficult to stabilize outside their native environment or to crystallize (Moraes et al. 2014). After the "resolution revolution" in cryo-EM, driven by improvements in transmission electron microscope optics, direct electron detectors, image processing algorithms, newer grids, and grid preparation tools (Kühlbrandt 2014), cryo-EM can now achieve resolutions comparable to X-ray crystallography. In fact, cryo-EM has become remarkably powerful for studying large, dynamic, and difficult to crystallize macromolecules and complexes, including transmembrane receptors and large multiprotein assemblies (Carvalho et al. 2018). Several advances have been achieved in protein crystallography, as well. Techniques like microcrystal electron diffraction (MicroED), serial femtosecond crystallography (SFX) using X-ray free electron lasers (XFELs), serial macromolecular crystallography, and the development of microfocus beamlines have allowed targeting of very small or weakly diffracting crystals and enabled time-resolved studies for extracting deeper biological insights from protein crystals. Another structural technique, cryo-electron tomography (cryo-ET) allows the study of membrane proteins in their native cellular environment or when reconstituted into proteoliposomes, providing insights into how membrane proteins integrate into the lipid bilayer. Cryo-ET combined with techniques like subtomogram averaging and integrated with prediction and other methods, allows for the visualization and modeling of macromolecular machines inside cells. In Fig. 4.1, we illustrate the number of structures deposited in the Protein Data Bank (PDB) during the decade 2015–2024, divided by experimental technique used for structure determination. While X-ray crystallography remains the dominant technique, there has been a remarkable increase in the number of protein structures determined by cryo-EM, especially in the latter half of this period. The sample preparation methods for membrane proteins have also advanced significantly over time. Improvements in expression systems (e.g., bacterial, yeast, insect, and mammalian cells) and the development of novel de-

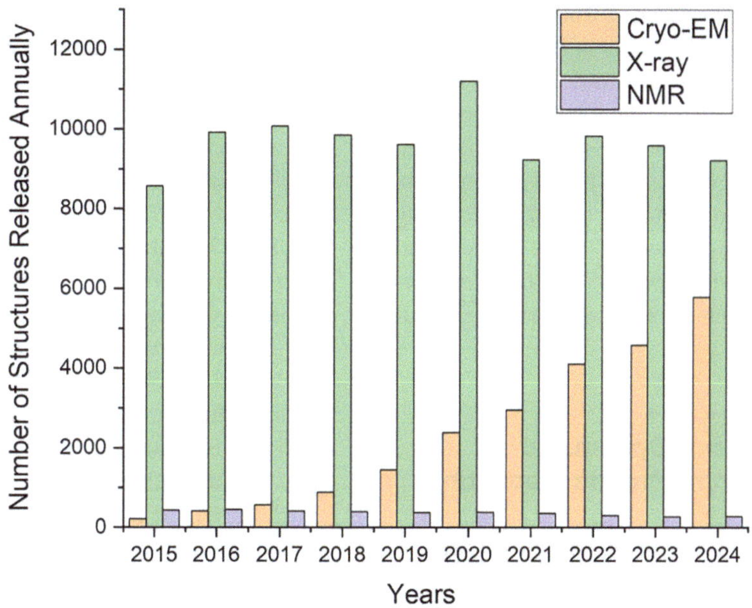

Fig. 4.1 PDB statistics of structures determined by structural technique. The growth of structures deposited in the Protein Data Bank per year for the years 2015–2024 is shown. X-ray crystallography has maintained the overall lead with a relatively stable number of structures per year during the past decade. However, the number of structures determined using cryo-EM shows a steep increase in the second half of the decade, while the number of NMR structures deposited has slightly declined over the years

tergents and membrane mimetics like amphipols, numerous copolymers (example, SMA, DIBMA, AASTY, Cubipol), and nanodiscs, have drastically improved protein solubilization, yield, and stability (Choy et al. 2021).

Modern structural biology employs integrative approaches that combine data from multiple experimental and theoretical methods to model complex biological systems and obtain comprehensive insights into molecular architecture and function (Schwalbe et al. 2024). For example, integrative methods can integrate information from techniques like X-ray crystallography, NMR, cryo-EM, cross-linking mass spectrometry, small-angle X-ray scattering (SAXS), and computational modeling (Rout and Sali 2019). In conclusion, structural biology has undergone a transformation due to significant advancements in various individual techniques along with the use of integrative approaches. In the following sections, we will delve deeper to examine how they have changed in recent years to supercharge the structure determination of membrane proteins.

4.2 Advances in Membrane Protein Preparation

The cell membrane serves as a crucial interface between the extracellular environment and the interior of the cell. Membrane proteins, which are embedded within the cell membrane, play essential roles in facilitating communication, transport, and signaling between the inside and outside of the cell. Remarkably, it is estimated that 20–30% of genes in a typical genome encode for membrane proteins (Krogh et al. 2001). To fully understand membrane protein function, it is essential to study them within their native lipid environment. Nonetheless, a common approach in structural biology involves the use of various membrane mimetics, which provide a substitute hydrophobic environment for structural studies.

This enables efficient scaling when protein copies in a cell are limited and aids in averaging to improve resolution and signal-to-noise ratio, among other parameters. In this section, we will explore the expression systems that are currently used for heterologous membrane protein expression, and the membrane mimetic systems, including detergents, amphipols, copolymers, extracellular vesicles, and lipid nanodiscs, which are used to stabilize membrane proteins for structural studies.

4.2.1 Expression Systems for Membrane Protein Production

The first step in structural studies of MPs is their expression in a functionally intact form using heterologous expression systems (Mancia and Love 2010, 2011). Achieving high yields of well-behaved membrane proteins requires an appropriate host system; therefore, the selection of a suitable host is critical for successful expression. Heterologous overexpression systems have facilitated the production of high-quality membrane proteins in sufficient quantities, supporting studies of both prokaryotic and eukaryotic targets (Mancia and Love 2011). Cell-free protein synthesis has also been used to produce membrane proteins of both prokaryotic and eukaryotic origin for structural analysis (Schneider et al. 2010; Shinoda et al. 2016), though it will not be examined further in this chapter. Currently, the main systems that are utilized for heterologous expression are the following:

(i) *Escherichia coli*: *Escherichia coli* is one of the most widely used systems for membrane protein overexpression due to its high expression efficiency, rapid growth rate, low cost, ease of scale-up, and simple transformation protocols (Pandey et al. 2016; Bruni and Kloss 2013). Additionally, engineered *E. coli* strains can help address challenges related to protein expression and solubility. However, *E. coli* expression does have notable limitations, including the absence of essential eukaryotic lipids in its membranes, lack of post-translational modifications, and limited protein folding chaperones. To overcome these issues, various strategies have been developed, such as co-expression of molecular chaperones, engineering of translational machinery, and protein tagging (Kesidis et al. 2020). According to one study, 477 combinations of promoters and *E. coli* strains have been tested for membrane protein expression optimization (Dilworth et al. 2018). Commonly used bacterial strains include BL21(DE3), C41(DE3), C43(DE3), BL21(DE3) CodonPlus, Rosetta2(DE3), and pLysS-containing BL21(DE3) (Kermani 2021). Furthermore, expression conditions, such as temperature, induction conditions, induction time, and optical density (OD) at the time of induction, significantly influence protein overexpression. Therefore, careful optimization of these parameters is crucial for successful expression.

(ii) *Yeast*: Yeast has been established as a robust system for the production of both membrane and soluble proteins (Routledge et al. 2016). It offers several advantages, including a relatively rapid growth rate, cost-effectiveness, simple culturing conditions, and scalability. Unlike prokaryotic systems, yeast possesses the machinery for eukaryotic post-translational modifications, such as glycosylation, protein folding, and disulfide bond formation (Yang and Zhang 2018). Yeast systems also support additional modifications, including palmitoylation and phosphorylation, which are important for the correct folding and function of many recombinant proteins. Although yeast grows more slowly than *E. coli*, its ability to perform eukaryotic-like modifications makes it a preferred system for the expression of eukaryotic membrane proteins. The two most commonly used yeast species for protein expression are *Saccharomyces cerevisiae* and *Pichia pastoris*. A major limitation of yeast expression systems is their glycosylation pattern, which significantly differs from that of higher eukaryotes and can adversely affect protein folding, function, and immunogenicity (Kesidis et al. 2020). Despite this, yeast systems are capable of producing high yields of membrane proteins, making them a valuable tool in structural and functional studies.

(iii) *Insect Cell/Baculovirus Expression System*: Insect cell expression is widely employed for the production of eukaryotic membrane proteins, primarily due to its post-translational modification (PTM) machinery, which more closely resembles that of mammalian cells compared to yeast (Junge et al. 2008). While insect cells can perform simple glycosylation, mammalian cells are capable of complex glycosylation, which may be critical for the function of certain proteins. The baculovirus expression vector system (BEVS) is a commonly used approach in insect cell systems (Kesidis et al. 2020). Baculoviruses can infect both insect and, in some modified systems, mammalian cells (Dukkipati et al. 2008; Goehring et al. 2014). One of the most widely adopted tools for baculovirus-based expression is the BAC-to-BAC system (Kesidis et al. 2020). In this method, the gene of interest is subcloned into a pFastBac vector, which is then used to produce a recombinant bacmid in *E. coli* strain DH10Bac. The resulting bacmid DNA is used to generate recombinant baculoviruses, which in turn are used to infect insect cells for protein production. Notably, the highest number of eukaryotic membrane protein structures solved to date have been produced using the insect cell expression system (He et al. 2014), underscoring its effectiveness in structural biology.

(iv) *Mammalian Cell Expression System*: Certain membrane proteins require near-native post-translational modifications (PTMs), lipid composition, and a cellular environment that closely mimics physiological conditions to achieve proper folding and functional activity. For some complex proteins, it is not feasible to obtain correctly folded and functional forms using prokaryotic or lower eukaryotic systems (Junge et al. 2008). Despite their advantages, mammalian expression systems come with significant limitations. The most notable challenges include low yields of the target protein, which can be insufficient for structural studies, as well as high production costs due to expensive media and complex culture requirements. As a result, relatively few membrane protein structures have been resolved using mammalian expression systems to date (Pandey et al. 2016). The two most commonly used mammalian cell lines for recombinant protein production are HEK (Human Embryonic Kidney) and CHO (Chinese Hamster Ovary) cells. Protein expression in these systems can be achieved through three main approaches: (a) Transient transfection—suitable for rapid, short-term expression; (b) Stable cell line generation—used for consistent long-term expression (e.g., (Elegheert et al. 2018)); and (c) Baculovirus-mediated transduction—an adaptation of the baculovirus system for use in mammalian cells (Dukkipati et al. 2008; Goehring et al. 2014). The use of suspension-adapted cells that also lack *N*-acetylglucosaminyltransferase I ($GnTI^-$) to restrict complex N-linked glycans into simple species only, has been extensively used in recent years to produce membrane proteins for structural studies (e.g., (Chaudhary et al. 2012)).

4.2.2 Detergents

Traditionally, and still most commonly, detergents have been used to solubilize and extract membrane proteins from their native biological membranes to analyze their structure, function, and dynamics in vitro. Detergents are amphiphilic molecules possessing a hydrophilic head group and a hydrophobic tail, allowing them to interact and disrupt biological membranes. Individual detergent molecules self-assemble into micellar structures above their critical micelle concentration (CMC), forming pseudo-spherical assemblies with hydrophobic tails buried in the core and hydrophilic head groups exposed to the surrounding aqueous solution (Stetsenko and Guskov 2017). The resulting protein–detergent complex (PDC) is soluble in an aqueous solution and maintains the protein in a stable state within micelles, as long as the concentration of detergent is kept above the CMC. Lower CMC values indicate greater micelle stability and lower risk of protein denaturation. Other important factors to consider include the hydrophile–lipophile balance, which indicates the detergent's ability to solubilize membrane proteins, and the aggregation number, which is the average number of detergent monomers in a micelle. The choice of detergent is critical for efficient extraction, solubility, and stability of membrane proteins throughout the downstream procedures and ultimately structure determination (Choy et al. 2021). For example, an unsuitable detergent may result in protein aggregation, avert the formation of crystal contacts during crystallization, or affect protein stability and/or particle distribution during cryo-EM sample preparation. There is still no universal detergent that is suitable for all membrane proteins, and initial screening remains essential to finding the optimal conditions for a specific protein. Therefore, detergent screens are widely used tools that typically involve screening a series of detergents in small-scale preparations to identify those that yield the largest amount of soluble, stable, active, and homogenous protein (Mancia and Love 2010).

Detergents are broadly classified into three groups based on the chemical nature and charge of their hydrophilic head group:

(i) *Ionic Detergents*: These detergents possess a charged hydrophilic head group, which can be either positive (cationic) or negative (anionic), and are known for their strong extraction capabilities. However, due to their high polarity, they can be harsh, disrupting both protein-lipid and protein-protein interactions, which may potentially lead to protein denaturation or inactivation. Their properties, particularly CMC values, are influenced by ionic strength. Examples include sodium dodecyl sulfate (SDS), sodium deoxycholate, and sodium taurocholate.

(ii) *Non-ionic Detergents*: These detergents have an uncharged hydrophilic head group carrying no charge. They are considered mild, with weaker extraction abilities but excellent optical properties that are useful for monitoring protein stability, while preserving the native state of the target protein and protein-protein interactions. Examples include alkyl maltosides, such as n-Dodecyl-β-D-Maltopyranoside (DDM), n-Decyl-β-D-Maltopyranoside (DM), and Lauryl Maltose Neopentyl Glycol (LMNG), and alkyl glucosides, including n-Octyl-β-D-Glucopyranoside (OG) and n-Nonyl-β-D-Glucopyranoside (NG). Polyoxyethylene glycols, such as C12E8 or Triton X-100, and Tween 20 are also non-ionic.

(iii) *Zwitterionic Detergents*: These detergents contain both positive and negative charges in their head group, resulting in an electrically neutral molecule. They are typically considered milder, offering gentle extraction and preserving the native state of the target protein with minimal protein denaturation. However, they are sensitive to strong ionic environments and may alter protein function or disrupt protein-protein interactions. Examples include CHAPS, CHAPSO, Lauryldimethylamine-N-oxide (LDAO), and sodium lauroyl sarcosinate.

4.2.2.1 Detergents in Structural Biology Methods

(i) *X-ray Crystallography*: In X-ray crystallography, membrane proteins are typically crystallized within a detergent micelle, where polar contacts between PDCs drive crystal lattice formation (Caffrey 2003). For crystallization in detergent micelles, detergents with relatively small polar headgroups are preferred to avoid covering the protein's polar surface needed for crystal contacts (Stetsenko and Guskov 2017). In general, shorter-chain detergents forming smaller micelles are favored as they can allow for tighter packing in the crystal lattice, potentially leading to higher resolution diffraction. High CMC detergents are also preferred in crystallography as they may improve crystal contact formation (Sonoda et al. 1993). However, excessive free detergent micelles can impede crystallization by promoting phase separation or yielding detergent crystals. DDM, DM, OG, and LDAO are the most used detergents for both purification and crystallization (Stetsenko and Guskov 2017; Parker and Newstead 2012). Glycosyl-substituted dicarboxylates (DCODs) are a novel detergent class designed for extraction, stabilization, and crystallization (Nguyen et al. 2018). Certain DCODs improve the diffraction resolution of crystals. Further, fluorinated diglucose detergents are a novel class of non-ionic fluorinated surfactants that are used to solubilize lipid vesicles and extract proteins from membranes (Boussambe et al. 2018).

(ii) *Cryo-electron Microscopy (cryo-EM)*: In cryo-EM, the choice of detergent may be responsible for complications in structure determination beyond protein stability, including increased ice thickness, reduced contrast, background noise due to empty detergent micelles (which can interfere with particle picking and alignment), and particle distribution issues (Kampjut et al. 2021; Michon et al. 2023). Interference by empty-detergent micelles is also applicable to negative-stain electron microscopy, often used for initial characterization of a purified membrane protein (Gewering et al. 2018). Detergents with low CMC and high-affinity PDCs are preferred because they enable using minimal total detergent concentration in the sample while keeping the protein soluble, which is crucial for optimal ice thickness and lower background noise in cryo-EM images (Kampjut et al. 2021; Li 2022). Free detergent can be easily removed or reduced without causing protein aggregation. LMNG, digitonin, and GDN (a synthetic analog of digitonin) use has significantly increased in cryo-EM due to their stabilizing properties and low CMC (Choy et al. 2021). For cryo-EM studies of eukaryotic membrane proteins, a lipid additive, cholesteryl hemisuccinate (CHS), is often used with detergents, such as DDM and LMNG, for solubilization and purification, where it functions as a membrane stabilizer, enhancing stability of membrane proteins in LMNG or conventional maltosides (Choy et al. 2021). Fluorinated detergents (e.g., FOM, FF8) are also used for solubilization and purification and have been shown to improve the distribution of membrane protein molecules in ice during cryo-EM grid preparation (Li 2022; Frotscher et al. 2015). Last, NDT-C11 (Neopentyl glycol-derived triglucoside-C11) is a novel detergent with high potential for single particle cryo-EM (Jiko et al. 2024).

(iii) *General Trends in Detergent Use*: Detergents have been instrumental in solving a vast number of membrane protein structures (Choy et al. 2021; Parker and Newstead 2012, 2016). Using a detergent mix has also become a common approach in screening and optimization (Parker and Newstead 2016). Novel detergents with improved properties for enhanced solubilization, stability, and structural studies are being developed. Sugar-based detergents, particularly DDM, DM, OG, and NG, and newer detergents, such as LMNG, GDN, and Digitonin, have been used for the majority of membrane protein structures deposited in the Protein Data Bank (Choy et al. 2021). Significant limitations of detergents include the potential delipidation of the target protein, as the loss of crucial lipids can reduce protein stability and function (Bechara et al. 2015; Stieger et al. 2021), along with concerns about maintaining native structural and functional integrity, including correct oligomeric states. Additionally, the presence of free micelles may interfere with downstream applications like crystallization or cryo-EM studies, while maintaining detergent concentrations above the CMC throughout purification can become costly for certain detergent classes.

4.2.3 Amphipols

Amphipols (APols) are a class of amphipathic polymers designed as alternatives to detergents for handling membrane proteins in aqueous solutions (Popot et al. 2011). These polymers are comprised of both hydrophilic groups and hydrophobic chains that adsorb specifically onto the hydrophobic transmembrane surface of membrane proteins, effectively keeping them water-soluble in the absence of detergent (Popot et al. 2011). The rationale of their design is that the hydrophobic octyl chains would bind to multiple contact points on membrane protein hydrophobic surfaces, resulting in a very high affinity interaction and preventing the dissociation of the polymer, even at extreme dilutions or doing so extremely slowly (Kleinschmidt and Popot 2014). The average length of the polymers is generally kept short (a few tens of units); therefore, several molecules are needed to keep a single membrane protein soluble. This facilitates homogenization of the APol belt size and minimizes the formation of bridges between adjacent membrane protein-APol complexes (Zoonens and Popot 2014). APols are highly soluble in water and form soluble complexes with a wide range of membrane proteins. Membrane proteins are generally more stable when trapped in APols compared to detergent solutions, exhibiting slower inactivation and stability at higher temperatures (Zoonens and Popot 2014).

Types of Amphipols The first-generation Apols were devised based on short amphipathic polymers carrying a large number of hydrophobic chains. Recent advancements led to the development of various other types of APols with modified basic structures. Further, labeling or functionalization of APols has conferred additional experimentally useful properties:

(i) *Classical APols*: Classical Apols were synthesized by forming amide bonds between octylamine and, in some cases, isopropylamine, and the carboxylic groups of low molecular weight polyacrylic acid precursors (Tribet et al. 1996). Examples include A8-35, A8-75, A34-35, and A34-75.

(ii) *Non-ionic, Glycosylated APols (NAPols)*: NAPols are amphipathic polymers that incorporate sugar groups and are non-ionic, indicating they possess no net electrical charge) (Sharma et al. 2012). The sugar groups (glycans) attached to the APol structure contribute to its solubility and stability. The absence of a net electrical charge on NAPols makes them compatible with a wide range of pH values and multivalent cations, which is important for certain biological applications (Bazzacco et al. 2012). Examples include NA8, NA10, NA11, NA29, NA45, and NA63.

(iii) *Sulfonated APols (SAPols)*: SAPols are derived from standard APols, such as A8-35, by replacing isopropyl groups with taurine moieties, which provide them with increased water solubility at lower pH (Dahmane et al. 2011). The nature of their hydrophilic moieties makes them pH-insensitive, and as a result, they are useful for studying membrane proteins in a variety of conditions, including low pH environments. Examples include SAPol-1, SAPol-2, SAPol-3, and SAPol-4.

(iv) *Phosphorylcholine-Based APols (PC-APols)*: PC-APols are random copolymers of n-octylacrylamide, N-isopropylacryla mide, and N-phosphorylcholine-N′-ethy lenedioxybis(ethyl)acrylamide (Diab et al. 2007). PC-APols are resistant to pH, high salt concentration, or the presence of divalent ions, making them suitable for various research applications, such as extracting and stabilizing MPs, immobilizing MPs onto solid supports for biosensors and assays, targeted drug delivery, and the development of biocompatible and antifouling biomedical materials (Diab et al. 2007; Lawton et al. 2017; Zhang and Chiao 2015). Examples include C45-68, C60-44, C60-30, and C22-43.

(v) *Cycloalkane-Modified APols*: These newer APols contain cyclic rather than linear aliphatic groups and have shown much greater efficiency at extracting proteins directly from the biological membrane compared to A8-35, bypassing the need for detergent extraction, and preserving protein-lipid interactions (Higgins et al. 2021). These amphipols have demonstrated improved performance in stabilizing membrane proteins for cryo-EM analysis (Marconnet et al. 2020, 2022). These include ArylAPols and CyclAPols with varying aromatic and cycloalkane side chains. An example in this group is the Ultrasolute Amphipols.

(vi) *Functionalized Amphipols*: APols, particularly A8–35, can be chemically modified for various applications (Della Pia et al. 2014; Perry et al. 2019). APols can be functionalized for specific purposes by grafting a small percentage of a tag or label onto the polymer without affecting its properties in solution (Le Bon et al. 2014a). Using a functionalized APol results in functionalizing the membrane protein of interest without the need for any chemical or genetic modification. This includes attaching:

   (a) *Affinity Tags*, such as biotin (BAPols) (Ferrandez et al. 2014), polyhistidine tags (HistAPols) (Giusti et al. 2015), imidazole cycles (ImidAPol) (Le Bon et al. 2018), and oligodeoxynucleotides (OligAPols) (Le Bon et al. 2014b).

   (b) *Fluorescent Probes* (FAPols), which can be used for visualization of membrane proteins, studying membrane protein-APol complex dynamics, and as reporters in various assays (Opačić et al. 2014).

   (c) *Isotopic labels*, such as ^{14}C, ^{3}H, ^{2}H, or Deuterated APols (DAPols), which are crucial for techniques like NMR, neutron scattering, and mass spectrometry to study protein structure and dynamics (Charvolin et al. 2014; Giusti et al. 2014; Tribet et al. 1997; Tehei et al. 2014).

(vii) *Universal Amphipols (UAPols)*: These are general precursors that can be easily functionalized with various probes or tags, simplifying the process of creating diverse functionalized APols. Examples include UAPol-NH2 and ThiAPols.

Applications of Amphipols APols have emerged as promising tools for folding denatured membrane proteins to their native, functional state, as demonstrated for bacteriorhodopsin and outer membrane proteins, such as OmpA and FomA (Watkinson et al. 2015; Pocanschi et al. 2013). APols stabilize membrane proteins against denaturation induced by heat or chaotropic agents, making them suitable for long-term storage and various biophysical studies (Dahmane et al. 2013). pH-insensitive APols, including SAPols and NAPols, enable solution NMR studies of membrane proteins in a detergent-free aqueous environment (Bazzacco et al. 2012). Deuterated amphipols (DAPols) are used to minimize signal overlaps and for contrast matching in various techniques, such as solution-state NMR and Small-Angle Neutron Scattering (SANS) studies of MPs and complex biomolecular systems (Giusti et al. 2014; Elter et al. 2014). APols have become valuable tools for cryo-EM studies of membrane proteins. Examples include structures of the TRPV1 ion channel (Huynh et al. 2014) and G-protein-coupled receptors (GPCRs) (Mary et al. 2014). Functionalized APols, bearing affinity tags, can serve as universal molecular adapters to immobilize membrane proteins onto solid supports (Charvolin et al. 2009), which can be used in biosensors, ligand-binding assays, surface plasmon resonance (SPR) studies, and creating membrane protein arrays. APols can be used to deliver membrane proteins, transmembrane peptides, and other hydrophobic molecules to lipid bilayers and cells in vitro and potentially in vivo (Nagy et al. 2001; Qi et al. 2012; Polovinkin et al. 2014). APols are used in proteomic workflows, including protein precipitation and processing (Ning et al. 2014), and are amenable to native electrospray ionization (ESI)-MS for studying membrane protein complexes (Watkinson et al. 2015). Furthermore, Apols have shown potential in stabilizing membrane proteins for vaccine formulations and enhancing their protective ability by increasing the immunoaccessibility of epitopes (Tifrea et al. 2011; Feinstein et al. 2014). In conclusion, Apols have evolved into mainstream tools in membrane biochemistry and biophysics. The continuous diversification of their chemical structures and the development of functionalized variants are expanding their range of applications, making them indispensable for a growing number of studies on membrane proteins.

4.2.4 Copolymers

Copolymers have emerged as noteworthy tools for studying membrane proteins, offering significant advantages over detergents by preserving native protein structure and lipid environment. The choice of a suitable copolymer depends on the specific characteristics of the protein and the experimental goals. The following examples are copolymers widely used for studying membrane proteins:

Styrene-Maleic Acid (SMA) Copolymer is an amphipathic copolymer composed of hydrophobic styrene and hydrophilic maleic acid groups, usually in a 2:1 or 3:1 ratio (Dörr et al. 2016; Bada Juarez et al. 2019). SMA works by inserting into the lipid membrane and spontaneously assembling into small discs of bilayer encircled by the polymer, known as SMA lipid particles (SMALPs) or native nanodiscs. SMALPs encapsulate membrane proteins within their native lipid patch, and unlike detergent-solubilized proteins, SMALPs are stable without requiring free polymer in buffers (Fiori et al. 2020). Advantages of SMA include detergent-free extraction, preservation of the native lipid environment and protein-lipid interactions, enhanced protein stability including thermal stability, retention of function and native conformation, compatibility with various cell types and analysis methods, including cryo-EM, X-ray crystallography, NMR, and mass spectrometry. Limitations include sensitivity to acidic pH and divalent cations, such as Mg^{2+} and Ca^{2+}, which can induce precipitation. Other limitations include the interference of the styrene ring with far-UV spectroscopy and the interference of free SMA with downstream applications like affinity purification and size-exclusion chromatography. Moreover, SMA products can have polydispersity, affecting the resolution of structural data. Modifications of the standard SMA for specialty applications include zwitterionic SMA (zSMA) for improved buffer compatibility (Fiori et al. 2017), styrene maleimide (SMI) with altered cation/pH sensitivity (Hall et al. 2018), SMA-Quaternary Ammonium (SMA-QA) for broader pH and high cation tolerance, SMA-sulfhydryl (SMA-SH) for various labeling procedures (Lindhoud et al. 2016), alkylamine-modified SMA (SMA-EA, SMA-ED, SMAd-A) (Ravula et al. 2017), styrene-maleamic acid (SMA-MA) for higher lipid content, styrene-taurine (SMA-Tau), enabling lipid phase transitions (Motov et al. 2024), SMA-benzylamine (SMA-BA) with variable lengths, native cell membrane nanoparticle (NCMN) polymers with enhanced tolerance to divalent ions and stability over a broad pH range (Kroeck et al. 2020), and substituted SMA analogues (R-SMA, X-SMA, R-SAA) to tune properties like solubilization efficacy and resistance to divalent ions. SMA applications include structural characterization, functional/biophysical studies, studying protein-lipid interactions, and high-throughput screening (HTS). SMA copolymers have been used for studies of GPCRs (Ayub et al. 2024).

Diisobutylene-Maleic Acid (DIBMA) is an amphipathic copolymer of diisobutylene and maleic acid groups, generally in a 1:1 ratio (Oluwole et al. 2017). DIBMA inserts into membranes to form DIBMA lipid particles (DIBMALPs), allowing detergent-free membrane protein purification while maintaining their native lipid environment. DIBMA forms larger diameter discs compared to SMA, providing a potentially less rigid environment. DIBMA encircles the lipid disc using one belt, unlike the two belts of SMA. Advantages of DIBMA include greater tolerance to divalent cations and a broader pH range than SMA, making it suitable for proteins requiring divalent cations. The absence of an aromatic group prevents UV absorbance interference in spectroscopic analyses. DIBMA is considered a gentler solubilizer with milder impact on lipid acyl-chain order, which may better mimic the native bilayer and allow for enhanced conformational changes and protein dynamics. DIBMALP's larger size can accommodate bigger proteins or complexes and offers higher lipid content (Voskoboynikova et al. 2021). Limitations include potentially lower yield, purity, less efficiency, and stability for some proteins compared to SMA. DIBMA can cause SDS-PAGE streaking and exhibits sensitivity to very low concentrations or high ionic strength buffers. Variants include glyco-DIBMA for forming discs under near-physiological conditions, sulfo-DIBMA (zwitterionic) with enhanced cation tolerance, glucosamine or amino-functionalized DIBMA to increase tolerance to divalent ions, PEGylated DIBMA (mPEG4-DIBMA) to reduce non-specific interactions with ligands, and DIBMA with different chain lengths (DIBMA 10, 12, 6). Functionalized variants like biotin-DIBMA and fluorescein-DIBMA allow for tag-free applications and visualization via fluorescent microscopy, respectively. DIBMA applications include broad protein encapsulation from various systems (including bacteria, yeast, and mammalian cells), SANS contrast matching for stealth nanodiscs (Guo et al. 2021), and structural characterization. DIBMA has been used to extract functional GPCRs (Chu et al. 2024) and human serotonin transporter (Dilworth et al. 2021).

Acrylic Acid-co-Styrene (AASTY) is an amphiphilic copolymer of acrylic acid and styrene, which is capable of solubilizing membrane proteins directly from bilayers into native lipid nanodiscs (Smith et al. 2020). It is often more aggressive in extraction due to higher acrylic acid content and lower molecular weight. Advantages of AASTY include effective solubilization, potential effectiveness at lower pH, amenability to structural studies like cryo-EM, and compatibility with downstream assays. Limitations include sensitivity to divalent cations, which can promote aggregation and precipitation, and incompatibility with spectroscopic procedures due to the presence of the styrene ring that causes UV absorbance interference (Timcenko et al. 2022). Advances involve RAFT polymerization for controlled properties, synthesis of copolymers with varied molecular weights and acrylic acid content for optimal performance (e.g., a 45:55 AA:STY ratio), and end-group modification. Applications include membrane protein extraction and solubilization, structural biology (cryo-EM), lipid studies, proteome-wide analysis, biophysical characterization, and functional studies (Walker et al. 2024).

Methacrylic Acid-co-Styrene (MAASTY) is an amphiphilic copolymer of methacrylic acid and styrene. It is based on the AASTY framework but differs from AASTY by a methyl group on the acid monomer (Pugh et al. 2024). Advantages include forming native nanodiscs, preserving endogenous lipids, facilitating high-resolution structural determination by cryo-EM, versatile solubilization of various lipids and proteins, and showing potential for capturing divalent cation-bound proteins (Pugh et al. 2024). Limitations for MAASTY include its polyanionic nature, sensitivity to divalent cations, and UV absorbance interference from styrene, making it incompatible for spectroscopy. It differs from AASTY in elution times and co-purified lipids. Advances include synthesis by RAFT polymerization for control, varied composition, and molecular weight. Applications include membrane protein extraction, structural characterization, studying interactions with divalent cations, and potential use in proteome-wide analyses (Pugh et al. 2024).

Stilbene-Maleic Anhydride/Stilbene-Maleic Acid (STMA) is a copolymer of stilbene and maleic acid, which forms homogeneous nanodiscs (Esmaili et al. 2020). Advantages of

STMA include a semi-rigid backbone, a broader usable pH range compared to SMA, formation of more homogeneous nanodiscs, effective protein extraction, and enhanced fluorescence. Limitations include potential for aggregation and incompatibilities with certain spectroscopic methods due to the stilbene group. Variants of STMA include methyl-substituted stilbenes such as poly(2-methylstilbene-alt-maleic anhydride) (2-STMA) and poly(4-methylstilbene-alt-maleic anhydride) (4-STMA)) (Smith et al. 2017). Applications in membrane protein research include solubilization, nanodisc formation, and structural/biophysical studies. STMA has been successfully used for the extraction of *E. coli* PagP (Esmaili et al. 2020).

Polymethacrylate (PMA) is an amphipathic polymethacrylate copolymer. Advantages of PMA include spontaneous lipid nanodisc formation, higher resistance to divalent cations compared to SMA, compatibility with standard purification processes, and the ability to form discs that are compatible with both circular dichroism and NMR methods. Limitations include challenges in controlling monomer sequence and molecular weight with traditional methods, leading to polydispersity, and the size of the disc formed that may not be optimal for all applications. PMAs with specific sidechain variations are stated to fragment membranes into discs of approximately 17 nm (Yasuhara et al. 2017). Variants involve different butyl to acetylcholine sidechain ratios, characterized by NMR. A library of amphiphilic polymethacrylates has been established, consisting of 25 polymers with variations in the hydrophobic unit, the degree of polymerization, and the corresponding number-averaged molecular weight. Examples include N-C4-24-4.0, N-C4-24-6.1, N-C4-25-1.7, and N-C4-26-13 (Yasuhara et al. 2017). Applications include the solubilization of membrane proteins to form nanodiscs and structural analysis using biophysical methods for proteins like GPCRs (Lavington and Watts 2021) and amyloid proteins (Yasuhara et al. 2017).

Alkyl Polyacrylic Acid (APAA) is a copolymer of alkyl and acrylic acid. Hydrophobic functionalization of polyacrylic acid with long-chain alkyl groups (hydrophobic tails) yields amphipathic polymers capable of extracting membrane proteins, and enhances their solubility and interaction with other molecules, including their drug delivery capabilities and gelation behavior (Hardin et al. 2019). Advantages of APAA include being a versatile platform for polymer nanodiscs, the ability to modulate nanodisc size including macro-nanodiscs, control over bilayer perturbation, suitability for UV spectroscopic studies in the absence of styrene, and alignment in an external magnetic field, which is particularly useful for NMR studies. Limitations include sensitivity to divalent cations and acidic pH, with tolerance varying depending on the alkyl chain length. Variants include pentyl, hexyl, and neopentyl derivatives, with RAFT synthesis offering control. Applications include membrane protein solubilization and structural characterization, suitability for UV spectroscopic studies, and the potential for use as adjuvants in vaccines (Chakraborty and Somasundaran 2006; Vial et al. 2006).

Butyl Methacrylate-Methacrylic Acid (BMAA) is a copolymer formed from butyl methacrylate and methacrylic acid. Advantages of BMAA include detergent-free extraction of membrane proteins, avoiding UV-absorbing and perturbing styrene units, comparable membrane solubilization efficacy to SMA, and different membrane disintegration patterns useful for studying microdomains differing in composition. Additionally, BMAA can be synthesized via atom transfer radical polymerization (ATRP), which provides precise control over the polymerization process and avoids the incorporation of an aliphatic chain present in some widely used RAFT agents (Janata et al. 2022). This is particularly advantageous as it could potentially affect the copolymer's performance in membrane solubilization assays. Limitations include sensitivity to acidic pH and divalent cations, leading to significant precipitation, similar to SMA. Variants involve tailoring molecular weight and BMA/MAA ratio via ATRP. For BMAA copolymers with BMA/MAA 1:1 composition, all copolymers, including the high-molecular-weight variant (MW of 14,000), were found to solubilize model membranes, while only the lowest-MW variant of the BMA/MAA 2:1 composition was effective. Applications of BMAA include membrane solubilization, isolation of membrane proteins, and studying membrane microdomains. BMAA has been used to solubilize Jurkat T-cell membranes (Janata et al. 2022).

Polymersomes are synthetic vesicles made of self-assembled block copolymers, serving as membrane mimetics, forming bilayered nanovesicles. Unlike disc-forming polymers, reconstitution into polymersomes (proteopolymersomes) requires detergents for initial protein extraction (Nallani et al. 2011). Advantages of polymersomes include superior stability compared to liposomes, lower permeability, and limited leakage. They offer membrane thickness of 5–50 nm, and their dense hydrophilic coronas enhance resistance to degradation. Vesicular architecture provides compartmentalization for studying transmembrane transport and signaling. Limitations include challenging functional reconstitution, polymers demixing from lipids in some cases, and fragility if highly fluid. Variants include diblock copolymers AB and triblock copolymers ABA, BAB, and ABC, featuring diverse block chemistries. In these configurations, A and C represent hydrophilic blocks, while B denotes the hydrophobic block. Advances include hybrid vesicles, blending phospholipids and block copolymers for robustness and biofunctionality, and methods to control size or create asymmetric membranes. Applications include artificial cell engineering, including artificial antigen-presenting cells (APCs) (Yang et al. 2020), structural and functional analysis of membrane transport proteins, membrane protein complexes, and membrane receptors, nanoreactors, and the potential for HTS in drug discovery. Examples of polymersome use include channel proteins for passive transport, such as aquaporin Z or synthetic porins (Habel et al. 2015), and protein complexes for active transport such as NADH:ubiquinone oxidoreductase (Complex I) (Leif et al. 1995), Proteorhodopsin (Gaitzsch et al. 2019), MHC-I molecule H-2Kb (Yang et al. 2020), and claudin-2 (Cldn2) (Nallani et al. 2011).

Cubipol copolymers enable single-step solubilization and stabilization of membrane proteins by capturing them in lipid–bilayer nanodiscs, providing room-temperature stability (Cubipol 2025). The Cubipol copolymer backbone integrates structural elements from existing backbones, such as SMA, DIBMA, and Ultrasolute Amphipol. Advantages of cubipol include structural preservation, low charge to minimize interference with protein-lipid interactions, and high resistance to extreme conditions such as variation in pH, ionic strength, and the presence of divalent ions. Variants of cubipol include cubipol biotin, designed for tag-free applications; cubipol amine, a positively charged variant that enhances interactions with negatively charged membrane surfaces; cubipol fluorescein, which is fluorescently labeled for visualization via fluorescent microscopy; cubipol glycerol characterized by low charge to minimize interference with protein–protein and protein–lipid interactions; cubipol PEG, which exhibits increased hydrophilicity, thereby reducing nonspecific interactions, improving solubility, and being particularly beneficial for proteins prone to aggregation; glyco-cubipol possessing the lowest charge and well-suited for charge-sensitive downstream applications; and sulfo-cubipol, which features low charge, is zwitterionic, and offers high resistance to extreme pH, ionic strength and divalent ion concentrations. Applications of cubipol include detergent-free membrane protein solubilization, isolation, and downstream applications like structural determination, functional studies, such as SPR, ELISA, and FACS, and drug discovery.

4.2.5 Extracellular Vesicles (EVs)

EVs are naturally cell-derived membrane vesicles. EVs are notable mediators of intercellular communication that change the recipient cell by transferring lipids, RNA, or protein cargo between cells. EVs, including exosomes and microvesicles, are produced through two main pathways: outward budding from the plasma membrane (microvesicles) or inward budding of endosomes to form multivesicular bodies (MVBs), which then release exosomes. During their biogenesis, membrane proteins are naturally incorporated into EVs when budding from the plasma membrane or the endosome (Doyle and Wang 2019). Some membrane proteins, especially those with transmembrane domains or lipid anchors, are selectively sorted and loaded into MVBs or their intraluminal vesicles (ILVs). This process is influenced by cellular signaling pathways and the presence of proteins like Endosomal Sorting Complex Required for Transport (ESCRT) complexes (Hurley 2010). EVs can also allow membrane proteins to form functional domains on their surface, like microdomains, which can be important for their function. EVs can be used to isolate membrane proteins and determine their structure in cell-derived membrane vesicles (Tao et al. 2023). Thus, EVs represent a platform for studying intact membrane proteins natively anchored with the correct topology in native biological membranes. Using the EV approach circumvents the need to conduct tedious detergent screens for solubilization, purification, and reconstitution required in classical membrane protein studies. Membrane protein-enriched extracellular vesicles (MPEEVs) are a type of EV specifically obtained to be enriched with a particular membrane protein.

Recombinant cell-surface membrane proteins displayed on EVs can be derived from cells in culture for structural studies. The interaction between these membrane proteins and ligands can also be analyzed directly on EVs. Further, EVs can be used for protein structure determination or immunization purposes (Delauzun et al. 2020).

The MPEEVs method has been applied to three integral type I membrane proteins, namely the Caenorhabditis elegans cell-cell fusion proteins AFF-1 and EFF-1, and the glycoprotein B (gB) from Herpes simplex virus type 1 (HSV1), resulting in low-resolution 3D reconstructions of EFF-1 and HSV1 gB in the membrane using electron tomography (Zeev-Ben-Mordehai et al. 1993). Recently, membrane vesicles directly derived from cells have enabled the structural determination of the ion channel Slo1 (high-conductance Ca^{2+}-activated K^{+} channel) at a resolution of 3.8 Å using vesicles derived from total membrane preparations, and at 2.7 Å using vesicles derived from plasma membrane preparations, employing cryo-EM (Tao et al. 2023).

Figure 4.2 shows the reconstructions of the ion channel Slo1 from total membrane vesicles (a), and plasma membrane vesicles (b), in which the ion channel is fully embedded within the lipid bilayer.

In addition, naturally generated lipid nanoparticles hold significant promise as engineerable therapeutic delivery vehicles (Peruzzi et al. 2024). However, active loading of protein cargo constitutes a challenge. Recently, it has been shown that by rationally enhancing the association of target proteins with lipid rafts in the membrane, the loading of proteins to EVs can also be enhanced. Moreover, select lipid tags that induce lipid modifications, such as myristoylation, palmitoylation, and farnesylation, can further increase EV loading of an engineered transcription factor, demonstrating that rational loading of cargo to EVs can be achieved (Peruzzi et al. 2024).

4.2.6 Lipid Nanodiscs

The development of lipid nanodiscs revolutionized the structural biology field by providing a tool for both the stabilization and structural characterization of membrane proteins. The most valuable property of this technology is that it provides a system with a lipid bilayer environment that can be tailored to mimic different biological membranes, including the plasma membrane, endoplasmic reticulum, or bacterial inner membrane. Lipid nanodiscs have

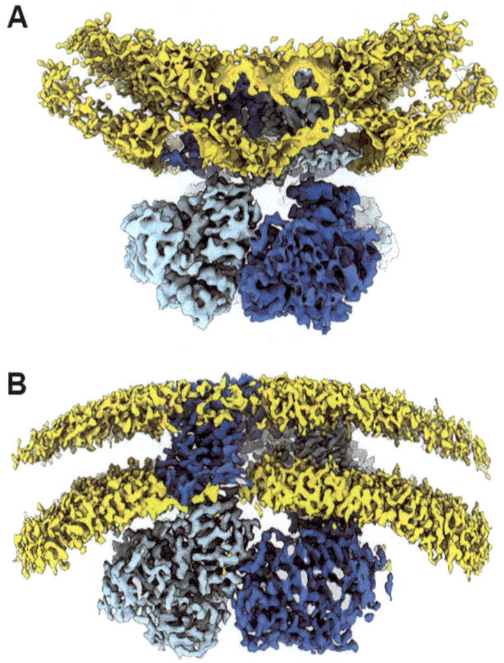

Fig. 4.2 Structure determination of the Slo1 ion channel from vesicles using single-particle cryo-EM. (**a**) Overall cryo-EM reconstruction of Slo1 from total membrane vesicles, and (**b**) Overall cryo-EM reconstruction of Slo1 from the plasma membrane vesicles. The improved resolution of the reconstruction from plasma membrane vesicles leads to additional detail in the cryo-EM map in (**b**). In both cases, the bilayer significantly contributes to the reconstructions and is shown in low contour in yellow color to showcase the curvature of the membrane. The Slo1 protomers are shown in blue and cyan. (The figure was published under Creative Commons Attribution License 4.0 (CC BY), and was adapted without modification from Tao et al. (2023))

enabled the structure determination of numerous membrane proteins, including many GPCRs (e.g., neurotensin receptor 1 in complex with G-proteins (Zhang et al. 2021a)), ion channels (e.g., transient receptor potential vanilloid 1 (TRPV1) ion channel (Gao et al. 2016)), and bacterial influx/efflux pumps (e.g., the multidrug efflux pump MexY (Gregor et al. 2025)).

A standard lipid nanodisc is composed of two copies of membrane scaffold protein (MSP) and a lipid bilayer (Bayburt and Sligar 2010; Denisov and Sligar 2016). The MSP protein is an engineered version of the α-helical human apolipoprotein A1, which has the ability to assemble itself into a discoidal structure. During the membrane protein nanodisc reconstitution, the MSP protein surrounds the membrane protein while the added lipids assemble into a phospholipid bilayer as detergent micelles are removed. As a result, the target membrane protein becomes embedded in a native-like bilayer environment (Cannon et al. 2023; Hagn et al. 2018). The length of the MSP protein determines the diameter of the nanodisc, and different MSP truncated forms have been created to enable the reconstitution of membrane proteins with different molecular weights, topologies, and oligomerization states (Hagn et al. 2018; Bruguera et al. 2022; Kumar et al. 2020). Among the widely used scaffold proteins are MSP1D1, MSP1E3D1, and MSP2N2. These MSPs yield nanodiscs of approximately 10 nm, 13 nm, and 16 nm in diameter, respectively (Cannon et al. 2023; Nouel Barreto et al. 2025). Furthermore, MSP nanodiscs can be filled using different lipids and in different ratios. However, the diameter of the nanodisc, lipid content, and lipid:MSP stoichiometry has to be carefully optimized, as these parameters can affect the conformation and function of the target protein (Nouel Barreto et al. 2025; Mi et al. 2017; Hoffmann et al. 2025; Arana et al. 2019; Zhang et al. 2021b; Dalal et al. 2024). For instance, the ABC transporter MsbA exhibits notable conformational differences when reconstituted into lipid nanodiscs compared to the structure determined in detergent micelles (Mi et al. 2017; Hoffmann et al. 2025; Arana et al. 2019). This data correlates with functional analysis, in which MsbA has been shown to have more ATPase activity in nanodiscs than in detergent micelles (Arana et al. 2019). Similarly, cryo-EM studies of the mechanosensitive channel MscS reconstituted into MSP1D1 nanodiscs demonstrated that lipid composition directly affects the protein gating. Specifically, short-chain lipids promote open-like conformations, whereas longer-chain lipids drive the protein towards closed or desensitized states (Zhang et al. 2021b).

Despite their versatility, the use of standard MSP-nanodisc is limited by heterogeneity and the lack of precise control over the number of proteins embedded in the system (Johansen et al. 2019; Nasr et al. 2017). To overcome these limitations, circularization of the MSP was introduced by covalently linking the N- and C-termini using sortase-mediated ligation (Nasr et al. 2017; Daniilidis et al. 2022). These circularized MSP-nanodiscs (cNDs) have offered significant improvements in stability by minimizing the dissociation of the membrane protein from the nanodisc, increasing monodispersity and size-control. Thus, enabling a more precise regulation of the number of membrane proteins incorporated (Johansen et al. 2019; Nasr et al. 2017; Daniilidis et al. 2022). Moreover, cNDs represent a next-generation vaccine platform capable of presenting antigens in a lipid bilayer environment. For example, Mabrouk et al. described for the first time the use of lipid cNDs containing four different SARS-CoV-2 spike proteins as a vaccine adjuvant system (Mabrouk et al. 2023). Additional advances to reduce the time-consuming process and increase the yield of cNDs have led to the development of engineered versions of MSP using the SpyCatcher–SpyTag technology (Zhang et al. 2021c). This approach involves a coupling system that forms an irreversible covalent bond in minutes (Zakeri et al. 2012). By fusing SpyCatcher and SpyTag to the N- and C-termini of MSPs, respectively, spontaneous circularization of the nanodisc occurs within cells, thereby eliminating the need for labor-intensive in vitro refolding and purification steps. This one-step method also provides a tenfold increase in cND yield compared to traditional methods and can produce nanodiscs with diameters ranging from 11 nm to 100 nm (Zhang et al. 2021c). However, although the utility of the SpyCatcher–SpyTag MSP-nanodisc was demonstrated through its application in determining the cryo-EM structures of the *Erwinia* ligand-gated ion channel (ELIC), further research is required to elucidate the versatility and constraints of this approach (Dalal et al. 2024).

In parallel to the improvements in nanodisc circularization, the development of monoclonal antibodies against MSP represents another significant innovation for membrane protein research. The anti-nanodisc antibody biND5 is the first reported antibody to specifically recognize a variety of MSP nanodiscs with high affinity. Since biND5 can recognize tag-free MSP, it enables potential applications that may require the immobilization of membrane proteins while maintaining their stabilization in a native-like lipid environment (Nakagawa et al. 2023).

Other lipid nanodiscs using DNA technology have been developed to mimic the complexity of multilayered membranes more precisely and have a customizable geometry (Zhao et al. 2018; Zhou et al. 2023; Zhang and Chapman 2020). Zhao et al. created nanodiscs up to ~70 nm in diameter using a DNA-origami barrel as a scaffold and confirmed the successful reconstitution of two membrane proteins (Zhao et al. 2018). Dong et al. reported a low-resolution cryo-EM structure of α-hemolysin integral membrane protein using a DNA origami-like strategy (Zhou et al. 2023).

In summary, lipid nanodiscs have revolutionized the study of membrane proteins by providing stability, size control, and a native-like lipid environment for structural and functional analyses. Continued developments over the last decade have helped to broaden the application of these platforms. Emerging DNA-based lipid nanodiscs further expand the toolkit, offering customizable architectures for the structure–function studies of integral proteins in more complex membrane environments.

Figure 4.3 presents a synthesis of the concepts introduced in this section, encapsulated into a schematic representation that summarizes a standard workflow that we employ for membrane protein purification.

4.3 Advances in Macromolecular Crystallography

Conventional macromolecular crystallography (MX) has been a widely employed technique for obtaining high-resolution structures of a vast number of proteins or biomolecules crystallized into well-ordered, large crystals (see Fig. 4.1). However, obtaining sufficiently large and well-ordered crystals suitable for diffraction studies presents numerous challenges, particularly for

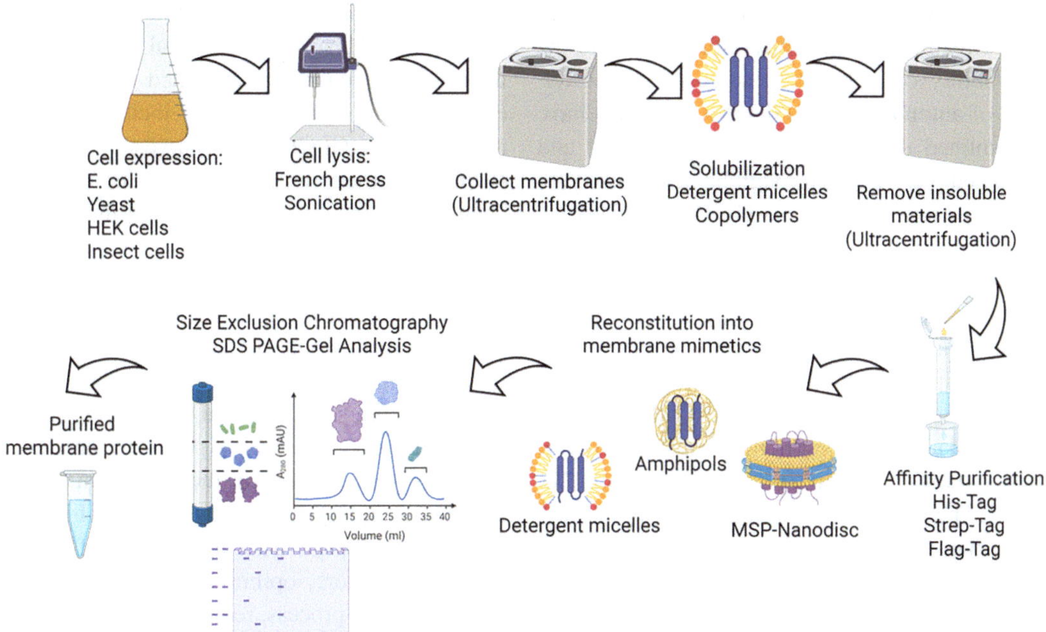

Fig. 4.3 Schematic representation of a membrane protein purification workflow. The process starts with protein expression in liquid cultures, followed by cell lysis using French press or sonication. Ultracentrifugation is used to collect membranes that are solubilized using detergent micelles or copolymers. Membrane proteins are then separated from insoluble materials by a second step of ultracentrifugation, followed by affinity purification using a genetically expressed protein tag. Different membrane mimetics (such as lipid nanodiscs, detergent micelles, or amphipols) may be used for membrane protein stabilization. The purity and stability of the final sample are confirmed by size-exclusion chromatography and SDS-PAGE analysis. (Image created with BioRender)

membrane proteins and large protein complexes. Challenges include optimizing crystallographic conditions, achieving the necessary crystal size and quality, and overcoming inherent protein instability or heterogeneity that can impede crystal formation. In this section, we will explore advances in macromolecular crystallography, which have enabled it to maintain the lead in structural output, despite the clear rise of cryo-EM in the past decade.

4.3.1 Lipidic Cubic Phase (LCP) Crystallization

The LCP (also referred to as *in meso*) method (Liu and Cherezov 2011; Caffrey 2015; Caffrey and Cherezov 2009) represents an advanced crystallization strategy designed for membrane proteins, which are often refractory to conventional crystallization methods due to low expression yields and inherent instability outside their non-native environments. The LCP is a biocontinuous lipid mesophase that self-assembles when a monoacylglycerol (MAG), such as monoolein, and an aqueous buffer are mixed at an appropriate ratio, resulting in a continuous cubic phase lipid bilayer (Caffrey 2015). The purified MPs diffuse into this cubic phase or mesophase bilayer upon reconstitution to provide a native membrane-like environment, retaining protein activity and lateral mobility. During crystallization setup, protein-laden LCP is equilibrated against a precipitant solution that often contains a polymer such as polyethylene glycol (PEG), salts, and a buffer to promote crystal nucleation and growth (Van 'T Hag et al. 2014; Zabara et al. 2017). Extensive crystallization trials are typically required to identify conditions conducive to crystal growth, followed by iterative optimizations to improve the crystal size and diffraction quality. Although crystallization screening is inherently labor-intensive, the availability of commercial screens and the advent of automated liquid dispensing robots have streamlined the process, enabling accurate and rapid setup with high-throughput screening.

Monoolein (or 9.9 MAG) (Kulkarni et al. 2011) remains the most widely employed lipid matrix to date for LCP or *in meso* crystallization trials at 20 °C. However, its inherent susceptibility to hydrolysis and transesterification, coupled with its inability to form a cubic phase below ambient temperatures (Qiu and Caffrey 2000), renders it unsuitable for labile membrane proteins that require handling at low temperature (4–6 °C). The rational design of alternative matrices by varying the hydrocarbon chain length led to the synthesis of a *cis*-monosaturated 7.9 MAG, enabling a stable cubic phase formation at low temperatures (Misquitta et al. 2004; Salvati Manni et al. 2015). Subsequently, the MS&FB group advanced the field by rationally designing and synthesizing a range of matrices, including 7.7, 7.8, 8.8, and 11.7 MAGs, to enable *in meso* crystallogenesis of target proteins (Caffrey 2015). Additionally, incorporation of additives such as cholesterol, monomethyl-DOPE, DOPE-mPEG350, and DOPC in combination with these MAGs has further improved the success of crystallogenesis. Most recently, two novel host lipids, 9.8 MAG, and 7.10 MAG have been developed, and the cubic phase formed with these lipids has successfully facilitated the crystallization of membrane proteins, including *E. coli* Apolipoprotein N-acetyl transferase (Lnt), β-barrel alginate transporter (AlgE), and the human adenosine A_{2A} receptor, compatible with cholesterol and POPC additives (Van Dalsen et al. 2021; Krawinski et al. 2024).

4.3.1.1 Applications of LCP

The LCP or *in meso* method was originally developed to address the intrinsic instability and poor solubility of MPs when extracted from their native environment. This approach enabled successful crystallization and subsequent structure determination of the first high-resolution crystal structure of bacteriorhodopsin by the Landau and Rosenbusch group (Landau and Rosenbusch 1996; Pebay-Peyroula et al. 1997). Since then, this approach has facilitated the structural elucidation of numerous MPs, including bacteriorhodopsins, GPCRs, light-harvesting complex II, photosynthetic reaction centers, transporters, channels, cytochrome oxidases, and many more (Caffrey 2015; Aherne et al. 2012; Winkler et al. 2019; Weinert et al. 2019; Zhu et al. 2020). Additionally, LCP has emerged as a versatile platform for structure-function studies, such as enzyme activity and ligand binding analysis (Li and Caffrey 2020), making it instrumental in structure-guided drug discovery. However, a persistent challenge with this approach lies in the formation of microcrystals during initial screenings, which are often too small to yield adequate X-ray diffraction data. Furthermore, optimizing crystal size is a labor-intensive process, requiring meticulous adjustments to crystallization conditions to achieve suitable dimensions for effective analysis. Recent advances have begun to address this challenge by leveraging these microcrystals or employing microcrystallography, for next-generation techniques such as SFX (see Sect. 3.2) and microcrystal electron diffraction (MicroED) (Zhu et al. 2020) (Sect. 5.1). Both techniques are capable of capturing high-resolution structural information from micro- or nanocrystals. In Micro-ED, protein crystals are deposited onto EM grids, blotted, and vitrified, followed by electron diffraction data collected akin to cryo-EM workflows. This integration of LCP with advanced techniques is reshaping the field of structural biology.

4.3.2 Microcrystallography and Serial Femtosecond Crystallography (SFX) with XFELs

In X-ray crystallography, prolonged exposure to X-rays can cause radiation damage (Garman 2010) and therefore affect the overall data completeness. To mitigate this damage, cryocooling was adopted to maintain cryogenic temperature (100 K) during the data collection. Yet, cryocooling requires screening for suitable cryoprotectants and often induces mosaicity and conformational heterogeneity (Vahedi-Faridi et al. 2003; Juers et al. 2018; Keedy et al. 2014). At the same time, room-temperature (RT) data collection approaches show promise for capturing dynamic conformations (Keedy et al. 2014). The advent of microcrystallography and serial femtosecond crystallography (SFX) using X-ray free-electron lasers (XFELs) has addressed many of these challenges. XFELs (Huang et al. 2021) produce ultrafast, high-intensity femtosecond pulses to collect diffraction data from micro- or nanocrystals before radiation damage occurs. This approach expands the possibility of data collection at room temperature (Owen et al. 2012) for the structural analysis of sensitive to radiation, and difficult to crystallize proteins, such as membrane proteins (MPs). As discussed in the previous section, MPs are not as readily crystallizable as soluble proteins due to their low expression and yield, amphiphilic nature, instability in non-native environments, and structural complexity (Carpenter et al. 2008). Given these obstacles, LCP (Liu and Cherezov 2011; Caffrey 2015; Caffrey and Cherezov 2009) has become a widely adopted method to crystallize membrane proteins mimicking the native bilayer environment (Sect. 3.1). Crystals produced by this approach tend to be in micron or sub-micron sizes (<20 μm) ideal for use with microfocus beamlines on synchrotron sources, such as ID13 and ID23-2 (Nanao et al. 2022) at ESRF, France, GM-CA (23-ID) at Advanced Photon Source (APS), FMX/AMX (17-ID-2/17-ID-1) at National Synchrotron Light Source II (NSLS II), USA, BL32XU beamline (Hirata et al. 2013), Spring-8, Japan, I24 and VMXi (Sanchez-Weatherby et al. 2019) beamlines at Diamond Light Source (DLS), UK, and PETRA III (Schroer et al. 2022) at DESY, Germany.

While cryocooling is extensively used to mitigate the radiation damage (Garman 2010) caused by ionizing radiation in microfocus beamlines, recent advances in XFELs have potentially eliminated this damage by utilizing ultrashort, extremely intense X-ray pulses lasting only for a few femtoseconds (fs). A list of global XFEL facilities along with their operating parameters is given in Table 4.1. The XFELs have enabled the collection of high-resolution diffraction data on micro- or nanocrystals at RT, preserving the native conformations before the onset of radiation-induced damage, a principle known as diffraction-before-destruction (Chapman et al. 2014). Using this approach, each diffraction pattern is collected from individual crystals, and a new crystal is required to collect the next pattern containing Bragg reflections, called crystal hits. This data collection strategy is now widely known as serial femtosecond crystallography (SFX) or sometimes nanocrystallography (Chapman et al. 2011).

4.3.2.1 Sample Delivery Methods

The experimental setup of SFX using XFELs involves the delivery of a continuous stream of micro- or nanocrystals into the X-ray beam by injecting crystal suspension across the beam with an injector device, such as jet-based (DePonte et al. 2008), droplet-based (Roessler et al. 1993), and fixed-target (Oghbaey et al. 2016) delivery systems. The jet-based Gas Dynamic Virtual Nozzle (GDVN) injectors (DePonte et al. 2008) remain one of the widely employed delivery systems in SFX applications to date. The GDVN injector utilizes a continuous focused coaxial gas stream to push microcrystal suspension through a 50–100 μm capillary to intersect the XFEL beam. However, GDVN typically requires a high sample flow rate (10–40 μl per min) (Oberthuer et al. 2017), which is too fast for the pulse repetition rate (120 Hz), leading to a substantial sample consumption (in 10–100 mg) during a full data collection. Ongoing advancements aim to enhance injector technology while minimizing sample consumption (Oberthuer et al. 2017; Zhao et al. 2019). Recent developments include new types of jet-based high-viscosity liquid injectors called Lipid Cubic Phase (LCP) injectors, which form a viscous gel matrix that can efficiently deliver the sample with minimal waste (Weierstall et al. 2014). LCP injectors efficiently deliver membrane proteins by enhancing their stability and providing a native membrane-like environ-

Table 4.1 Global XFEL facilities and their operating parameters

XFEL facilities	Location	Operational since	X-ray energy range (keV)	Pulse rate (pulses/sec)	Minimum laser wavelength (nm)	Maximum electron energy (GeV)	Photos per pulse
European XFEL	Schenefeld, Germany	2017	0.25–25	27,000	0.05	17.5 GeV	10^{22}
Linac Coherent Light Source (LCLS)	Stanford, USA	2009	0.25–25	120	0.15	14.3 GeV	~10^{12}
LCLS II, CuRF	Stanford, USA	2019	0.25–25	120	0.05	15 GeV	~10^{12}
LCLS II, SCRF	Stanford, USA	2020 (expected full operation from 2025)	0.25–25	1,000,000	0.25	5 GeV	~10^{12}–10^{13})
SACLA	Harima, Japan	2011	4–15	60	0.08	8	~$2x10^{11}$
SwissFEL	Villigen, Switzerland	2016	0.1–12.4	100	0.1	5.8	~10^{11}
PAL-XFEL	Pahong, South Korea	2017	0.25–15	60	0.06	10	~10^{12}
SXFEL/SHINE	Shanghai, China	SXFEL: 2021, SHINE: 2025 (expected)	0.4–25	1,000,000	0.05	8	~10^{12}

ment. While primarily employed for membrane proteins, soluble proteins can also be crystallized in LCP or mixed with pre-grown crystals before sample delivery (Fromme et al. 2015). Other high-viscous jet-based injectors have also been utilized (Kovácsová et al. 2017; Sugahara et al. 2017; Sugahara et al. 2016) but are sensitive to sample buffers, except for grease (Sugahara et al. 2015) and Vaseline (Botha et al. 2015) matrices.

As an alternative to jet-based injectors, crystals can be delivered via droplets generated from microcrystal resuspension using piezoelectric or acoustic ejection systems (Roessler et al. 1993; Mafuné et al. 2016; Fuller et al. 2017). This method enables drop-on-demand (DoD) data collection where droplet ejection is precisely synchronized with the XFEL pulse (120 Hz), significantly reducing the sample consumption compared to the jet-based injectors. A recent modification to this method involves more controlled sample delivery using acoustic droplet ejection combined with a conveyor belt drive system, referred to as Drop-on-tape (DoT), enabling simultaneous time-resolved SFX and spectroscopic measurements of photochemical and chemical reactions across a broad range of time scales (Fuller et al. 2017).

In addition to liquid jet and droplet-based delivery systems, fixed-target systems, such as microfabricated chips made from silicon (Oghbaey et al. 2016; Mehrabi et al. 2020), silicon nitride (Hunter et al. 2014), or polymers (Mueller et al. 2015), offer precise spatial and temporal control of crystal presentation. Typically, these chips contain patterned wells or indentations designed to trap crystals from the microcrystal suspension, and later, chips are blotted to produce a fine, thin liquid film around the crystals. Sample loading and blotting are typically performed in humid conditions to minimize background scattering and prevent sample dehydration. The chips are sometimes sealed with X-ray transparent membranes such as Mylar to prevent dehydration. During data collection, high-precision motors rapidly raster these chips through the XFEL beam.

4.3.2.2 Applications of SFX with XFEL

SFX diffraction data collected from serially delivered crystals often contain many blank images, as numerous X-ray pulses miss the crystals. Only a subset of these tens or hundreds of thousands of images contains Bragg reflections, called crystal hits. Pre-processing programs like Cheetah (Barty et al. 2014) are used to initially evaluate the data quality based on the useful accepted diffraction spots suitable for structure determination. These hits are then further processed using programs like CrystFEL (White et al. 2016), ccrbx.xfel (Sauter et al. 2013; Hattne et al. 2014), and nXDS (Kabsch 2014). To date, most protein structures solved using XFELs have relied on the molecular replacement (MR) method, which uses a known homologous structure to derive initial phases. However, in cases where MR is not possible, de novo phasing is utilized. Unlike synchrotron sources, de novo phasing at XFELs is more difficult due to inaccuracies in structure factor amplitudes arising from XFEL pulse fluctuations, variations in crystal size and mosaicity, and partial reflections. The first structure solved through de novo phasing at XFELs used the strong anomalous signal from gadolinium-derivatized (gadoteridol) lysozyme crystals with single-wavelength anomalous dispersion (SAD) scattering measurements (Barends et al. 2014). Subsequently, several studies have reported structures determined through de novo phasing at XFELs, including the Hg-derivative of luciferin-regenerating enzyme (LRE) obtained by single isomorphous replacement with anomalous scattering (SIRAS) (Yamashita et al. 2015); a novel crystal structure of BinAB, a naturally occurring paracrystalline larvicide, determined in complex with three heavy atom derivatives, mercury, gadolinium, and iodine, phased by multiple isomorphous replacement (MIR) (Colletier et al. 2016); and the membrane protein bacteriorhodopsin (bR) derivatized with iodine-labeled (HDD13a, I3C) and detergent, solved using SAD, SIR, and SIRAS methods (Nakane et al. 2016).

SFX has been instrumental in determining membrane protein structures using LCP as the crystallization as well as the crystal delivery medium. As of April 2025, there are a total of 836 protein entries in the protein data bank (PDB) obtained at XFELs, of which 235 structures belong to transmembrane proteins. Eighty percent of structures (704 entries) were solved at a 1.0–2.5 Å resolution range. One of the first membrane protein structures determined with LCP using SFX-XFEL was collected on G-protein coupled receptors (GPCR) microcrystals of human serotonin receptor 5-HT_{2B} (Liu et al. 2013). Following this, several GPCR (β_2AR, A_{2A}AR, glucagon receptor, and 5-HT_{2B}) structures (Weierstall et al. 2014), the angiotensin receptor (Zhang et al. 2015), human delta-opioid receptor (Fenalti et al. 2015), angiotensin II receptor (Zhang et al. 2017a), and glucagon class B GPCR (Zhang et al. 2017b) were solved using SFX-XFEL. Over the last 5 years (2020–2025), a total of 102 membrane protein structures have been solved using SFX-XFELs. Notable examples include the structure of CRTH2 (a GPCR protein) bound to a Prostaglandin D_2 (PGD_2) derivative (Liu et al. 2021a), the ATP-binding cassette (ABC) transporter CmABCB1 from *Cyanidioschyzon merolae* (Pan et al. 2022), and the light-driven chloride pump halorhodopsin (HR) from *Nonlabens marinus* (*Nm*HR), which is involved in ATP production and maintaining osmotic balance (Mous et al. 2022). Additionally, another structure of *Nm*HR-3 was determined using time-resolved SFX (TR-SFX) (Hosaka et al. 2022). Other highlights include the TR-SFX structure of rhodopsin (Gruhl et al. 2023), the structure of human corticotrophin-releasing hormone receptor 1 (CRF1R) (Kim et al. 2023), and the structure of cytochrome oxidase (CcO), with the latter studied through resonance Raman spectroscopy coupled with SFX (Ishigami et al. 2023). Furthermore, TR-SFX was used to study the structural response of CcO upon flash photolysis of carbon monoxide to track the dynamics in the active site CcO (Safari et al. 2023). The TR-SFX structures of the light-driven sodium pump rhodopsin 2 from *Krokinobacter eikastus* (KR2) (Wranik et al. 2023) and the oxygen-evolving photosystem II

(PSII), providing crucial insights into the PSII structural dynamics during S-state transition, were also determined (Li et al. 2024).

4.3.3 Cryo-Crystallography and In Situ Diffraction

Cryocooling, also referred to as cryo-crystallography, is one of the most widely used and reliable techniques in structural biology for protecting protein crystals from radiation damage (Garman 2010) during X-ray diffraction data collection (Pflugrath 2015). While it has become a cornerstone of MX, its use inherently limits the observation of conformational heterogeneity (Keedy et al. 2014) and dynamic motions occurring within the crystal lattice. To address these limitations, emerging crystallographic approaches, such as serial crystallography (SX) and time-resolved serial femtosecond crystallography (TR-SFX), have been employed, eliminating the need for crystal cryocooling, and allowing capture of structural snapshots in transient conformational states at ambient temperatures (Fromme et al. 2015; Liu et al. 2013; Aquila et al. 2012). However, they require large numbers of microcrystals for data collection on XFEL facilities, as one crystal is shot only once to obtain a diffraction image. Hence, microcrystallography has emerged as a powerful approach to address the challenges in producing and handling microcrystals, supported by various advancements.

Nevertheless, cryocooling still remains an essential method for experiments that rely on single-crystal diffraction at advanced microfocus beamlines (third- and fourth-generation synchrotrons) to collect high-resolution diffraction data. However, additional cryoprotectant screening is often labor-intensive. Recent advancements in cryocooling have significantly streamlined this process. Technologies like NANUQ (https://www.mitegen.com/nanuqtm-cryocooling-device/), which is an automated cryocooling device employing ultra-fast plunging and cold-gas removal, enables rapid freezing of crystals, reducing the need for cryoprotectants. This minimizes ice formation and crystal damage commonly associated with manual handling. In addition, the integration of laser-based photoablation (Zander et al. 2016) for automated crystal harvesting has enhanced efficiency, reducing manual intervention and time. This approach enables precise removal of mother liquor, the residual liquid post-crystallization, prior to mounting. It streamlines the cryocooling workflow through the utilization of a high-precision laser ablation technique and innovative mounting strategies that are challenging or unattainable with traditional tools. More recently, innovations in crystal mounting devices, such as X-ray-compatible transparent microfluidic chips, have further improved sample handling. These devices offer excellent optical transparency and low X-ray background, enabling array-based trapping of crystals suitable for robotic automation at synchrotron beamlines. Furthermore, the developments of methods like Cryo2RT have expanded MX capabilities by facilitating X-ray diffraction data collection at RT from cryocooled crystals, thereby allowing structural studies of protein dynamics across a range of temperatures (Huang et al. 2024).

In parallel to these advances, in situ diffraction has emerged as an alternative strategy for X-ray data collection that minimizes the need for cryocooling, crystal harvesting, and mounting at cryogenic temperatures. Initially developed for high-throughput screening of well-diffracting crystals directly within the crystallization plates without physical disturbance, it has since evolved into a powerful method for full dataset collection. Moreover, RT data collection using SX and TR-SFX has enabled the detailed study of protein dynamics, reaction kinetics, and transient conformational states of functional proteins and protein-ligand complexes, with growing applications in drug discovery. To support such studies, advanced synchrotron facilities now offer both RT and cryogenic data collection setups (Broecker et al. 2018) that are compatible with 96-well crystallization plates and *in meso* in situ plates, particularly advantageous for membrane protein targets. In this approach, Mylar-based film sandwich plates compatible with automated imaging and producing minimal background scattering are employed.

These plates support various crystallization formats, including microbatch, vapor-diffusion, and LCP or *in meso* setups.

Recent improvements in in situ crystallography have enabled de novo structure determination of membrane proteins by combining systematic diffraction screening and rapid data collection at cryogenic temperatures from *in meso*-grown microcrystals (Huang et al. 2018). Experimental phasing is achieved by accumulating anomalous signals from isomorphous crystals identified after collection, enhancing the efficiency of native and heavy-atom phasing strategies. Other advancements include the development of new 3D-printed holders to facilitate mounting of in situ samples on a goniometer at cryogenic or RT conditions for *in meso* in situ fixed target SX (Huang et al. 2020). These setups also support co-crystallization strategies such as heavy-atom or ligand soaking before data collection. In parallel, on-chip protein crystallization using dialysis methods has enabled in situ single-crystal or microcrystal-based SX experiments at RT (Jaho et al. 2021). This method integrates microfabrication techniques into a chip design containing a semi-permeable regenerated cellulose dialysis membrane, allowing precise and reversible control over the crystallization process. Once protein crystals grow to suitable sizes, the microchips can be directly mounted using a 3D-printed holder for data collection. More recently, in situ diffraction data were directly collected from microcrystals grown in a 96-well plate for the challenging protein autotaxin, solved in the low-symmetry space group P1, achieving a resolution of 3 Å at the microfocus beamline ID23-2 at ESRF (Foos et al. 2024). This demonstrates the feasibility of high-quality data collection from microcrystals without manual harvesting. Furthermore, advancements in automation, such as the use of crystal-on-crystal devices for in situ diffraction at RT, have enabled high-resolution data collection that is fully compatible with both synchrotron and XFEL-based beamlines (Ren et al. 2020). Collectively, these advancements establish in situ crystallography as a streamlined, high-throughput method for structural studies, especially of membrane proteins, enabling efficient data collection at semi- or fully automated synchrotron or XFEL platforms.

4.3.4 Phasing Techniques

Macromolecular crystallography (MX) involves X-ray diffraction data collection by measuring the diffraction intensities of scattered X-rays within the crystal lattice. While these intensities provide the amplitude of scattered waves, they lack corresponding phase information essential for generating electron density maps (Taylor 2010; Hendrickson 2023). To overcome this "phase problem," two main approaches are typically utilized: molecular replacement (MR) and experimental phasing. In the MR approach, phases are derived from known structure coordinates of homologous proteins, and are often favored due to their efficiency and minimal experimental overhead. On the other hand, experimental phasing is a more difficult and labor-intensive process that leverages either intrinsic anomalous scatterers or heavy atoms incorporated exogenously to generate differences in diffraction intensities. One of the preferred and commonly used methods of experimental phasing is single-wavelength anomalous dispersion (SAD) (Hendrickson 2014), which utilizes data collected with the anomalous scatterers close to their absorption edge. These anomalous scatterers include heavy atoms, such as Hg, Au, and Ir, incorporated by soaking (Pike et al. 2016) or through selenomethionine (SeMet) substitution (Strub et al. 2003) into proteins. However, these approaches have notable limitations, as soaking can induce crystal isomorphism and affect diffraction quality, while SeMet substitution may impair protein expression yield. As an alternative, native-SAD phasing emerged as a powerful strategy that circumvents the need for derivatization to solve the phase problem.

Native-SAD utilizes the weak anomalous signals from endogenous light atoms inherently present within the protein. A prime example is sulfur-SAD (S-SAD), where S atoms from cysteine or methionine can be utilized to enable phase determination. While the S K-edge is 2.472 keV, corresponding to a wavelength of approximately 5.02 Å, most synchrotron beamlines operate at wavelengths ranging $\lambda = 0.7$ to 2.06 Å, well below the sulfur K-edge. Consequently, S-SAD diffraction data are typically collected at the maximum available wavelengths ($\lambda = 1.77$ to 2.06 Å) with weak anomalous signals. To overcome this limitation, high-multiplicity data sets are often acquired by collecting redundant measurements from the same or multiple isomorphous crystals to enable experimental phasing. Leveraging this approach, fast native-SAD phasing strategies, combining Se-SAD and S-SAD, have successfully enabled the structure determination of 11 proteins, including a human membrane protein, using low-dose, high multiplicity data collected from a single crystal (Weinert et al. 2015). While low-dose high multiplicity data collection can significantly improve anomalous signals, air scattering and absorption still introduce background noise that attenuates the already weak anomalous signals, posing a significant challenge for native-SAD phasing. Recent advancements, however, such as the implementation of helium-filled beam paths, have significantly reduced both background scatter and air absorption, thereby enhancing the quality of anomalous signals (Karasawa et al. 2022).

Further advancements in the native-SAD phasing have extended beyond S-SAD, with other endogenous light elements such as Ca, K, Cl, and P, with K-edges attainable at synchrotron wavelengths. Notably, beamline I23 at Diamond Light Source has enabled the structure determination of 22 macromolecules, highlighting the expanding applicability of native-SAD (El Omari et al. 2023). Similarly, the use of iodine-labeled detergent bicelles, such as HAD13a, which contains an I3C head group, facilitated robust SAD, SIR, and SIRAS phasing in SFX experiments (Nakane et al. 2016). In another promising approach, halide-SAD phasing was successfully applied to four distinct MPs by utilizing iodide's affinity for positively charged residues at the membrane interface (Melnikov et al. 2017). Parallel efforts have employed selenourea as a promising soaking reagent for LCP-grown crystals, providing a high-density, non-disruptive heavy atom source that effectively expands the phasing toolkit for hydrophobic, membrane-embedded proteins such as PlsY (Luo et al. 2022). Together, these methods mark significant developments in experimental phasing, broadening its accessibility, reliability, and applicability in the field of structural biology.

4.3.5 Time-Resolved Crystallography

Time-Resolved Crystallography (TRX) offers a powerful approach to visualize transient intermediates and conformational dynamics that occur during biochemical reactions. Unlike MX or static crystallography, which provides static structural snapshots, TRX introduces time as a fourth dimension, capturing a sequence of structural frames to reveal conformational changes in real time. Additionally, most MX or static experiments utilize a monochromatic X-ray beam to collect diffraction data using a single cryocooled crystal, producing high-resolution but time-averaged structural information. In contrast, TRX collects data at ambient or RT using a polychromatic beam through Laue diffraction (Moffat et al. 1984; Hajdu et al. 1987; Moffat 2019), with exposure times faster than the timescale of the reaction, allowing simultaneous capture of numerous reflections. The potential of Laue diffraction was first demonstrated in studies that mapped early structural changes in the GTP-bound form of the G-protein Ha-Ras p21, minutes after photolytic release of caged GTP (Schlichting et al. 1990). This work traced the progression of GTP hydrolysis over a span of 14 min. Subsequently, TRX measurements on a myoglobin crystal bound to carbon monoxide (Mb:CO) (Srajer et al. 1996) and photoactive yellow protein (PYP) (Genick et al. 1997), initiated by laser-induced photolysis, enabled the capture of structural snapshots on the millisecond (ms) and nanosecond (ns) time scale. Further advancements in synchrotron sources and X-ray detectors have facilitated extensive TRX studies on these stable, well-diffracting Mb:CO and PYP protein crystals, ideal for Laue diffraction, achieving time resolutions down to the picosecond (ps) range (Bourgeois et al. 2003; Schmidt et al. 2004; Ren et al. 2012; Schotte et al. 2012; Jung et al. 2013). However, only a limited subset of proteins fulfills the stringent requirements for successful TRX studies using Laue diffraction. To expand TRX to more systems, time-resolved X-ray solution scattering (TR-XSS) (Cammarata et al. 2008; Cho et al. 2010; Björling et al. 2016) has been employed to measure protein secondary structural rearrangements in solution. While this method avoids crystallization constraints, it lacks the atomic-level resolution offered by Laue diffraction-based studies. The advent of XFELs has further revolutionized TRX studies by delivering ultrashort, intense X-ray pulses. These pulses probe a continuous stream of hundreds of thousands of micro- or nanocrystals delivered via various sample injection methods (see Sect. 3.2.1). XFEL-based TRX studies have enabled the capture of rapid structural conformations at extremely short pulse durations (fs range) using serially presented microcrystals exposed to focused XFEL pulses, a technique known as SFX (For more details, refer to Sect. 3.2.2) or time-resolved SFX; TR-SFX.

A critical step in TRX studies is the initiation of the reaction to be probed by an X-ray pulse. This is typically achieved using a pump-probe method, in which a "pump" pulse (often a laser) triggers the reaction, followed by an X-ray "probe" pulse at a defined time delay to obtain diffraction images. The first TR-SFX study was performed using an optical pump laser synchronized with an XFEL pulse on a large membrane protein photosystem I-Ferredoxin complex (Aquila et al. 2012), following the successful development of SFX methods elucidating the structure of photosystem I nanocrystals (Chapman et al. 2011). In this TR-SFX study, photosystem I-ferredoxin co-crystals were delivered via a liquid jet-based device, irradiated with an optical pump laser, and subsequently probed by XFEL pulses, capturing light-induced changes at time delays of 5–10 μs (Aquila et al. 2012). Using a similar approach, TR-SFX has enabled ultrafast trans-to-cis isomerization trajectories of chromophores within naturally light-sensitive proteins such as PYP (Pande et al. 2016; Pandey et al. 2020), chromophore twisting in the reversibly photoswitchable fluorescent protein rsEGFP2 (Coquelle et al. 2018), and snapshots of structural photoresponse in phytochrome proteins (Claesson et al. 2020).

An alternative approach for studying non-photosensitive proteins is to attach a photolabile group (a photocage) to either the ligand or the protein, rendering the system inactive until a light pulse triggers photocage cleavage and releases the active molecule. One example of this approach is the study of ATP-driven dimerization of the nucleotide-binding domain (NBD) of MsbA flippase enzyme. In this TR-XSS experiment, a light-sensitive caged ATP analog, NPE-ATP, was used, and upon irradiation, it released the active ATP, which bound to NBDs and triggered dimerization (Josts et al. 2018). In another study, enzyme-ligand-induced structural dynamics were visualized in fluoroacetate dehalogenase. TR-SFX measurements at time-points ranging from 30 ms to 30 s after photolytic release of fluoroacetate from pHP-caged fluoroacetate revealed four sequential events: binding of substrate at the active site, covalent ester intermediate formation, positioning of a water molecule for intermediate hydrolysis, followed by the product release (Mehrabi et al. 2019).

The development of mix-and-inject serial crystallography (MISC) further expanded time-resolved studies of non-light-sensitive systems by mixing microcrystals with ligands or substrates in a liquid jet. Notably, this method captured the binding and cleavage of the third-generation antibiotic ceftriaxone in Mycobacterium tuberculosis β-lactamase (BlaC) microcrystals, determined at near-atomic resolution across five time points from 30 ms to 2 s after mixing (Olmos et al. 2018). Subsequently, MISC studies on enzyme isocyanide hydratase (ICH) microcrystals revealed transient states including a thioimidate covalent intermediate (Dasgupta et al. 2019). These data suggest that modification of active-site cysteine induces helical movement, enabling water entry into the ICH active site to hydrolyze the intermediate. This establishes a mechanistic link between helical mobility and catalytic efficiency in ICH. More recently, TR-SFX enabled structural kinetics of boronate-based inhibition of β-lactamase using a mix-and-diffuse method, where boric acid was mixed and allowed to diffuse into microcrystals, resulting in 22 data sets acquired with a resolution of 1.4–2.04 Å (Prester et al. 2024). In parallel, several other techniques, such as freeze-quenching or tapping and temperature-jump (T-jump) have been employed to conduct TR-SFX studies. In freeze-quenching, a reaction is triggered and then rapidly quenched by cooling the crystals using cryogens at precise time intervals to trap transient states. A recent development, a crystal plunger Spitrobot, enables reaction quenching via cryo-trapping with ms time-resolution. Functionally similar to a cryo-EM vitrification device, it operates in a humidity- and temperature-controlled environment and utilizes crystals mounted on micromesh sample holders (Mehrabi et al. 2023). In T-jump experiments, an infrared (IR) laser excites the O-H stretching in water molecules, resulting in rapid heating of the sample, causing thermal activation. Using this approach, structural motions in lysozyme crystals were visualized at nanosecond time resolutions (Wolff et al. 2023). Together, these advancements underscore the versatility and growing impact of TRX in capturing dynamic structural events across a wide range of biological systems.

4.3.6 Fourth-Generation Synchrotrons

Fourth-generation synchrotron sources offer an advanced platform for producing ultra-bright, highly coherent X-rays by accelerating electrons close to the speed of light in circular paths inside storage rings, enabling cutting-edge research in the field of MX and related disciplines. These exceptional beam properties are made possible by the implementation of innovative magnetic multi-bend achromat (MBA) lattice designs, which drastically reduce the horizontal electron beam emittance, thereby minimizing both its transverse size and divergence. As a result, the brilliance and spatial coherence of the emitted X-ray photon beam increase significantly. Brilliance, defined as the number of photons emitted per second per unit source area (photon flux) and phase-space volume over a particular bandwidth, is a key parameter, as higher brilliance leads to a greater coherent flux and markedly improves signal-to-noise ratios (SNR). This enhanced beam quality is particularly advantageous for MX studies, allowing researchers to acquire data from small or more weakly diffracting crystals. In addition to higher energy X-ray sources, fourth-generation synchrotron facilities integrate advanced detectors and computational algorithms, further enhancing faster data collection and analysis capabilities.

MAX IV, located in Lund, Sweden, became the world's first operational fourth-generation synchrotron light source, commissioned in 2016 (Tavares et al. 2018; Grabski and Al-Dmour 2021). Its 3 GeV storage ring (with 528 m circumference) is optimized for ultra-bright hard X-rays and features a 20-fold seven-bend achromat (7BA) lattice achieving an impressively low emittance of 0.33 nm·rad. MAX IV currently operates 16 beamlines, among them, BioMAX, operational since 2017, is dedicated to high-throughput MX experiments and supports advanced phasing techniques such as SAD and MAD (Gonzalez et al. 2025). The MicoMAX beamline focuses on serial crystallography (SX), including TRX. Additionally, MAX IV hosts the FragMAX platform for fragment-based drug discovery, and the FemtoMAX beamline facilitates protein diffraction experiments at ultrafast time resolutions.

Following MAX IV, the second fourth-generation synchrotron source, SIRIUS, began operation in 2019. It is operated by the Brazilian Synchrotron Light Laboratory (LNLS) in Campinas, Brazil. It features a 518 m circumference electron storage ring based on a 20-cell five-bend achromat (5BA) lattice, achieving a bare lattice horizontal emittance of 0.25 nm·rad for a 3 GeV beam (Liu et al. 2019a). The facility is designed to host up to 38 beamlines, with around 20 already in operation. Among these, the MANACÁ beamline is tailored for MX using micro- or nanocrystals, supporting room-temperature and remote data collection. The SAPUCAIA beamline is dedicated to small-angle X-ray scattering (SAXS) experiments, further broadening the structural biology capabilities at SIRIUS.

The third fourth-generation synchrotron source, the Extremely Brilliant Source (ESRF-EBS), represents the experimental realization of the novel Hybrid Multi-Bend Achromat (HMBA) storage ring magnetic lattice concept at the European Synchrotron Radiation Facility (ESRF), Grenoble, France (Raimondi et al. 2023). This upgraded storage ring, featuring 32 straight sections, 28 of which are used to produce X-ray radiation, houses a 6 GeV light source and achieves an impressively low horizontal emittance between 0.12–0.13 nm·rad. The energy consumption of the HMBA lattice is smaller compared to previous designs, contributing to improved beam stability.

In addition to the existing sources or storage rings, several other advanced facilities are under construction or in the planning stages. These include the Advanced Photon Source Upgrade (APS-U) (https://www.aps.anl.gov/APS-Upgrade/About-the-APS-Upgrade) and Advanced Light Source Upgrade (ALS-U) (https://als.lbl.gov/als-u/overview/) in the United States, which are set to enhance the capabilities of their respective synchrotron light sources. PETRA IV (Schroer et al. 2018) and BESSY III (Goslawski et al. 2022) in Germany, ALBA II (Biscari et al. 2024) in Spain, SOLEIL-2 (Abiven et al. 2023) in France, and Elettra-2 (Karantzoulis et al. 2024) in Italy will further advance synchrotron technology in Europe, while Diamond II (Ghasem et al. 2024) in the United Kingdom, are poised to offer cutting-edge beamlines. Other sources like CLS-2 (Boland et al. 2021) in Canada and Korea-4GSR (Jang et al. 2022) at the Pohang Accelerator Laboratory (PAL) in South Korea are also being developed as a fourth-generation synchrotron light source, adding to the growing global network of high-performance research infrastructure. In Asia, China's flagship synchrotron project, the High Energy Photon Source (HEPS) (Jiao et al. 2018), officially launched its joint commissioning phase in March 2025 and is expected to become fully operational by the end of this year. HEPS has achieved an electron beam emittance of 0.93 nm·rad in its storage ring, enabling the generation of exceptionally bright and high-quality synchrotron radiation suitable for advanced scientific research.

4.4 Advances in Single-Particle Cryo-EM

Cryogenic electron microscopy (cryo-EM) has revolutionized structural biology by enabling the visualization of biomolecules in a frozen hydrated state at near-atomic resolution (Li et al. 2013). After the "resolution revolution" in cryo-EM, driven by improvements in microscope optics, direct electron detectors, image processing algorithms, newer grids, and grid preparation tools, cryo-EM can now achieve resolutions comparable to X-ray crystallography. Single-particle cryo-EM has become the preferred technique for the study of membrane proteins, given the bottleneck that crystallization presented for membrane protein structural biology and large, heterogeneous, or dynamic multiprotein assemblies, due to the ability to extract multiple conformations from the same dataset. In this section, we will explore recent advances in single-particle cryo-EM instrumentation and software that have contributed to the rise of cryo-EM utilization for structure determination of membrane proteins and multiprotein complexes.

4.4.1 Direct Electron Detectors and Image Processing

A critical component for the rise of cryo-EM has been the development of a direct electron detector (DED) that can capture images with high sensitivity, improved signal-to-noise ratio, and faster acquisition time compared to film and Charge-Coupled Devices (CCDs) (Wu et al. 2016). Modern DEDs, such as Gatan K2/K3 (Sun et al. 2021), FEI's Falcon 4 (Kuijper et al. 2015; Nakane et al. 2020), and Direct Electron's Apollo (Peng et al. 2023) coupled with an energy filter to provide high-quality data, also in combination with powerful image processing software, such as cryoSPARC, Relion, and Scipion, are now able to support structural resolution to the near-atomic level (DiIorio and Kulczyk 2022).

Much of the detector technology is based on developments in microelectronics, which have allowed the design of direct detectors with fine pixels and fast readout. Traditional methods of recording images in the electron microscope began with film and subsequently migrated to CCDs. The use of film was labor-intensive due to the extensive processing required to convert it to a digital format for further analysis. Additionally, CCD technology faced limitations in both sensitivity and resolution, impacting its effectiveness in capturing images. The introduction of direct electron detectors marked a significant leap forward (McMullan et al. 2014). These detectors directly capture electrons from the microscope's electron beam, eliminating the need for scintillator and fiber optic coupling often found in older CCD-based systems. DEDs offer several advantages over CCDs, including high detective quantum efficiency (DQE), high frame rates, and low noise. They enable super-resolution data collection and electron counting, which further enhances the DQE, making them highly effective for capturing detailed and high-quality images. Hence, the impact of direct electron detectors on structural biology has been profound. These detectors have enabled the visualization of complex biomolecules, such as proteins and nucleic acids, at near-atomic resolution. The advanced capabilities of DEDs have catalyzed significant breakthroughs across various fields, including drug discovery, structural genomics, and the elucidation of disease mechanisms, underscoring the transformative impact of this technology (Nakane et al. 2020).

The development of direct electron detectors continues to evolve, with ongoing research focused on further enhancing their sensitivity, resolution, and speed. Innovations, such as hybrid pixel detectors and advanced image processing algorithms, are expected to expand the boundaries of cryo-EM even further. The development of DEDs optimized for 200 keV (Milazzo et al. 2010) or even 100 keV (Karia et al. 2024) is particularly noteworthy. These detectors offer a cost-effective alternative to the more common 300 keV DEDs, while still providing resolution that meets the requirements for many projects. A recent study using a LaB6 source instead of a field emission gun (FEG) at 120 keV obtained remarkable resolution, paving the way to even further cost reduction through the introduction of a newly developed Gatan Alpine DED camera optimized for microscopes operating at 80–200 keV (Venugopal et al. 2025).

Another significant advancement in cryo-EM is the development of sophisticated image processing software. This innovation has established single-particle cryo-EM as the preferred method for structural biologists seeking to determine high-resolution structures of diverse macromolecules, in particular MPs. Among the data processing software, cryoSPARC (Punjani et al. 2017) and Relion (Scheres 2016) are widely used. The newly developed Scipion enables users to create an entire workflow by integrating the most relevant image processing packages (Jiménez-Moreno et al. 2021), allowing for easier customization of the workflow to meet research needs. These software suites allow both new and expert users to perform image processing and 3D reconstructions (DiIorio and Kulczyk 2022), using the same basic workflow: movies acquired by the microscope detectors undergo correction for beam-induced motion and contrast transfer function (CTF) estimation (Ripstein and Rubinstein 2016). Next, individual particle images are selected and extracted from averaged movie frames for iterative 2D and 3D classification, followed by 3D reconstruction, refinement, and validation. Because various software packages employ different algorithms and require varying levels of expertise to operate, the 3D maps they generate often differ in quality and resolution. Thus, users regularly transfer data between a variety of programs for optimal results (Sorzano et al. 2021).

Figure 4.4 provides a detailed overview of a single-particle cryo-EM workflow starting from a purified protein and incorporating sample preparation, sample optimization, data collection, data processing, and atomic model building.

4.4.2 Near-Atomic Resolution

Cryo-EM structure determination has rapidly evolved in recent years, from producing low-resolution images, often colloquially referred to as "blob-ology," which lacked the ability to resolve individual atoms or fine details, to achieving near-atomic resolution, where individual atoms within biological macromolecules or complexes can be visualized. This transformation has been facilitated by technological advances in transmission electron microscopes, the development of superior direct electron detectors and user-friendly image processing software, and the ever-increasing accessibility of computational power.

In recent years, the resolution of structures determined using cryo-EM has improved significantly, showing a continuous shift towards higher attainable resolution. Whereas most of the solved protein structures are obtained at a resolution of 3–4 Å, a growing number of structures are in the 2–3 Å range, and a few are at a resolution better than 2 Å. At resolutions better than 4 Å, atomic model building becomes feasible. However, the direct visualization of true atomic positions in protein structure determination necessitates a resolution better than 1.5 Å (Yip et al. 2020). It is well established that cryo-EM is a powerful method for determining the 3D structures of biological macromolecules. However, before 2013, most atomic structures deposited in the protein data bank (PDB) were determined by X-ray crystallography (Cheng 2015). The attainment of near-atomic resolution for the TRP channel TRPV1 at 3.4 Å resolution marked the onset of the so-called "resolution revolution," breaking the side-chain resolution barrier for membrane proteins without the need for crystallization (Liao et al. 2013). This study focused on TRPV1 domain interactions and topology, providing a structural blueprint for elucidating unique aspects of TRP channel function. Since then, cryo-EM has made significant advances, offering advantages in comparison to X-ray crystallography, such as the reduced sample concentration requirements and the capability to study flexible proteins and macromolecular complexes (Nogales and Scheres 2015). In order to obtain the highest possible resolution of a biologically stable sample, it is critical to fine-tune the microscope, camera, and energy filter. A 1.7 Å resolution cryo-EM reconstruction of the human membrane protein β3 GABAA receptor homopentamer and a 1.22 Å mouse apoferritin structure were recently achieved using a combination of the latest cryo-electron microscope (Titan Krios G4), a Falcon 4 DED, and a stable energy filter (Nakane et al. 2020).

Resolving water molecules is an important metric for assessing the quality of a true atomic resolution map. In a recent study, the number of water molecules in different cryo-EM maps was compared. The results indicated that water molecules identified in one map were located within 1.0 Å of their positions in other maps (Zhang et al. 2020). This observation demonstrated that the latest 300 keV microscope and DED can accurately describe atomic positions in a macromolecular complex with a resolution ranging between 1.2 and 1.3 Å. Moreover, atomic resolution or near-atomic resolution can elucidate not only the fine details of molecular interactions, but also assist in fine-tuning the method utilized, setting the stage for extending the resolution further (Grant and Grigorieff 2015).

Atomic resolution has also facilitated the study of charge and bond properties of hydrogen (Maki-Yonekura et al. 2023) and led to the determination of the structure of the Frh complex unambiguously (Allegretti et al. 2014). These achievements underscore the boundless potential for obtaining novel biological insights as the resolution of cryo-EM reconstructions continues to advance. In addition, a recent study that utilized a 200 keV microscope to determine the structure of *Escherichia coli* beta-galactosidase at a reso-

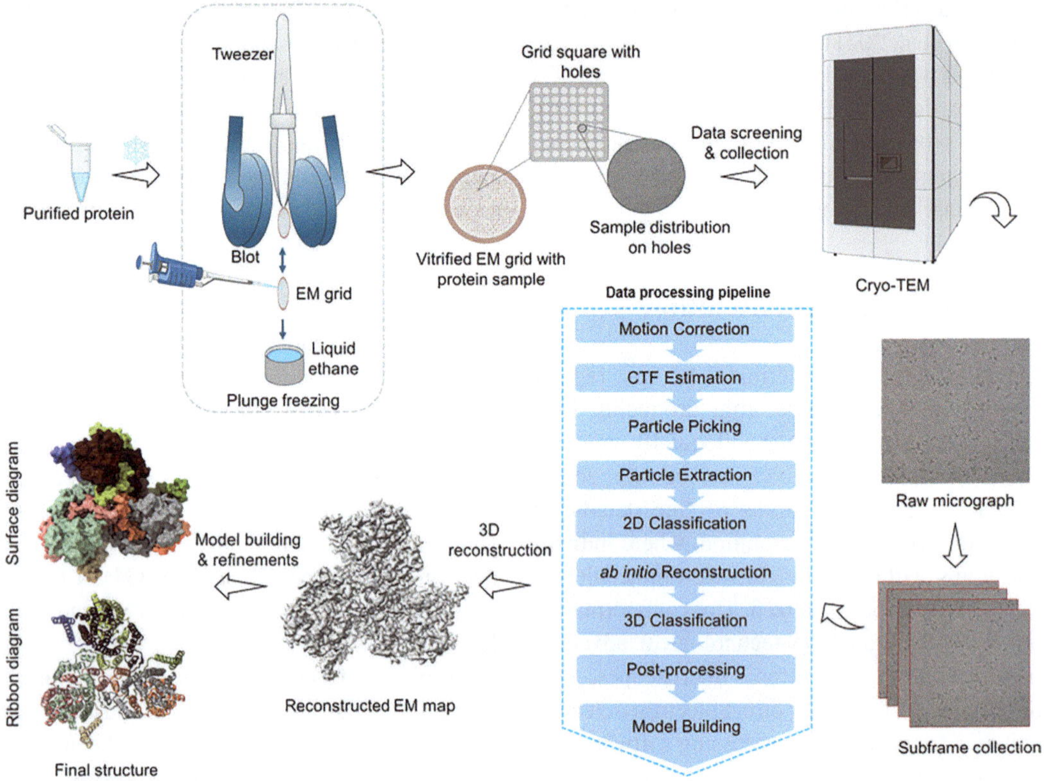

Fig. 4.4 Schematic overview of a single-particle cryo-EM workflow, illustrating key steps from sample preparation to final 3D model reconstruction. The workflow includes sample loading onto cryo-EM grids, vitrification, grid screening and data collection using a cryo-TEM, followed by data processing and 3D reconstruction. The final map (EMDB ID: EMD-30923) and corresponding atomic model (PDB ID: 7DYR) are shown for illustration purposes only (Pettersen et al. 2004). The EM density map was visualized using UCSF Chimera, and the ribbon and surface representation of the fitted model were prepared using Chimera X (Goddard et al. 2018)

lution of 1.8 Å (Merk et al. 2020), has demonstrated the value of electron microscopes operating at 200 keV for imaging protein complexes at atomic resolution using cryo-EM. This development broadens accessibility towards cryo-EM research, as the availability of 200 keV microscopes is significantly higher than that of the high end 300 keV microscopes.

4.4.3 Grid Coating and Functionalization

Although single-particle cryo-EM has enhanced our understanding of biologically important molecules in near-atomic resolution, specific technical challenges remain. A significant bottleneck lies in identifying optimal conditions for sample vitrification, which directly influences the behavior of the sample on the grid as well as overall image quality (Naydenova et al. 2020; Wang et al. 2020a). To address these challenges, coated and functionalized grids have been created to improve particle distribution, minimize preferred orientation, and reduce beam-induced motion (Naydenova et al. 2019, 2020; Wang et al. 2020a).

Ideally, the samples should be vitrified in a thin film of ice with a thickness that closely approximates the size of the particles to ensure maximum contrast (Wang et al. 2020a; Neselu et al. 2023; Fujita et al. 2023). Unfortunately, as detailed in the following section, the sample is often exposed to the air–water interface (AWI) (see Sect. 4.4) immediately before vitrification. This exposure can lead to denaturation, aggregation, or preferred orientation (Naydenova et al. 2019; D'Imprima et al. 2019). Previous studies have demonstrated that coating the metal grid with a thin carbon layer provides an anchoring surface, limiting the interaction between the sample and the AWI (Naydenova et al. 2019; D'Imprima et al. 2019; Liu et al. 2019b, 2021b). However, the conventional carbon grid increases background noise, suffers from bending, and is prone to the accumulation of contaminants, ultimately delivering poor image quality (D'Imprima et al. 2019; Pantelic et al. 2011; Russo and Passmore 2014). In contrast, graphene carbon grids reduce background noise and also possess higher conductivity, reducing the motion of particles due to charging effects (Naydenova et al. 2019; Pantelic et al. 2011; Wang et al. 2020b). Additionally, graphene oxide (GO), a modified version of graphene with oxygen-containing functional groups, has emerged as a new support grid material (Liu et al. 2021b; Pantelic et al. 2011; Pantelic et al. 2010). Unlike standard graphene grids, GO grids are naturally hydrophilic, promoting better protein adsorption (Pantelic et al. 2011; van de Put et al. 2015).

Further advancements in GO-based grid preparation have been made to ensure the formation of uniform GO monolayers and enhance protein adsorption capabilities. For instance, different crosslinking strategies, such as epoxidation, polyethyleneimine (PEI) polymeric coatings, and thermal treatments, have been employed (Fujita et al. 2023; Liu et al. 2021b; Patel et al. 2021). The epoxidized-GO films improve protein adsorption, particle distribution, and reduce beam-induced motion, leading to higher contrast images (Fujita et al. 2023). The PEI coating prevents GO flaking, increases GO coverage and adhesion capability, yet studies of this method remain limited (Patel et al. 2021). The UV/thermal treatment reduces the space between GO sheets, contains fewer functional groups than conventional GO films, and enhances conductivity. As a result, these reduced GO grids are less charging-prone, leading to improved image quality. The impact of these modified grids was confirmed in cryo-EM reconstructions of small protein complexes, reaching near-atomic resolution (Liu et al. 2021b).

Alternatively, GO functionalized grids with nickel-nitrilotriacetic (NTA) and two-dimensional (2D) streptavidin crystals have been used to capture His-tagged and biotinylated/strep-tagged proteins, respectively (Wang et al. 2020a; Liu et al. 2019b; Benjamin et al. 2016; Fan et al. 2019). Though, the latter requires additional computational analysis to subtract the streptavidin signal (Han et al. 2016; Zhao et al. 2024). These grids have been shown to effectively capture target proteins directly from lysate and keep them away from the AWI in favorable orientations for high-resolution cryo-EM imaging (Benjamin et al. 2016). More recently, a method to attach SpyTag to GO grids covalently has been developed (Wang et al. 2020a). Leveraging the rapid and irreversible SpyTag–SpyCatcher coupling system, these grids specifically and robustly bind proteins fused to SpyCatcher (Zakeri et al. 2012; Wang et al. 2020a). This approach facilitates selective immobilization, ensures sample concentration on grids, and improves particle distribution. A 3.3Å reconstruction of the human mitochondrial TRAP1 protein proves the effectiveness of these grids for high-resolution structural studies (Wang et al. 2020a).

In addition, GO antibody-based grids have emerged as a versatile strategy for capturing low-abundance proteins and reducing AWI interface problems. In this approach, monoclonal antibodies are immobilized on the grid surface via ester Pyr-NHS linkers, enabling the determination of atomic resolution cryo-EM structures (Zhao et al. 2024; Yu et al. 2014, 2016). The grids can be used to directly purify endogenous molecules on the grid without necessitating an affinity tag (Zhao et al. 2024; Yu et al. 2016). However, extensive optimization may be required, which can be a time-consuming process.

Finally, in contrast to GO, gold grids exhibit higher mechanical stability, reduced charging effects, and higher conductivity. These properties make gold grids less prone to beam-induced motion, contributing to improved overall reconstruction quality (Russo and Passmore

2014). Yet, the holes on conventional gold grids still permit some ice deformation and particle movement. A new design using gold grids with ultra-small holes (290 nm diameter) arranged in a hexagonal pattern (HexAufoil) has been appears to minimize ice buckling caused by cryo-cooling and beam-induced particle movement to sub-Ångstrom levels (Naydenova et al. 2020; Naydenova and Russo 2022). These innovative grids have been employed to obtain cryo-EM structures of light-harvesting 2 complex from *Marichromatium purpuratum* and human ABCB11 bile salt export pump at 2.4 Å and 2.8 Å resolution, respectively (Gardiner et al. 2021; Gruget et al. 2025).

4.4.4 The Air–Water Interface (AWI)

The air–water interface (AWI) is the chaotic boundary where the solution meets the air. With its distinct physicochemical properties that differentiate it from the bulk phases of water and air and a characteristic width of 3-10 Å, the AWI acts as a hotspot for various chemical and physical processes (D'Imprima et al. 2019). This unique region is marked by alterations in the dielectric constant and the presence of a structurally anisotropic region, among its other unique properties, which can accelerate chemical reactions involving reactive oxygen species (ROS). These characteristics pose a significant bottleneck in the preparation of high-quality vitrified specimens for cryo-EM. During conventional vitrification, a thin film of sample solution of <100 nm is formed, creating an interface between the hydrophobic air and the hydrophilic aqueous solution. Macromolecules tend to diffuse and be adsorbed at this AWI within a millisecond. However, current vitrification devices typically operate on longer timescales, making contact with the AWI inevitable (Liu and Wang 2023).

Impacts of the AWI At the AWI, protein molecules tend to align along the interface with hydrophilic regions facing inward towards the water environment and hydrophobic regions facing outward towards the air environment. Direct interactions between proteins and the interface lead to partial or complete protein unfolding and denaturation. The high surface-to-volume ratio at the AWI can also damage proteins. Adsorption to the AWI frequently induces challenges, such as uneven particle distribution, dissociation of complexes, and preferred particle orientation, which arises from the selective adsorption of protein molecules in specific orientations (D'Imprima et al. 2019).

Strategies to Address Challenges Posed by the AWI There are several approaches to mitigate the deleterious effects of the AWI during cryo-EM sample preparation:

A. *Incorporating Additives into the Solution*:
   (a) *Surfactants (Detergents)*: Adding detergents to the sample at or below their critical micelle concentration (CMC) (Yadav and Vinothkumar 2024) reduces the surface tension of the aqueous solution, alleviating AWI adsorption (Kang et al. 2023). As amphipathic molecules, surfactants preferentially adsorb at the AWI, forming a protective layer that prevents direct contact between the target macromolecules and the AWI, thereby preserving the integrity of the protein particles (Kampjut et al. 2021; Chen et al. 2019; Li et al. 2021). Commonly used detergents include Tween 80, CHAPSO, DDM, NP40, and fluorinated surfactants (like FL-FOS, FL-FOM).
   (b) *Late Embryogenesis Abundant (LEA) Proteins*: Late Embryogenesis Abundant (LEA) proteins, which help organisms withstand desiccation, have be used as additives during plunge freezing to mitigate AWI damage to fragile multi-subunit complexes. They are thought to form a protective barrier at the AWI (Abe et al. 2024).
   (c) *Metallo-Supramolecular Branched Polymers*: Application of a palladium-polyethylene glycol (PEG) metallo-

supramolecular branched polymer (MSBP) in the sample preparation has been shown to significantly reduce the AWI effect. The hyper-branched polymers surround the proteins and shield the particles from the AWIs on both sides of the grid (Xu et al. 2024).

B. *Employing Fast Vitrification Devices*:
   (a) Reducing the time between sample deposition on grids and freezing can minimize the time proteins interact with the AWI. The target is to freeze the macromolecules in solution before they significantly adsorb at the interface. Devices like Chameleon or the cryoWriter aid in reducing the spot-to-plunge times down to as little as 100 ms (Rima et al. 2022; Noble et al. 2018; Levitz et al. 2022).

C. *Utilizing Supporting Films*:
   (a) *Continuous Thin Films*: Using thin layers of continuous supporting films, such as amorphous carbon or graphene, can adsorb macromolecules, effectively keeping them away from the AWI (Liu and Wang 2023; Kang et al. 2023). The supporting film induces proteins to adsorb to the carbon/graphene–water interface rather than the AWI, protecting them from denaturation.
   (b) *Functionalized Supporting Films*: Supporting films can be functionalized with affinity ligands to specifically anchor the target particles away from the AWI.
      (i) *Graphene and its Derivatives (e.g., graphene oxide)*: Hydrophilized graphene has emerged as a new generation of support due to its negligible background noise and mechanical robustness, which can also be functionalized to meet specific needs (Nickl et al. 2023). For example, proteins can be covalently anchored to functionalized graphene sheets to prevent denaturation and improve particle orientations (Chio et al. 2024).
      (ii) *Graphene with Surfactant Self-Assembled Monolayers (GSAMs)*: This superstructure combines an impenetrable surfactant monolayer on a graphene membrane (Zheng et al. 2024), aiming to neutralize both air-water and graphene-water interfaces, enriching particle orientation and improving the particle utilization ratio, defined as the ratio of the particles used in the final reconstruction to the total number of particles extracted.
      (iii) *Streptavidin Monolayer Grids*: These holey-carbon cryo-EM grids contain 2D streptavidin crystals that are grown onto a biotinylated lipid monolayer. The high-affinity interaction between streptavidin and biotin allows for the binding of biotinylated samples, thereby protecting them from the AWI. Additionally, this strategy concentrates the samples available in limited quantities and allows purification of protein complexes directly on the grids (Cookis et al. 2023).
      (iv) *Support Film Using 2D Crystals of Hydrophobin HFBI*: Support films using 2D crystals of hydrophobin HFBI can adsorb sample particles via electrostatic interactions, protecting them from the AWI (Fan et al. 2021). The hydrophilic side of the HFBI film adsorbs protein, allowing the formation of sufficiently thin ice for high-quality data collection. The particle orientation distribution can be regulated by adjusting the buffer pH.
      (v) *An immobilized Antibody-Based Affinity Grid (IAAG)*: IAAG has been developed based on the high-affinity PA-tag/NZ-1 antibody epitope tag system, which employs Pyr-NHS as a linker to immobilize NZ-1 Fab on the graphene oxide or carbon-covered grid surface. The IAAG grid effectively enriches PA-tagged target proteins and overcomes preferred

orientation issues. It has been utilized for on-grid purification of low-abundance target complexes from cell lysates (Zhao et al. 2024).

D. *Other Approaches*:
   (a) *Specimen Preparation on a Tube-Reconstruction of Average Subtracted Tubular Regions (SPOT-RASTR)*: This method uses lipid nanotubes embedded in bulk ice to which the sample particles adhere to, thus protecting them from the AWI and promoting a wider range of particle orientations (Esfahani et al. 2024).
   (b) *Electrospray-Assisted Cryo-EM Sample Preparation (ESI-cryoPrep)*: This blotting-free method sprays charged droplets containing macromolecules onto the grid, confining them within the middle of the ice layer and preventing adsorption to interfaces (Yang et al. 2024).
   (c) *Laser Flash Melting and Re-vitrification*: The flash melting and re-vitrification of cryo-EM samples using microsecond laser pulses reduces the tendency of the particles to align in a preferred orientation (Straub et al. 2025). Flash melting appears to exert small forces on the proteins, which changes their orientation and causes them to adopt an uneven spatial distribution after re-vitrification. This technique can potentially also detach proteins adsorbed at the AWI.
   (d) *Nanoparticle Stabilization*: Gold nanoparticles functionalized with thiol groups (AuTCOOH) can stabilize proteins by providing a supportive scaffold, keeping them away from the AWI (Moyano and Rotello 2011).

To summarize, AWI denaturation significantly impacts protein structure determination in cryo-EM. The implementation of strategies to mitigate the challenges associated with the AWI enhances the quality of the structural data obtained. However, the optimal approach for overcoming AWI-related challenges is sample-specific and may necessitate optimization.

4.4.5 Particle Size Augmentation

The structural analysis of small proteins via single-particle cryo-EM has limitations, particularly for membrane proteins without significant extramembrane domains (Wu and Rapoport 2021). Small membrane proteins are reassembled into detergent micelles or nanodiscs for stabilization and cryo-EM structure determination. However, the different membrane mimetics often hinder the visualization of the small protein (Bloch et al. 2021). Additionally, the absence of large soluble domains acting as a fiducial marker, combined with the inherently low signal-to-noise ratio of cryo-EM images, complicates the proper alignment of the particles (Uchański et al. 2021). These factors collectively limit the resolution achievable in the cryo-EM analysis of small membrane proteins (Xie et al. 2025). To address these obstacles, fiducial-assisted approaches such as binding partners and fusion strategies have been developed to increase the size of particles and improve particle alignment during image processing (Bloch et al. 2021; Zhang et al. 2022a).

The fusion approach using BRIL has been widely applied for the structural characterization of GPCRs. BRIL is a 12 kDa engineered variant of apocytochrome b562 consisting of four α-helices, typically fused into the intracellular loop-3 (ICL3) of the GPCR protein (Xie et al. 2025; Zhang et al. 2022a; Mukherjee et al. 2020; Tsutsumi et al. 2020; Miyagi et al. 2020). Tsutsumi et al. reported the first cryo-EM structure of human Frizzled5 receptor fused to BRIL at 3.7 Å resolution and utilized an anti-BRIL antigen-binding fragment/nanobody complex to further augment the overall particle's size. Yet, the flexibility of the complex hampered obtaining higher resolution reconstructions (Tsutsumi et al. 2020). As further demonstrated by Zhang et al., BRIL alone might not be a sufficient extramembrane feature for particle alignment, and thus achieve high-resolution reconstructions of small membrane proteins, even when engineered to be more rigid (Xie et al. 2025; Zhang et al. 2022a). Successful BRIL insertion typically requires anchoring between two α-helices of the target protein. Therefore, prior structural knowledge of the protein of interest is needed to avoid distortions (Mukherjee et al. 2020; Robertson et al. 2022). More recently, the application of a rigid poly-alanine linker has made it possible to fuse BRIL to the C-terminus of NorA, a small bacterial Major Facilitator Superfamily (MFS) transporter (Xie et al. 2025). Using computational modeling, this fiducial-assisted design ensured particle size augmentation via a rigid connection, reduced steric occlusions, and was minimally invasive to the native structure and function of the protein. Importantly, this approach allowed resolving high-resolution cryo-EM structures of NorA and other MFS transporters in multiple conformations, expanding the applicability of BRIL-based fiducials to a broader spectrum of small membrane proteins.

In addition to fusion strategies, target-binding methodologies, including antigen-binding fragments (Fabs) or nanobodies (Nb), have been successfully used to characterize the structure of small membrane proteins with cryo-EM (Bloch et al. 2021; Robertson et al. 2022; Lin et al. 2020; Kim et al. 2019). For instance, Roberson et al. have reported cryo-EM structures of four inactive-state GPCRs, at resolutions as high as 2.4 Å using Nanobody6 (Nb6), which binds to the ICL3 loop (Robertson et al. 2022). Unlike BRIL fusion, nanobody binding does not require a modification of the native sequence of the target protein. However, given their small size (15 kDa), Nbs alone may be insufficient for accurate particle alignment in very small proteins. In such cases, a second binding partner is often required to increase the molecular size further. By contrast, Fabs, due to their larger size (50 kDa), are often sufficient as standalone fiducials to solve high-resolution cryo-EM structures (Kang et al. 2018; Butterwick et al. 2018; Bloch et al. 2020). Their distinct shape will also contribute more across projections, unlike small Nbs, which might appear similar from multiple angles during image processing. Nevertheless, despite their usual high binding affinity towards the target, the flexibility in the hinge region of Fabs can blur the 2D class images if it is not constrained. Using Fabs has made it possible to obtain cryo-EM structures of several prokaryotic and eukaryotic small membrane proteins at high resolution. Some examples include the insect olfactory receptor Orco, the yeast glucosyltransferase ALG6, the *Plasmodium falciparum* transporter PfCRT, the human serotonin transporter bound to ibogaine, and the bacterial drug efflux pump NorA (Butterwick et al. 2018; Bloch et al. 2020; Kim et al. 2019; Coleman et al. 2019; Brawley et al. 2022). In this latter case, Fabs were used for both size segmentation and particle alignment, leading to the creation of a peptide efflux pump inhibitor (Brawley et al. 2022). While effective for size augmentation, it is essential to note that both Nbs and Fabs can restrict the protein structure into a particular conformational state, biasing the resulting structure (Zimmermann et al. 2018).

More complex strategies combining the use of Fabs, Nb, and/or scaffold proteins have been used to overcome size limitations, symmetry, and preferential orientation problems in cryo-EM (Wu and Rapoport 2021; Uchański et al. 2021). Uchański et al. introduced the "megabodies" tool, in which an Nb was grafted onto the circularly permuted scaffold proteins *Helicobacter pylori* HopQ and the *Escherichia coli* K12 glucosidase YgjK to produce molecules of 50–100 kDa (Uchański et al. 2021). This design is customizable since the Nb can be replaced to target different proteins. The megabody approach has been used to determine a high-resolution cryo-EM structure of the small membrane protein hedgehog acetyltransferase HHAT, the human SPCA1a, and the human α1β3γ2 $GABA_A$ receptor (Coupland et al. 2021; Chen et al. 2023; Laverty et al. 2019). Recently, megabodies have also been proven effective in structure determination of human serum albumin by X-ray crystallography (De Felice et al. 2024). Notably, this strategy has been adapted to target MSP-based nanodiscs, highlighting its potential as a generalizable tool for cryo-EM structure determination of small membrane proteins (Uchański et al. 2021). In parallel, "legobodies" have emerged as a second modular strategy for structure determination of small proteins (Wu and Rapoport 2021; Kang and Chen 2023; Fan et al. 2024; Reimund et al. 2025). In this approach, an Nb/Fab complex is bound to the target along with a maltose-binding protein (MBP) as a scaffold protein. To further stabilize the complex and enhance rigidity, MBP is bound to domain C of *Staphylococcus aureus* protein A, followed by sequential fusions of domain D of protein A and protein G (Wu and Rapoport 2021; Ackle et al. 2025). The latter reinforces the complex integrity by interacting with the heavy chains of the variable and conserved regions of the Fab (Wu and Rapoport 2021). While legobodies significantly enhance the size of small particles, limitations include steric interference, residual flexibility, and extensive engineering before finding a specific and rigid Nb. Yet, Ackle et al. provide straightforward guidelines for adapting nanobodies to these enlargement strategies to facilitate the structural analysis of small proteins using cryo-EM (Ackle et al. 2025).

4.4.6 Next-Generation Sample Preparation Systems

A large number of technical capabilities have advanced at a breathtaking pace over the past decade in the field of cryo-EM, including microscopes, DEDs, energy filters, and software. However, until very recently, the method used for vitrifying samples has remained essentially unchanged since it was first developed almost 40 years ago (Dubochet et al. 1988). The method consists of applying a small volume of 2–3 μL of sample to the surface of an EM grid, which is typically constructed of a metal mesh, such as Cu and Au, which is covered by a support film, typically C or Au, and is often fenestrated. Most of the sample is blotted away using filter paper, leaving a very thin layer in the order of 10–100 nm, which is supported by the holey substrate. The grid is then rapidly plunged into a cryogen, which is typically liquid ethane or an ethane/propane mix that enables fast cooling of the sample. While some instruments, such as FEI Vitrobot, Gatan CP3, and Leica GP, have been developed to streamline the mechanics of vitrification, some disadvantages remain. For example, most of the sample is removed by the filter paper and the thickness of the vitrified layer is uneven and often unpredictable.

A prototype of a new sample vitrification device called Spotiton V0.5 was developed in 2012 (Jain et al. 2012). Spotiton used an inkjet head to deliver small droplets onto the grid and then immediately dropped it into a cryogen for freezing. One of the advantages of Spotiton is that it uses only 30 nl of sample, which is significantly less than the conventional freezing method that typically uses 3 μl of sample. Additionally, the piezo-electric inkjet mechanism preserves the structural integrity of macromolecules. The device became more consistent with the development of nanowire grids 4 years later (Razinkov et al. 2016). One of the main features is the use of a camera to verify the vitrification process, allowing the operator to retain only grids of acceptable quality, thereby saving time during the screening process. Spotiton has matured as a technology with the addition of automated grid handling, an integrated glow discharger, twin cameras for monitoring the vitrification process, and the generation of a detailed freezing session report at the end of each sample preparation (Dandey et al. 2018). The nanowire grid has also been fine-tuned and enhanced with the latest Spotiton machine (Wei et al. 2018). The system is currently in routine and continuous use in labs under the commercial name "Chameleon" and has been used to successfully vitrify a wide variety of samples.

The development of Spotiton sparked heavy interest and additional development in the field of cryo-EM sample preparation. A number of instruments have been developed or are currently in development, aiming to optimize and streamline cryo-EM sample preparation and fine-tune it for different applications. Vitrojet, the spiritual successor of the Vitrobot, is one of the new instruments that employ pin printing to deposit a sample onto a regular cryo-EM grid. Similar to Spotiton, Vitrojet offers optical ice thickness measurement. However, the cryofixation is achieved through jet vitrification (Henderikx et al. 2024). Vitrojet can handle a variety of samples and features two main advantages. First, the jet vitrification can produce samples with higher spatial and time resolution. Second, higher cooling rates from jet vitrification may offer an advantage in cell vitrification for cryo-electron tomography (cryo-ET), enabling the vitrification of larger samples without the formation of crystalline ice, unlike the regular plunge freezing method (Berger et al. 2021).

There are several other methods that produce excellent results for targeted applications. First is the microfluidic spraying-plunging method that enables time-resolved cryo-EM for target proteins (Feng et al. 1993). Second, ultrasonic specimen application and through-grid wicking were combined in a high-speed specimen-preparation device that was named "Back-it-up" or BIU (Tan and Rubinstein 2020). BIU sprays sample solution onto one side of a holey cryo-EM grid and then wicks it away through the grid by a glass-fiber filter held against the opposite side, often called the "back" of the grid, producing a film suitable for vitrification. This process can be completed in tens of milliseconds and will further reduce the tendency for preferred orientation due to the super-fast wicking time. Third, the Easy-Grid system offers better vitrification of large, mammalian cells compared to standard plunge-freezing (Gemin et al. 2024). Lastly, the "Shake-It-Off" method provides an economical option for labs that want to experiment with blot-less technology (Rubinstein et al. 2019).

As discussed in Sect. 4.3, the grid plays a crucial role in cryo-EM. The type of grid used may be determined by the type of grid preparation instrument. One example is the nanowire grid used with Spotition/Chameleon. Alternatively, the grid may be chosen based on specific properties. For example, UltrAuFoil Holey Gold Films and HexAufoil extra small hole support grids can reduce beam-induced movement of the sample, resulting in superior image quality and enhanced information retention for reconstructions (Naydenova et al. 2020). In another example, as was previously mentioned, grids with graphene oxide (Palovcak et al. 2018) or a pure graphene layer have been shown to improve preferential orientation and sample denaturation due to adsorption to the air–water interface (AWI). Lastly, the graphene sandwich specimen preparation sets a new direction to explore in cryo-EM analysis of biological specimens (Liu and Wang 2024; Wu et al. 2021).

4.4.7 Time-Resolved Cryo-EM

Over the last decade, cryo-EM has become an indispensable tool for determining high-resolution structures of biologically relevant macromolecules. However, traditional single-particle cryo-EM remains inherently limited to the analysis of static structural states, often failing to capture the dynamic conformational changes that govern biological activity in complex systems such as proteins with discrete intermediate states. Time-resolved (TR) cryo-EM overcomes this limitation by enabling investigators to capture and resolve short-lived conformational states of biomolecular complexes during biochemical reactions or processes. These transient states, sometimes present for only milliseconds, are commonly responsible for regulating protein function and activity, making them critical targets for dynamic structure elucidation (Mäeots and Enchev 2022; Frank 2017; Amann et al. 2023).

TR cryo-EM works by combining rapid sample mixing for precise initiation of biochemical reactions—such as substrate binding, conformational switching, or complex assembly—followed by precisely timed vitrification to "trap" biomolecular complexes in discrete intermediate states (Mäeots and Enchev 2022; Frank 2017; Amann et al. 2023). One of the first instances of this approach was introduced by Berriman and Unwin in 1994 to study transient states of the Torpedo acetylcholine receptor (nAChR) (Berriman and Unwin 1994). In their study, the authors used an atomizer to propel 1 μm droplets of acetylcholine onto grids that had been pre-applied with nAChR, initiating the reaction directly on the grid and allowing for 1–100 ms temporal resolution. Despite its spatial resolution, sample uniformity and stoichiometric limitations, this study laid the groundwork for future iterations of this method.

In recent years, several approaches have emerged to achieve high mixing efficiency and temporal resolution for TR cryo-EM sample preparation. In 2020, Dandey et al. introduced a novel approach utilizing the Spotiton system described in Sect. 4.6, employing a piezoelectric dispenser to deposit two different 50 pL sample droplets onto a grid within 5–11 ms of one another, thereby facilitating a mixing interval of 90–500 ms before sample vitrification. The authors demonstrated the utility of this method for four processes including ribosomal 70S complex assembly, RNA polymerase promoter binding, calcium-induced conformational changes during MthK potassium channel gating, and dynamin-mediated membrane constriction (Dandey et al. 2020).

In the same year, Mäeots et al. developed a modular microfluidic polydimethylsiloxane (PDMS)-based chip comprised of a 3D micromixer to promote efficient homogenization of two sample solutions, and a tunable ejection delay line that allowed for controlled incubation time between mixing and vitrification. This design allows residence times ranging from 10 to thousands of ms, depending on the chosen ejection delay line length and flow rate. Similar to Dandey et al., the authors utilized a piezoelectric sprayer to dispense homogenized sample droplets onto cryo-EM grids. To demonstrate its utility, the authors captured early intermediates of RecA filament formation on single-stranded DNA, revealing a structural continuum of assembly states over 20–1000 ms (Mäeots et al. 2020). Battacharjee et al. developed a similar TR cryo-EM method featuring a reusable SiO_2-coated PDMS-based microfluidic chip, allowing strict reaction stoichiometry conservation, with predefined mixing intervals of 10, 25, 140, and 900 ms. Using this method, the authors investigated HflX-mediated ribosome recycling, enabling them to capture and visualize three distinct intermediate states of HflX-mediated splitting of the 70S *E. coli* ribosome in the presence of GTP, over 10–140 milliseconds (Bhattacharjee et al. 2024).

An alternative approach includes laser beam-induced rapid in situ melting and revitrification for microsecond TR cryo-EM (Harder et al. 2022; Voss et al. 2021). In these examples, the authors note that the 5 μs timescale of their method is too short to detect any major protein conformational changes despite minimal photoinduced sample damage. Instead, this method shows promise toward investigating the disassembly of large complexes following electron-beam irradiation, as they observed with the bacterial chaperonin, GroEL, as a model.

As far as the biological application of TR cryo-EM goes, many studies, including those described above, employ this method to study conformational changes of large biomolecular structures, such as ribosomal (Loveland et al. 2020; Carbone et al. 2021; Petrychenko et al. 2021), and proteasomal complexes (Zhang et al. 2022b). This can be attributed to the ease of their purification, high spatial resolution, and slow reaction rates at cold temperatures (4–10 °C) in the presence of inhibitors. In these cases, reaction rates ranging from seconds to minutes permit the manual preparation of TR cryo-EM samples through blot plunge-freezing after timed incubation. This approach enables the capture of an ensemble of intermediate states across multiple time points.

Considering recent advancements in the field, the focus of TR cryo-EM appears to be on attaining higher temporal resolution to capture short-lived conformational changes of biological systems in their native states, without employing reaction rate slowing techniques. Hence, it is anticipated that there will be significant expansion and exploration of methodologies employing rapid mixing microfluidic dispersion platforms in the forthcoming decade.

4.4.8 AI and Machine Learning in Cryo-EM

Single-particle cryo-EM has emerged as a robust method in structural biology, enabling high-resolution structure determination of biological macromolecules without the need for crystallization or high amounts of purified sample. Despite its transformative advantages, obtaining cryo-EM data presents several challenges, such as sample quality and stability to withstand vitrification stress during the plunge-freezing process. Furthermore, samples are susceptible to radiation damage during electron exposure, resulting in beam-induced motion that can degrade image quality by introducing noise. Continuous advancements in microscope hardware, sample preparation techniques, and data processing software have helped mitigate many of these limitations. For cryo-EM data processing, software suites such as cryoSPARC (Punjani et al. 2017) and RELION (Scheres 2012) are widely used and have become central to modern workflows. Additionally, recent progress in Artificial Intelligence (AI) and Machine Learning (ML) has further transformed cryo-EM by significantly enhancing the efficiency and accuracy of data processing.

(i) *Deep learning (DL) in Particle Picking*: One of the crucial steps in a cryo-EM workflow is particle picking from motion-corrected micrographs with estimated contrast transfer function (CTF) parameters. While manual picking is labor-intensive and inefficient, the development of numerous semi-automatic and deep learning-based particle picking algorithms has now enabled high-throughput and reliable particle identification (Bepler et al. 2019). Among these particle pickers, Topaz is one of the widely used tools. It employs a convolutional neural network (CNN) and general-purpose positive-unlabeled (PU) learning methods to detect particles with high accuracy. Topaz has demonstrated superior performance, retrieving more particles with lower false-positive rates than conventional DoG (Voss et al. 2009) and CNN-based pickers such as crYOLO (Wagner et al. 2019) and DeepPicker (Wang et al. 2016). crYOLO, which is based on the "You Only Look Once" (YOLO) deep learning object detection framework, is particularly noted for its speed and accuracy. It processes up to five micrographs per second and delivers high recall and precision, especially when it is trained with an initial dataset of 200–2500 particles (Wagner et al. 2019). Another powerful tool, WARP (Tegunov and Cramer 2019), integrates deep learning for accurate particle picking and image denoising, while also offering real-time evaluation and pre-processing during data collection. WARP corrects both global and local motions, estimates local defocus using novel algorithms, and provides real-time monitoring of key imaging parameters for each recorded micrograph. In addition to these, a growing number of deep-learning particle pickers, such as DeepPicker (Wang et al. 2016), DeepEM (Zhu et al. 2017), R-CNN (Xiao and Yang 2017), PIXER (Zhang et al. 2019), DeepCryoPicker (Al-Azzawi et al. 2020), TransPicker (Zhang et al. 2021d), CryoSegNet (Gyawali et al. 2024), CASSPER (George et al. 2021), and CryoTransformer (Dhakal et al. 2024), have been implemented, continuing to drive developments in computational processing of cryo-EM data. Overall, these tools enhance accuracy, reduce manual picking, and further improve the efficiency and performance of particle picking in cryo-EM workflows.

(ii) *DL in 3D map Reconstruction*: Following particle picking, particle extraction from the processed micrographs enables subsequent 2D classification, wherein particles are grouped into classes based on similarities in shape and orientation. This critical step serves both to assess particle quality and to eliminate poorly defined or non-particle images, retaining only well-defined particle images for subsequent analysis. Software packages such as cryoSPARC and RELION implement advanced algorithms for 2D class averaging, producing high-quality 2D classes that represent particles in different orientations. These 2D classes serve as the basis for ab initio 3D reconstruction without prior structural knowledge about the sample's 3D state. CryoSPARC (Punjani et al. 2017) employs DL-based algorithms such as stochastic gradient descent (SGD) and branch-and-bound maximum likelihood optimization to perform ab initio 3D reconstructions. Additionally, the integration of SGD with Bayesian marginalization enables ab initio 3D classification, facilitating fully automated structural reconstructions from heterogeneous datasets. Similarly, RELION uses an SGD algorithm to generate a de novo 3D initial model from the 2D particles. Other programs used for ab initio 3D reconstructions include SIMPLE (Caesar et al. 2020), cisTEM (Grant et al. 2018), and CryoDRGN (Levy et al. 2025; Zhong et al. 2021), each offering distinct computational strategies for initial model generation.

(iii) *AI and DL in De Novo Atomic Model Building*: The development of DL algorithms has profoundly transformed the field of structural biology, particularly the field of cryo-EM. This advancement has contributed to a substantial increase in the number of macromolecular structures resolved using cryo-EM techniques. A pivotal breakthrough in this area was the development of ML-based approaches such as AlphaFold2 (Jumper et al. 2021). AlphaFold2 is a neural network-based DL model capable of predicting protein structures directly from amino acid sequences with near-atomic accuracy, even in the absence of homologous structural templates. Building on this progress, several other DL-based structure prediction tools have since emerged, including RoseTTaFold (Baek et al. 2021), ESMFold (Lin et al. 2023), and I-TASSER-MTD (Zhou et al. 2022a). Predicted models from these algorithms guide the process of cryo-EM model building. In parallel, several DL-based tools have been developed to perform model building and refinements within cryo-EM density maps. One such example is EMBUILD (He et al. 2022), which integrates AlphaFold predicted structural models through a multi-step pipeline involving FFT-based global fitting, domain-based semi-flexible refinement, and graph-based iterative assembly guided by main-chain probability maps. Similarly, DEMO-EM (Zhou et al. 2022b) is an automated program that constructs multi-domain structural models from medium-to-low resolution cryo-EM maps through a progressive refinement pipeline that combines rigid-body domain fitting and flexible assembly simulations. Additional recently developed DL-based de novo modeling tools that offer efficient and accurate model building from cryo-EM maps include DeepTracer-ID (Chang et al. 2022), Pathwalking (Hryc and Baker 2022), Cryo2Struct (Giri and Cheng 2024), and ModelAngelo (Jamali et al. 2024). Collectively, these innovations underscore the transformative role of AI and ML in cryo-EM, and future developments in this field are expected to further enhance resolution, automation, and the overall interpretability of cryo-EM data.

4.5 Emerging Modalities for the Study of Membrane Proteins

In addition to single-particle cryo-EM, several other modalities of structural biology that utilize a cryo-transmission electron microscope have been developed. In this section, we will explore recent advances in cryo-electron tomography (cryo-ET), a technique that enables the study of macromolecules within the native cellular environment, and microcrystal electron diffraction, which enables the determination of structures from nanocrystals inaccessible to traditional X-ray methods. In addition, we will examine how integrative methods combine different techniques to overcome technical challenges and provide biological insights that may not have been otherwise accessible.

4.5.1 Cryo-Electron Tomography for In Situ Structure Determination

Determining the structure and function of membrane proteins remains a significant challenge in biochemistry. Traditionally, these proteins are recombinantly expressed and then purified using various detergent compositions designed to mimic the native lipid environment. However, this approach raises concerns about whether the purified protein structure truly reflects its native, physiologically-relevant state. In recent years, cryo-electron tomography (cryo-ET) has emerged as a powerful technique for visualizing macromolecular complexes in near-native states moderate to high resolution. With recent advances in cryo-ET, it is now possible to study membrane proteins directly within the lipid bilayer of intact cells, without the need for purification (Kudryashev 2024). This enables the structural analysis of membrane proteins in their native environment, embedded in the membrane and interacting with physiological partners, thereby eliminating many of the ambiguities associated with traditional purification-based methods. Cryo-ET is also a powerful technique for visualizing host–pathogen interactions in infectious diseases. It provides detailed insights into how pathogens replicate, survive, and interact with host cellular structures at the molecular level (Lemos et al. 2025). Major advancements over the past two decades, such as improved sample preparation methods, direct electron detectors, subtomogram averaging, and enhanced image processing algorithms, have significantly contributed to the success and widespread application of cryo-ET in "in situ" cellular structural biology.

Sample preparation for cryo-ET is highly challenging, complex, and time-consuming. The types of samples suitable for cryo-ET range from viruses and bacteria to eukaryotic cells, small organisms, tissues, and spheroids. One of the most critical steps in the workflow is ultra-rapid freezing to ensure that water transitions into a glass-like, vitreous state, preserving cellular structures in their near-native condition (Kuba et al. 2021). This process, known as vitrification, is typically achieved by plunge freezing for thin samples (less than a few micrometers thick), whereas high-pressure freezing is used for thicker specimens. Cryoprotectants such as glycerol can also assist in successful vitrification. For imaging, biological samples are commonly observed using a 300 kV cryo-transmission electron microscope (cryo-TEM), which allows electrons to traverse the vitrified sample. However, electrons have a mean free path of approximately 300–400 nm before undergoing inelastic scattering (Rice et al. 2018). When imaging thicker samples, such as eukaryotic cells, significant noise can result, leading to reduced resolution (Keller and Fernández-Busnadiego 2024). While thin samples like viruses and bacteria can often be imaged directly, thicker eukaryotic specimens require thinning to allow proper transmission of electrons through the sample. Two primary techniques are used for sample thinning: cryo-focused ion beam (cryo-FIB) milling and cryo-ultramicrotomy.

Cryo-ultramicrotomy involves mechanically sectioning high-pressure frozen samples into thin slices using a diamond knife (Schiøtz et al. 2024). Although this method is faster than cryo-FIB milling, it is associated with artifacts such as knife marks, compression, and mechanical distortion, which can compromise high-resolution reconstructions of protein complexes. As a result, cryo-FIB milling is now considered the method of choice for preparing thin cellular samples for cryo-ET, due to its superior preservation of ultrastructure.

Cryo-FIB milling has been the key technique for thinning vitrified biological specimens for over a decade (Wagner et al. 2020). In this method, a focused beam of gallium ions is used to precisely remove cellular material, producing thin sections of approximately 100–200 nm, known as lamellae. These lamellae are suitable for high-resolution cryo-electron tomography. For larger or multicellular specimens, such as tissues or organoids, additional strategies have been developed. One such method is the lift-out technique, in which a piece of the vitrified sample is first extracted and transferred onto an EM grid, where it undergoes cryo-FIB milling (Schaffer et al. 2019). Another approach, Vibratome-assisted High-pressure freezing Ultramicrotomy Thinning (VHUT–Cryo-FIB), involves slicing the sample using a vibratome before high-pressure freezing. After vitrification, the sample is further trimmed using cryo-ultramicrotomy, followed by FIB milling to achieve the final desired thickness (Zhang et al. 2021e). To precisely target regions of interest within the specimen for cryo-FIB milling, correlative light and electron microscopy (cryo-CLEM) is often employed. This technique overlays fluorescent light microscopy signals with electron microscopy imaging, enabling accurate localization of specific structures or proteins before milling (Keller and Fernández-Busnadiego 2024).

(i) *Data Collection, Tomogram Reconstruction, and Processing*: In cryo-ET, projection images of a specimen are acquired at different tilt angles, typically ranging from −60° to +60°. These images are computationally aligned and merged to generate a 3D reconstruction called a tomogram. However, as the tilt angle increases, the electron path length through the sample becomes longer, resulting in loss of contrast and resolution. Due to these limitations, the full ±90° tilt range is generally not achievable. This incomplete angular coverage leads to gaps in the sampled data in Fourier space, creating a region known as the missing wedge. The missing wedge results in artifacts and loss of information along the Z-axis, causing reconstructed structures to appear stretched or blurred vertically (Lučić et al. 2005). To address these artifacts, sophisticated algorithms and software tools that attempt to compensate for the missing wedge by reconstructing or inferring missing information in Fourier space have been developed, thus improving the accuracy of tomograms (Ding et al. 2019). Additionally, vitrified biological samples are highly sensitive to radiation damage. To minimize this, a low total electron dose, typically around 120 electrons per square angstrom ($e^-/Å^2$), is used during data acquisition. This low-dose imaging contributes to a poor signal-to-noise ratio (SNR), further complicating data analysis. Whole cells and complex biological samples imaged by cryo-ET are inherently dense and crowded, filled with organelles, membranes, and macromolecular complexes. The combination of low SNR and overlapping densities poses a significant challenge to the unequivocal identification of specific features, such as individual protein complexes, within the tomogram. These challenges necessitate the use of advanced image processing, denoising techniques, and machine learning-based tools to extract meaningful structural information from noisy data (Frangakis 2021).

(ii) *Tomogram Analysis and Particle Picking*: Denoising not only enhances the visual quality of cryo-ET tomograms but also significantly improves the accuracy of segmentation, which is essential for interpreting the spatial organization and function of subcellular components. Recent advances in machine learning and deep learning have led to the development of powerful software tools, such as Dragonfly (Heebner et al. 2022), Amira (Thermo Fisher Scientific), and DeepPict (de Teresa-Trueba et al. 2023), which greatly reduces the need for labor-intensive manual segmentation. Segmentation enables not only the identification and analysis of tomographic features but also supports quantitative analyses, such as measuring inter-organelle distances and calculating volumetric or spatial distributions (Salfer et al. 2020). Beyond visualization and quantification, segmentation plays a critical role in streamlining the particle picking process by providing spatial context for subtomogram averaging, which is crucial for resolving macromolecular structures in situ. Tools such as PickYOLO (Genthe et al. 2023) and MemBrain (Lamm et al. 2022) leverage deep learning to automatically identify and extract particles, particularly membrane-associated proteins, directly from raw tomograms. These advancements are transforming cryo-ET workflows by increasing efficiency, consistency, and scalability in structural analysis.

(iii) *Subtomogram Averaging*: In cryo-ET, individual macromolecular complexes, such as ribosomes or membrane proteins, are often difficult to analyze in detail due to the low electron dose and the densely crowded cellular environment, which results in low signal-to-noise ratios. However, when identical or similar structures are present repeatedly within a tomogram, small 3D volumes containing these features—called subtomograms—can be extracted. These subtomograms are then iteratively aligned and averaged, a process known as subtomogram averaging, to enhance the signal and improve the resolution of the target complex (Basanta et al. 2020; Ni et al. 2022). This technique enables the structural analysis of macromolecules in situ, within their native cellular context. Gaining in-depth knowledge of the molecular interactions between proteins embedded in membranes provides critical insights into the structural organization and functional mechanisms of biological membranes. This cellular-context approach allows us to explore dynamic processes, such as signaling, transport, and membrane remodeling, with high spatial precision.

Figure 4.5 provides a detailed overview of a cryo-electron tomography workflow focusing on the five main aspects of tomography, including sample vitrification, cryo-FIB, cryo-CLEM, data collection using a cryo-TEM, tomogram reconstruction, denoising, segmentation, and subtomogram averaging.

4.5.2 Microcrystal Electron Diffraction

Microcrystal electron diffraction (MicroED) is a technique used in structural biology to determine the atomic structure of proteins and other macromolecules from very small crystals measuring less than 0.5 μm in size. It is a cryo-EM technique that uses diffraction mode in a transmission electron microscope (TEM), with the microcrystal continuously rotated while diffraction data is collected. The data collected via MicroED can be processed using software typically used for X-ray crystallography. This technique, named by the Gonen laboratory in 2013 (Gonen 2013), was accompanied by a detailed protocol, and a 2.9 Å lysozyme structure was achieved in the same year (Shi et al. 2013). It shares certain similarities with electron crystallography, which was established more than 50 years ago (Wisedchaisri et al. 1993). MicroED offers benefits for the study of intrinsically disordered proteins, direct localization of charged ions within protein structures, and determination of nearly damage-free protein structures (Rodriguez et al. 2017). During recent advancements of the technique, an additional improvement in resolution was achieved, with lysozyme having now been resolved at 1.5 Å, and resolutions of 1 Å or better for small organic molecules (Jones et al. 2018).

X-ray crystallography often requires large crystals, while X-ray free-electron lasers (XFELs) use a significantly brighter beam than traditional synchrotrons, with data obtained from crystals as small as 1 μm, but multiple crystals are commonly required to obtain a structure due to radiation damage. MicroED, on the other hand, can handle crystals that are too small for traditional X-ray methods. A high-resolution lysozyme, achieved at a resolution of 2.5 Å, was obtained using an enhanced data collection protocol known as the "continuous rotation" method for MicroED (Nannenga et al. 2014). This was accomplished a year later, following the proof-of-concept demonstrated by a group of researchers from Netherlands the previous year (Nederlof et al. 2013). Prior to this, the data were collected as single still images. Electron diffraction has many similarities to X-ray crystallography, though several distinctions between the two techniques can be made: (1) Electrons interact stronger with matter compared to X-rays, and impart less energy to the sample,

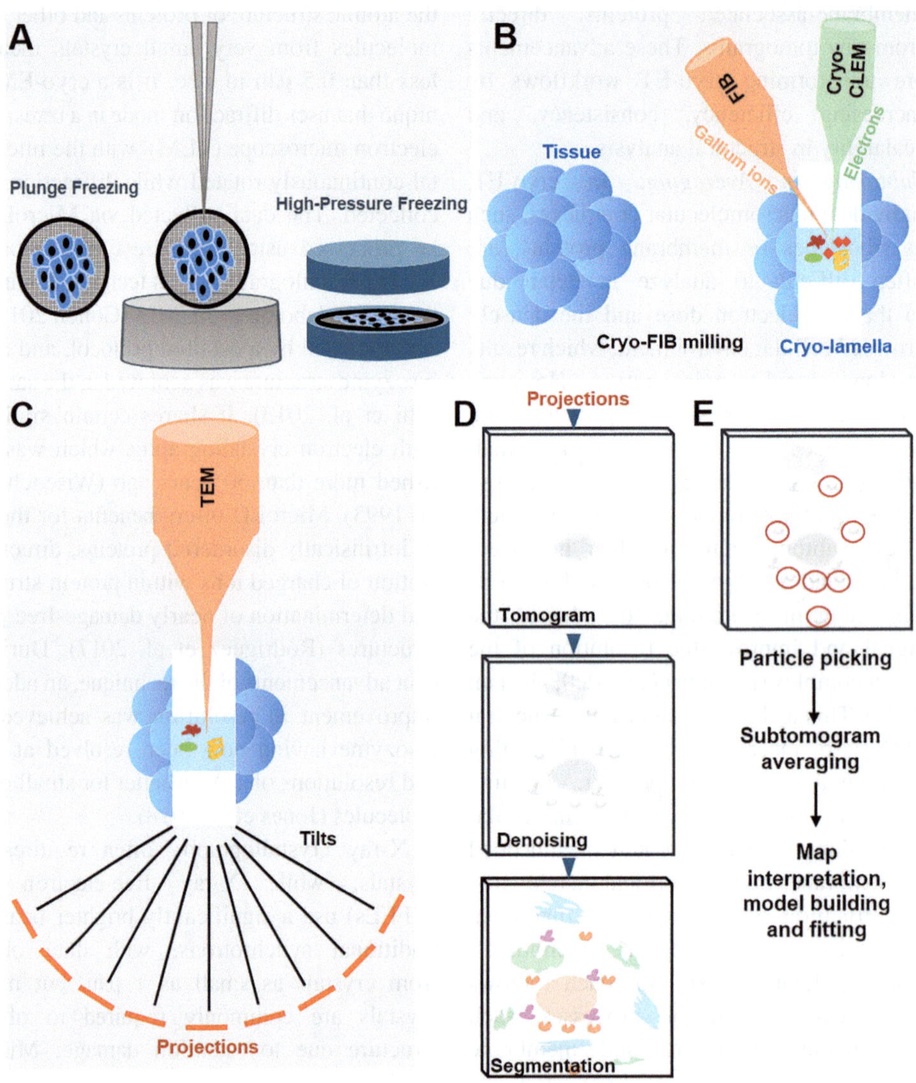

Fig. 4.5 Schematic overview of a cryo-ET workflow. (**a**) Sample can be vitrified either by plunge freezing or high-pressure freezing, depending on the sample thickness. (**b**) Correlative light and electron microscopy (cryo-CLEM) can be used for area determination for cryo-FIB milling to prepare a thin lamella. (**c**) At different tilts, many projections of the sample are collected using a cryo-TEM. (**d**) Each generated tomogram is denoised to facilitate quantitative analysis. Cellular organelles can be analyzed by membrane segmentation in terms of volume, surface, and distance. (**e**) In some cryo-ET tomograms, protein complexes can be picked by manual particle picking and/or automated and deep learning methods, and subtomogram averaging will enable a reconstruction of a near-native structure

which enables reaching high-resolution from much smaller crystals compared to traditional X-ray crystallography. (2) The wavelength of electrons is significantly smaller compared to wavelengths used for X-ray crystallography, leading to a significantly larger Ewald sphere. (3) In some cases, phase information can be recovered by collecting images of the crystals (Nannenga and Gonen 2016).

MicroED is capable of working with highly ordered microcrystals or nanocrystals, requiring only a few crystals to obtain the necessary data for structure determination (Nannenga and Gonen 2019). This is particularly advantageous when the growth of large, well-ordered crystals poses a significant bottleneck for crystallographic approaches, especially with challenging samples, such as membrane proteins and protein complexes. After a decade of technological and computational improvements, MicroED is now also beginning to demonstrate utility in contemporary drug discovery (Danelius et al. 2023).

Finally, it has been demonstrated that sonication, vigorous pipetting, or vortexing can fragment large, imperfect crystals into small, single-crystal fragments suitable for data collection and atomic structure determination using MicroED. This was particularly beneficial in cases where untreated crystals were mosaic, produced diffraction patterns with multiple lattices, or were otherwise unsuitable for standard crystallographic experiments. The study involved a total of eight test crystals, achieving resolutions between 1.1–2.9 Å (de la Cruz et al. 2017). Cryo-FIB offers an alternative method for studying large protein crystals by thinning them to 200 nm lamellas, enabling meaningful data generation by MicroED while maintaining near atomic resolution of the crystal lattice (Duyvesteyn et al. 2018).

Figure 4.6 provides a detailed overview of a MicroED workflow, detailing critical steps beginning with growing crystals with typical crystallography approaches to crystal screening, followed by cryo-grid preparation, setting up the cryo-TEM for a MicroED collection, optimizing data collection, and processing data to acquire electron density maps for atomic model building.

4.5.3 Integrative Structural Biology Methods

All organisms rely on protein–protein interactions to transmit signals, respond to stimuli, and regulate growth. Understanding these interactions at the molecular level is made possible through structural biology. In the past, gaining structural insight into complex proteins was limited by technological constraints. However, recent advances have significantly expanded our ability to study proteins in their native environments. Integrative structural biology has emerged as a powerful approach by integrating multiple complementary techniques to study complex biological systems. Many cellular processes depend on weak or transient interactions between proteins, which are often lost during traditional purification steps aimed at obtaining homogeneous protein samples. While techniques like super-resolution or confocal microscopy allow in situ visualization, they typically offer only low-resolution data.

To overcome these limitations, researchers have begun combining low-resolution co-localization data with high-resolution structural information, forming the foundation of integrative structural biology (Ziegler et al. 2021). Classical high-resolution methods, such as X-ray crystallography, nuclear magnetic resonance (NMR) spectroscopy, and cryo-electron microscopy (cryo-EM), are now being integrated with other techniques like mass spectrometry (MS), small-angle X-ray scattering (SAXS), AlphaFold-predicted atomic models, and cryo-electron tomography (cryo-ET) to build more complete and physiologically relevant models of macromolecular assemblies. For example, the nuclear pore complex (NPC) is an enormous macromolecular assembly, approximately 52 megadaltons in size, composed of around 550 protein subunits from 30 different types (Sali 2021). Due to its size, complexity, and dynamic nature, resolving the complete structure of the NPC using classical structural biology methods alone is nearly impossible (Rout and Sali 2019). However, integrative structural biology makes this achievable by combining diverse sources of information. Stoichiometry data from protein biochemistry, structural frameworks from cryo-ET, and residue–residue distance constraints from cross-linking mass spectrometry can be integrated to construct a precise and functionally relevant model of the NPC.

Modeling large, multi-subunit protein complexes from cryo-EM density maps presents considerable challenges due to the complexity and size of these assemblies. Traditional methods often struggle with accurately interpreting the

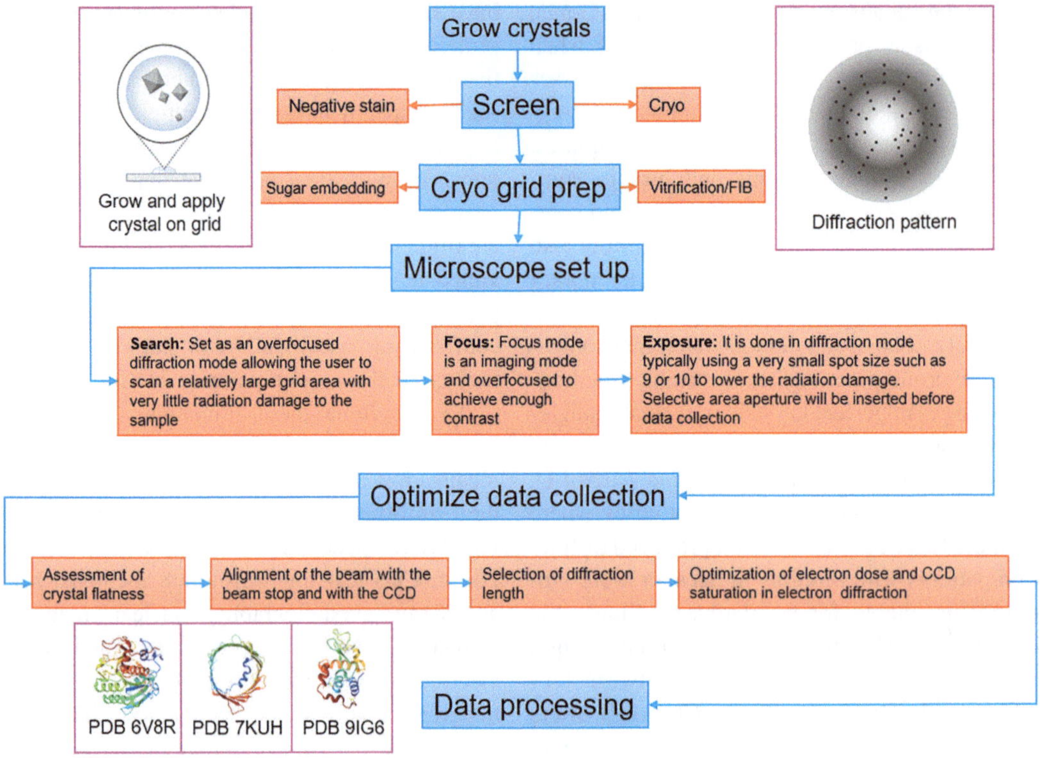

Fig. 4.6 Schematic overview of a MicroED workflow. The process starts with the production of microcrystals using standard macromolecular crystallization procedures. The microcrystals are screened by either negative stain or cryo-EM methods. A sample that is suitable for electron diffraction studies is selected for data collection after careful microscope setup for MicroED. The grid can be prepared either via sugar embedding or vitrification (with or without FIB milling). Microscope setup for low-dose electron diffraction includes three settings: search, focus, and exposure. The search mode is set as an overfocused diffraction mode, allowing the user to scan a relatively large grid area with very little radiation damage to the sample. The focus is identified and then overfocused from there to yield a highly contrasted image. Exposure is in diffraction mode, typically using a very small spot size such as 9 or 10. The focus mode is only used once or twice in a session for untilted specimens. For tilted specimens, it is recommended to refocus and realign directly adjacent to the target crystal before every ED collected. Subsequently, optimization for data collection takes place. Crystal flatness needs to be assessed, the beam must align well with the beam stop and the detector, and the proper diffraction length needs to be selected to achieve optimal data. Finally, the electron dose must be carefully selected to prevent CCD saturation so the spot intensities can be accurately determined. The resulting MicroED data is processed using software like DIALS, MOSFLM, and XDS. Three examples of protein structures determined using this method and deposited to the Protein Data Bank are shown

intricate details within these maps. However, recent advancements in deep learning have introduced innovative approaches to overcome these obstacles. One such approach involves segmenting the cryo-EM density map into 3D pixels, known as voxels. The voxels are analyzed using deep learning models trained to predict specific protein features, such as secondary structures and amino acid types. This voxel-wise analysis facilitates the identification of structural motifs within the density map. Subsequently, the predicted features serve as inputs for a Hidden Markov Model (HMM). Utilizing the Viterbi algorithm, the HMM determines the most probable sequence of hidden states, effectively aligning the predicted amino acid sequence with the observed density features. This alignment yields an initial model of the protein backbone. To refine this preliminary model, tools like AlphaFold are employed. AlphaFold leverages advanced deep learning techniques to predict protein structures with remarkable accuracy (Giri and Cheng 2025). By integrating the initial backbone model derived from cryo-EM data with AlphaFold's predictive capabilities, more reliable and precise atomic models of complex protein structures can be generated. This integrative methodology exemplifies the synergy between experimental data and computational modeling, enhancing our ability to elucidate the structures of intricate biomolecular assemblies.

Another powerful integrative approach involves combining cryo-correlative light and electron microscopy (cryo-CLEM) with cryo-electron tomography (cryo-ET) to achieve a more complete understanding of biological structures (Klein et al. 2021). In cryo-CLEM, a fluorescence microscope is integrated with an electron microscope under cryogenic conditions. This setup enables the correlation of dynamic or molecule-specific signals from the light microscope with high-resolution structural data obtained from the electron microscope. By linking molecular identity to ultrastructure, cryo-CLEM enables precise localization of specific proteins or events within the complex cellular context, enhancing the interpretability and relevance of cryo-ET data.

4.6 Membrane Protein Structural Biology and Drug Discovery Applications

Historically, drug development has relied on forward pharmacology, where naturally derived or synthetic compounds are tested for desired biological activity, sometimes without a complete understanding of their mechanism of action or biological target—acetaminophen being a rather famous example (Takenaka 2001; Gerriets et al. 2025). This is especially true before the introduction of X-ray crystallography for protein structure elucidation, the first example of which came by solving the structure of myoglobin in 1958 (Kendrew et al. 1958). To elucidate a drug's mechanism of action and to develop better drugs for particular diseases, it is crucial to have thorough knowledge of a drug target's structure and binding mechanics. With X-ray crystallography dominating the field of structural biology for the better part of half a century, this technique has favored structure elucidation for soluble proteins, which are often easier to crystallize compared to their membrane-bound counterparts (https://www.rcsb.org/stats/summary; Renaud et al. 2016). Over the past few decades, more tools have become available for protein structure determination including cryo-EM (Renaud et al. 2016; Ceska et al. 2019; Van Drie and Tong 2020; Renaud et al. 2018), which was used to determine the structure of bacteriorhodopsin in 1975 (Henderson and Unwin 1975) and nuclear magnetic resonance (NMR), which was employed to solve the structure of proteinase inhibitor IIA in 1985 (Williamson et al. 1985). Despite these historic achievements, early tools of structural biology faced limitations in throughput, structural resolution, and the study of membrane-bound proteins, thus hindering the comprehensive utilization of structure-based drug design. Over the past decade, numerous technical advancements have significantly reshaped the landscape of drug development with the modern advent of cryo-EM. These advancements, including enhancements in throughput, spatial resolution, and the ability to study membrane-bound targets using nanodiscs, have enabled structure determination at resolutions as high as 1.2 Å. These achievements facilitate accurate visualization of small molecule binding and provide mechanistic insights that were previously unattainable (Renaud et al. 2016; Ceska et al. 2019; Van Drie and Tong 2020; Renaud et al. 2018).

In the past decade alone, there have been major breakthroughs involving cryo-EM in the context of drug development. For example, a study published in 2018 by Amporndanai et al. demonstrated that cryo-EM could be used to determine the structure of membrane-bound cytochrome bc_1, a parasitic drug target that was previously studied via crystallographic methods through its mammalian orthologs, often resulting in failed drug candidacy (Amporndanai et al. 2018). This was largely attributed to the significant difference between the structures of mammalian cytochrome bc_1 and those of parasitic orthologs, which are notoriously challenging to purify and isolate in the amounts required for X-ray crystallography. Additionally, inhibitor-bound structures of cytochrome bc_1 were also determined by cryo-EM, further highlighting its ability to study previously inaccessible drug targets. The same was observed in a study published in 2017 by Morris et al., where cryo-EM was used to attain the structure of human and *P. falciparum* 20S proteasome bound to a variety of inhibitory ligands for use in expanding the arsenal of antimalarials (Morris and Da Fonseca 2017).

The use of cryo-EM also extends to drug development in the context of treating cancer. In 2020, Saur et al. published the first cryo-EM structure of pyruvate kinase PKM2, a promising oncologic drug target (Saur et al. 2020). In their study, the authors determined the structure of PKM2, as well as several other desirable drug targets, in the presence of fragment-sized ligands, promoting cryo-EM for use in fragment-based drug development. Similarly, Merk et al. employed cryo-EM to determine ultra-high resolution structures of small metabolic enzymes, such as isocitrate, glutamate, and lactate dehydrogenase(s), showcasing active site localization for various small molecule inhibitors (Merk et al. 2016).

The clinical gap in understanding membrane-bound drug targets in high-profile etiologies, such as Alzheimer's disease (AD), is also being addressed by cryo-EM. In 2015, Bai et al. presented a full-atom structure for all four components of γ-secretase, a multi-subunit transmembrane complex, dysfunction of which has been implicated in the progression of the disease (Bai et al. 2015). Of these subunits, the authors showed that AD-derived mutations in presenilin 1 (PS1) affect the interactions between neighboring subunits, allowing for additional flexible conformations and a plastic proteolytic active site. These findings serve as a molecular basis for elucidating the function of γ-secretase, and by extension, AD in general. These methods also extend toward membrane proteins involved in cell signaling, with Zhang et al. employing cryo-EM to resolve the structure of a ligand-bound G-protein coupled receptor (GPCR) without a stabilizing nanobody, which was previously required for studying proteins of this class via X-ray crystallography (Zhang et al. 2021f). Many of these studies, including that by Zhang et al., feature glucagon-like peptide-1 (GLP-1) receptors, a type of GPCR and a major target in the treatment of type 2 diabetes, the structural study of which would be inherently challenging without the use of cryo-EM (Liang et al. 2018; Zhang et al. 2017c).

Of the studies discussed above, one of the most exciting future directions for cryo-EM involves the "Build and Retrieve" (BaR) method developed by Su et al. in 2021. Despite its throughput, the need to produce pure and homogeneous protein samples for cryo-EM slows down the process of data acquisition. Using the BaR method, it is now feasible to identify and determine cryo-EM structures for multiple membrane proteins, soluble proteins, and protein complexes concurrently, using a single heterogenous protein sample derived from cellular lysate. In this single study, the authors were able to determine the structure of eight bacterial proteins from three separate heterogenous samples.

While challenges such as the need for substantial computational resources for data analysis remain, cryo-EM offers a considerable edge over earlier tools in structural biology. Collectively, these developments position cryo-EM as an essential technique poised to further accelerate the discovery and optimization of novel therapeutics.

Figure 4.7 provides three examples of drug discovery applications using cryo-EM: cytochrome bc_1 in complex with an anti-malarial compound, pyruvate kinase PKM2 in complex with a promising oncologic drug target, and the glucagon-like peptide-1 receptor (GLP-1R) in complex with a GLP-1R antagonist. We specifically highlight the densities of the bound drug to the receptor molecule, showcasing how they interact in each case.

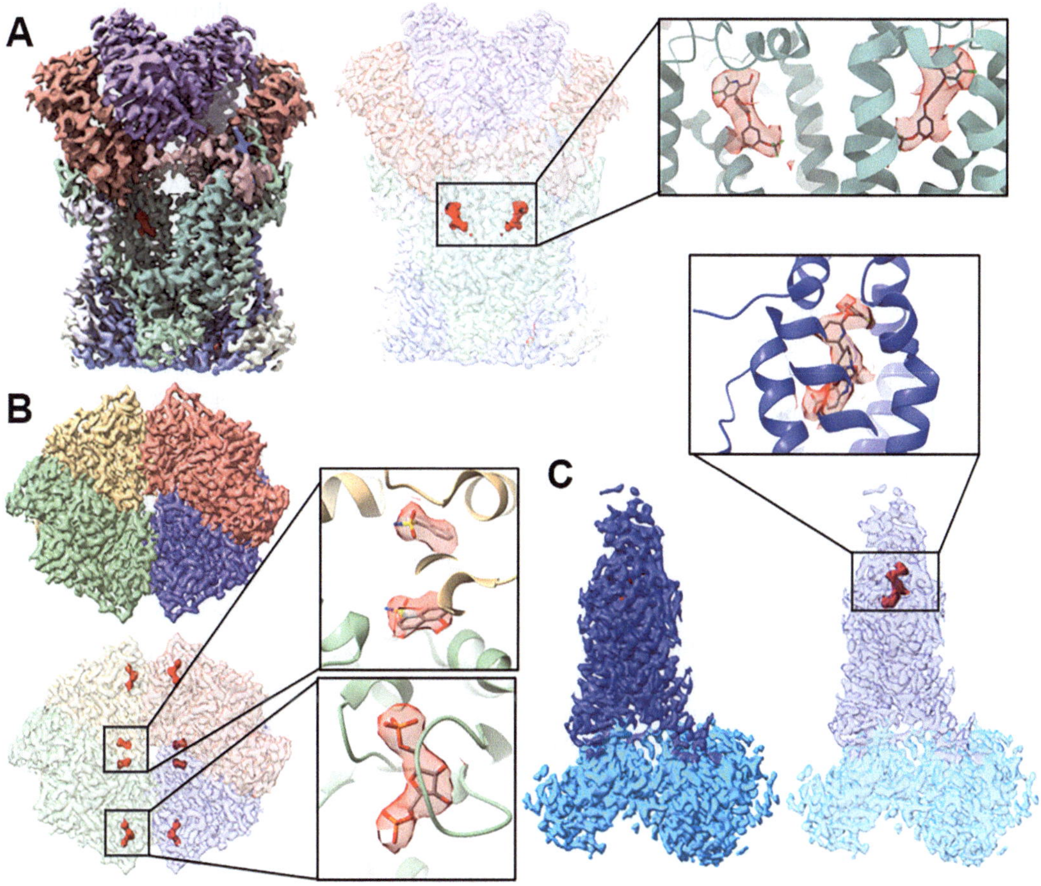

Fig. 4.7 Examples of cryo-EM in drug discovery applications. (**a**) Cryo-EM density map of cytochrome bc_1 in complex with anti-malarial compound GSK-932121, highlighted in red (PDB 6FO0, EMDB-4286). (**b**) Cryo-EM density map of pyruvate kinase 2 in complex with 5-hydroxynaphthalene-1-sulfonamide (Cmp5) (top panel), and fructose-1,6-bisphosphate (bottom panel), highlighted in red (PDB 6TTF, EMDB-10575). (**c**) Cryo-EM density map of glucagon-like peptide-1 receptor (GLP-1R) in complex with GLP-1R antagonist PF-06882961 (Danuglipron), highlighted in red (PDB: 7LCJ, EMDB-23275). For each receptor–ligand complex, a magnified frame is shown depicting the ligand in stick model, overlayed by a semi-transparent electron density map in red, surrounded by a cartoon model of the receptor. In each image, receptor subunits are depicted in various pastel colors

4.7 Conclusions and Outlook

The convergence of diverse, high-resolution structural techniques has ushered in a new era of structural biology, enabling the detailed visualization of biomolecules at near-atomic resolution or in some cases true atomic resolution. Despite this progress, sample preparation remains a critical bottleneck, particularly for membrane proteins, which are inherently challenging due to their hydrophobic nature and dynamic behavior. These proteins are vital targets in structural biology due to their central roles in essential cellular processes, such as signal transduction, ion transport, and cell communication, accounting for more than 50% of current drug targets. Unlocking their structures is fundamental both for gaining mechanistic insights and advancing structure-based drug design. To meet these challenges, significant progress has been made in expression and solubilization strategies. Membrane proteins are now routinely produced in a variety of host systems, such as

E. coli, *yeast*, Sf9 insect cells, and HEK293 mammalian cells, and solubilized using an array of detergents, amphipathic polymers (amphipols), copolymers, and lipid nanodiscs. In X-ray crystallography, the use of LCP crystallization methods to produce microcrystals suitable for microcrystallography, SFX, and time-resolved crystallography using XFELs has made it possible to study previously intractable targets. The emergence of fourth-generation synchrotron sources, delivering brighter and more focused beams, has significantly enhanced the quality and throughput of macromolecular data collection, making previously intractable problems solvable.

In parallel, cryo-EM has experienced a revolution in capability and accessibility due to advances, such as direct electron detectors, smart grid functionalization, and machine learning and AI-based data processing algorithms. The development of sample preparation innovations, including next-generation plunge-freezing devices and automated vitrification platforms, is reducing variability and increasing reproducibility. Particularly for membrane proteins, these improvements have enabled structure determination with minimal sample, revealing intricate mechanisms and offering structural templates critical for rational drug design.

The field of structural biology is rapidly moving beyond static structures to a dynamic understanding of macromolecular behavior. Single-particle cryo-EM is increasingly able to characterize native, heterogeneous and labile complexes. Cryo-ET with subtomogram averaging has enabled in situ visualization of macromolecules in their native cellular context. Furthermore, MicroED offers atomic resolution from nano-crystals inaccessible to traditional X-ray methods. In the next decade, integrative approaches, combining crystallography, cryo-EM, NMR, mass spectrometry, and computational modeling, will further enhance our ability to offer comprehensive views of biomolecular structure, dynamics, and function. Such enhanced capabilities will be critical for uncovering transient states, conformational ensembles, and allosteric regulation, to drive our understanding of fundamental biology and pave the way for designing the next generation of precision therapeutics targeting previously "undruggable" proteins.

Acknowledgments We would like to acknowledge support by the National Institute of General Medical Sciences (NIGMS) and the National Institute of Aging (NIA) via NIH grants R35GM150831 and R03AG081787 to VIP.

References

Abe KM, Li G, He Q, Grant T, Lim CJ (2024) Small LEA proteins mitigate air-water interface damage to fragile cryo-EM samples during plunge freezing. Nat Commun 15:7705

Abiven Y-M, Berrier S-E, Buteau A, Chado I, Fonda E, Frahi E et al (2023) SOLEIL II: towards a major transformation of the facility. Schaa VR, Götz A, Venter J, White K, Robichon M, Rowland V, editors;5 pages, 4.907 MB

About the APS upgrade [Internet]. Available from: https://www.aps.anl.gov/APS-Upgrade/About-the-APS-Upgrade

Ackle F, Thavarasah S, Earp JC, Seeger MA (2025). Rigid enlargement of sybodies with antibody fragments for cryo-EM analyses of small membrane proteins. Sci Rep 15:9460

Advanced Light Source Upgrade [Internet]. Available from: https://als.lbl.gov/als-u/overview/

Aherne M, Lyons JA, Caffrey M (2012) A fast, simple and robust protocol for growing crystals in the lipidic cubic phase. J Appl Crystallogr 45:1330–1333

Al-Azzawi A, Ouadou A, Max H, Duan Y, Tanner JJ, Cheng J (2020) DeepCryoPicker: fully automated deep neural network for single protein particle picking in cryo-EM. BMC Bioinformatics 21:509

Allegretti M, Mills DJ, McMullan G, Kühlbrandt W, Vonck J (2014) Atomic model of the F420-reducing [NiFe] hydrogenase by electron cryo-microscopy using a direct electron detector. eLife 3:e01963

Amann SJ, Keihsler D, Bodrug T, Brown NG, Haselbach D (2023) Frozen in time: analyzing molecular dynamics with time-resolved cryo-EM. Structure 31:4–19

Amporndanai K, Johnson RM, O'Neill PM, Fishwick CWG, Jamson AH, Rawson S et al (2018) X-ray and cryo-EM structures of inhibitor-bound cytochrome bc_1 complexes for structure-based drug discovery. IUCrJ 5:200–210

Aquila A, Hunter MS, Doak RB, Kirian RA, Fromme P, White TA et al (2012) Time-resolved protein nanocrystallography using an X-ray free-electron laser. Opt Express 20:2706–2716

Arana MR, Fiori MC, Altenberg GA (2019) Functional and structural comparison of the ABC exporter MsbA studied in detergent and reconstituted in nanodiscs. Biochem Biophys Res Commun 512:448–452

Ayub H, Murray RJ, Kuyler GC, Napier-Khwaja F, Gunner J, Dafforn TR et al (2024) GPCRs in the round: SMA-like copolymers and SMALPs as a platform for investigating GPCRs. Arch Biochem Biophys 754:109946

Bada Juarez JF, Harper AJ, Judge PJ, Tonge SR, Watts A (2019) From polymer chemistry to structural biology: the development of SMA and related amphipathic polymers for membrane protein extraction and solubilisation. Chem Phys Lipids 221:167–175

Baek M, DiMaio F, Anishchenko I, Dauparas J, Ovchinnikov S, Lee GR et al (2021) Accurate prediction of protein structures and interactions using a three-track neural network. Science 373:871–876

Bai X, Yan C, Yang G, Lu P, Ma D, Sun L et al (2015) An atomic structure of human γ-secretase. Nature 525:212–217

Barends TRM, Foucar L, Botha S, Doak RB, Shoeman RL, Nass K et al (2014) De novo protein crystal structure determination from X-ray free-electron laser data. Nature 505:244–247

Barty A, Kirian RA, Maia FRNC, Hantke M, Yoon CH, White TA et al (2014) Cheetah: software for high-throughput reduction and analysis of serial femtosecond X-ray diffraction data. J Appl Crystallogr 47:1118–1131

Basanta B, Chowdhury S, Lander GC, Grotjahn DA (2020) A guided approach for subtomogram averaging of challenging macromolecular assemblies. J Struct Biol X 4:100041

Bayburt TH, Sligar SG (2010) Membrane protein assembly into nanodiscs. FEBS Lett 584:1721–1727

Bazzacco P, Billon-Denis E, Sharma KS, Catoire LJ, Mary S, Le Bon C et al (2012) Nonionic homopolymeric amphipols: application to membrane protein folding, cell-free synthesis, and solution nuclear magnetic resonance. Biochemistry 51:1416–1430

Bechara C, Nöll A, Morgner N, Degiacomi MT, Tampé R, Robinson CV (2015) A subset of annular lipids is linked to the flippase activity of an ABC transporter. Nat Chem 7:255–262

Benjamin CJ, Wright KJ, Bolton SC, Hyun S-H, Krynski K, Grover M et al (2016) Selective capture of histidine-tagged proteins from cell lysates using TEM grids modified with NTA-Graphene Oxide. Sci Rep 6:32500

Bepler T, Morin A, Rapp M et al (2019) Positive-unlabeled convolutional neural networks for particle picking in cryo-electron micrographs. Nat Methods 16, 1153–1160

Berger C, Ravelli RBG, López-Iglesias C, Kudryashev M, Diepold A, Peters PJ (2021) Structure of the *Yersinia* injectisome in intracellular host cell phagosomes revealed by cryo FIB electron tomography. J Struct Biol

Berriman J, Unwin N (1994) Analysis of transient structures by cryo-microscopy combined with rapid mixing of spray droplets. Ultramicroscopy 56:241–252

Bhattacharjee S, Feng X, Maji S, Dadhwal P, Zhang Z, Brown ZP et al (2024) Time resolution in cryo-EM using a PDMS-based microfluidic chip assembly and its application to the study of HflX-mediated ribosome recycling. Cell 187:782–796.e23

Biscari C, Aigner E, Attenkofer K, Casas J, Ferrer S, Matilla O et al (2024) ALBA synchrotron heading towards its upgrade. Synchrotron Radiat News 37:18–23

Björling A, Berntsson O, Lehtivuori H, Takala H, Hughes AJ, Panman M et al (2016) Structural photoactivation of a full-length bacterial phytochrome. Sci Adv 2:e1600920

Bloch JS, Pesciullesi G, Boilevin J, Nosol K, Irobalieva RN, Darbre T et al (2020) Structure and mechanism of the ER-based glucosyltransferase ALG6. Nature 579:443–447

Bloch JS, Mukherjee S, Kowal J, Filippova EV, Niederer M, Pardon E et al (2021) COMPLEX development of a universal nanobody-binding Fab module for fiducial-assisted cryo-EM studies of membrane proteins. Proc Natl Acad Sci 118:e2115435118

Boland M, Baribeau C, Bertwistle D, Hunchak P, Le Pimpec F, Patel J et al (2021) Concept Design for the CLS2 Accelerator Complex. Liu (Ed.) L, M. (Ed.) B John, Neuenschwander (Ed.) R, Renan (Ed.) P, W. (Ed.) S Volker R, editors. Proc 12th Int Part Accel Conf;IPAC2021:3 pages, 2.287 MB

Botha S, Nass K, Barends TRM, Kabsch W, Latz B, Dworkowski F et al (2015) Room-temperature serial crystallography at synchrotron X-ray sources using slowly flowing free-standing high-viscosity microstreams. Acta Crystallogr D Biol Crystallogr 71:387–397

Bourgeois D, Vallone B, Schotte F, Arcovito A, Miele AE, Sciara G et al (2003) Complex landscape of protein structural dynamics unveiled by nanosecond Laue crystallography. Proc Natl Acad Sci USA 100:8704–8709

Boussambe GNM, Guillet P, Mahler F, Marconnet A, Vargas C, Cornut D et al (2018) Fluorinated diglucose detergents for membrane-protein extraction. Methods 147:84–94

Brawley DN, Sauer DB, Li J, Zheng X, Koide A, Jedhe GS et al (2022) Structural basis for inhibition of the drug efflux pump NorA from Staphylococcus aureus. Nat Chem Biol 18:706–712

Broecker J, Morizumi T, Ou W-L, Klingel V, Kuo A, Kissick DJ et al (2018) High-throughput in situ X-ray screening of and data collection from protein crystals at room temperature and under cryogenic conditions. Nat Protoc 13:260–292

Bruguera ES, Mahoney JP, Weis WI (2022) Reconstitution of purified membrane protein dimers in lipid nanodiscs with defined stoichiometry and orientation using a split GFP tether. J Biol Chem 298:101628

Bruni R, Kloss B (2013) High-throughput cloning and expression of integral membrane proteins in *Escherichia coli*. Curr Protoc Protein Sci Editor Board John E Coligan Al 74:29.6.1–29.6.34

Butterwick JA, del Mármol J, Kim KH, Kahlson MA, Rogow JA, Walz T et al (2018) Cryo-EM structure of the insect olfactory receptor Orco. Nature 560: 447–52

Caesar J, Reboul CF, Machello C, Kiesewetter S, Tang ML, Deme JC et al (2020) SIMPLE 3.0. Stream single-particle cryo-EM analysis in real time. J Struct Biol X 4:100040

Caffrey M (2003) Membrane protein crystallization. J Struct Biol 142:108–132

Caffrey M (2015) A comprehensive review of the lipid cubic phase or in meso method for crystallizing membrane and soluble proteins and complexes. Acta Crystallogr Sect F Struct Biol Commun 71:3–18

Caffrey M, Cherezov V (2009) Crystallizing membrane proteins using lipidic mesophases. Nat Protoc 4:706–731

Cammarata M, Levantino M, Schotte F, Anfinrud PA, Ewald F, Choi J et al (2008) Tracking the structural dynamics of proteins in solution using time-resolved wide-angle X-ray scattering. Nat Methods 5:881–886

Cannon KS, Sarsam RD, Tedamrongwanish T, Zhang K, Baker RW (2023) Lipid nanodiscs as a template for high-resolution cryo-EM structures of peripheral membrane proteins. J Struct Biol 215:107989

Carbone CE, Loveland AB, Gamper HB, Hou Y-M, Demo G, Korostelev AA (2021) Time-resolved cryo-EM visualizes ribosomal translocation with EF-G and GTP. Nat Commun 12:7236

Carpenter EP, Beis K, Cameron AD, Iwata S (2008) Overcoming the challenges of membrane protein crystallography. Curr Opin Struct Biol 18:581–586

Carvalho V, Pronk JW, Engel AH (2018) Characterization of membrane proteins using Cryo-Electron Microscopy. Curr Protoc Protein Sci 94:e72

Ceska T, Chung C-W, Cooke R, Phillips C, Williams PA (2019) Cryo-EM in drug discovery. Biochem Soc Trans 47:281–293

Chakraborty S, Somasundaran P (2006) Sequestration of drugs using poly(acrylic acid) and alkyl modified poly(acrylic acid) nanoparticles. Soft Matter 2:850–854

Chang L, Wang F, Connolly K, Meng H, Su Z, Cvirkaite-Krupovic V et al (2022) DeepTracer-ID: De novo protein identification from cryo-EM maps. Biophys J 121:2840–2848

Chapman HN, Fromme P, Barty A, White TA, Kirian RA, Aquila A et al (2011) Femtosecond X-ray protein nanocrystallography. Nature 470:73–77

Chapman HN, Caleman C, Timneanu N (2014) Diffraction before destruction. Philos Trans R Soc B Biol Sci 369:20130313

Charvolin D, Perez J-B, Rouvière F, Giusti F, Bazzacco P, Abdine A et al (2009) The use of amphipols as universal molecular adapters to immobilize membrane proteins onto solid supports. Proc Natl Acad Sci 106:405–410

Charvolin D, Picard M, Huang L-S, Berry EA, Popot J-L (2014) Solution behavior and crystallization of cytochrome bc1 in the presence of Amphipols. J Membr Biol 247:981–996

Chaudhary S, Pak JE, Gruswitz F, Sharma V, Stroud RM (2012) Overexpressing human membrane proteins in stably transfected and clonal human embryonic kidney 293S cells. Nat Protoc 7:453–466

Chen J, Noble AJ, Kang JY, Darst SA (2019) Eliminating effects of particle adsorption to the air/water interface in single-particle cryo-electron microscopy: bacterial RNA polymerase and CHAPSO. J Struct Biol X 1:100005

Chen Z, Watanabe S, Hashida H, Inoue M, Daigaku Y, Kikkawa M et al (2023) Cryo-EM structures of human SPCA1a reveal the mechanism of Ca2+/Mn2+ transport into the Golgi apparatus. Sci Adv 9:eadd9742

Cheng Y (2015) Single-particle Cryo-EM at crystallographic Resolution. Cell 161:450–457

Chio US, Palovcak E, Smith AAA, Autzen H, Muñoz EN, Yu Z et al (2024) Functionalized graphene-oxide grids enable high-resolution cryo-EM structures of the SNF2h-nucleosome complex without crosslinking. Nat Commun 15:2225

Cho HS, Dashdorj N, Schotte F, Graber T, Henning R, Anfinrud P (2010) Protein structural dynamics in solution unveiled via 100-ps time-resolved x-ray scattering. Proc Natl Acad Sci 107:7281–7286

Choy BC, Cater RJ, Mancia F, Pryor EEJ (2021) A 10-year meta-analysis of membrane protein structural biology: detergents, membrane mimetics, and structure determination techniques. Biochim Biophys Acta Biomembr 1863:183533

Chu C, Vargas C, Barbosa MC, Sommerhage S, Schröder GF, Keller S, et al (2024) Capturing G protein-coupled receptors into native lipid-bilayer nanodiscs using new diisobutylene/maleic acid (DIBMA) copolymers. bioRxiv 2024.01.20.576420. Available from: https://www.biorxiv.org/content/10.1101/2024.01.20.576420v1

Claesson E, Wahlgren WY, Takala H, Pandey S, Castillon L, Kuznetsova V et al (2020) The primary structural photoresponse of phytochrome proteins captured by a femtosecond X-ray laser. Elife 9:e53514

Coleman JA, Yang D, Zhao Z, Wen P-C, Yoshioka C, Tajkhorshid E et al (2019) Serotonin transporter–ibogaine complexes illuminate mechanisms of inhibition and transport. Nature 569:141–145

Colletier J-P, Sawaya MR, Gingery M, Rodriguez JA, Cascio D, Brewster AS et al (2016) De novo phasing with X-ray laser reveals mosquito larvicide BinAB structure. Nature 539:43–47

Cookis T, Sauer P, Poepsel S, Han B-G, Herbst DA, Glaeser R et al (2023) Streptavidin-affinity grid fabrication for Cryo-Electron Microscopy sample preparation. J Vis Exp JoVE. https://doi.org/10.3791/66197

Coquelle N, Sliwa M, Woodhouse J, Schirò G, Adam V, Aquila A et al (2018) Chromophore twisting in the excited state of a photoswitchable fluorescent protein captured by time-resolved serial femtosecond crystallography. Nat Chem 10:31–37

Coupland CE, Andrei SA, Ansell TB, Carrique L, Kumar P, Sefer L et al (2021) Structure, mechanism, and inhibition of Hedgehog acyltransferase. Mol Cell 81:5025–5038.e10

Cubipol [Internet]. Cube Biotech. [cited 2025 May 5]. Available from: https://cube-biotech.com/products/membrane-protein-stabilization/copolymer-nanodisc-products/cubipol/

D'Imprima E, Floris D, Joppe M, Sánchez R, Grininger M, Kühlbrandt W (2019) Protein denaturation at the air-water interface and how to prevent it. eLife 8:e42747

Dahmane T, Giusti F, Catoire LJ, Popot J-L (2011) Sulfonated amphipols: synthesis, properties, and applications. Biopolymers 95:811–823
Dahmane T, Rappaport F, Popot J-L (2013) Amphipol-assisted folding of bacteriorhodopsin in the presence or absence of lipids: functional consequences. Eur Biophys J 42:85–101
Dalal V, Arcario MJ, Petroff JT, Tan BK, Dietzen NM, Rau MJ et al (2024) Lipid nanodisc scaffold and size alter the structure of a pentameric ligand-gated ion channel. Nat Commun 15:25
Dandey VP, Wei H, Zhang Z, Tan YZ, Acharya P, Eng ET et al (2018) Spotiton: new features and applications. J Struct Biol 202:161–9
Dandey VP, Budell WC, Wei H, Bobe D, Maruthi K, Kopylov M et al (2020) Time-resolved cryo-EM using Spotiton. Nat Methods 17:897–900
Danelius E, Patel K, Gonzalez B, Gonen T (2023) MicroED in drug discovery. Curr Opin Struct Biol 79:102549. Available from: https://www.sciencedirect.com/science/article/pii/S0959440X23000234
Daniilidis M, Brandl MJ, Hagn F (2022). The advanced properties of circularized MSP nanodiscs facilitate high-resolution NMR studies of membrane proteins. J Mol Biol 434:167861
Dasgupta M, Budday D, de Oliveira SHP, Madzelan P, Marchany-Rivera D, Seravalli J et al (2019) Mix-and-inject XFEL crystallography reveals gated conformational dynamics during enzyme catalysis. Proc Natl Acad Sci USA 116:25634–25640
De Felice S, Romanyuk Z, Chinellato M, Zoia G, Linciano S, Kumada Y et al (2024) Crystal structure of human serum albumin in complex with megabody reveals unique human and murine cross-reactive binding site. Protein Sci 33:e4887
de la Cruz MJ, Hattne J, Shi D, Seidler P, Rodriguez J, Reyes FE et al (2017) Atomic-resolution structures from fragmented protein crystals with the cryoEM method MicroED. Nat Methods 14:399–402
de Teresa-Trueba I, Goetz SK, Mattausch A, Stojanovska F, Zimmerli CE, Toro-Nahuelpan M et al (2023) Convolutional networks for supervised mining of molecular patterns within cellular context. Nat Methods 20:284–294
Delauzun V, Amigues B, Gaubert A, Leone P, Grange M, Gauthier L et al (2020) Extracellular vesicles as a platform to study cell-surface membrane proteins. Methods 180:35–44
Della Pia EA, Hansen RW, Zoonens M, Martinez KL (2014) Functionalized amphipols: a versatile toolbox suitable for applications of membrane proteins in synthetic biology. J Membr Biol 247:815–826
Denisov IG, Sligar SG (2016) Nanodiscs for structural and functional studies of membrane proteins. Nat Struct Mol Biol 23:481–486
DePonte DP, Weierstall U, Schmidt K, Warner J, Starodub D, Spence JCH et al (2008) Gas dynamic virtual nozzle for generation of microscopic droplet streams. J Phys Appl Phys 41:195505
Dhakal A, Gyawali R, Wang L, Cheng J (2024) CryoTransformer: a transformer model for picking protein particles from cryo-EM micrographs. Elofsson A, editor. Bioinformatics 40:btae109
Diab C, Tribet C, Gohon Y, Popot J-L, Winnik FM (2007) Complexation of integral membrane proteins by phosphorylcholine-based amphipols. Biochim Biophys Acta BBA – Biomembr 1768:2737–2747
DiIorio MC, Kulczyk AW (2022) A robust single-particle Cryo-Electron Microscopy (cryo-EM) processing workflow with cryoSPARC, RELION, and Scipion. J Vis Exp JoVE
Dilworth MV, Piel MS, Bettaney KE, Ma P, Luo J, Sharples D et al (2018) Microbial expression systems for membrane proteins. Methods 147:3–39
Dilworth MV, Findlay HE, Booth PJ (2021) Detergent-free purification and reconstitution of functional human serotonin transporter (SERT) using diisobutylene maleic acid (DIBMA) copolymer. Biochimica et Biophysica Acta (BBA) 1863:183602
Ding G, Liu Y, Zhang R, Xin HL (2019) A joint deep learning model to recover information and reduce artifacts in missing-wedge sinograms for electron tomography and beyond. Sci Rep 9:12803
Dörr JM, Scheidelaar S, Koorengevel MC, Dominguez JJ, Schäfer M, van Walree CA et al (2016) The styrene–maleic acid copolymer: a versatile tool in membrane research. Eur Biophys J 45:3–21
Doyle LM, Wang MZ (2019) Overview of extracellular vesicles, their origin, composition, purpose, and methods for exosome isolation and analysis. Cells 8:727
Dubochet J, Adrian M, Chang JJ, Homo JC, Lepault J, McDowall AW et al (1988) Cryo-electron microscopy of vitrified specimens. Q Rev Biophys 21:129–228
Dukkipati A, Park HH, Waghray D, Fischer S, Garcia KC (2008) BacMam system for High-level expression of recombinant soluble and membrane glycoproteins for structural studies. Protein Expr Purif 62:160–170
Duyvesteyn HME, Kotecha A, Ginn HM, Hecksel CW, Beale EV, de Haas F et al (2018) Machining protein microcrystals for structure determination by electron diffraction. Proc Natl Acad Sci USA 115:9569–9573
El Omari K, Duman R, Mykhaylyk V, Orr CM, Latimer-Smith M, Winter G et al (2023) Experimental phasing opportunities for macromolecular crystallography at very long wavelengths. Commun Chem 6:219
Elegheert J, Behiels E, Bishop B, Scott S, Woolley RE, Griffiths SC et al (2018) Lentiviral transduction of mammalian cells for fast, scalable and high-level production of soluble and membrane proteins. Nat Protoc 13:2991–3017
Elter S, Raschle T, Arens S, Viegas A, Gelev V, Etzkorn M et al (2014) The use of Amphipols for NMR structural characterization of 7-TM proteins. J Membr Biol 247:957–964
Esfahani BG, Randolph PS, Peng R, Grant T, Stroupe ME, Stagg SM (2024) SPOT-RASTR-A cryo-EM specimen preparation technique that overcomes problems with preferred orientation and the air/water interface. PNAS Nexus 3:pgae284

Esmaili M, Brown CJ, Shaykhutdinov R, Acevedo-Morantes C, Wang YL, Wille H et al (2020) Homogeneous nanodiscs of native membranes formed by stilbene–maleic-acid copolymers. Nanoscale 12:16705–16709

Fan X, Wang J, Zhang X, Yang Z, Zhang J-C, Zhao L et al (2019) Single particle cryo-EM reconstruction of 52 kDa streptavidin at 3.2 Angstrom resolution. Nat Commun 10:2386

Fan H, Wang B, Zhang Y, Zhu Y, Song B, Xu H et al (2021) A cryo-electron microscopy support film formed by 2D crystals of hydrophobin HFBI. Nat Commun 12: 7257

Fan L, Zhuang Y, Wu H, Li H, Xu Y, Wang Y et al (2024) Structural basis of psychedelic LSD recognition at dopamine D1 receptor. Neuron 112:3295–3310.e8

Feinstein EH, Tifrea D, Sun G, Popot J-L, de la Maza LM, Cocco MJ (2014) Long-term stability of a vaccine formulated with the Amphipol-trapped major outer membrane protein from Chlamydia trachomatis. J Membr Biol 247:1053–1065

Fenalti G, Zatsepin NA, Betti C, Giguere P, Han GW, Ishchenko A et al (2015) Structural basis for bifunctional peptide recognition at human δ-opioid receptor. Nat Struct Mol Biol 22:265–268

Feng X, Fu Z, Kaledhonkar S, Jia Y, Shah B, Jin A et al (1993) A fast and effective microfluidic spraying-plunging method for High-Resolution single-particle Cryo-EM. Struct Lond Engl 2017(25):663–670.e3

Ferrandez Y, Dezi M, Bosco M, Urvoas A, Valerio-Lepiniec M, Le Bon C et al (2014) Amphipol-mediated screening of molecular orthoses specific for membrane protein targets. J Membr Biol 247:925–940

Fiori MC, Jiang Y, Altenberg GA, Liang H (2017) Polymer-encased nanodiscs with improved buffer compatibility. Sci Rep 7:7432

Fiori MC, Zheng W, Kamilar E, Simiyu G, Altenberg GA, Liang H (2020) Extraction and reconstitution of membrane proteins into lipid nanodiscs encased by zwitterionic styrene-maleic amide copolymers. Sci Rep 10:9940

Foos N, Florial J-B, Eymery M, Sinoir J, Felisaz F, Oscarsson M et al (2024) *In situ* serial crystallography facilitates 96-well plate structural analysis at low symmetry. IUCrJ 11:780–791

Frangakis AS (2021) It's noisy out there! A review of denoising techniques in cryo-electron tomography. J Struct Biol 213:107804

Frank J (2017) Time-resolved cryo-electron microscopy: recent progress. J Struct Biol 200:303–306

Fromme R, Ishchenko A, Metz M, Chowdhury SR, Basu S, Boutet S et al (2015) Serial femtosecond crystallography of soluble proteins in lipidic cubic phase. IUCrJ 2:545–551

Frotscher E, Danielczak B, Vargas C, Meister A, Durand G, Keller S (2015) A fluorinated detergent for membrane-protein applications. Angew Chem Int Ed 54:5069–5073

Fujita J, Makino F, Asahara H, Moriguchi M, Kumano S, Anzai I et al (2023) Epoxidized graphene grid for highly efficient high-resolution cryoEM structural analysis. Sci Rep 13:2279

Fuller FD, Gul S, Chatterjee R, Burgie ES, Young ID, Lebrette H et al (2017) Drop-on-demand sample delivery for studying biocatalysts in action at X-ray free-electron lasers. Nat Methods 14:443–449

Gaitzsch J, Hirschi S, Freimann S, Fotiadis D, Meier W (2019) Directed insertion of light-activated proteorhodopsin into asymmetric polymersomes from an ABC block copolymer. Nano Lett 19:2503–2508

Gao Y, Cao E, Julius D, Cheng Y (2016) TRPV1 structures in nanodiscs reveal mechanisms of ligand and lipid action. Nature 534:347–351

Gardiner AT, Naydenova K, Castro-Hartmann P, Nguyen-Phan TC, Russo CJ, Sader K et al (2021) The 2.4 Å cryo-EM structure of a heptameric light-harvesting 2 complex reveals two carotenoid energy transfer pathways. Sci Adv 7:eabe4650

Garman EF (2010) Radiation damage in macromolecular crystallography: what is it and why should we care? Acta Crystallogr D Biol Crystallogr 66:339–351

Gemin O, Armijo V, Hons M, Bissardon C, Linares R, Bowler MW et al (2024) EasyGrid: a versatile platform for automated cryo-EM sample preparation and quality control. bioRxiv 2024.01.18.576170. Available from: https://www.biorxiv.org/content/10.1101/2024.01.18.576170v2

Genick UK, Borgstahl GE, Ng K, Ren Z, Pradervand C, Burke PM et al (1997) Structure of a protein photocycle intermediate by millisecond time-resolved crystallography. Science 275:1471–1475

Genthe E, Miletic S, Tekkali I, Hennell James R, Marlovits TC, Heuser P (2023) PickYOLO: fast deep learning particle detector for annotation of cryo electron tomograms. J Struct Biol 215:107990

George B, Assaiya A, Roy RJ, Kembhavi A, Chauhan R, Paul G et al (2021) CASSPER is a semantic segmentation-based particle picking algorithm for single-particle cryo-electron microscopy. Commun Biol. 4:200

Gerriets V, Anderson J, Patel P, Nappe TM (2025) Acetaminophen. StatPearls [Internet]. Treasure Island (FL): StatPearls Publishing; [cited 2025 Apr 28]. Available from: http://www.ncbi.nlm.nih.gov/books/NBK482369/

Gewering T, Januliene D, Ries AB, Moeller A (2018) Know your detergents: a case study on detergent background in negative stain electron microscopy. J Struct Biol 203:242–246

Ghasem H, Kraljevic NB, Singh B, Martin IPS (2024) Lattice design for the diamond-II light source storage ring. Phys Rev Accel Beams 27:110704

Giri N, Cheng J (2024) De novo atomic protein structure modeling for cryoEM density maps using 3D transformer and HMM. Nat Commun 15:5511

Giri N, Cheng J (2025) Atomic protein structure modeling from Cryo-EM using multi-modal deep learning and AlphaFold3. bioRxiv 2025.03.16.643561. Available from: http://biorxiv.org/lookup/doi/10.1101/2025.03.16.643561

Giusti F, Rieger J, Catoire LJ, Qian S, Calabrese AN, Watkinson TG et al (2014) Synthesis, characterization and applications of a perdeuterated Amphipol. J Membr Biol 247:909–924

Giusti F, Kessler P, Hansen RW, Della Pia EA, Le Bon C, Mourier G et al (2015) Synthesis of a polyhistidine-bearing amphipol and its use for immobilizing membrane proteins. Biomacromolecules 16:3751–3761

Goddard TD et al (2018) UCSF ChimeraX: meeting modern challenges in visualization and analysis. Protein Sci 27:14–25

Goehring A, Lee C-H, Wang KH, Michel JC, Claxton DP, Baconguis I et al (2014) Screening and large-scale expression of membrane proteins in mammalian cells for structural studies. Nat Protoc 9:2574–2585

Gonen T (2013) The collection of high-resolution electron diffraction data. Methods Mol Biol 955:153–69. Available from: https://www.ncbi.nlm.nih.gov/pmc/articles/PMC3856639/

Gonzalez A, Krojer T, Nan J, Bjelčić M, Aggarwal S, Gorgisyan I et al (2025) Status and perspective of protein crystallography at the first multi-bend achromat based synchrotron MAX IV. J Synchrotron Radiat 32:779–791

Goslawski P, Abo-Bakr M, Arlandoo M, Bengtsson J, Holldack K, Jankowiak A et al (2022) BESSY III status report and lattice design process. Zimmermann F, Tanaka H, Sudmuang P, Klysubun P, Sunwong P, Chanwattana T et al, editors;3 pages, 2.143 MB

Grabski M, Al-Dmour E (2021) Commissioning and operation status of the MAX IV 3 GeV storage ring vacuum system. J Synchrotron Radiat 28:718–731

Grant T, Grigorieff N (2015) Measuring the optimal exposure for single particle cryo-EM using a 2.6 Å reconstruction of rotavirus VP6. eLife 4:e06980

Grant T, Rohou A, Grigorieff N (2018) cisTEM, user-friendly software for single-particle image processing. eLife 7:e35383

Gregor WD, Maharjan R, Zhang Z, Chiaraviglio L, Sastry N, Cui M et al (2025) Cryo-EM structure of the Pseudomonas aeruginosa MexY multidrug efflux pump. mBio 16:e03826–e03824

Gruget C, Reddy BG, Moore JM (2025) A structural and mechanistic model for BSEP dysfunction in PFIC2 cholestatic disease. Commun Biol 8:1–12

Gruhl T, Weinert T, Rodrigues MJ, Milne CJ, Ortolani G, Nass K et al (2023) Ultrafast structural changes direct the first molecular events of vision. Nature 615:939–944

Guo R, Sumner J, Qian S (2021) Structure of diisobutylene maleic acid copolymer (DIBMA) and its lipid particle as a "stealth" membrane-mimetic for membrane protein research. ACS Appl Bio Mater 4:4760–4768

Gyawali R, Dhakal A, Wang L, Cheng J (2024) CryoSegNet: accurate cryo-EM protein particle picking by integrating the foundational AI image segmentation model and attention-gated U-Net. Brief Bioinform 25: bbae282

Habel J, Hansen M, Kynde S, Larsen N, Midtgaard SR, Jensen GV et al (2015) Aquaporin-based biomimetic polymeric membranes: approaches and challenges. Membranes 5:307–351

Hagn F, Nasr ML, Wagner G (2018) Assembly of phospholipid nanodiscs of controlled size for structural studies of membrane proteins by NMR. Nat Protoc 13:79–98

Hajdu J, Machin PA, Campbell JW, Greenhough TJ, Clifton IJ, Zurek S et al (1987) Millisecond X-ray diffraction and the first electron density map from Laue photographs of a protein crystal. Nature 329:178–181

Hall SCL, Tognoloni C, Charlton J, Bragginton ÉC, Rothnie AJ, Sridhar P et al (2018) An acid-compatible co-polymer for the solubilization of membranes and proteins into lipid bilayer-containing nanoparticles. Nanoscale 10:10609–10619

Han B-G, Watson Z, Kang H, Pulk A, Downing KH, Cate J et al (2016) Long shelf-life streptavidin support-films suitable for electron microscopy of biological macromolecules. J Struct Biol 195:238–244

Harder OF, Voss JM, Olshin PK, Drabbels M, Lorenz UJ (2022) Microsecond melting and revitrification of cryo samples: protein structure and beam-induced motion. Acta Crystallogr Sect Struct Biol 78:883–889

Hardin NZ, Ravula T, Mauro GD, Ramamoorthy A (2019) Hydrophobic functionalization of polyacrylic acid as a versatile platform for the development of polymer lipid nanodisks. Small Weinh Bergstr Ger 15:e1804813

Hattne J, Echols N, Tran R, Kern J, Gildea RJ, Brewster AS et al (2014) Accurate macromolecular structures using minimal measurements from X-ray free-electron lasers. Nat Methods 11:545–548

He Y, Wang K, Yan N (2014) The recombinant expression systems for structure determination of eukaryotic membrane proteins. Protein Cell 5:658–672

He J, Lin P, Chen J, Cao H, Huang S-Y (2022) Model building of protein complexes from intermediate-resolution cryo-EM maps with deep learning-guided automatic assembly. Nat Commun 13:4066

Heebner JE, Purnell C, Hylton RK, Marsh M, Grillo MA, Swulius MT (2022) Deep learning-based segmentation of cryo-electron tomograms. J Vis Exp 189:e64435 https://doi.org/10.3791/64435

Henderikx RJM, Mann D, Domanska A, Dong J, Shahzad S, Lak B et al (2024) VitroJet: new features and case studies. Acta Crystallogr Sect Struct Biol 80:232–246

Henderson R, Unwin PNT (1975) Three-dimensional model of purple membrane obtained by electron microscopy. Nature 257:28–32

Hendrickson WA (2014) Anomalous diffraction in crystallographic phase evaluation. Q Rev Biophys 47:49–93

Hendrickson WA (2023) Facing the phase problem. IUCrJ 10:521–543

Higgins AJ, Flynn AJ, Marconnet A, Musgrove LJ, Postis VLG, Lippiat JD et al (2021) Cycloalkane-modified amphiphilic polymers provide direct extraction of membrane proteins for CryoEM analysis. Commun Biol 4:1337

Hirata K, Kawano Y, Ueno G, Hashimoto K, Murakami H, Hasegawa K et al (2013) Achievement of protein micro-crystallography at SPring-8 beamline BL32XU. J Phys Conf Ser 425:012002

Hoffmann L, Baier A, Jorde L, Kamel M, Schäfer J-H, Schnelle K et al (2025) The ABC transporter MsbA in a dozen environments. Structure 33:916–923.e4
Hosaka T, Nomura T, Kubo M, Nakane T, Fangjia L, Sekine S-I et al (2022) Conformational alterations in unidirectional ion transport of a light-driven chloride pump revealed using X-ray free electron lasers. Proc Natl Acad Sci USA 119:e2117433119
Hryc CF, Baker ML (2022) Beyond the backbone: the next generation of pathwalking utilities for model building in CryoEM density maps. Biomolecules 12:773
Huang C-Y, Olieric V, Howe N, Warshamanage R, Weinert T, Panepucci E et al (2018) In situ serial crystallography for rapid de novo membrane protein structure determination. Commun Biol 1:124
Huang C-Y, Meier N, Caffrey M, Wang M, Olieric V (2020) 3D-printed holders for in meso in situ fixed-target serial X-ray crystallography. J Appl Crystallogr 53:854–859
Huang N, Deng H, Liu B, Wang D, Zhao Z (2021) Features and futures of X-ray free-electron lasers. Innov Camb Mass 2:100097
Huang C-Y, Aumonier S, Olieric V, Wang M (2024) Cryo2RT: a high-throughput method for room-temperature macromolecular crystallography from cryo-cooled crystals. Acta Crystallogr Sect Struct Biol 80:620–628
Hunter MS, Segelke B, Messerschmidt M, Williams GJ, Zatsepin NA, Barty A et al (2014) Fixed-target protein serial microcrystallography with an x-ray free electron laser. Sci Rep 4:6026
Hurley JH (2010) The ESCRT complexes. Crit Rev Biochem Mol Biol 45:463–487
Huynh KW, Cohen MR, Moiseenkova-Bell VY (2014) Application of amphipols for structure-functional analysis of TRP channels. J Membr Biol 247:843–851
Ishigami I, Sierra RG, Su Z, Peck A, Wang C, Poitevin F et al (2023) Structural insights into functional properties of the oxidized form of cytochrome c oxidase. Nat Commun 14:5752
Jaho S, Junius N, Borel F, Sallaz-Damaz Y, Salmon J-B, Budayova-Spano M (2021) Crystallization of proteins on chip by microdialysis for in situ X-ray diffraction studies. J Vis Exp JoVE
Jain T, Sheehan P, Crum J, Carragher B, Potter CS (2012) Spotiton: a prototype for an integrated inkjet dispense and vitrification system for cryo-TEM. J Struct Biol 179:68–75
Jamali K, Käll L, Zhang R, Brown A, Kimanius D, Scheres SHW (2024) Automated model building and protein identification in cryo-EM maps. Nature 628: 450–457
Janata M, Čadová E, Angelisová P, Charnavets T, Hořejší V, Raus V (2022) Tailoring butyl methacrylate/methacrylic acid copolymers for the solubilization of membrane proteins: the influence of composition and molecular weight. Macromol Biosci 22:e2200284
Jang GS, Shin S, Yoon M, Ko J, Yoon YD, Lee J et al (2022) Low emittance lattice design for Korea-4GSR. Nucl Instrum Methods Phys Res Sect A 1034:166779
Jiao Y, Xu G, Cui X-H, Duan Z, Guo Y-Y, He P et al (2018) The HEPS project. J Synchrotron Radiat 25:1611–1618
Jiko C, Li J, Moon Y, Tanaka Y, Gopalasingam CC, Shigematsu H et al (2024) NDT-C11 as a viable novel detergent for single particle Cryo-EM. ChemPlusChem 89:e202400242
Jiménez-Moreno A, Del Caño L, Martínez M, Ramírez-Aportela E, Cuervo A, Melero R et al (2021) Cryo-EM and single-particle analysis with Scipion. J Vis Exp JoVE
Johansen NT, Tidemand FG, Nguyen TTTN, Rand KD, Pedersen MC, Arleth L (2019) Circularized and solubility-enhanced MSPs facilitate simple and high-yield production of stable nanodiscs for studies of membrane proteins in solution. FEBS J 286:1734–1751
Jones CG, Martynowycz MW, Hattne J, Fulton TJ, Stoltz BM, Rodriguez JA et al (2018) The CryoEM method MicroED as a powerful tool for small molecule structure determination. ACS Cent Sci 4:1587–1592
Josts I, Niebling S, Gao Y, Levantino M, Tidow H, Monteiro D (2018) Photocage-initiated time-resolved solution X-ray scattering investigation of protein dimerization. IUCrJ 5:667–672
Juers DH, Farley CA, Saxby CP, Cotter RA, Cahn JKB, Holton-Burke RC et al (2018) The impact of cryosolution thermal contraction on proteins and protein crystals: volumes, conformation and order. Acta Crystallogr Sect Struct Biol 74:922–938
Jumper J, Evans R, Pritzel A, Green T, Figurnov M, Ronneberger O et al (2021) Highly accurate protein structure prediction with AlphaFold. Nature 596:583–589
Jung YO, Lee JH, Kim J, Schmidt M, Moffat K, Srajer V et al (2013) Volume-conserving trans-cis isomerization pathways in photoactive yellow protein visualized by picosecond X-ray crystallography. Nat Chem 5:212–220
Junge F, Schneider B, Reckel S, Schwarz D, Dötsch V, Bernhard F (2008) Large-scale production of functional membrane proteins. Cell Mol Life Sci 65:1729–1755
Kabsch W (2014) Processing of X-ray snapshots from crystals in random orientations. Acta Crystallogr D Biol Crystallogr 70:2204–2216
Kampjut D, Steiner J, Sazanov LA (2021) Cryo-EM grid optimization for membrane proteins. iScience 24:102139
Kang Y, Chen L (2023) Structural basis for the binding of DNP and purine nucleotides onto UCP1. Nature 620: 226–231
Kang Y, Kuybeda O, de Waal PW, Mukherjee S, Van Eps N, Dutka P et al (2018) Cryo-EM structure of human rhodopsin bound to an inhibitory G protein. Nature 558:553–558
Kang JS, Zhou X, Liu Y-T, Wang K, Zhou ZH (2023) Theoretical framework and experimental solution for the air–water interface adsorption problem in cryoEM. Biophys Rep 9:215–229
Karantzoulis E, Di Mitri S, Barbo F, Barletta W, Bassanese S, Bracco R et al (2024) Design strategies and technology of Elettra 2.0 for a versatile offer to the

user community. Nucl Instrum Methods Phys Res Sect Accel Spectrometers Detect Assoc Equip 1060:169007

Karasawa A, Andi B, Fuchs MR, Shi W, McSweeney S, Hendrickson WA et al (2022) Multi-crystal native-SAD phasing at 5 keV with a helium environment. IUCrJ 9:768–777

Karia D, Koh AF, Yang W, Cushing VI, Basanta B, Mihaylov DB et al (2024) Sub-3 Å resolution protein structure determination by single-particle cryo-EM at 100 keV. bioRxiv 2024.09.05.611417. Available from: https://www.biorxiv.org/content/10.1101/2024.09.05.611417v2

Keedy DA, van den Bedem H, Sivak DA, Petsko GA, Ringe D, Wilson MA et al (2014) Crystal cryocooling distorts conformational heterogeneity in a model Michaelis complex of DHFR. Structure 22:899–910

Keller J, Fernández-Busnadiego R (2024) In situ studies of membrane biology by cryo-electron tomography. Curr Opin Cell Biol 88:102363

Kendrew JC, Bodo G, Dintzis HM, Parrish RG, Wyckoff H, Phillips DC (1958) A three-dimensional model of the myoglobin molecule obtained by X-ray analysis. Nature 181:662–666

Kermani AA (2021) A guide to membrane protein X-ray crystallography. FEBS J 288:5788–5804

Kesidis A, Depping P, Lodé A, Vaitsopoulou A, Bill RM, Goddard AD et al (2020) Expression of eukaryotic membrane proteins in eukaryotic and prokaryotic hosts. Methods 180:3–18

Kim J, Tan YZ, Wicht KJ, Erramilli SK, Dhingra SK, Okombo J et al (2019) Structure and drug resistance of the Plasmodium falciparum transporter PfCRT. Nature 576:315–320

Kim H, Lim T, Ha GE, Lee J-Y, Kim J-W, Chang N et al (2023) Structure-based drug discovery of a corticotropin-releasing hormone receptor 1 antagonist using an X-ray free-electron laser. Exp Mol Med 55:2039–2050

Klein S, Wimmer BH, Winter SL, Kolovou A, Laketa V, Chlanda P (2021) Post-correlation on-lamella cryo-CLEM reveals the membrane architecture of lamellar bodies. Commun Biol 4:137

Kleinschmidt JH, Popot J-L (2014) Folding and stability of integral membrane proteins in amphipols. Arch Biochem Biophys 564:327–343

Kovácsová G, Grünbein ML, Kloos M, Barends TRM, Schlesinger R, Heberle J et al (2017) Viscous hydrophilic injection matrices for serial crystallography. IUCrJ 4:400–410

Krawinski P, Smithers L, Van Dalsen L, Boland C, Ostrovitsa N, Pérez J et al (2024) 7.10 MAG. A novel host monoacylglyceride for *in meso* (lipid cubic phase) crystallization of membrane proteins. Cryst Growth Des 24:2985–3001

Kroeck KG, Qiu W, Catalano C, Trinh TKH, Guo Y (2020) Native cell membrane nanoparticles system for membrane protein-protein interaction analysis. J Vis Exp JoVE:e61298

Krogh A, Larsson B, Von Heijne G, Sonnhammer ELL (2001) Predicting transmembrane protein topology with a hidden markov model: application to complete genomes 11 Edited by F. Cohen. J Mol Biol 305:567–580

Kuba J, Mitchels J, Hovorka M, Erdmann P, Berka L, Kirmse R et al (2021) Advanced cryo-tomography workflow developments – correlative microscopy, milling automation and cryo-lift-out. J Microsc 281:112–124

Kudryashev M (2024) The big chill: growth of *in situ* structural biology with cryo-electron tomography. QRB Discov 5:e10

Kühlbrandt W (2014) The resolution revolution. Science 343:1443–1444

Kuijper M, van Hoften G, Janssen B, Geurink R, De Carlo S, Vos M et al (2015) FEI's direct electron detector developments: Embarking on a revolution in cryo-TEM. J Struct Biol 192:179–187

Kulkarni CV, Wachter W, Iglesias-Salto G, Engelskirchen S, Ahualli S (2011) Monoolein: a magic lipid? Phys Chem Chem Phys PCCP 13:3004–3021

Kumar A, Basak S, Rao S, Gicheru Y, Mayer ML, Sansom MSP et al (2020) Mechanisms of activation and desensitization of full-length glycine receptor in lipid nanodiscs. Nat Commun 11:3752

Lamm L, Righetto RD, Wietrzynski W, Pöge M, Martinez-Sanchez A, Peng T et al (2022) MemBrain: a deep learning-aided pipeline for detection of membrane proteins in Cryo-electron tomograms. Comput Methods Prog Biomed 224:106990

Landau EM, Rosenbusch JP (1996) Lipidic cubic phases: a novel concept for the crystallization of membrane proteins. Proc Natl Acad Sci USA 93:14532–14535

Laverty D, Desai R, Uchański T, Masiulis S, Stec WJ, Malinauskas T et al (2019) Cryo-EM structure of the human α1β3γ2 GABAA receptor in a lipid bilayer. Nature 565:516–520

Lavington S, Watts A (2021) Detergent-free solubilisation & purification of a G protein coupled receptor using a polymethacrylate polymer. Biochim Biophys Acta Biomembr 1863:183441

Lawton JM, Habib M, Ma B, Brooks RA, Best SM, Lewis AL et al (2017) The effect of cationically-modified phosphorylcholine polymers on human osteoblasts in vitro and their effect on bone formation in vivo. J Mater Sci Mater Med 28:144

Le Bon C, Popot J-L, Giusti F (2014a) Labeling and functionalizing amphipols for biological applications. J Membr Biol 247:797–814

Le Bon C, Della Pia EA, Giusti F, Lloret N, Zoonens M, Martinez KL et al (2014b) Synthesis of an oligonucleotide-derivatized amphipol and its use to trap and immobilize membrane proteins. Nucleic Acids Res 42:e83

Le Bon C, Marconnet A, Masscheleyn S, Popot J-L, Zoonens M (2018) Folding and stabilizing membrane proteins in amphipol A8-35. Methods 147:95–105

Leif H, Sled VD, Ohnishi T, Weiss H, Friedrich T (1995) Isolation and characterization of the proton-translocating NADH:ubiquinone oxidoreductase from *Escherichia coli*. Eur J Biochem 230:538–548

Lemos M, Augusto I, De Souza W, Miranda K (2025) Advances, challenges, and applications of cryo-electron tomography workflows for three-dimensional cellular imaging of infectious pathogens. J Microsc:jmi.13408
Levitz TS, Weckener M, Fong I, Naismith JH, Drennan CL, Brignole EJ et al (2022) Approaches to using the chameleon: robust, automated, fast-plunge cryoEM specimen preparation. Front Mol Biosci 9:903148
Levy A, Raghu R, Feathers JR, Grzadkowski M, Poitevin F, Johnston JD, Vallese F, Clarke OB, Wetzstein G, Zhong ED (2025) CryoDRGN-AI: neural ab initio reconstruction of challenging cryo-EM and cryo-ET datasets. Nat Methods 22:1486–1494
Li S (2022) Detergents and alternatives in cryo-EM studies of membrane proteins. Acta Biochim Biophys Sin 54:1049–1056
Li D, Caffrey M (2020) Structure and functional characterization of membrane integral proteins in the lipid cubic phase. J Mol Biol 432:5104–5123
Li X, Mooney P, Zheng S, Booth CR, Braunfeld MB, Gubbens S et al (2013) Electron counting and beam-induced motion correction enable near-atomic-resolution single-particle cryo-EM. Nat Methods 10:584–590
Li B, Zhu D, Shi H, Zhang X (2021) Effect of charge on protein preferred orientation at the air-water interface in cryo-electron microscopy. J Struct Biol 213:107783
Li H, Nakajima Y, Nango E, Owada S, Yamada D, Hashimoto K et al (2024) Oxygen-evolving photosystem II structures during S1-S2-S3 transitions. Nature 626:670–677
Liang Y-L, Khoshouei M, Glukhova A, Furness SGB, Zhao P, Clydesdale L et al (2018) Phase-plate cryo-EM structure of a biased agonist-bound human GLP-1 receptor–Gs complex. Nature 555:121–125
Liao M, Cao E, Julius D, Cheng Y (2013) Structure of the TRPV1 ion channel determined by electron cryo-microscopy. Nature 504:107–112
Lin X, Li M, Wang N, Wu Y, Luo Z, Guo S et al (2020) NANOBODY structural basis of ligand recognition and self-activation of orphan GPR52. Nature 579:152–7
Lin Z, Akin H, Rao R, Hie B, Zhu Z, Lu W et al (2023) Evolutionary-scale prediction of atomic-level protein structure with a language model. Science 379:1123–1130
Lindhoud S, Carvalho V, Pronk JW, Aubin-Tam M-E (2016) SMA-SH: modified styrene–maleic acid copolymer for functionalization of lipid nanodiscs. Biomacromolecules 17:1516–1522
Liu W, Cherezov V (2011) Crystallization of membrane proteins in lipidic mesophases. J Vis Exp JoVE.:2501
Liu N, Wang H-W (2023) Better Cryo-EM specimen preparation: how to deal with the air–water interface? J Mol Biol 435:167926
Liu N, Wang H-W (2024) Graphene in cryo-EM specimen optimization. Curr Opin Struct Biol 86:102823
Liu W, Wacker D, Gati C, Han GW, James D, Wang D et al (2013) Serial femtosecond crystallography of G protein-coupled receptors. Science 342:1521–1524
Liu L, Neuenschwander RT, Rodrigues ARD (2019a) Synchrotron radiation sources in Brazil. Philos Transact A Math Phys Eng Sci 377:20180235
Liu N, Zhang J, Chen Y, Liu C, Zhang X, Xu K et al (2019b) Bioactive functionalized monolayer graphene for high-resolution Cryo-Electron Microscopy. J Am Chem Soc 141:4016–4025
Liu H, Deepak RNVK, Shiriaeva A, Gati C, Batyuk A, Hu H et al (2021a) Molecular basis for lipid recognition by the prostaglandin D2 receptor CRTH2. Proc Natl Acad Sci USA 118:e2102813118
Liu N, Zheng L, Xu J, Wang J, Hu C, Lan J et al (2021b) Reduced graphene oxide membrane as supporting film for high-resolution cryo-EM. Biophys Rep 7:227–238
Loveland AB, Demo G, Korostelev AA (2020) Cryo-EM of elongating ribosome with EF-Tu•GTP elucidates tRNA proofreading. Nature 584:640–645
Lučić V, Förster F, Baumeister W (2005) Structural studies by electron tomography: from cells to molecules. Annu Rev Biochem 74:833–865
Luo Z, Gu W, Wang Y, Tang Y, Li D (2022) Selenourea for experimental phasing of membrane protein crystals grown in lipid cubic phase. Crystals 12:976
Mabrouk MT, Zidan AA, Aly N, Mohammed MT, Ghantous F, Seaman MS et al (2023) Circularized Nanodiscs for multivalent mosaic display of SARS-CoV-2 spike protein antigens. Vaccines 11:1655
Mäeots M-E, Enchev RI (2022) Structural dynamics: review of time-resolved cryo-EM. Acta Crystallogr Sect Struct Biol 78:927–935
Mäeots M-E, Lee B, Nans A, Jeong S-G, Esfahani MMN, Ding S et al (2020) Modular microfluidics enables kinetic insight from time-resolved cryo-EM. Nat Commun 11:3465
Mafuné F, Miyajima K, Tono K, Takeda Y, Kohno J, Miyauchi N et al (2016) Microcrystal delivery by pulsed liquid droplet for serial femtosecond crystallography. Acta Crystallogr Sect Struct Biol 72:520–523
Maki-Yonekura S, Kawakami K, Takaba K, Hamaguchi T, Yonekura K (2023) Measurement of charges and chemical bonding in a cryo-EM structure. Commun Chem 6:98
Mancia F, Love J (2010) High-throughput expression and purification of membrane proteins. J Struct Biol 172:85–93
Mancia F, Love J (2011) High throughput platforms for structural genomics of integral membrane proteins. Curr Opin Struct Biol 21:517–522
Marconnet A, Michon B, Le Bon C, Giusti F, Tribet C, Zoonens M (2020) Solubilization and stabilization of membrane proteins by cycloalkane-modified amphiphilic polymers. Biomacromolecules 21:3459–3467
Marconnet A, Michon B, Prost B, Solgadi A, Le Bon C, Giusti F et al (2022) Influence of hydrophobic groups attached to amphipathic polymers on the solubilization of membrane proteins along with their lipids. Anal Chem 94:14151–14158
Mary S, Damian M, Rahmeh R, Mouillac B, Marie J, Granier S et al (2014) Amphipols in G protein-coupled

receptor pharmacology: what are they good for? J Membr Biol 247:853–860

McMullan G, Faruqi AR, Clare D, Henderson R (2014) Comparison of optimal performance at 300 keV of three direct electron detectors for use in low dose electron microscopy. Ultramicroscopy 147:156–163

Mehrabi P, Schulz EC, Dsouza R, Müller-Werkmeister HM, Tellkamp F, Miller RJD et al (2019) Time-resolved crystallography reveals allosteric communication aligned with molecular breathing. Science 365:1167–1170

Mehrabi P, Müller-Werkmeister HM, Leimkohl JP, Schikora H, Ninkovic J, Krivokuca S et al (2020) The HARE chip for efficient time-resolved serial synchrotron crystallography. J Synchrotron Radiat 27:360–370

Mehrabi P, Sung S, Von Stetten D, Prester A, Hatton CE, Kleine-Döpke S et al (2023) Millisecond cryo-trapping by the spitrobot crystal plunger simplifies time-resolved crystallography. Nat Commun 14:2365

Melnikov I, Polovinkin V, Kovalev K, Gushchin I, Shevtsov M, Shevchenko V et al (2017) Fast iodide-SAD phasing for high-throughput membrane protein structure determination. Sci Adv 3:e1602952

Merk A, Bartesaghi A, Banerjee S, Falconieri V, Rao P, Davis MI et al (2016) Breaking Cryo-EM Resolution barriers to facilitate drug discovery. Cell 165:1698–1707

Merk A, Fukumura T, Zhu X, Darling JE, Grisshammer R, Ognjenovic J et al (2020) 1.8 Å resolution structure of β-galactosidase with a 200 kV CRYO ARM electron microscope. IUCrJ 7:639–643

Mi W, Li Y, Yoon SH, Ernst RK, Walz T, Liao M (2017) Structural basis of MsbA-mediated lipopolysaccharide transport. Nature 549:233–237

Michon B, López-Sánchez U, Degrouard J, Nury H, Leforestier A, Rio E et al (2023) Role of surfactants in electron cryo-microscopy film preparation. Biophys J 122:1846–1857

Milazzo A-C, Moldovan G, Lanman J, Jin L, Bouwer JC, Klienfelder S et al (2010) Characterization of a direct detection device imaging camera for transmission electron microscopy. Ultramicroscopy 110:744–747

Misquitta Y, Cherezov V, Havas F, Patterson S, Mohan JM, Wells AJ et al (2004) Rational design of lipid for membrane protein crystallization. J Struct Biol 148:169–175

Miyagi H, Asada H, Suzuki M, Takahashi Y, Yasunaga M, Suno C et al (2020) XRAY the discovery of a new antibody for BRIL-fused GPCR structure determination. Sci Rep 10:11669

Moffat K (2019) Laue diffraction and time-resolved crystallography: a personal history. Philos Trans R Soc Math Phys Eng Sci 377:20180243

Moffat K, Szebenyi D, Bilderback D (1984) X-ray Laue diffraction from protein crystals. Science 223:1423–1425

Moraes I, Evans G, Sanchez-Weatherby J, Newstead S, Stewart PDS (2014) Membrane protein structure determination – the next generation. Biochim Biophys Acta 1838:78–87

Morris EP, Da Fonseca PCA (2017) High-resolution cryo-EM proteasome structures in drug development. Acta Crystallogr Sect Struct Biol 73:522–533

Motov VV, Kot EF, Kislova SO, Bocharov EV, Arseniev AS, Boldyrev IA et al (2024) On the properties of styrene–maleic acid copolymer–lipid nanoparticles: a solution NMR perspective. Polymers (Basel) 16: 3009

Mous S, Gotthard G, Ehrenberg D, Sen S, Weinert T, Johnson PJM et al (2022) Dynamics and mechanism of a light-driven chloride pump. Science 375:845–851

Moyano DF, Rotello VM (2011) Nano meets biology: structure and function at the nanoparticle Interface. Langmuir 27:10376–10385

Mueller C, Marx A, Epp SW, Zhong Y, Kuo A, Balo AR et al (2015) Fixed target matrix for femtosecond time-resolved and in situ serial micro-crystallography. Struct Dyn Melville N 2:054302

Mukherjee S, Erramilli SK, Ammirati M, Alvarez FJD, Fennell KF, Purdy MD et al (2020) BRIL Synthetic antibodies against BRIL as universal fiducial marks for single−particle cryoEM structure determination of membrane proteins. Nat Commun 11:1598

Nagy JK, Kuhn Hoffmann A, Keyes MH, Gray DN, Oxenoid K, Sanders CR (2001) Use of amphipathic polymers to deliver a membrane protein to lipid bilayers. FEBS Lett 501:115–120

Nakagawa F, Kikkawa M, Chen S, Miyashita Y, Hamaguchi-Suzuki N, Shibuya M et al (2023) Anti-nanodisc antibodies specifically capture nanodiscs and facilitate molecular interaction kinetics studies for membrane protein. Sci Rep 13:11627

Nakane T, Hanashima S, Suzuki M, Saiki H, Hayashi T, Kakinouchi K et al (2016) Membrane protein structure determination by SAD, SIR, or SIRAS phasing in serial femtosecond crystallography using an iododetergent. Proc Natl Acad Sci USA 113:13039–13044

Nakane T, Kotecha A, Sente A, McMullan G, Masiulis S, Brown PMGE et al (2020) Single-particle cryo-EM at atomic resolution. Nature 587:152–156

Nallani M, Andreasson-Ochsner M, Tan C-WD, Sinner E-K, Wisantoso Y, Geifman-Shochat S et al (2011) Proteopolymersomes: in vitro production of a membrane protein in polymersome membranes. Biointerphases 6:153–157

Nanao M, Basu S, Zander U, Giraud T, Surr J, Guijarro M et al (2022) ID23-2: an automated and high-performance microfocus beamline for macromolecular crystallography at the ESRF. J Synchrotron Radiat 29:581–590

Nannenga BL, Gonen T (2016) MicroED opens a new era for biological structure determination. Curr Opin Struct Biol 40:128–135

Nannenga BL, Gonen T (2019) The cryo-EM method microcrystal electron diffraction (MicroED). Nat Methods 16:369–379

Nannenga BL, Shi D, Leslie AGW, Gonen T (2014) High-resolution structure determination by continuous-rotation data collection in MicroED. Nat Methods 11:927–930

NANUQ cryocooling device [Internet]. Available from: https://www.mitegen.com/nanuqtm-cryocooling-device/

Nasr ML, Baptista D, Strauss M, Sun Z-YJ, Grigoriu S, Huser S et al (2017) Covalently circularized nanodiscs for studying membrane proteins and viral entry. Nat Methods 14:49–52

Naydenova K, Russo CJ (2022) Integrated wafer-scale manufacturing of electron cryomicroscopy specimen supports. Ultramicroscopy 232:113396

Naydenova K, Peet MJ, Russo CJ (2019) Multifunctional graphene supports for electron cryomicroscopy. Proc Natl Acad Sci 116:11718–11724

Naydenova K, Jia P, Russo CJ (2020) Cryo-EM with sub–1 Å specimen movement. Science 370:223–226

Nederlof I, van Genderen E, Li YW, Abrahams JP (2013) A Medipix quantum area detector allows rotation electron diffraction data collection from submicrometre three-dimensional protein crystals. Acta Crystallogr D Biol Crystallogr 69:1223–1230

Neselu K, Wang B, Rice WJ, Potter CS, Carragher B, Chua EYD (2023). Measuring the effects of ice thickness on resolution in single particle cryo-EM. J Struct Biol X 7:100085

Nguyen K-A, Peuchmaur M, Magnard S, Haudecoeur R, Boyère C, Mounien S et al (2018) Glycosyl-substituted dicarboxylates as detergents for the extraction, overstabilization, and crystallization of membrane proteins. Angew Chem Int Ed Engl 57:2948–2952

Ni T, Frosio T, Mendonça L, Sheng Y, Clare D, Himes BA et al (2022) High-resolution in situ structure determination by cryo-electron tomography and subtomogram averaging using emClarity. Nat Protoc 17: 421–444

Nickl P, Hilal T, Olal D, Donskyi IS, Radnik J, Ludwig K et al (2023) A new support film for Cryo Electron Microscopy protein structure analysis based on covalently functionalized Graphene. Small Weinh Bergstr Ger 19:e2205932

Ning Z, Hawley B, Seebun D, Figeys D (2014) APols-aided protein precipitation: a rapid method for concentrating proteins for proteomic analysis. J Membr Biol 247:941–947

Noble AJ, Wei H, Dandey VP, Zhang Z, Tan YZ, Potter CS et al (2018) Reducing effects of particle adsorption to the air–water interface in cryo-EM. Nat Methods 15:793–795

Nogales E, Scheres SHW (2015) Cryo-EM: a unique tool for the visualization of macromolecular complexity. Mol Cell 58:677–689

Nouel Barreto A, Cuello LG, Zoghbi ME (2025) ABC transporter activity is affected by the size of lipid nanodiscs. FEBS Lett 599:502–511

Oberthuer D, Knoška J, Wiedorn MO, Beyerlein KR, Bushnell DA, Kovaleva EG et al (2017) Double-flow focused liquid injector for efficient serial femtosecond crystallography. Sci Rep 7:44628

Oghbaey S, Sarracini A, Ginn HM, Pare-Labrosse O, Kuo A, Marx A et al (2016) Fixed target combined with spectral mapping: approaching 100% hit rates for serial crystallography. Acta Crystallogr Sect Struct Biol 72:944–955

Olmos JL, Pandey S, Martin-Garcia JM, Calvey G, Katz A, Knoska J et al (2018) Enzyme intermediates captured "on the fly" by mix-and-inject serial crystallography. BMC Biol 16:59

Oluwole AO, Danielczak B, Meister A, Babalola JO, Vargas C, Keller S (2017) Solubilization of membrane proteins into functional lipid-bilayer nanodiscs using a diisobutylene/maleic acid copolymer. Angew Chem Int Ed Engl 56:1919–1924

Opačić M, Giusti F, Popot J-L, Broos J (2014) Isolation of *Escherichia coli* Mannitol Permease, EIImtl, trapped in Amphipol A8-35 and fluorescein-labeled A8-35. J Membr Biol 247:1019–1030

Owen RL, Axford D, Nettleship JE, Owens RJ, Robinson JI, Morgan AW et al (2012) Outrunning free radicals in room-temperature macromolecular crystallography. Acta Crystallogr D Biol Crystallogr 68:810–818

Palovcak E, Wang F, Zheng SQ, Yu Z, Li S, Betegon M et al (2018) A simple and robust procedure for preparing graphene-oxide cryo-EM grids. J Struct Biol 204:80–84

Pan D, Oyama R, Sato T, Nakane T, Mizunuma R, Matsuoka K et al (2022) Crystal structure of CmABCB1 multi-drug exporter in lipidic mesophase revealed by LCP-SFX. IUCrJ 9:134–145

Pande K, Hutchison CDM, Groenhof G, Aquila A, Robinson JS, Tenboer J et al (2016) Femtosecond structural dynamics drives the trans/cis isomerization in photoactive yellow protein. Science 352:725–729

Pandey A, Shin K, Patterson RE, Liu X-Q, Rainey JK (2016) Current strategies for protein production and purification enabling membrane protein structural biology. Biochem Cell Biol 94:507–527

Pandey S, Bean R, Sato T, Poudyal I, Bielecki J, Cruz Villarreal J et al (2020) Time-resolved serial femtosecond crystallography at the European XFEL. Nat Methods 17:73–78

Pantelic RS, Meyer JC, Kaiser U, Baumeister W, Plitzko JM (2010) Graphene oxide: a substrate for optimizing preparations of frozen-hydrated samples. J Struct Biol 170:152–156

Pantelic RS, Suk JW, Hao Y, Ruoff RS, Stahlberg H (2011) Oxidative Doping renders graphene hydrophilic, facilitating its use as a support in biological TEM. Nano Lett 11:4319–4323

Parker JL, Newstead S (2012) Current trends in α-helical membrane protein crystallization: an update. Protein Sci Publ Protein Soc 21:1358–1365

Parker JL, Newstead S (2016) Membrane protein crystallisation: current trends and future perspectives. Adv Exp Med Biol 922:61–72

Patel A, Toso D, Litvak A, Nogales E (2021) Efficient graphene oxide coating improves cryo-EM sample preparation and data collection from tilted grids. bioRxiv 2021.03.08.434344. Available from: https://www.biorxiv.org/content/10.1101/2021.03.08.434344v1

PDB Data Distribution by Experimental Method and Molecular Type [Internet]. RCSB PDB. [cited 2025

Apr 28]. Available from: https://www.rcsb.org/stats/summary

Pebay-Peyroula E, Rummel G, Rosenbusch JP, Landau EM (1997) X-ray structure of bacteriorhodopsin at 2.5 angstroms from microcrystals grown in lipidic cubic phases. Science 277:1676–1681

Peng R, Fu X, Mendez JH, Randolph PS, Bammes BE, Stagg SM (2023) Characterizing the resolution and throughput of the Apollo direct electron detector. J Struct Biol X 7:100080

Perry TN, Souabni H, Rapisarda C, Fronzes R, Giusti F, Popot J-L et al (2019) BAmSA: visualising transmembrane regions in protein complexes using biotinylated amphipols and electron microscopy. Biochim Biophys Acta Biomembr 1861:466–477

Peruzzi JA, Gunnels TF, Edelstein HI, Lu P, Baker D, Leonard JN et al (2024) Enhancing extracellular vesicle cargo loading and functional delivery by engineering protein-lipid interactions. Nat Commun 15:5618

Petrychenko V, Peng B-Z, De AP, Schwarzer AC, Peske F, Rodnina MV, Fischer N (2021) Structural mechanism of GTPase-powered ribosome-tRNA movement. Nat Commun 12:5933

Pettersen EF et al (2004) UCSF Chimera—a visualization system for exploratory research and analysis. J Comput Chem 25:1605–1612

Pflugrath JW (2015) Practical macromolecular cryocrystallography. Acta Crystallogr Sect F Struct Biol Commun 71:622–642

Pike ACW, Garman EF, Krojer T, von Delft F, Carpenter EP (2016) An overview of heavy-atom derivatization of protein crystals. Acta Crystallogr Sect Struct Biol 72:303–318

Pocanschi CL, Popot J-L, Kleinschmidt JH (2013) Folding and stability of outer membrane protein A (OmpA) from *Escherichia coli* in an amphipathic polymer, amphipol A8-35. Eur Biophys J 42:103–118

Polovinkin V, Gushchin I, Sintsov M, Round E, Balandin T, Chervakov P et al (2014) High-resolution structure of a membrane protein transferred from Amphipol to a lipidic mesophase. J Membr Biol 247:997–1004

Popot J-L, Althoff T, Bagnard D, Banères J-L, Bazzacco P, Billon-Denis E et al (2011) Amphipols from A to Z. Annu Rev Biophys 40:379–408

Prester A, Perbandt M, Galchenkova M, Oberthuer D, Werner N, Henkel A et al (2024) Time-resolved crystallography of boric acid binding to the active site serine of the β-lactamase CTX-M-14 and subsequent 1,2-diol esterification. Commun Chem 7:152

Pugh CF, Præstegaard KF, Feilen LP, Živković D, Borthwick NJ, de Lichtenberg C et al (2024) MAASTY: A (dis)ordered copolymer for structural determination of human membrane proteins in native nanodiscs. bioRxiv 2024.08.19.608676. Available from: https://www.biorxiv.org/content/10.1101/2024.08.19.608676v1

Punjani A, Rubinstein JL, Fleet DJ, Brubaker MA (2017) cryoSPARC: algorithms for rapid unsupervised cryo-EM structure determination. Nat Methods 14:290–296

Qi L, Wu L, Zheng S, Wang Y, Fu H, Cui D (2012) Cell-penetrating magnetic nanoparticles for highly efficient delivery and intracellular imaging of siRNA. Biomacromolecules 13:2723–2730

Qiu H, Caffrey M (2000) The phase diagram of the monoolein/water system: metastability and equilibrium aspects. Biomaterials 21:223–234

Raimondi P, Benabderrahmane C, Berkvens P, Biasci JC, Borowiec P, Bouteille J-F et al (2023) The extremely brilliant source storage ring of the European Synchrotron Radiation Facility. Commun Phys 6:82

Ravula T, Hardin NZ, Ramadugu SK, Ramamoorthy A (2017) pH tunable and divalent metal ion tolerant polymer lipid nanodiscs. Langmuir 33:10655–10662

Razinkov I, Dandey V, Wei H, Zhang Z, Melnekoff D, Rice WJ et al (2016) A new method for vitrifying samples for cryoEM. J Struct Biol 195:190–198

Reimund M, Dearborn AD, Graziano G, Lei H, Ciancone AM, Kumar A et al (2025) Structure of apolipoprotein B100 bound to the low-density lipoprotein receptor. Nature 638:829–35

Ren Z, Srajer V, Knapp JE, Royer WE (2012) Cooperative macromolecular device revealed by meta-analysis of static and time-resolved structures. Proc Natl Acad Sci USA 109:107–112

Ren Z, Wang C, Shin H, Bandara S, Kumarapperuma I, Ren MY et al (2020) An automated platform for in situ serial crystallography at room temperature. IUCrJ 7:1009–1018

Renaud J-P, Chung C, Danielson UH, Egner U, Hennig M, Hubbard RE et al (2016) Biophysics in drug discovery: impact, challenges and opportunities. Nat Rev Drug Discov 15:679–698

Renaud J-P, Chari A, Ciferri C, Liu W, Rémigy H-W, Stark H et al (2018) Cryo-EM in drug discovery: achievements, limitations and prospects. Nat Rev Drug Discov 17:471–492

Rice WJ, Cheng A, Noble AJ, Eng ET, Kim LY, Carragher B et al (2018) Routine determination of ice thickness for cryo-EM grids. J Struct Biol 204:38–44

Rima L, Zimmermann M, Fränkl A, Clairfeuille T, Lauer M, Engel A et al (2022) cryoWriter: a blotting free cryo-EM preparation system with a climate jet and cover-slip injector. Faraday Discuss 240:55–66

Ripstein ZA, Rubinstein JL (2016) Processing of Cryo-EM movie data. Methods Enzymol 579:103–124

Robertson MJ, Papasergi-Scott MM, He F, Seven AB, Meyerowitz JG, Panova O et al (2022) NANOBODY Structure determination of inactive-state GPCRs with a universal nanobody. Nat Struct Mol Biol 29:1188–95

Rodriguez JA, Eisenberg DS, Gonen T (2017) Taking the measure of MicroED. Curr Opin Struct Biol 46:79–86

Roessler CG, Agarwal R, Allaire M, Alonso-Mori R, Andi B, Bachega JFR et al (1993) Acoustic injectors for drop-on-demand serial femtosecond crystallography. Struct Lond Engl 2016(24):631–640

Rout MP, Sali A (2019) Principles for integrative structural biology studies. Cell 177:1384–1403

Routledge SJ, Mikaliunaite L, Patel A, Clare M, Cartwright SP, Bawa Z et al (2016) The synthesis of recombinant membrane proteins in yeast for structural studies. Methods 95:26–37

Rubinstein JL, Guo H, Ripstein ZA, Haydaroglu A, Au A, Yip CM et al (2019) Shake-it-off: a simple ultrasonic cryo-EM specimen-preparation device. Acta Crystallogr Sect Struct Biol 75:1063–70

Russo CJ, Passmore LA (2014). Ultrastable gold substrates for electron cryomicroscopy. Science 346:1377–1380

Safari C, Ghosh S, Andersson R, Johannesson J, Båth P, Uwangue O et al (2023) Time-resolved serial crystallography to track the dynamics of carbon monoxide in the active site of cytochrome c oxidase. Sci Adv 9:eadh4179

Salfer M, Collado JF, Baumeister W, Fernández-Busnadiego R, Martínez-Sánchez A (2020) Reliable estimation of membrane curvature for cryo-electron tomography. Saucerman JJ, editor. PLOS Comput Biol 16:e1007962

Sali A (2021) From integrative structural biology to cell biology. J Biol Chem 296:100743

Salvati Manni L, Zabara A, Osornio YM, Schöppe J, Batyuk A, Plückthun A et al (2015) Phase behavior of a designed cyclopropyl analogue of monoolein: implications for low-temperature membrane protein crystallization. Angew Chem Int Ed Engl 54:1027–1031

Sanchez-Weatherby J, Sandy J, Mikolajek H, Lobley CMC, Mazzorana M, Kelly J et al (2019) VMXi: a fully automated, fully remote, high-flux in situ macromolecular crystallography beamline. J Synchrotron Radiat 26:291–301

Santos R, Ursu O, Gaulton A, Bento AP, Donadi RS, Bologa CG et al (2017) A comprehensive map of molecular drug targets. Nat Rev Drug Discov 16:19–34

Saur M, Hartshorn MJ, Dong J, Reeks J, Bunkoczi G, Jhoti H et al (2020) Fragment-based drug discovery using cryo-EM. Drug Discov Today 25:485–490

Sauter NK, Hattne J, Grosse-Kunstleve RW, Echols N (2013) New Python-based methods for data processing. Acta Crystallogr D Biol Crystallogr 69:1274–1282

Schaffer M, Pfeffer S, Mahamid J et al (2019) A cryo-FIB lift-out technique enables molecular-resolution cryo-ET within native Caenorhabditis elegans tissue. Nat Methods 16:757–762

Scheres SHW (2012) RELION: implementation of a Bayesian approach to cryo-EM structure determination. J Struct Biol 180:519–530

Scheres SHW (2016) Processing of structurally heterogeneous Cryo-EM data in RELION. Methods Enzymol 579:125–157

Schiøtz OH, Klumpe S, Plitzko JM, Kaiser CJO (2024) Cryo-electron tomography: *en route* to the molecular anatomy of organisms and tissues. Biochem Soc Trans 52:2415–2425

Schlichting I, Almo SC, Rapp G, Wilson K, Petratos K, Lentfer A et al (1990) Time-resolved X-ray crystallographic study of the conformational change in Ha-Ras p21 protein on GTP hydrolysis. Nature 345: 309–315

Schmidt M, Pahl R, Srajer V, Anderson S, Ren Z, Ihee H et al (2004) Protein kinetics: structures of intermediates and reaction mechanism from time-resolved x-ray data. Proc Natl Acad Sci USA 101:4799–4804

Schneider B, Junge F, Shirokov VA, Durst F, Schwarz D, Dötsch V et al (2010) Membrane protein expression in cell-free systems. Methods Mol Biol Clifton NJ 601:165–186

Schotte F, Cho HS, Kaila VRI, Kamikubo H, Dashdorj N, Henry ER et al (2012) Watching a signaling protein function in real time via 100-ps time-resolved Laue crystallography. Proc Natl Acad Sci USA 109:19256–19261

Schroer CG, Agapov I, Brefeld W, Brinkmann R, Chae Y-C, Chao H-C et al (2018) PETRA IV: the ultralow-emittance source project at DESY. J Synchrotron Radiat 25:1277–1290

Schroer CG, Wille H-C, Seeck OH, Bagschik K, Schulte-Schrepping H, Tischer M et al (2022) The synchrotron radiation source PETRA III and its future ultra-low-emittance upgrade PETRA IV. Eur Phys J Plus 137:1312

Schwalbe H, Audergon P, Haley N, Amaro CA, Agirre J, Baldus M et al (2024) The future of integrated structural biology. Structure 32:1563–1580

Sharma KS, Durand G, Gabel F, Bazzacco P, Le Bon C, Billon-Denis E et al (2012) Non-ionic amphiphilic homopolymers: synthesis, solution properties, and biochemical validation. Langmuir 28:4625–4639

Shi D, Nannenga BL, Iadanza MG, Gonen T (2013) Three-dimensional electron crystallography of protein microcrystals. elife 2:e01345

Shinoda T, Shinya N, Ito K, Ishizuka-Katsura Y, Ohsawa N, Terada T et al (2016) Cell-free methods to produce structurally intact mammalian membrane proteins. Sci Rep 6:30442

Smith AAA, Autzen HE, Laursen T, Wu V, Yen M, Hall A et al (2017) Controlling styrene maleic acid lipid particles through RAFT. Biomacromolecules 18:3706–3713

Smith AAA, Autzen HE, Faust B, Mann JL, Muir BW, Howard S et al (2020) Lipid nanodiscs via ordered copolymers. Chem 6:2782–2795

Sonoda Y, Newstead S, Hu N-J, Alguel Y, Nji E, Beis K et al (1993) Benchmarking membrane protein detergent stability for improving throughput of high-resolution X-ray structures. Struct Lond Engl 2011(19):17–25

Sorzano COS, Jiménez-Moreno A, Maluenda D, Ramírez-Aportela E, Martínez M, Cuervo A et al (2021) Image processing in Cryo-Electron Microscopy of single particles: the power of combining methods. In: Methods Mol Biol Clifton NJ, vol 2305, pp 257–289

Srajer V, Teng T, Ursby T, Pradervand C, Ren Z, Adachi S et al (1996) Photolysis of the carbon monoxide complex of myoglobin: nanosecond time-resolved crystallography. Science 274:1726–1729

Stetsenko A, Guskov A (2017) An overview of the top ten detergents used for membrane protein crystallization. Crystals 7:197

Stieger B, Steiger J, Locher KP (2021) Membrane lipids and transporter function. Biochim Biophys Acta (BBA) – Mol Basis Dis 1867:166079

Straub MS, Harder OF, Mowry NJ, Barrass SV, Hruby J, Drabbels M et al (2025) Laser flash melting cryo-EM samples to overcome preferred orientation. Nat Methods 22:1880–1886

Strub M-P, Hoh F, Sanchez J-F, Strub JM, Böck A, Aumelas A et al (2003) Selenomethionine and Selenocysteine double labeling strategy for crystallographic phasing. Structure 11:1359–1367

Su C-C, Lyu M, Morgan CE, Bolla JR, Robinson CV, Yu EW (2021) A 'build and retrieve' methodology to simultaneously solve cryo-EM structures of membrane proteins. Nat Methods 18:69–75

Subramaniam S, Kleywegt GJ (2022) A paradigm shift in structural biology. Nat Methods 19:20–23

Sugahara M, Mizohata E, Nango E, Suzuki M, Tanaka T, Masuda T et al (2015) Grease matrix as a versatile carrier of proteins for serial crystallography. Nat Methods 12:61–63

Sugahara M, Song C, Suzuki M, Masuda T, Inoue S, Nakane T et al (2016) Oil-free hyaluronic acid matrix for serial femtosecond crystallography. Sci Rep 6:24484

Sugahara M, Nakane T, Masuda T, Suzuki M, Inoue S, Song C et al (2017) Hydroxyethyl cellulose matrix applied to serial crystallography. Sci Rep 7:703

Sun M, Azumaya CM, Tse E, Bulkley DP, Harrington MB, Gilbert G et al (2021) Practical considerations for using K3 cameras in CDS mode for high-resolution and high-throughput single particle cryo-EM. J Struct Biol 213:107745

Takenaka T (2001) Classical vs reverse pharmacology in drug discovery. BJU Int 88(Suppl 2):7–10; discussion 49–50

Tan YZ, Rubinstein JL (2020) Through-grid wicking enables high-speed cryoEM specimen preparation. Acta Crystallogr Sect Struct Biol 76:1092–1103

Tao X, Zhao C, MacKinnon R (2023) Membrane protein isolation and structure determination in cell-derived membrane vesicles. Proc Natl Acad Sci USA 120:e2302325120

Tavares PF, Al-Dmour E, Andersson Å, Cullinan F, Jensen BN, Olsson D et al (2018) Commissioning and first-year operational results of the MAX IV 3 GeV ring. J Synchrotron Radiat 25:1291–1316

Taylor GL (2010) Introduction to phasing. Acta Crystallogr D Biol Crystallogr 66:325–338

Tegunov D, Cramer P (2019) Real-time cryo-electron microscopy data preprocessing with Warp. Nat Methods 16:1146–1152

Tehei M, Perlmutter JD, Giusti F, Sachs JN, Zaccai G, Popot J-L (2014) Thermal fluctuations in Amphipol A8-35 particles: a neutron scattering and molecular dynamics study. J Membr Biol 247:897–908

Thompson TB, Kovall RA (2019) Structural biology: gaining atomic level insight into the biological function of macromolecules. Exp Biol Med 244:1507–1509

Tifrea DF, Sun G, Pal S, Zardeneta G, Cocco MJ, Popot J-L et al (2011) Amphipols stabilize the *Chlamydia* major outer membrane protein and enhance its protective ability as a vaccine. Vaccine 29:4623–4631

Timcenko M, Autzen AAA, Autzen HE (2022) Characterization of divalent cation interactions with AASTY nanodiscs. ACS Appl Polym Mater 4:1071–1083

Tribet C, Audebert R, Popot J-L (1996) Amphipols: polymers that keep membrane proteins soluble in aqueous solutions. Proc Natl Acad Sci 93:15047–15050

Tribet C, Audebert R, Popot J-L (1997) Stabilization of hydrophobic colloidal dispersions in water with amphiphilic polymers: application to integral membrane proteins. Langmuir 13:5570–5576

Tsutsumi N, Mukherjee S, Waghray D, Janda CY, Jude KM, Miao Y et al (2020) BRIL Structure of human Frizzled5 by fiducial-assisted cryo-EM supports a heterodimeric mechanism of canonical Wnt signaling. eLife 9:e58464

Uchański T, Masiulis S, Fischer B, Kalichuk V, López-Sánchez U, Zarkadas E et al (2021) Megabodies expand the nanobody toolkit for protein structure determination by single-particle cryo-EM. Nat Methods 18: 60–8

Vahedi-Faridi A, Lovelace J, Bellamy HD, Snell EH, Borgstahl GEO (2003) Physical and structural studies on the cryocooling of insulin crystals. Acta Crystallogr D Biol Crystallogr 59:2169–2182

Van 'T Hag L, Darmanin C, Le TC, Mudie S, Conn CE, Drummond CJ (2014) In meso crystallization: compatibility of different lipid bicontinuous cubic mesophases with the cubic crystallization screen in aqueous solution. Cryst Growth Des 14:1771–1781

Van Dalsen L, Smithers L, Boland C, Weichert D, Caffrey M (2021) 9.8 MAG: a new host lipid for *in meso* (lipid cubic phase) crystallization of integral membrane proteins. Cryst Growth Des 21:490–500

van de Put MWP, Patterson JP, Bomans PHH, Wilson NR, Friedrich H, van Benthem RATM et al (2015) Graphene oxide single sheets as substrates for high resolution cryoTEM. Soft Matter 11:1265–1270

Van Drie JH, Tong L (2020) Cryo-EM as a powerful tool for drug discovery. Bioorg Med Chem Lett 30:127524

Venugopal H, Mobbs J, Taveneau C, Fox DR, Vuckovic Z, Gulati S et al (2025) High-resolution cryo-EM using a common LaB6 120-keV electron microscope equipped with a sub–200-keV direct electron detector. Sci Adv 11:eadr0438

Vial F, Oukhaled AG, Auvray L, Tribet C (2006) Long-living channels of well defined radius opened in lipid bilayers by polydisperse, hydrophobically-modified polyacrylic acids. Soft Matter 3:75–78

Voskoboynikova N, Margheritis EG, Kodde F, Rademacher M, Schowe M, Budke-Gieseking A et al (2021) Evaluation of DIBMA nanoparticles of variable size and anionic lipid content as tools for the structural and functional study of membrane proteins. Biochimica et Biophysica Acta (BBA) 1863:183588

Voss NR, Yoshioka CK, Radermacher M, Potter CS, Carragher B (2009) DoG picker and TiltPicker: software tools to facilitate particle selection in single particle electron microscopy. J Struct Biol 166:205–213

Voss JM, Harder OF, Olshin PK, Drabbels M, Lorenz UJ (2021) Rapid melting and revitrification as an approach to microsecond time-resolved cryo-electron microscopy. Chem Phys Lett 778:138812

Wagner T, Merino F, Stabrin M, Moriya T, Antoni C, Apelbaum A et al (2019) SPHIRE-crYOLO is a fast and accurate fully automated particle picker for cryo-EM. Commun Biol 2:218

Walker G, Brown C, Ge X, Kumar S, Muzumdar MD, Gupta K et al (2024) Oligomeric organization of membrane proteins from native membranes at nanoscale spatial and single-molecule resolution. Nat Nanotechnol 19:85–94

Wang F, Gong H, Liu G, Li M, Yan C, Xia T et al (2016) DeepPicker: a deep learning approach for fully automated particle picking in cryo-EM. J Struct Biol 195:325–336

Wang F, Liu Y, Yu Z, Li S, Feng S, Cheng Y et al (2020a) General and robust covalently linked graphene oxide affinity grids for high-resolution cryo-EM. Proc Natl Acad Sci 117:24269–24273

Wang F, Yu Z, Betegon M, Campbell MG, Aksel T, Zhao J et al (2020b) Amino and PEG-amino graphene oxide grids enrich and protect samples for high-resolution single particle cryo-electron microscopy. J Struct Biol 209:107437

Watkinson TG, Calabrese AN, Giusti F, Zoonens M, Radford SE, Ashcroft AE (2015) Systematic analysis of the use of amphipathic polymers for studies of outer membrane proteins using mass spectrometry. Int J Mass Spectrom 391:54–61

Wagner FR, Watanabe R, Schampers R, Singh D, Persoon H, Schaffer M et al (2020) Preparing samples from whole cells using focused-ion-beam milling for cryo-electron tomography. Nat Protoc 15:2041–2070

Wei H, Dandey VP, Zhang Z, Raczkowski A, Rice WJ, Carragher B et al (2018) Optimizing "self-wicking" nanowire grids. J Struct Biol 202:170–174

Weierstall U, James D, Wang C, White TA, Wang D, Liu W et al (2014) Lipidic cubic phase injector facilitates membrane protein serial femtosecond crystallography. Nat Commun 5:3309

Weinert T, Olieric V, Waltersperger S, Panepucci E, Chen L, Zhang H et al (2015) Fast native-SAD phasing for routine macromolecular structure determination. Nat Methods 12:131–133

Weinert T, Skopintsev P, James D, Dworkowski F, Panepucci E, Kekilli D et al (2019) Proton uptake mechanism in bacteriorhodopsin captured by serial synchrotron crystallography. Science 364:61–65

White TA, Mariani V, Brehm W, Yefanov O, Barty A, Beyerlein KR et al (2016) Recent developments in CrystFEL. J Appl Crystallogr 49:680–689

Williamson MP, Havel TF, Wüthrich K (1985) Solution conformation of proteinase inhibitor IIA from bull seminal plasma by 1H nuclear magnetic resonance and distance geometry. J Mol Biol 182:295–315

Winkler MBL, Kidmose RT, Szomek M, Thaysen K, Rawson S, Muench SP et al (2019) Structural insight into eukaryotic sterol transport through Niemann-pick type C proteins. Cell 179:485–497.e18

Wisedchaisri G, Reichow SL, Gonen T (1993) Advances in structural and functional analysis of membrane proteins by electron crystallography. Struct Lond Engl 2011(19):1381–1393

Wolff AM, Nango E, Young ID, Brewster AS, Kubo M, Nomura T et al (2023) Mapping protein dynamics at high spatial resolution with temperature-jump X-ray crystallography. Nat Chem 15:1549–1558

Wranik M, Kepa MW, Beale EV, James D, Bertrand Q, Weinert T et al (2023) A multi-reservoir extruder for time-resolved serial protein crystallography and compound screening at X-ray free-electron lasers. Nat Commun 14:7956

Wu X, Rapoport TA (2021) LEGOBODY Cryo-EM structure determination of small proteins by nanobody-binding scaffolds (Legobodies). Proc Natl Acad Sci 118:e2115001118

Wu S, Armache J-P, Cheng Y (2016) Single-particle cryo-EM data acquisition by using direct electron detection camera. Microscopy 65:35–41

Wu K, Wu D, Zhu L, Wu Y (2021) Application of monolayer Graphene and its derivative in Cryo-EM sample preparation. Int J Mol Sci 22:8940

Xiao Y, Yang G (2017) A fast method for particle picking in cryo-electron micrographs based on fast R-CNN. Rome p 020080. Available from: https://pubs.aip.org/aip/acp/article/586344

Xie P, Li Y, Lamon G, Kuang H, Wang DN, Traaseth NJ (2025) BRIL A fiducial-assisted strategy compatible with resolving small MFS transporter structures in multiple conformations using cryo-EM. Nat Commun 16:7

Xu Y, Qin Y, Wang L, Zhang Y, Wang Y, Dang S (2024) Metallo-supramolecular branched polymer protects particles from air-water interface in single-particle cryo-electron microscopy. Commun Biol 7: 1–10

Yadav S, Vinothkumar KR (2024) Factors affecting macromolecule orientations in thin films formed in cryo-EM. Acta Crystallogr Sect Struct Biol 80:535–550

Yamashita K, Pan D, Okuda T, Sugahara M, Kodan A, Yamaguchi T et al (2015) An isomorphous replacement method for efficient de novo phasing for serial femtosecond crystallography. Sci Rep 5:14017

Yang Z, Zhang Z (2018) Engineering strategies for enhanced production of protein and bio-products in *Pichia pastoris*: a review. Biotechnol Adv 36:182–195

Yang W, Deng H, Zhu S, Lau J, Tian R, Wang S et al (2020) Size-transformable antigen-presenting cell-mimicking nanovesicles potentiate effective cancer immunotherapy. Sci Adv 6:eabd1631

Yang Z, Fan J, Wang J, Fan X, Ouyang Z, Wang H-W et al (2024) Electrospray-assisted cryo-EM sample preparation to mitigate interfacial effects. Nat Methods 21:1023–1032

Yasuhara K, Arakida J, Ravula T, Ramadugu SK, Sahoo B, Kikuchi J et al (2017) Spontaneous lipid nanodisc fomation by amphiphilic polymethacrylate copolymers. J Am Chem Soc 139:18657–18663

Yip KM, Fischer N, Paknia E, Chari A, Stark H (2020) Atomic-resolution protein structure determination by cryo-EM. Nature 587:157–161

Yu G, Vago F, Zhang D, Snyder JE, Yan R, Zhang C et al (2014) Single-step antibody-based affinity cryo-electron microscopy for imaging and structural analysis of macromolecular assemblies. J Struct Biol 187:1–9

Yu G, Li K, Huang P, Jiang X, Jiang W (2016). Antibody-based affinity Cryoelectron Microscopy at 2.6-Å resolution. Structure 24:1984–1990

Zabara A, Meikle TG, Trenker R, Yao S, Newman J, Peat TS et al (2017) Lipidic cubic phase-induced membrane protein crystallization: interplay between lipid molecular structure, mesophase structure and properties, and crystallogenesis. Cryst Growth Des 17:5667–5674

Zakeri B, Fierer JO, Celik E, Chittock EC, Schwarz-Linek U, Moy VT et al (2012) Peptide tag forming a rapid covalent bond to a protein, through engineering a bacterial adhesin. Proc Natl Acad Sci 109:E690–E697

Zander U, Hoffmann G, Cornaciu I, Marquette J-P, Papp G, Landret C et al (2016) Automated harvesting and processing of protein crystals through laser photoablation. Acta Crystallogr Sect Struct Biol 72:454–466

Zeev-Ben-Mordehai T, Vasishtan D, Siebert CA, Whittle C, Grünewald K (1993) Extracellular vesicles: a platform for the structure determination of membrane proteins by Cryo-EM. Struct England 2014(22):1687–1692

Zhang Z, Chapman ER (2020) Programmable nanodisc patterning by DNA origami. Nano Lett 20:6032–6037

Zhang H, Chiao M (2015) Anti-fouling coatings of poly(dimethylsiloxane) devices for biological and biomedical applications. J Med Biol Eng 35:143–155

Zhang H, Unal H, Gati C, Han GW, Liu W, Zatsepin NA et al (2015) Structure of the angiotensin receptor revealed by serial femtosecond crystallography. Cell 161:833–844

Zhang H, Han GW, Batyuk A, Ishchenko A, White KL, Patel N et al (2017a) Structural basis for selectivity and diversity in angiotensin II receptors. Nature 544:327–332

Zhang H, Qiao A, Yang D, Yang L, Dai A, de Graaf C et al (2017b) Structure of the full-length glucagon class B G-protein-coupled receptor. Nature 546:259–264

Zhang Y, Sun B, Feng D, Hu H, Chu M, Qu Q et al (2017c) Cryo-EM structure of the activated GLP-1 receptor in complex with a G protein. Nature 546:248–253

Zhang J, Wang Z, Chen Y, Han R, Liu Z, Sun F et al (2019) PIXER: an automated particle-selection method based on segmentation using a deep neural network. BMC Bioinformatics 20:41

Zhang K, Pintilie GD, Li S, Schmid MF, Chiu W (2020) Resolving individual atoms of protein complex by cryo-electron microscopy. Cell Res 30:1136–1139

Zhang M, Gui M, Wang Z-F, Gorgulla C, Yu JJ, Wu H et al (2021a) Cryo-EM structure of an activated GPCR–G protein complex in lipid nanodiscs. Nat Struct Mol Biol 28:258–267

Zhang Y, Daday C, Gu R-X, Cox CD, Martinac B, de Groot BL et al (2021b) Visualization of the mechanosensitive ion channel MscS under membrane tension. Nature 590:509–14

Zhang S, Ren Q, Novick SJ, Strutzenberg TS, Griffin PR, Bao H (2021c) One-step construction of circularized nanodiscs using SpyCatcher-SpyTag. Nat Commun 12:5451

Zhang C, Li H, Wan X, Chen X, Yang Z, Feng J et al (2021d) TransPicker: a transformer-based framework for particle picking in cryoEM micrographs. In: 2021 IEEE Int Conf Bioinforma biomed BIBM. IEEE, Houston pp 1179–1184. Available from: https://ieeexplore.ieee.org/document/9669524/

Zhang J, Zhang D, Sun L, Ji G, Huang X, Niu T et al (2021e) VHUT-cryo-FIB, a method to fabricate frozen hydrated lamellae from tissue specimens for in situ cryo-electron tomography. J Struct Biol 213:107763

Zhang X, Johnson RM, Drulyte I, Yu L, Kotecha A, Danev R et al (2021f) Evolving cryo-EM structural approaches for GPCR drug discovery. Structure 29:963–974.e6

Zhang K, Wu H, Hoppe N, Manglik A, Cheng Y (2022a) BRIL fusion protein strategies for cryo-EM study of G protein-coupled receptors. Nat Commun 13:4366

Zhang S, Zou S, Yin D, Zhao L, Finley D, Wu Z et al (2022b) USP14-regulated allostery of the human proteasome by time-resolved cryo-EM. Nature 605:567–574

Zhao Z, Zhang M, Hogle JM, Shih WM, Wagner G, Nasr ML (2018) DNA-corralled nanodiscs for the structural and functional characterization of membrane proteins and viral entry. J Am Chem Soc 140:10639–10643

Zhao F-Z, Zhang B, Yan E-K, Sun B, Wang Z-J, He J-H et al (2019) A guide to sample delivery systems for serial crystallography. FEBS J 286:4402–4417

Zhao Q, Hong X, Wang Y, Zhang S, Ding Z, Meng X et al (2024) An immobilized antibody-based affinity grid strategy for on-grid purification of target proteins enables high-resolution cryo-EM. Commun Biol. 7:715

Zheng L, Xu J, Wang W, Gao X, Zhao C, Guo W et al (2024) Self-assembled superstructure alleviates air-water interface effect in cryo-EM. Nat Commun 15:7300

Zhong ED, Bepler T, Berger B, Davis JH (2021) CryoDRGN: reconstruction of heterogeneous cryo-EM structures using neural networks. Nat Methods 18:176–185

Zhou X, Zheng W, Li Y, Pearce R, Zhang C, Bell EW et al (2022a) I-TASSER-MTD: a deep-learning-based platform for multi-domain protein structure and function prediction. Nat Protoc 17:2326–2353

Zhou X, Li Y, Zhang C, Zheng W, Zhang G, Zhang Y (2022b) Progressive assembly of multi-domain protein structures from cryo-EM density maps. Nat Comput Sci 2:265–275

Zhou Y, Dong J, Wang Q (2023) Fabricating higher-order functional DNA origami structures to reveal biological processes at multiple scales. NPG Asia Mater 15:1–24

Zhu Y, Ouyang Q, Mao Y (2017) A deep convolutional neural network approach to single-particle recognition in cryo-electron microscopy. BMC Bioinformatics 18:348

Zhu L, Bu G, Jing L, Shi D, Lee M-Y, Gonen T et al (2020) Structure determination from lipidic cubic phase embedded microcrystals by MicroED. Structure 28:1149–1159.e4

Ziegler SJ, Mallinson SJB, St John PC, Bomble YJ (2021) Advances in integrative structural biology: towards understanding protein complexes in their cellular context. Comput Struct Biotechnol J 19:214–225

Zimmermann I, Egloff P, Hutter CA, Arnold FM, Stohler P, Bocquet N et al (2018) Synthetic single domain antibodies for the conformational trapping of membrane proteins. eLife 7:e34317

Zoonens M, Popot J-L (2014) Amphipols for each season. J Membr Biol 247:759–796

Exploring Biophysics at the Membrane with Single-Molecule TIRF Microscopy

5

R. Charles Kissell, James T. Pentikis, Austin M. Baggetta, and Leigh D. Plant

Abstract

Cellular functions result from the action of heteromeric protein machines that typically exhibit greater utility and dynamism than the sum of their parts. Therefore, the structural composition of these multimeric machines must be determined to delineate their functional attributes, including how information is transduced between the extracellular and intracellular environments, and between cells. While evaluating the activity of a protein complex at the population level provides valuable insights, measuring the activity of individual molecules offers a powerful approach for uncovering functional dynamics and reaction kinetics that are often obscured in ensemble measurements. Here we describe the background and application of *T*otal *I*nternal *R*eflection *F*luorescence, or TIRF microscopy. We introduce the photo-physics that underpin TIRF and describe its utility in the study of membrane protein biophysics. We focus on applications of TIRF designed to determine the stoichiometry and real-time movement of protein complexes, especially ion channels and signaling receptors, in biological membranes.

Keywords

TIRF microscopy · Ion channel · Fluorescence photobleaching · Particle tracking · Stoichiometry · Membrane protein

R. C. Kissell · J. T. Pentikis · A. M. Baggetta · L. D. Plant (✉)
Department of Pharmaceutical Sciences, The Center for Drug Discovery, School of Pharmacy & Pharmaceutical Sciences, Bouvé College of Health Sciences, Northeastern University, Boston, MA, USA
e-mail: l.plant@northeastern.edu

5.1 The Theoretical Basis of TIRF

5.1.1 The Law of Refraction

The *law of refraction* was first described by Persian scientist Ibn Sahl in 984 CE (Rashed 1990). It states that when a propagating light wave encounters an interface between two translucent media it will change velocity. The extent to which the wave velocity changes depends on the difference in the refractive index (n) of the two translucent media. The result is a change in the angle of travel of the light wave called refraction, or reflection, if the light wave bounces off of the interface and continues to propagate through the first medium (Mondal and Diaspro 2014). As light enters a medium with a higher refractive index, it will slow down, and its wavelength becomes shorter.

A. Rosenhouse-Dantsker (ed.), *Decoding Ion Channels Structure and Function*, Advances in Experimental Medicine and Biology 1497, https://doi.org/10.1007/978-3-032-07523-9_5

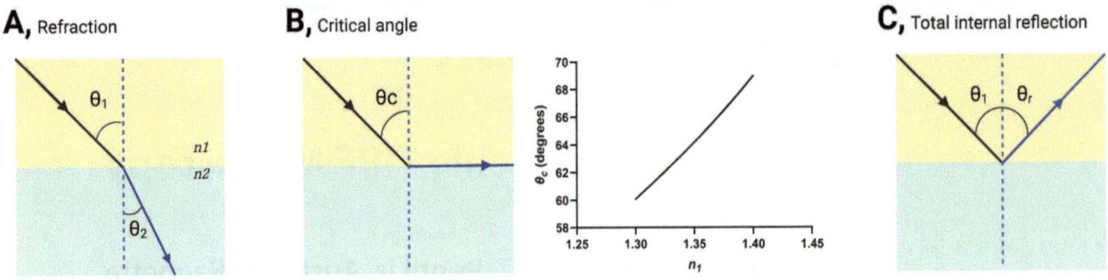

Fig. 5.1 Visualization of basic principles of light wave photophysics. An incident beam travels through a translucent medium (yellow) and at an angle (ϑ) to the normal (dashed line) and arrives at an interface with a second translucent medium (green). (**a**) Refraction occurs when light passes through different media, causing the path of the light to change at an angle that is related to the refractive indices (*n*) of the media. (**b**) The critical angle (ϑc), relative to the normal, allows the incident beam to propagate along the interface between the two media. The value of ϑc depends on the refractive index of *n1*. (**c**) When the incidence angle ϑ1 exceeds ϑc, the incident beam is totally internally reflected from the interface between the media and propagates in medium at an angle reflection, ϑr, relative to the normal

When light enters a lower refractive index material, the speed of light increases, and the wavelength becomes longer, Fig. 5.1.

The law of refraction was defined mathematically by Dutch mathematician Willebrord Snellius in 1621, who derived Snell's Law (Eq. 5.1) (Hentschel and Hentschel 2001):

$$n_1 \sin\theta_1 = n_2 \sin\theta_2 \tag{5.1}$$

$$\theta_1 = \theta_r \tag{5.2}$$

$$\theta_C = \sin^{-1}\frac{n_2}{n_1} \tag{5.3}$$

where n_1 and n_2 are the indices of refraction and θ_1 and θ_2 are the angles of incidence of two media and θ_r is the angle of reflection, respectively measured against the normal (see Fig. 5.1).

Reflection occurs when light bounces off an interface between two media, and the angle of incidence is equal to the angle of reflection (Eq. 5.2). Total internal reflection occurs when light attempts to travel from a medium with a higher refractive index (n_1) to one with a lower refractive index ($n_1 > n_2$) (Lakowicz 2006). The transition from refraction to reflection can be observed in an experimental device where the incidence angle of a propagating light wave can be adjusted with respect to an interface between two translucent media. In this transition, the angle of incidence will reach a critical angle (θ_C), where the angle of refraction allows the light to propagate along the interface between the two media ($\theta_2 = 90°$). When the angle of incidence θ_1 surpasses the critical angle θ_C, refraction is no longer possible, and all the light is reflected into the original medium. This phenomenon is called total internal reflection, and it can be observed in systems where the angle of incidence is adjustable, such as in optical fiber setups or experimental prisms that form the basis of most TIRF microscopes.

5.1.2 The Evanescent Field

Under the conditions of total internal reflection, the incident light wave does not propagate into the second medium, but energy is still transmitted across the interface via an evanescent field (also called an evanescent wave). An evanescent field has the same frequency as the incident light and can carry sufficient energy to excite fluorophores near the interface. However, unlike a propagating incident wave, the intensity of the evanescent field decays exponentially with distance perpendicular to the interface (Matsuoka et al. 2012). The evanescent field intensity *I(z)* decays according to Eq. 5.4:

$$I(z) = I_0 \exp\left(-\frac{z}{\delta}\right) \tag{5.4}$$

$$\delta = \frac{\lambda_0}{4\pi} \times \left(n_1^2 \sin^2\theta_1 - n_2^2\right)^{-1/2} \tag{5.5}$$

where I_0 is the intensity of light at the interface ($z = 0$), z is the perpendicular distance from the interface, and δ is the penetration depth at which the field intensity drops to I_0/e (37% of I_0). The penetration depth is defined by Eq. (5.5) (Lakowicz 2006).

In a typical application of TIRF microscopy, the incident light is introduced through a high-index medium such as a glass coverslip ($n_1 = 1.5$). For simplification, we will approximate the cytoplasm of cells on top of the coverslip as having comparable optical properties to water, and therefore the experimental buffer solution ($n_2 = 1.33$). As described by Eq. 5.5, if 470 nm wavelength incident light strikes the glass/water interface at an incidence angle of 70°, the penetration depth of the evanescent field is 57.4 nm. For reference, the average thickness of a mammalian cell membrane is approximately 5 nm (Angle et al. 2014), while the diameter of the whole cell is 15,000 nm (Seidel et al. 2023). In contrast, the axial resolution of most applications of confocal microscope is 0.5–1.5 μm, depending on the format of the objective lens used. In a TIRF microscope, the decrease in intensity of the evanescent field with distance from the interface is visualized in Fig. 5.2. The evanescent wave intensity $I(z)$ decays exponentially as axial distance from the interface increases. The "z" length represented in the x axis is the perpendicular distance from the surface of the cover slip. Fluorophores located within the evanescent field are selectively excited and emit light that can be detected and measured during an experiment. Crucially, the limited penetration depth of the evanescent wave minimizes background signals from fluorophores outside this region, including those deeper within the cytoplasm of the cell. In addition to delimiting the location of the fluorophores under investigation to the plasma membrane of the cell, total internal reflection fluorescence improves the signal-to-background noise ratio of the data, facilitating single molecule level analysis.

In summary, total internal reflection occurs when light passes from a medium with a higher refractive index to a lower one ($n1 > n2$) at an angle greater than the critical angle (as defined by Snell's Law). Although the light does not propagate into the second medium, the evanescent field extends a short distance into it. In TIRF microscopy, this evanescent field excites fluorophores in a thin region near the cell–glass interface, yielding high-contrast, low-background fluorescence signals suitable for detailed studies of membrane-localized processes.

5.2 TIRF and Single Molecule Stoichiometry Studies

5.2.1 Overview

The cell membrane hosts a wide array of proteins whose interactions underlie foundational cellular processes, including the transduction of information between the extracellular and intracellular environments. Determining the organization and dynamics of these protein machines is critical to understanding their role in cellular physiology, pathophysiology, and pharmacology (Leake et al. 2006; Jain et al. 2011). Furthermore, it is desirable to study these interactions on a single-molecule basis, as the specific stoichiometries of protein complexes in living cells can be challenging to determine directly using macroscopic, or ensemble methods (Jain et al. 2012). To this end, TIRF microscopy provides a means of observing and studying fluorescent-tagged single proteins in living cells in a high signal-to-background modality (Mashanov et al. 2004; Monici 2005).

An early, prominent application of TIRF microscopy in single-molecule studies was the determination of protein subunit stoichiometry within a functioning protein complex, as demonstrated in a landmark study of the bacterial flagellar motor (Leake et al. 2006). MotB, a component of the *E. Coli* flagellar motor, was labeled with green fluorescent protein (GFP), and the number of MotB molecules in a single motor was calculated from measurements of the GFP fluorescence intensity relative to the initial fluorescence. The method of subunit counting in membrane proteins was further developed by Ulbrich and Isacoff, who described a technique for counting the number of photobleaching steps

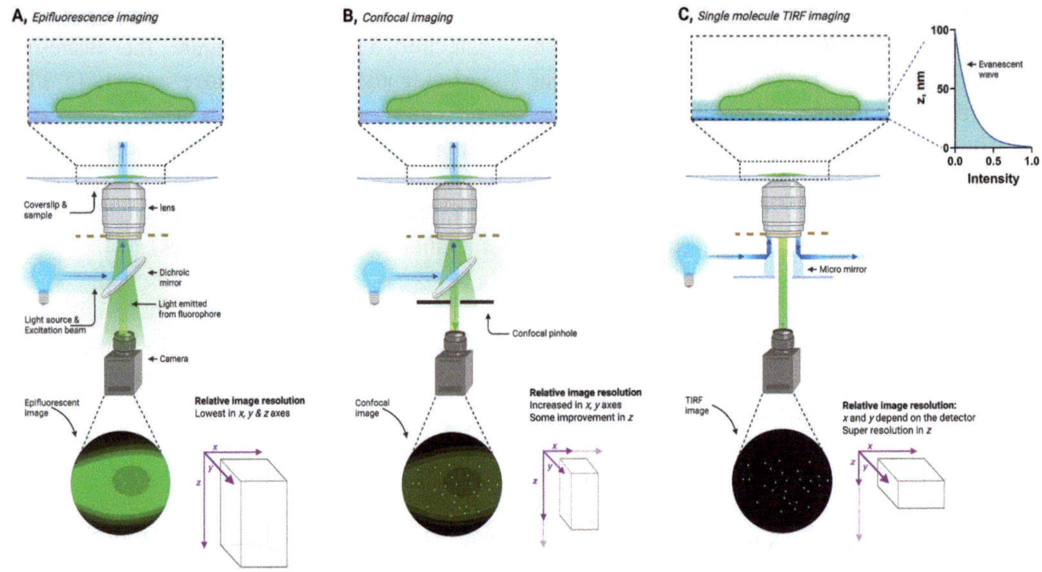

Fig. 5.2 The principles of epifluorescence, confocal, and single-molecule TIRF imaging. Cells expressing a membrane protein of interest tagged with a fluorescent protein such as eGFP can be imaged using several modes of optical microscopy. (**a**) Schematic representation of the principles of epifluorescent microscopy. The excitation beam (here, blue light) is focused on a field of cells using the objective lens. The excitation beam penetrates the cells and excites the fluorophore through the depth of the sample. Emitted light (green) that is in focus, is collected by the lens and directed to a camera, or the eyepieces of a microscope. The excitation beam is reflected towards the sample by a dichroic mirror that lets the emission light pass. The relative resolution of the fluorescent image is low in the *x*, *y*, and *z* axes. (**b**) Schematic representation of simple confocal microscope. The excitation of the sample fluorophores is like the epifluorescence experiment described in (**a**). A pinhole aperture in the emitted light path prevents out of focus light from reaching the detector, increasing the resolution of the fluorescent image in the *x* and *y* axes. Improvements in the resolution of the *z* axis depend on the configuration of the system. (**c**) Schematic of a micro mirror TIRF microscopy. The excitation (laser) beam is reflected into the back focal plane of the objective lens at an angle that allow for total internal reflection when the beam meets the aqueous interface of the sample. An evanescent wave with the same wavelength as the excitation beam penetrates less than 100 nm into the sample, exciting fluors at the cell membrane. The high signal-to-noise emission beam is focused on a detector allowing the study of single fluorescent particles at the cell surface. The exiting excitation beam is directed away from the sample by a second micro mirror. The relative resolution of the fluorescent image in the *x* and *y* axes is determined by the detector while super-resolution is achieved in the *z* axis

of GFP fused to a membrane protein of interest (Ulbrich and Isacoff 2007). Photobleaching occurs when a fluorophore ceases to fluoresce upon exposure to the excitation light. Thus, the number of photobleaching events, observed as stepwise decreases in the fluorescence intensity, may be used to deduce the number of fluorophores present in a single protein complex.

In Ulbrich and Isacoff's study of canonically tetrameric cyclic nucleotide-gated (CNG) channels, tagged with GFP and expressed in *Xenopus laevis* oocytes, it was observed that there were as many samples with four bleaching steps as with three steps. Further, some samples only showed one or two steps. Overall, such data form a binomial distribution of bleaching steps. A binomial distribution results from inherent properties of the system, principally the single molecule behavior of fluorescent proteins. For example, it was determined that approximately 10% of protein tetramers would appear as trimers due to the temporal proximity of consecutive bleaching steps. In other words, two bleaching steps very close together or even coincident in time appear as one step, depending on the resolution of the recording apparatus. Despite correcting for this contingency, a significant number of fluorescence spots still bleached in less than four discrete steps.

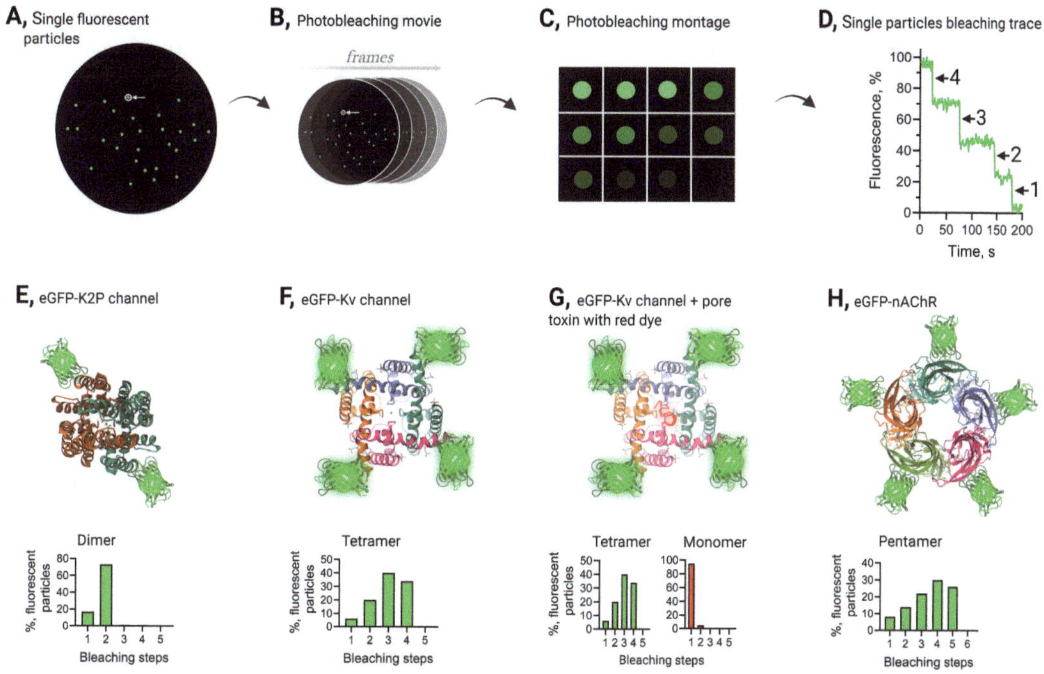

Fig. 5.3 The principles of single particle photobleaching to determine the stoichiometry of membrane proteins. (**a**) A representation of idealized single fluorescent particle TIRF microscopy data showing how single fluorescent particles (for example clusters of subunits tagged with eGFP) imaged at the surface of a live cell. An example of a single particle is highlighted with a white circle. (**b**) The fluorescence is photobleached by continual illumination and the data are recorded as a 'movie' of stacked images. (**c**) An idealized photomontage showing the stepwise in fluorescence intensity of the example particle. (**d**) A plot of the relative fluorescence intensity of an example particle over time. The signal decreases in 4-equal steps, indicating 4 eGFPs, i.e., that the single particle comprises 4-subunits. (**e**), Idealized histogram for eGFP-tagged K2P subunits show 90% of the particles photobleaching in 2-steps, consistent with a canonical dimeric channel. (**f**) Idealized histogram for eGFP-tagged Kv subunits show a maximum of 4-steps in a bimodal distribution, consistent with a canonical tetrameric channel. (**g**) The experiment in (**f**) is repeated with a pore-blocking toxin labeled with a red dye. The 2-color, simultaneous photobleaching data show one toxin (single red step) per tetrameric channel. (**h**) eGFP-labeled nicotinic acetylcholine receptor photobleaches in 5-steps, consistent with a canonical pentamer. The histogram shows a bimodal distribution and a maximum of five bleaching steps.

After fitting the number of bleaching steps to a calculated distribution of tetrameric channels containing 1, 2, 3, or 4 functional GFP tags, it was reasoned that approximately 20% of the GFPs are not in a fluorescent state, causing some proportion of GFP-tagged tetramers to give fewer than four bleaching steps. The observed distribution of bleaching steps matches the predicted distribution if 20% of GFPs are not fluorescing. For complexes composed of more than five subunits, these parameters propagate, requiring increasingly large datasets to allow for an accurate reflection of the expected offsets. In high order structures, such as the bacterial flagellar motor, the subunit number can be estimated from the fluorescence intensity of a single bleaching step relative to initial fluorescence. Fluorescent tags besides GFP and its variants are also used for stoichiometry studies and have offsets that depend on the properties of that protein. These parameters must be considered to accurately assess TIRF microscopy data for subunit counting of protein complexes. Figure 5.3 illustrates these principles, using idealized datasets for membrane protein complexes with distinct subunit stoichiometries.

5.2.2 The Behavior of Single Fluorescent Proteins

The probability that any given fluorescent protein (FP) is not observed is less than one. There are several reasons for this, including: (1) an FP might be pre-bleached, (2) an FP might not be folded, (3) an FP could exist in a dark state (more on *fluorescence blinking* in the next section), and (4) the quantum yield (see below) of an FP is always <1. We can define this probability as $1—\pi$, where π is the probability of observing each possible bleaching event successfully (Hines 2013). The value of π is a variable that depends on experimental conditions and varies between research settings. These parameters complicate single fluorophore analysis because different π values will lead to different distributions of observed bleaching events. Together, these factors present a challenge in determining which distribution provides better evidence of a conclusion regarding subunit stoichiometry. For example, if a dataset is too small and not adequately powered for analysis, the true number of bleaching events (n), might be larger than the highest number of observed bleaching events. This could be the result if the true tails of the distribution are not observed due to the finite number of observations (i.e., only $n—1$ or fewer bleaching events are observed). Alternatively, artifacts from the data collection algorithm could result in an overestimation of n. Indeed, inferring the observed bleaching steps is a challenge because there is no unique combination of π and n that could have produced a specific dataset. To address these concerns, Bayesian inference is employed to accurately estimate parameter confidence (Hines 2013; Plant et al. 2014).

5.3 Fluorescent Protein Photophysics

The phenomenon of fluorescence is central to most microscopy and spectroscopy applications, including TIRF. Fluorescence is a category of luminescence, that is, the emission of light from a substance that occurs from changes in electronically excited states (Lakowicz 2006). Fluorescence occurs when an electron in an excited singlet state returns to the ground state of its paired electron. The transition from the excited state can occur when the paired electrons have opposite spin and is accompanied by the release of a photon. This transition is a very rapid process, typically 10 ns or less.

The first molecule observed to possess fluorescent properties was quinine. Interest in the use of quinine as an antimalarial drug led to the development of the first spectrofluorometer in the 1950s by the National Institutes of Health in the United States (Udenfriend 1995). Since that time, the ever-growing toolbox of experimental techniques that are founded on fluorescence measurements led to power the identification of an increasingly large number of fluorescent molecules, many of which have applications in TIRF microscopy.

Alexander Jablonski is recognized as the father of fluorescence spectroscopy for his many contributions toward our understanding and utilization of fluorescent molecules. One of these is the Jablonski diagram, which illustrates the electronic processes that occur between the absorption and emission of light (Fig. 5.4) (Jabłoński 1935). The singlet electronic states are represented by S_0, S_1, and S_2, which indicate the ground, first, and second states, respectively. Fluorophores can exist in several discrete vibrational energy levels (0, 1, 2, etc.) at each electronic state. A fluorophore at room temperature is typically in the S_0 state. Following absorption of light, a fluorophore is excited to a higher vibrational level of either the S_1 or S_2 state. Transitions between states last approximately 10^{-15} s. This is usually followed by a slower but still nearly instantaneous 10^{-12} s transition to the lowest vibrational level of S_1. By comparison, fluorescence, as mentioned earlier, typically lasts 10^{-8} s. Because these transitions occur on time scale orders of magnitude faster than fluorescence emission, the emission generally results from the lowest vibrational level of S_1. The fluorophore typically returns to a higher vibrational ground state level before reaching thermal equilibrium at the lowest vibrational level of the ground state. An example Jablonski diagram is given in Fig. 5.4.

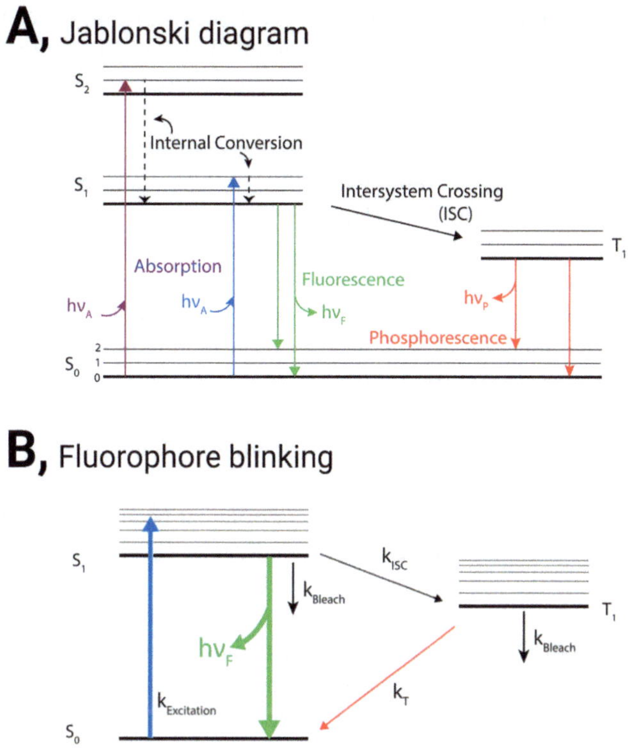

Fig. 5.4 Jablonski diagrams of electronic transition states and fluoropore blinking (**a**) Example Jablonski Diagram demonstrates the excitation of a fluorophore by absorption of a photon (*hv*), resulting in multiple possible excited states S_1 and S_2. Following excitation, the fluorophore will return to the ground state, S_0 which is accompanied by the release of a photon (hv_F), corresponding to fluorescence. (**b**) Differences in the relative rates of excitation $k_{Excitation}$, fluorescence bleaching k_{Bleach}, intersystem crossing k_{ISC}, and transition from triplet state to ground state k_T are proposed to cause some fluorophores to "blink"; a phenomenon that manifests as temporary cessation of fluorescence

Fluorescence blinking is believed to occur primarily when the fluorescent molecule undergoes intersystem crossing (ISC) to the triplet state T_1 (Lakowicz 2006). This process results in a lack of fluorescence until the molecule returns to the ground singlet state. This blinking activity is primarily determined by the rate constants for intersystem crossing (k_{ISC}) and for return to the ground state S_0 from the triplet state (k_T). The similar yet distinct process of photobleaching (k_{Bleach}) is, by contrast, irreversible and occurs due to destruction of the fluorophore. It is proposed that photobleaching occurs from either the S_1 or T_1 state to the ground state (Fig. 5.4).

5.4 Fluorescent Proteins and Dyes

Imaging cells with TIRF microscopy, or any fluorescence microscopy technique, requires selection of fluorophores appropriate for the system being used and the goals of the experiment being performed. Typically, experimenters will either employ a fluorescent protein or an organic dye as the fluorophore of choice. Outlined below is a synopsis of key differences between these classes of fluorophores, including rationale for their inclusion in relevant experimental contexts.

Green fluorescent protein (GFP), derived originally from the jellyfish species *Aequorea vic-*

toria, has seen a great number of modifications and improvements, with enhanced GFP, or eGFP, being the most utilized fluorescent protein for a range of applications. Various mutational and directed evolutionary strategies have also diversified GFP to produce variants that fluoresce with distinct wavelengths (colors) (Tsien 1998; Heim et al. 1994; Heim and Tsien 1996). These fluorescent proteins are often a popular choice for cellular biologists who aim to couple a fluorescence signal to the presence or function of a protein of interest within a cell (Stepanenko et al. 2008; Chalfie et al. 1994). Since early demonstrations of this principle, the amount of information that can be probed through fluorescent proteins has been expanded using fluorescent fusion proteins (Heim et al. 1994; Prasher et al. 1992). Fluorescence fusion proteins specifically tag one protein of interest by incorporating a fluorescent protein into the sequence, typically at the N- or C-terminus.

It is common practice for TIRF microscopy experiments to be performed with ion channels or G-protein coupled receptors that are fused to a fluorescent protein. There are several advantages to this approach including that each subunit is tagged and that the protein or protein complexes of interest can be studied in living cells, giving direct visualization of cellular processes in action. However, there are multiple scenarios in which fluorescent proteins may not be suitable for these experimental setups. In some cases, the chosen fluorophore may be incompatible with the protein of interest, and the fusion of the fluorophore to the protein causes impairment of protein function or localization (Pedelacq et al. 2006; Kuey et al. 2019). Such occurrences obstruct the observation of the underlying physiological phenomenon of interest, necessitating an alternative approach.

If the fusion of a fluorescent protein to a target protein is not possible, fluorescent dye-based approaches may be more suitable. Some of the more commonly used dyes include the Alexa dyes, which are improvements over previous organic dye compounds with respect to brightness and photostability (Panchuk-Voloshina et al. 1999). Organic dyes may be conjugated directly to an antibody specific to the protein of interest in the cell membrane, or antibody recognition may occur indirectly via a dye-conjugated "secondary" antibody that is specific for an untagged "primary" that recognizes the protein of interest (Masi et al. 2010). In either case, antibodies will be targeted to an endogenous protein or unmodified protein in a heterologous expression system, without the possibility of a fluorescent protein interfering with the function or trafficking of the protein of interest. This approach is often necessary for studying primary cell cultures, where fusion of fluorescent proteins to the protein of interest is not possible. Furthermore, antibodies may potentially offer very high specificity for a protein of interest and may be designed to distinguish between multiple states of a protein or different posttranslational modifications it has undergone (Hattori and Koide 2018; Arur and Schedl 2014). In this way, coupling a fluorescent signal to antibody-binding of a protein of interest could offer more specific information on the state of the protein compared to indiscriminate fluorescence from a fluorescent protein fused to the protein of interest during synthesis. Lastly, when designing an experiment using fluorescence microscopy, researchers must consider the possibility of spectral overlap, particularly as more proteins or molecules in a cell are being labeled for imaging. Organic dyes typically have narrower excitation and emission profiles than fluorescent proteins and can provide another avenue for researchers to tag many molecules of interest while maintaining sufficiently distinct emission profiles for fluorescent imaging to remain viable.

Different fluorescent proteins (FPs) possess different excitation and emission spectra. An emission spectrum is the distribution of wavelengths emitted by a fluorophore at a single constant excitation wavelength. In contrast, an excitation spectrum describes the distribution of emission intensity observed upon scanning a range of excitation wavelengths (Lakowicz 2006). Other photophysical properties vary with different fluorophores as well and can influence the choice of an FP to be used for a particular experimental application. For example, the extinction coefficient describes how efficiently a fluorophore absorbs light, and the quantum yield describes the efficiency of converting light into emitted photons. It is also worth considering that some FPs do not exist as monomers, such as tdTomato, which exists as a tandem dimer. This property has direct bearing on experiments designed to measure the stoichiometry of a protein or protein complex of interest as it may be biased by the multimerization propensity of the FP (Shaner et al. 2004).

5.4.1 A Note on Quantum Yield

The quantum yield of a fluorescent protein is a measure of its efficiency in converting absorbed photons into emitted photons and is defined as the ratio of the number of photons emitted as fluorescence n_E to the number of photons absorbed by the fluorophore n_A. Thus, quantum yield Φ can be determined as:

$$\Phi = \frac{n_E}{n_A} \tag{5.6}$$

Key points about quantum yield:

1. Value Range: The quantum yield ranges between 0 and 1 (or 0–100%), where a higher value indicates greater efficiency of fluorescence.
   - A quantum yield of 1 means every absorbed photon results in fluorescence.
   - A quantum yield close to 0 indicates most absorbed energy is lost as heat or other non-radiative processes.
2. Influencing Factors:
   - Protein Environment: The quantum yield can vary depending on the fluorophore's local environment, such as pH, ion concentration, and interactions with nearby molecules.
   - Intrinsic Properties: It is also determined by the intrinsic properties of the fluorescent protein, such as its structure and chromophore stability.
3. Relevance in Experiments: A fluorescent protein's quantum yield is directly correlated withits brightness. Brightness is also influenced by the extinction coefficient, which describes how strongly the protein absorbs light (Tables 5.1 and 5.2).

5.4.2 Notes for Selecting Fluorophores

(i) *Excitation & Emission Wavelengths (nm):*
   - Excitation refers to the wavelength at which the protein absorbs light.
   - Emission refers to the wavelength at which the protein fluoresces (emits light).

(ii) *Quantum Yield:*
   - If multiple fluorophores are required, consideration should be given to their respective quantum yields, and how the signal may be impacted if they differ. It is advisable to avoid large disparities in quantum yield, as optimal imaging parameters will vary depending on brightness.

(iii) *Monomer/Dimer Status:*
   - *Monomer:* Ideal for fusion proteins and cellular imaging.
   - *Tandem Dimer (e.g., tdTomato):* Engineered to act as a single unit but cannot split into monomers. Can have enhanced brightness over a monomer FP.
   - *Tetramer (e.g., DsRed):* Less suitable for tagging due to oligomerization, which can affect protein function. However, such FPs can have enhanced brightness over a monomer FP.

(iv) *Fluorescent Protein Selection Tips:*
   - *Photostability:* Some fluorophores bleach faster than others (e.g., TagRFP-T is a variant of TagRFP that is optimized for photostability).
   - *Brightness:* mTFP1 and mTurquoise2 are among the brightest in their respective spectral ranges.
   - *FRET Compatibility:* Certain pairs (e.g., eCFP & eYFP) are commonly used for Förster Resonance Energy Transfer (FRET) experiments.

Table 5.1 Key properties of common fluorescent proteins

Fluorescent protein	Excitation wavelength (nm)	Emission wavelength (nm)	Quantum yield	Monomer/ dimer
Enhanced green fluorescent protein (EGFP)	488	507	0.60	Monomer
Enhanced yellow fluorescent protein (EYFP)	514	527	0.61	Monomer
mVenus	515	528	0.57	Monomer
eCFP (enhanced cyan fluorescent protein)	433	475	0.40	Monomer
mCherry	587	610	0.22	Monomer
mOrange	548	562	0.69	Monomer
mStrawberry	574	596	0.29	Monomer
mPlum	590	649	0.10	Monomer
mTFP1	462	492	0.85	Monomer
mTurquoise2	434	474	0.93	Monomer
DsRed	558	583	0.70	Tetramer
TagRFP-T	555	584	0.41	Monomer
tdTomato	554	581	0.69	Tandem dimer

(v) *Special Considerations:*
- *mCherry* is commonly used due to its good balance of brightness and photostability.
- *mPlum* has a very low quantum yield but is one of the few far-red monomeric proteins.
- *tdTomato* is often preferred over DsRed because it retains brightness while behaving more like a monomer.

5.5 Applications of TIRF Microscopy

5.5.1 Overview

G protein-coupled receptors (GPCRs) are membrane proteins that are essential for intercellular communications due to their direct role in mediating signals from many hormones, neurotransmitters, and xenobiotics (Sungkaworn et al. 2017). Central to their function is, as their name suggests, their interaction with heterotrimeric guanidine triphosphate (GTP)-binding proteins (usually shortened to "G protein"). When a GPCR binds a G protein, the GPCR exchanges guanidine diphosphate (GDP) bound to the G protein for GTP. As a result, the G protein dissociates into two subunits: G_{α} and $G_{\beta\gamma}$. These subunits interact with other regulatory proteins, directly changing the environment of the cell. For example, when acetylcholine (ACh) binds to the cardiac muscarinic ACh receptor (mAChR), $G_{\beta\gamma}$ dissociates from the receptor and activates cardiac potassium channels, decreasing heart rate (Logothetis et al. 1987; Trautwein et al. 1958). Studying the mechanistic nature of these interactions can enhance our understanding of a wide range of physiological and pathophysiological processes, thus informing novel methods of modulating these processes in the treatment of disease.

Approximately 34% of Food and Drug Administration (FDA)-approved drugs target GPCRs, highlighting their importance in pathophysiology and pharmaceutical research (Hauser et al. 2018). While our understanding of GPCR structure-function and physiology has improved in recent decades, many outstanding questions remain. One example is the spatial distribution of GPCRs on the cell membrane. Fresh insights have been provided by leveraging TIRF microscopy in conjunction with a specialized camera built to detect single photons in a technique called photoactivated localization microscopy (PALM). In their study of G protein-coupled α_{2A} adrenergic

Table 5.2 Summary of equations

Equation	Formula	
(5.1)	$n_1 \sin\theta_1 = n_2 \sin\theta_2$	n, refractive index θ_1, angle of incidence θ_2, angle of refraction
(5.2)	$\theta_1 = \theta_2$	θ_1, angle of incidence θ_2, angle of refraction
(5.3)	$\theta_C = \sin^{-1}\frac{n_2}{n_1}$	θ_c, critical angle n_1, refractive index of medium where light originates n_2, refractive index of medium light is entering
(5.4)	$I(z) = I_0 \exp\left(-\frac{z}{\delta}\right)$	$I(z)$, evanescent field intensity I_0, intensity of light at the interface z, perpendicular distance from the interface δ, penetration depth
(5.5)	$\delta = \frac{\lambda_0}{4\pi} \times \left(n_1^2 \sin^2\theta_1 - n_2^2\right)^{-1/2}$	δ, penetration depth λ_0, wavelength of incident illumination n_1, refractive index of medium where light originates n_2, refractive index of medium light is entering θ_2, angle of refraction
(5.6)	$\Phi = \frac{n_E}{n_A}$	Φ, quantum yield n_E is the number of photons emitted as fluorescence n_A is the number of photons absorbed by fluorophore
(5.7)	$d = \lambda/2\, n \sin(\theta)$	d, minimally resolvable distance λ, wavelength of illumination n, refractive index θ, half-angle of cone of light entering detector
(5.8)	$d = 0.61\lambda/\mathrm{NA}$	d, minimally resolvable distance λ, wavelength of illumination NA, numerical aperture
(5.9)	$I(r) = (2\, J_1(kr)/kr)^2$	$I(r)$, intensity of light at radius r r, radius k, wave number J_1, Bessel function of the first kind, first order

receptors (α_{2A}-AR), Sungkaworn et al. measured the mobility of the receptors within the plasma membrane, the frequency of receptor-G protein complex formation, and association/dissociation kinetics of this complex to demonstrate the presence of "hot spots," areas on the membrane in which GPCRs and G proteins are localized and preferentially couple (Sungkaworn et al. 2017). This finding reveals novel potential strategies for modulating GPCR function pharmacologically, such as through modification of complex formation kinetics, GPCR mobility, or receptor-G protein coupling at hot spots. Such avenues of research would not be possible without single-molecule imaging enabled by TIRF.

5.5.2 Single Molecule Pull-Down Assay (SiMPull)

The single molecule pull-down (SiMPull) assay is an experimental technique designed to characterize protein interactions at the level of a single complex (Jain et al. 2011, 2012). The SiMPull Assay resembles a more typical co-immunoprecipitation assay (Barrios-Rodiles et al. 2005), but offers greater sensitivity and the ability to image individual complexes (Jain et al. 2011, 2012). The technique functions by first creating a flow chamber using passivated microscope slides and cover slips. The additional inclusion of biotinylated antibodies specific for the "bait" protein of interest allows for cell lysates to be passed through the flow chamber, after which "prey" proteins are isolated in individual complexes on the cover slip. Proteins of interest, which are either tagged with genetically encoded fluorescent proteins or antibody-labeling, can then be visualized by TIRF microscopy (Jain et al. 2011, 2012).

The SiMPull assay has multiple applications for studying protein complexes. Combined with TIRF microscopy, the stoichiometry of isolated protein complexes can be analyzed by photobleaching of unique fluorophores on each component of the protein complex (Jain et al. 2011, 2012, 2014). Because complexes are individually isolated, as opposed to a "bulk" pull-down typical of a co-immunoprecipitation assay, information can also be gained on the proportion of complexes forming with different binding partners (Jain et al. 2011, 2012). The sensitive nature of the SiMPull assay additionally allows for the study of protein complexes present only in relatively low abundance or from single cell lysates (Jain et al. 2011, 2012; Zhao et al. 2023; Stolpner and Dickinson 2022).

5.5.3 Instant TIRF-SIM

Most super-resolution microscopy techniques present a tradeoff between temporal and spatial resolution. While a technique such as STORM can achieve lateral resolution of around 20 nm (Rust et al. 2006), image acquisition is lengthy and fails to capture dynamic processes in live cells. Instant TIRF-SIM, however, can achieve sub diffraction imaging at a resolution of approximately 115 nm while maintaining the ability to capture images at a rate of 100 Hz (Guo et al. 2018). Structured illumination microscopy, or SIM, functions by combining an illumination pattern of the experimenter's design with the emitted light from a fluorescent sample (Gustafsson 2000). Multiple illumination patterns are used, and the resulting images are reconstructed into one super-resolution image of the original structure of interest. Improved temporal resolution of SIM has been achieved by instantaneous algorithmic processing of images, as opposed to digitally. The resulting technique, iSIM, can perform super-resolution microscopy in real time with an axial resolution of about 350 nm (York et al. 2013; Curd et al. 2015). Further improvements in axial resolution are obtained via TIRF-based illumination in the case of instant TIRF-SIM, bringing axial resolution to about 120 nm (Guo et al. 2018). Instant TIRF-SIM is a particularly desirable optical technique for imaging membrane-bound structures in live cells as well as cellular processes. Unlike other super-resolution microscopy techniques that lose the ability to resolve events on short time scales, instant TIRF-SIM offers a significant increase in lateral resolution while maintaining the functionality of near-instantaneous image acquisition.

5.5.4 Single Particle Tracking

The movement of microscopic particles has been of interest to scientists since the initial characterization of pollen granules moving in water by the botanist Robert Brown in the early nineteenth century (Brown et al. 1866). It was concluded from his extensive observations that the movement of these particles was not due to evaporation of the aqueous media or due to local currents, but rather likely arose from the movement of the particle itself (Brown et al. 1866). This observation would later be named after him as Brownian motion and was vital to prove the existence of atoms, which was awarded the Nobel Prize in Physics in 1926 (Einstein 1905; R. E 1911; Bian et al. 2016). These initial observations led to the birth of a new framework, i.e., that the movement of particles within a medium was due to the kinetic fluctuation of atoms acting upon the particle of interest from multiple directions, ultimately leading to the observed diffusive motion (Einstein 1905; R. E 1911; Bian et al. 2016). Thus, particles undergoing Brownian motion exhibit an apparent "random walk" through their environment as they move away from their point of origin (Bian et al. 2016). The phenomenon of Brownian motion has been influential in multiple disciplines, including microrheology (Valentine et al. 2001), economics (Bachelier et al. 2006), and biophysics (Saffman and Delbrück 1975). The rest of this section will be dedicated to understanding the study of particle movement in biological systems.

To study the movement of single particles in biological systems, scientists have used single particle tracking (SPT), a methodology that allows for the simultaneous identification and tracking of large numbers of individual particles across time. For biological systems, a "particle" is often any protein of interest, and SPT has become highly popular in biology as it provides unparalleled insight into the temporal dynamics of protein movement in living cells. SPT has been fruitful for studying the diffusivity of proteins in the lipid bilayer of the plasma membrane. Researchers can make predictions about the microenvironment a protein of interest exists in when observed using SPT (Saffman and Delbrück 1975; Simon et al. 2024; Jaqaman et al. 2008). For example, it has been repeatedly observed that membrane proteins do not only undergo Brownian diffusion, but also show periods of confinement, or sub-diffusion, as well as periods of super-diffusion (Sungkaworn et al. 2017; Gross and Watt 2018; Simson et al. 1995; Wilson et al. 1996; Goulian and Simon 2000). This finding was originally of interest because this anomalous diffusion was a deviation from the predictions of the fluid mosaic model, which predicted that membrane proteins would undergo Brownian motion in the plasma membrane (Singer and Nicolson 1972). It was subsequently shown that membrane proteins moved between distinct microenvironments in the plasma membrane. And, in turn, specific membrane microenvironments influenced the diffusivity of proteins. SPT helped to update the fluid mosaic model to include interactions between the plasma membrane components and the underlying cytoskeleton, as well as the idea of membrane "fences" and "fenceposts/pickets" created by the cytoskeleton and other transmembrane proteins (Sungkaworn et al. 2017; Jacobson et al. 1995; Nicolson 2013; Freeman et al. 2018). These results showcase how SPT has been influential to our understanding of membrane protein movement in the plasma membrane of living cells.

SPT requires the labeling of a protein of interest with a molecule that allows for microscopic identification. Historically, colloidal gold particles or fluorescent latex beads had been used to label proteins for tracking (Simson et al. 1995), but recently the use of fluorescent genetically-encoded proteins, including photo-activatable or photo-switchable proteins (Bayle et al. 2021), or photostable fluorescent dyes, has become the main way to visualize proteins as they diffuse in the plasma membrane (Simon et al. 2024; Calebiro et al. 2013). The development of Halo (Los et al. 2008), SNAP (Keppler et al. 2003), and CLIP (Gautier et al. 2008) tags have enabled labeling of intracellular proteins with cell permeable dyes that are photostable and bright (Simon et al. 2024). Recent experiments have been conducted to look at the colocalization of membrane proteins by combining multiple protein tags in the same cell (Sungkaworn et al. 2017; Grimes et al. 2023), furthering what types of questions can be asked with these new tools. SPT also relies on a high signal-to-noise ratio to detect the labeled protein over background noise (Simon et al. 2024; Jaqaman et al. 2008; Michalet 2010). This need for a high signal-to-noise ratio has perfectly situated TIRF microscopy as an ideal method for conducting SPT of membrane proteins (Sungkaworn et al. 2017; Jaqaman et al. 2008; Calebiro et al. 2013; Grimes et al. 2023). Once a fluorescent label has been chosen and particle movies have been acquired, individual particle tracks can be constructed (Jaqaman et al. 2008). An idealized version of this process is shown in Fig. 5.5. The protocols employed to generate single particle tracks and subsequent track analyses to measure diffusion coefficients are outside the scope of this section, but great reviews have been written on these topics (Simon et al. 2024; Lelek et al. 2021; Shen et al. 2017).

The combination of TIRF microscopy with SPT has recently enabled fundamental insights into how G-protein coupled receptors (GPCRs) temporally and spatially interact with their G-protein, which is important for downstream signaling (Sungkaworn et al. 2017). SPT has

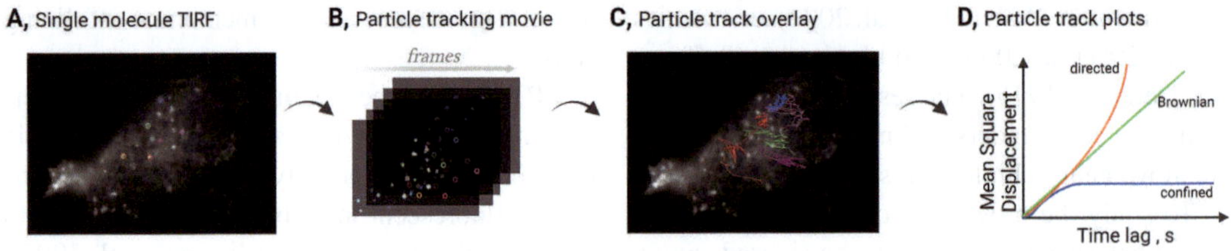

Fig. 5.5 Single particle tracking for ion channel motion in the cell membrane. (**a**) Grayscale TIRF image of live cell expressing an mCherry-tagged 5-HT_{1A} receptor. Example single fluorescent particles that will be tracked over time are labeled with colored circles. (**b**) Periodic data collection measures the movement of particles over time. (**c**), The motion track of each example particle is superimposed on the cell. (**d**) Idealized graph showing the Mean Square Displacement (MSD) of example tracks with three motion behaviors highlighted: Confined (the channel is restricted to a specific patch of the membrane, blue); Brownian (consistent with Brownian motion; green); Directed (implies a mechanism facilitating motion; red)

even provided insight into the interactions between the β2-adrenergic receptor (β2AR) and β-arrestin 2 (βArr2) as well as the involvement of phosphatidylinositol-4,5-bisphosphate (PIP_2) in agonist-dependent β2AR and βArr2 interaction (Grimes et al. 2023). Since β-arrestins are important for GPCR internalization and have a variety of other signaling functions (Pierce et al. 2001), these results make clear that the temporal interactions due to lateral diffusion in the plasma membrane of β2AR and βArr2 have a large influence on receptor fate. Both studies showcase that membrane protein diffusion has consequences for downstream signaling kinetics and dynamics. These fundamental discoveries would not have been possible without the combination of TIRF microscopy and SPT enabling accurate characterization of protein diffusion.

For many researchers who use SPT to study membrane proteins, visualizing particles moving in real time comes as an initial surprise, much like the surprise Robert Brown experienced when making his initial observations of pollen granules (Brown et al. 1866). The static images in textbooks and classrooms come to life, and the realization that proteins are moving within all our cells at every moment is a sobering thought. Future studies combining SPT and TIRF microscopy will undoubtedly continue to provide fundamental insights into membrane protein signaling. For example, it has been shown that GPCRs exhibit different levels of basal, or constitutive, activity (activity in the absence of an agonist) (Rosenbaum et al. 2009; Zeghal et al. 2023). Could basal activity differences be attributed to differences in diffusion kinetics and the probability of GPCR and G-protein interactions (Sungkaworn et al. 2017)? With the widespread attention that SPT is gaining, and with the improved signal-to-noise measurements that TIRF microscopy allows, this and many other questions can be addressed. An application of SPT via TIRF for observing the signaling components associated with a single GPCR, mCherry-tagged $5HT_{1A}$, is demonstrated in Fig. 5.5. In this example, particle trajectories can be analyzed to determine whether they are randomly diffused through the membrane or controlled via some other process.

5.5.5 Single-Molecule Detection

The ability to measure biochemical phenomena on a single-molecule basis has long been an analytical goal. Measurement of single-molecule fluorescence requires highly sensitive equipment as well as minimal background fluorescence. In practice, it is impossible to eliminate background signal due to sample impurities, emission from the optical components of the equipment, scattering of light at the incident wavelength, and the detection of aberrant, or "dark" current present in processing equipment such as camera silicon chips. One strategy of reducing background signal is to restrict the volume of the sample. Limiting the observed sample volume is one of the major benefits of TIRF microscopy. Because the evanescent wave only penetrates 100 nm into the sample, a much lower biological volume is studied in comparison to traditional fluorescence microscopy (Lakowicz 2006). This facet of TIRF microscopy is particularly important for single-molecule imaging, as restricting the observed volume reduces the diffraction of fluorescent signals upon imaging (Liu et al. 2015).

5.6 TIRF and Related Fluorescence Microscopy Techniques

Fluorescence microscopy techniques have long been used to visualize cellular structures or physiological events that are coupled with fluorescent signals. A commonality of these techniques is the visualization of a sample by excitation of the fluorophores within it, but different approaches to sample illumination across fluorescence microscopy techniques yield great differences in resolution, signal-to-noise ratio, and other factors that can limit the amount of information extracted from a sample. TIRF microscopy operates under similar principles to the more common epifluorescence and confocal microscopy techniques, but the use of a rapidly decaying evanescent wave to excite fluorophores enables uniquely high resolution in the z-plane (see Fig. 5.2). Accordingly, TIRF microscopy is particularly well suited to and has many applications for the study of membrane-bound proteins.

5.6.1 Epifluorescence Microscopy

Epifluorescence microscopy, otherwise known as widefield fluorescence microscopy, is a commonplace technique for the imaging of biological samples, owing to its speed, versatility, and convenience, and reliance on relatively inexpressive equipment (Webb and Brown 2013; Erdogan 2011). As shown in Fig. 5.2, the technique involves a light source, typically a mercury or xenon bulb, or more recently, light-emitting diodes (LEDs), that excite fluorophores within the entire sample indiscriminately of depth (Webb and Brown 2013; Ladouceur and Brown 2021). Optical filters are used to limit the wavelength(s) of light that pass through the sample as well as the fluorescence emitted by the fluorophores. Filtering out light outside of the experimentally relevant excitation and emission frequencies improves the signal-to-noise ratio of the technique. After passing through the emission filter, light is collected through the objective lens of the microscope and can be passed to a detector for image capture and analysis (Webb and Brown 2013). Epifluorescence microscopy is both economical and has high throughput. Images of samples can be acquired rapidly and in quick succession, permitting time-lapse imaging of biological samples (Zou and Bai 2019). Furthermore, because this technique captures bulk signals, it can detect low-level emission from weakly fluorescing particles in a sample or samples only sparsely populated with fluorophores, albeit at the expense of spatial resolution (Webb and Brown 2013).

One of the greatest deficits associated with epifluorescence microscopy is the large amount of background signal present from exciting fluorophores outside of the focal plane, a consequence of light passing through the entirety of the sample (Hupfel et al. 2021). Depending on the intended application, this is not always problematic, but the spatial resolution of images collected by epifluorescence microscopy is inherently poorer than what is achievable with other fluorescence microscopy techniques, such as TIRF microscopy. Due to excessive background fluorescence, axial resolution of epifluorescence microscopes often exceeds 500 nm–1.5 μm, depending on the objective lens. In contrast, the enhanced signal to noise characteristics of TIRF microscopy permits axial resolutions closer to 100 nm (Fish 2022). Compared to TIRF microscopy, illumination of the entire sample makes it particularly difficult to resolve individual fluorescent particles with epifluorescence microscopy (Lang et al. 2006).

Because TIRF microscopy selectively excites fluorophores at the cell surface, much of the background noise caused by fluorescence outside of the focal plane in epifluorescence microscopy is not present.

5.6.2 Confocal Microscopy

Confocal microscopy is a similar technique to widefield microscopy but has been optimized for imaging thicker samples or multiple depths within one sample. The composition of a confocal microscope is functionally similar to that of instruments used for widefield microscopy, although importantly, a pinhole aperture is included between the sample and the detector (Elliott 2020). The function of this pinhole aperture is to limit the light passing to the detector from regions outside of the focal plane (Kitamura 2021). Accordingly, beams of light that are "in focus" will reach the detector, while light that is not coming from the focal plane will be filtered out. In fluorescence microscopy applications, excitation of fluorophores occurring outside of the focal plane produces significant background signals (noise), so elimination of any of this out-of-focus fluorescence will enhance the resolution of the images produced (Elliott 2020; Kitamura 2021).

Confocal microscopy is often used to image thicker samples repeatedly, with the potential to reconstruct a three-dimensional image of the sample following scanning (Elliott 2020). This is a unique feature of confocal microscopy, although the acquisition of such three-dimensional reconstructions is a lengthy process. Furthermore, as scanning is slow, this imaging technique is better suited for imaging larger, static structures, and can be less well suited for interrogating dynamic cellular events (Honzel et al. 2024; Ulfhake et al. 1992).

Compared to TIRF microscopy, confocal microscopy is better suited for samples or experiments that would benefit from scanning. However, even though the axial resolution of confocal imaging is significantly improved over epifluorescence microscopy techniques, TIRF microscopy remains the superior imaging technique for maximizing axial resolution, particularly for the imaging of structures in the cell membrane (Lang et al. 2006; Honzel et al. 2024). Approaching optimal usage conditions, confocal microscopes can achieve a lateral resolution of around 180 nm and an axial resolution of about 500 nm (Fouquet et al. 2015), compared to the axial resolution of about 100 nm that is possible through TIRF microscopy (Fish 2009). In addition to slower imaging precluding the observation of more dynamic cellular processes, larger voxel depth from confocal microscopes may also make it more difficult to resolve individual fluorescent particles within a sample. Compared to confocal microscopy, TIRF microscopy is better suited for both live cell imaging as well as higher resolution imaging of cell membrane-bound structures.

5.6.3 Mathematical Limitations of Optical Imaging Techniques

Conventional TIRF microscopy, as well as widefield epifluorescence and confocal microscopy, are diffraction-limited imaging techniques. This is an inherent limitation of these approaches because of the nature of light being used for imaging (Abbe 1873). The diffraction limit refers to the phenomenon of light scattering upon interaction with a fluorophore or other structure being imaged, making it impossible to resolve structures within 100–200 nm of each other (Galbraith and Galbraith 2011). Described first by Ernst Abbe in 1873 (Abbe 1873), the diffraction limit is related to the wavelength of the light being used for imaging and the numerical aperture of the microscope. This relationship is represented by the following equation:

$$d = \frac{\lambda}{2} n \sin(\theta) \quad (5.7)$$

where d refers to the distance in nm between points in the optical system that are theoretically able to be resolved, λ is the wavelength of light in nm, n is the refraction index of the medium through which imaging is occurring, and θ refers to the half-angle of the largest possible cone of light that can enter the detector of the microscope. Together, the term $n \sin(\theta)$ represents the numerical aperture of the microscope, a value that expresses the angles from which a microscope's detector will accept light.

While the diffraction limit describes a theoretical and physical barrier to a microscope's resolution, the Rayleigh criterion is a similar but slightly more practical concept (Born and Wolf 1999). The diffraction limit is more applicable to understanding the general limitations of resolution within an optical system, and the Rayleigh criterion is more closely related to the ability of two points to be resolved. The minimum resolvable distance between the two points is defined by the following equation:

$$d = 0.61\lambda/\text{NA} \tag{5.8}$$

where d refers to the minimum resolvable distance in nm between points in the optical field, λ is the wavelength of light in nm, and the numerical aperture NA is a value that expresses the angles from which a microscope's detector will accept light. According to the Rayleigh criterion, two fluorophores less than a distance (d) away from each other will appear as a single, blurry point in the optical field during imaging.

Another means of quantifying a microscope's limits of resolution is the point spread function, which describes the difference between light emitted from a point within a sample and how it is imaged by a microscope's detector (Shaw and Rawlins 1991). In microscopy applications, the point spread function is typically represented by an Airy disk, which is a blurred area around the point being imaged surrounded by several concentric circles. The intensity of the light around the point, as a function of distance from the point, is defined by the Airy function:

$$I(r) = (2\, J_1(kr)/kr)^2 \tag{5.9}$$

The term $I(r)$ refers to the intensity of light at radius r from the center of the point from which light is originating. k is the wave number, defined by the equation $2\pi/\lambda$, describing spatial frequency in the form of wavelengths per nanometer. The $J_1(kr)$ term is derived from the Bessel function of the first kind and conveys information about wave propagation and oscillation. The point spread function is affected by the numerical aperture of the optical system being used, and therefore can be expected to change with different wavelengths of light or different refraction indexes from media being used for imaging (Shaw and Rawlins 1991). Additionally, microscopes are not often constructed in a way that reflect ideal conditions, so any further imperfections in the path light travels through will further limit resolution (Smith 2008).

The diffraction limit, Rayleigh criterion, and point spread function quantify the limitations of conventional microscopy techniques to resolve structures at the molecular scale. For traditional microscopy methods, these limitations exclude the possibility of single-molecule imaging. While some of these limitations are imposed by the physical nature of light itself, more recent advances in imaging techniques have allowed for super-resolution imaging—designed in a way that can bypass these physical limitations.

5.6.4 Super-Resolution Microscopy

While TIRF microscopy can provide sub-diffraction limit imaging in the "z" plane, it does not provide sub-diffraction limit imaging in the x-y axes. Here the practical limit for a given TIRF microscope set up is constrained by the magnification factors of the excitation light path and the pixel size of the imaging device. In typical optical microscopes, the diffraction limit was long viewed as an inherent restraint to optical microscopy techniques. This has begun to change with the rise of multiple super-resolution microscopy techniques, which are capable of imaging beyond the theoretical diffraction limit of an optical system. Stochastic optical reconstruction microscopy (STORM) and stimulated emission depletion microscopy (STED) represent two unique methods of imaging that offer lateral resolution surpassing the diffraction limit.

STORM microscopy, first described in 2006, is a technique that relies on manipulating the behavior of fluorophores in order to capture sections of an image individually before reconstruction into a cohesive whole (Rust et al. 2006). Samples being imaged contain photoswitchable and photostable fluorophores, which may be controlled by the experimenter. These fluorophores are cyclically switched on or off, with only a small proportion fluorescing at any given time point. In a series of steps, an amount of fluorophores are turned on or off, and fluorophores present in the field of view are localized by Gaussian fitting of the associated point spread functions. After several steps, each fluorophore in the field of view has been localized, and a reconstructed image can be formed with a lateral resolution of about 20 nm. It is the fact that fluorophores in close proximity may be switched on or off and imaged separately that allows for sub diffraction limit resolution.

STED microscopy bypasses the diffraction limit by selectively inhibiting fluorescence at the outer edges of a fluorophore's point spread function (Hell and Wichmann 1994). An excitation laser functions as is typical for fluorescence microscopy techniques, illuminating the sample and any fluorophores within its beam path. A second ring-shaped depletion laser, in which the excitation laser is nested within, prevents fluorescence signals from being detected in an area around the fluorophore. In this manner, the fluorescence signals being detected may be resolved at resolutions on the order of 20–50 nm.

In addition to the benefits towards lateral resolution gained by employing STORM or STED techniques, further improvements of axial resolution can be achieved by combining these techniques with TIRF-based illumination (Rust et al. 2006; Han et al. 2013; Gould et al. 2011; Sivankutty et al. 2020). Although these techniques are compatible, it may not always be desirable for experimenters to use these combined techniques in all situations. Limitations of the described super-resolution microscopy techniques include throughput, sample preparation, and imaging requirements (Tam and Merino 2015; Valli et al. 2021). For example, STORM requires lengthy image acquisition times as well as additional analysis and image processing before a final reconstructed image is produced. Low temporal resolution makes it nearly impossible to image dynamic live-cell events. A limiting factor of STED microscopy is the number of unique fluorophores that can be used, as the depletion laser must be matched to the emission spectra of any fluorophores to effectively inhibit fluorescence. Further constraints on compatible fluorophores exist due to the high intensity of the emission and depletion lasers, demanding particularly photostable fluorophores. TIRF microscopy on its own may not provide equivalent lateral resolution but also lacks issues with throughput or constraints on multicolor imaging. The heightened axial resolution of TIRF microscopy and its compatibility with multiple research applications cement as one of the best options for live-cell imaging and imaging of cell membrane-bound structures.

5.7 Practical Considerations of TIRF Microscopy

5.7.1 Overview

One of the most significant advantages of TIRF microscopy is the ability to excite fluorophores in the cell membrane with minimal background fluorescence compared to other methods. This is because of the rapid decay of evanescent wave past the cover slip-cell membrane interface. Thus, while TIRF microscopy is not suited for studying deep within the cell or within tissue, it is a powerful tool for resolving events taking place in the cell membrane with single-molecule precision.

Thus far, we have discussed TIRF in terms of theoretical, ideal conditions. But in practice, additional considerations must be made when interpreting TIRF data. Let us consider Eq. 5.5, which calculates the penetration depth δ at which the evanescent wave has 37% of its intensity at the interface. But the wave continues to decay past δ, asymptotically approaching an intensity of 0 (visualized in Fig. 5.2). Because of this gradual drop-off (as opposed to a binary on/off condition),

δ is defined relative to the system's signal-to-noise ratio. If we extrapolate Eq. 5.4, we find that after three penetration depths, the wave intensity decays to less than 5% of the initial intensity ($I(3\delta) = 0.0498$), which is in line with typical noise levels (Oheim et al. 2019). Thus $3^*\delta$, or 172.3 nm following our previous example, could be considered the effective penetration depth to arbitrarily account for noise inherent to the system. Still, in practice, δ is in fact a variable that is challenging to accurately define.

Following Snell's law, the depth of the evanescent wave δ is one specific value that can be precisely calculated. However, this relies on several assumptions, primarily: (1) The incident light is perfectly collimated—that is, all of the incident light is completely parallel to itself and all strikes the interface at precisely the same angle, (2) there is no scattering of light by the optics of the TIRF setup, and (3) the refractive index of the cover slip and cell milieu is precisely known and does not contain any local variations (Oheim et al. 2019).

In practice, all three of these assumptions are usually invalid. This does not preclude the ability to employ TIRF to excite fluorophores in a spatially confined manner, but it can affect the conclusions one may draw from TIRF microscopy data. For example, beginning with the first assumption, incident light typically strikes the interface at a range of angles due to imperfect collimation, resulting in a range of evanescent wave depths. One can approximate this range of depths following Eq. 5.5. Using our previous example, if the incident angle is in fact a range from 65° to 75° due to imperfect collimation, the evanescent wave depth then ranges from 53.8 nm to 63.3 nm. The effect that the incident angle has on penetration depth is nonlinear and visualized in Fig. 5.5. The true range of incident angles will vary with each setup, and it is up to the experimenter to decide how much variation is acceptable given the objectives. For example, quantification of the motion of the bacterial flagellar motor, which has a height approximately 90 nm above the coverslip (Leake et al. 2006), would require a very narrow range of wave depths for accurate conclusions to be drawn from fluorescence measurements. But for studies of experimental constructs situated within a cell membrane, such variation is less consequential.

The second and third assumptions are less quantifiable and less controllable. The scattering of incident light by the optics guiding it to the sample can result in normally propagating light, which does not rapidly decay in intensity, reaching the sample. This enables the excitation of fluorophores beyond the intended area of illumination. The relevance of this parameter to TIRF data analysis is an ongoing matter of debate and worthy of consideration before beginning TIRF microscopy experiments.

Uncertainty regarding the true refractive indices (RI), especially of the second medium, is a major source of debate regarding TIRF microscopy data interpretation. We have previously referred to this medium as "water" or "cell milieu," but the identity depends on experimental conditions and often cannot be precisely known. This is particularly true when studying an intact cell. Indeed, the RI varies within a cell such that near-membrane endosomes possess a measurably lower RI (1.344) compared to that of mitochondria (1.378) (Brunstein et al. 2017). While this may seem like a minor distinction, the relationship between n_1 and δ is not linear, as demonstrated in Fig. 5.6. Following our previous example using Snell's law and constraining the incident angle to 70°, this corresponds to a penetration depth ranging from 67.4 nm to 80.6 nm (the true range is likely larger as not all cell components were measured). These fluctuations and uncertainties present a serious challenge to the reproducibility and viability of reported TIRF measurements. Fortunately, some methods have been proposed to address these concerns.

Techniques for the direct measurement of the angle of incident light θ and the calibration of $\delta(\theta)$ can benefit the accurate interpretation of TIRF data. Typically, this requires a known axial fluorophore distribution $F(z)$; by measuring intensity along the axial dimension $I(z)$, the experimental θ may be calculated. Once θ is determined, fluorescence intensity measurements at various heights may be used to estimate $\delta(\theta)$. Similarly, a technique involving imaging fluorescent molecules embedded in agarose beads at a range of heights

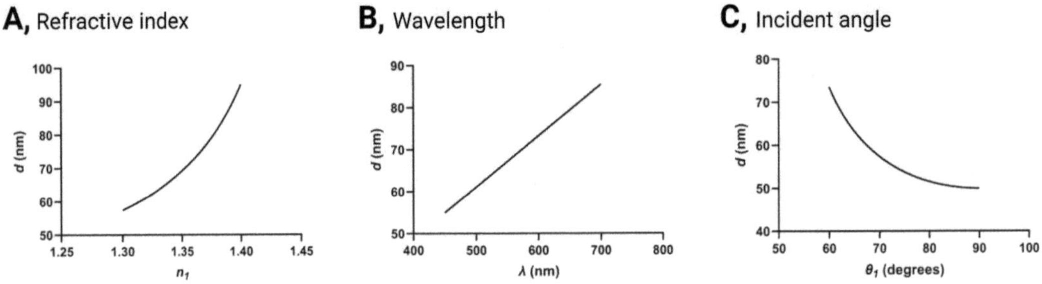

Fig. 5.6 Evanescent wave depth is related to refractive index, wavelength, and incident angle. The depth of the evanescent field is positively related to the refractive index of the medium in a nonlinear manner (**a**); The wavelength of light the incident beam, in a direct manner (**b**); and the incidence angle used for illumination in a nonlinear, inverse manner (**c**)

has been described (Oheim et al. 2019). This technique enables accurate measurement of the effective excitation range of the evanescent wave, as well as the quantitative determination of the relative evanescent and nonevanescent components of excitation.

5.7.2 Composition of a TIRF Microscope

The construction of a TIRF microscope can be customized to experimental needs. For example, studies involving multiple fluorophores require multiwavelength excitation. This is necessary when interactions between two different proteins (necessitating two different fluorophores) are being studied. Multiwavelength TIRF typically relies on an objective-based setup whereby total internal reflection is initiated by focusing excitation laser beams through a high numerical aperture (>1.4) microscope objective lens. In this arrangement, both excitation and emission light transmit through the objective, necessitating additional optics to separate them. This can involve a combination of dichroic mirrors and emission filters, in which excitation light is directed into the microscope objective through the back aperture by the dichroic mirror (Larson et al. 2014). However, separating the excitation and emission lights becomes more difficult with more lasers and distinct fluorophores. More dichroic mirrors can be employed but may reduce the efficiency of collecting emitted photons.

To circumvent these concerns, micromirrors have been developed for use in TIRF, hereafter referred to as "mmTIRF." These are small broadband mirrors that direct the excitation beams into and out of the back aperture of the microscope objective. This setup is more physically stable than dichroic mirrors and does not rely on the flatness of the dichroic or motors employed to manipulate them (typical in commercial systems). As a result, mmTIRF is optimized for observation of multiple fluorophores with high photon collection efficiency. Furthermore, this setup enables precise adjustment of the incident angle of each excitation incident beam and thus the penetration depth of each evanescent field (Larson et al. 2014). The benefits of this setup are particularly suited for studying colocalization events, which can demonstrate biomolecular interactions between two proteins.

5.7.3 Collection of TIRF Data

A popular expression system for diverse proteins of interest is the Human Embryonic Kidney (HEK) cell line (Gada et al. 2023), although various other options have equal practical utility. For TIRF experiments, adherent cell lines such as HEK-293 T are commonly used, as the cells will inherently attach to the bottom surface of the vessel they are grown in. After depositing, or seeding, cells on glass coverslips, gene(s) encoding the protein(s) of interest are expressed in the cells. One commonly used method to achieve this is transfection, in which plasmid DNA is combined with a polymeric endocytic agent such as polyethylenimine (PEI) then incubated with the cells for some hours. Regardless of the method of gene expression, the intended result is cells expressing the construct(s) of interest adhered to a thin piece of glass that may be placed in a TIRF apparatus for study.

5.7.4 Analysis of TIRF Data

Following collection of TIRF microscopy data, the collected images must be processed and analyzed to draw conclusions regarding any observations made. The general workflow of data analysis for TIRF microscopy is described below, although variations may occur based on the needs of the experimenter or the specific software being used for data analysis.

Before data are analyzed, it must first be processed to maximize the signal-to-noise ratio of imaged specimens. For any given image, the background fluorescence intensity should be subtracted in order to prevent any autofluorescence or light scattering in the sample from interfering with the interpretation of results (Mattheyses et al. 2010). Further improvements to the signal-to-noise ratio can be made by the process of "denoising." Whether due to imperfections in the detector, background fluorescence, or random fluctuations in observations made, applying filters or algorithms to the data can improve the quality of the final data (Carlton et al. 2010). During longer periods of imaging, such as in time-lapse applications, samples may also move slightly over the course of data collection. In this case, images can be aligned to a reference image in order to ensure correct spatial localization of imaged fluorophores over the course of the imaging process (Pitkeathly et al. 2012).

Depending on the purpose of data collection, multiple types of analyses may be desired. Fluorescence intensity analysis, at its most basic, provides information on the localization of fluorophores within a sample being imaged. Observations of fluorescence intensity can be used to describe the cell surface density of fluorophores (Fox et al. 2013), the presence of protein complex formation (Masi et al. 2010), and offer detailed information about the movement of individual fluorophores over time (Rao et al. 2022), either as a means of single particle tracking or studying larger-scale dynamic processes in live cells. Another common use of TIRF imaging is to describe the stoichiometry of membrane complexes (Yamamura et al. 2015). Each protein subunit within a complex can be labeled either by genetic encoding of a fluorophore or antibody labeling, after which exposure to high intensity light will cause photobleaching. As fluorescence intensity is measured, photobleaching will cause emitted fluorescence to decrease in discrete "steps" (see Fig. 5.3). If each protein is labeled with a unique fluorophore, observing the number of photobleaching steps enables the experimenter to determine the number of each protein subunit present in each complex. To accurately determine the stoichiometry of these protein complexes, multiple observations are made to create a histogram describing the number of photobleaching steps. This histogram is often compared to an idealized binomial distribution, after which conclusions can be drawn regarding the stoichiometry of membrane complexes and the relative abundance of each unique complex should multiple variants be able to form from the same protein subunits.

5.8 Conclusions and Outlook

TIRF microscopy is a powerful, flexible and innovative technique for studying cell physiology via single fluorescent molecules. It offers unprecedented imaging resolution at the cell membrane which enables observation of single membrane proteins including ion channels and receptors. With diverse applications, from probing stoichiometry of membrane protein complexes, to tracking the movement of individual particles in live cells, TIRF microscopy opens novel avenues of research that may be pursued to further our understanding of cellular processes at molecular resolution.

Both advancements in the design of fluorescent probes and further integration of TIRF microscopy with techniques such as super-resolution imaging will continue to improve the power and practicality of TIRF microscopy. This opens novel avenues of research that may be pursued to further our understanding of cellular processes at molecular resolution, with direct applications to established drug targets and under-explored signaling pathways alike.

Looking ahead, TIRF microscopy is poised to remain an indispensable tool in experimental medicine and biology, particularly as advances in fluorophore design, detector sensitivity, and image analysis algorithms continue to enhance its spatial and temporal resolution. Its unique capacity to selectively illuminate and monitor events near the plasma membrane makes it especially valuable for dissecting intricate cellular processes such as receptor signaling, single particle motion, and protein complex dynamics. As integration with complementary techniques such as super-resolution microscopy continues to grow, TIRF microscopy will play a central role in bridging the gap from molecular events to physiological function, solidifying its place in the next generation of biomedical discovery.

Acknowledgements The authors dedicate this chapter to Sergei Noskov in acknowledgement of his foundational contributions to ion channel biophysics. We extend our appreciation to our colleagues in the Plant and Logothetis labs for their constant encouragement and feedback. The authors thank their families and one author's cherished cat, Senator Amy, for their unwavering support in making this chapter possible.

References

Abbe E (1873) Beiträge zur Theorie des Mikroskops und der mikroskopischen Wahrnehmung. Arch Mikrosk Anat 9(1):413–468

Angle MR et al (2014) Penetration of cell membranes and synthetic lipid bilayers by nanoprobes. Biophys J 107(9):2091–2100

Arur S, Schedl T (2014) Generation and purification of highly specific antibodies for detecting post-translationally modified proteins in vivo. Nat Protoc 9(2):375–395

Bachelier L et al (2006) Louis Bachelier's Theory of speculation: the origins of modern finance. Princeton University Press. xv, Princeton, p 188

Barrios-Rodiles M et al (2005) High-throughput mapping of a dynamic signaling network in mammalian cells. Science 307(5715):1621–1625

Bayle V et al (2021) Single-particle tracking photoactivated localization microscopy of membrane proteins in living plant tissues. Nat Protoc 16(3)

Bian X, Kim C, Karniadakis GE (2016) 111 years of Brownian motion. Soft Matter 12(30):6331

Born M, Wolf E (1999) Principles of optics: electromagnetic theory of propagation, interference and diffraction of light, 7th edn. Cambridge University Press, Cambridge

Brown R, Brown R, Bennett JJ, Hardwicke R (1866) The miscellaneous botanical works of Robert Brown. Published for the Ray society by R. Hardwicke, London, p 1

Brunstein M, Roy L, Oheim M (2017) Near-membrane refractometry using supercritical angle fluorescence. Biophys J 112(9):1940–1948

Calebiro D et al (2013) Single-molecule analysis of fluorescently labeled G-protein-coupled receptors reveals complexes with distinct dynamics and organization. Proc Natl Acad Sci USA 110(2):743

Carlton PM et al (2010) Fast live simultaneous multiwavelength four-dimensional optical microscopy. Proc Natl Acad Sci USA 107(37):16016–16022

Chalfie M et al (1994) Green fluorescent protein as a marker for gene expression. Science 263(5148):802–805

Curd A et al (2015) Construction of an instant structured illumination microscope. Methods 88:37–47

Einstein A (1905) Über die von der molekularkinetischen Theorie der Wärme geforderte Bewegung von in ruhenden Flüssigkeiten suspendierten Teilchen. Ann Phys 322(8)

Elliott AD (2020) Confocal microscopy: principles and modern practices. Curr Protoc Cytom 92(1):e68

Erdogan T (2011) Optical filters for wavelength selection in fluorescence instrumentation. Curr Protoc Cytom 2:2 4 1–2 4 25

Fish KN (2009) Total internal reflection fluorescence (TIRF) microscopy. Curr Protoc Cytom. Chapter 12: p. Unit12 18

Fish KN (2022) Total internal reflection fluorescence (TIRF) microscopy. Curr Protoc 2(8):e517

Fouquet C et al (2015) Improving axial resolution in confocal microscopy with new high refractive index mounting media. PLoS One 10(3):e0121096

Fox PD, Loftus RJ, Tamkun MM (2013) Regulation of Kv2.1 K(+) conductance by cell surface channel density. J Neurosci 33(3):1259–1270

Freeman SA et al (2018) Transmembrane pickets connect Cyto- and Pericellular skeletons forming barriers to receptor engagement. Cell 172(1-2)

Gada KD et al (2023) Imaging membrane proteins using total internal reflection fluorescence microscopy (TIRFM) in mammalian cells. Bio Protoc 13(4):e4614

Galbraith CG, Galbraith JA (2011) Super-resolution microscopy at a glance. J Cell Sci 124(Pt 10):1607–1611

Gautier A et al (2008) An engineered protein tag for multiprotein labeling in living cells. Chem Biol 15(2)

Gould TJ, Myers JR, Bewersdorf J (2011) Total internal reflection STED microscopy. Opt Express 19(14):13351–13357

Goulian M, Simon SM (2000) Tracking single proteins within cells. Biophys J 79(4):2188

Grimes J et al (2023) Plasma membrane preassociation drives β-arrestin coupling to receptors and activation. Cell 186(10)

Gross D, Watt W (2018) Cell surface clustering and mobility of the Liganded LDL receptor Meas. Spectroscopic Membrane Probes. CRC

Guo M et al (2018) Single-shot super-resolution total internal reflection fluorescence microscopy. Nat Methods 15(6):425–428

Gustafsson MG (2000) Surpassing the lateral resolution limit by a factor of two using structured illumination microscopy. J Microsc 198(Pt 2):82–87

Han R et al (2013) Recent advances in super-resolution fluorescence imaging and its applications in biology. J Genet Genomics 40(12):583–595

Hattori T, Koide S (2018) Next-generation antibodies for post-translational modifications. Curr Opin Struct Biol 51:141–148

Hauser AS et al (2018) Pharmacogenomics of GPCR drug targets. Cell 172(1-2)

Heim R, Tsien RY (1996) Engineering green fluorescent protein for improved brightness, longer wavelengths and fluorescence resonance energy transfer. Curr Biol 6(2):178–182

Heim R, Prasher DC, Tsien RY (1994) Wavelength mutations and posttranslational autoxidation of green fluorescent protein. Proc Natl Acad Sci USA 91(26):12501–12504

Hell SW, Wichmann J (1994) Breaking the diffraction resolution limit by stimulated emission: stimulated-emission-depletion fluorescence microscopy. Opt Lett 19(11):780–782

Hentschel K, Hentschel K (2001) Das Brechungsgesetz in der Fassung von Snellius. Arch Hist Exact Sci 55(4):297

Hines KE (2013) Inferring subunit stoichiometry from single molecule photobleaching. J Gen Physiol 141(6):737–746

Honzel E et al (2024) A comparison of confocal and epifluorescence microscopy for quantification of RNAScope and immunohistochemistry fluorescent images. J Neurosci Methods 403:110050

Hupfel M et al (2021) Wavelet-based background and noise subtraction for fluorescence microscopy images. Biomed Opt Express 12(2):969–980

Jabłoński A (1935) Über den Mechanismus der Photolumineszenz von Farbstoffphosphoren. Z Phys 94(1–2):38–46

Jacobson K, Sheets ED, Simson R (1995) Revisiting the fluid mosaic model of membranes. Science NY 268(5216):1441

Jain A et al (2011) Probing cellular protein complexes using single-molecule pull-down. Nature 473(7348):484–488

Jain A et al (2012) Single-molecule pull-down for studying protein interactions. Nat Protoc 7(3):445–452

Jain A et al (2014) Stoichiometry and assembly of mTOR complexes revealed by single-molecule pulldown. Proc Natl Acad Sci USA 111(50):17833–17838

Jaqaman K et al (2008) Robust single-particle tracking in live-cell time-lapse sequences. Nat Methods 5(8)

Keppler A et al (2003) A general method for the covalent labeling of fusion proteins with small molecules in vivo. Nat Biotechnol 21(1)

Kitamura A (2021) Pinhole closure improves spatial resolution in confocal scanning microscopy. Methods Mol Biol 2274:385–389

Kuey C et al (2019) Unintended perturbation of protein function using GFP nanobodies in human cells. J Cell Sci 132(21)

Ladouceur AM, Brown CM (2021) Fluorescence microscopy light source review. Curr Protoc 1(9):e243

Lakowicz JR (2006) Principles of fluorescence spectroscopy, vol xxvi, 3rd edn. Springer, New York, p 954

Lang E, Baier J, Kohler J (2006) Epifluorescence, confocal and total internal reflection microscopy for single-molecule experiments: a quantitative comparison. J Microsc 222(Pt 2):118–123

Larson J et al (2014) Design and construction of a multiwavelength, micromirror total internal reflectance fluorescence microscope. Nat Protoc 9(10):2317–2328

Leake MC et al (2006) Stoichiometry and turnover in single, functioning membrane protein complexes. Nature 443(7109):355–358

Lelek M et al (2021) Single-molecule localization microscopy. Nat Rev Methods Primers 1:39

Liu Z, Lavis LD, Betzig E (2015) Imaging live-cell dynamics and structure at the single-molecule level. Mol Cell 58(4)

Logothetis DE et al (1987) The βγ subunits of GTP-binding proteins activate the muscarinic K+ channel in heart. Nature 325:6102

Los GV et al (2008) HaloTag: a novel protein labeling technology for cell imaging and protein analysis. ACS Chem Biol 3(6)

Mashanov GI et al (2004) The spatial and temporal dynamics of pleckstrin homology domain binding at the plasma membrane measured by imaging single molecules in live mouse myoblasts. J Biol Chem 279(15):15274–15280

Masi A et al (2010) Optical methods in the study of protein-protein interactions. Adv Exp Med Biol 674:33–42

Matsuoka S et al (2012) Single-molecule imaging of stochastic signaling events in living cells. Cold Spring Harb Protoc 2012(3):267–278

Mattheyses AL, Simon SM, Rappoport JZ (2010) Imaging with total internal reflection fluorescence microscopy for the cell biologist. J Cell Sci 123(Pt 21):3621–3628

Michalet X (2010) Mean square displacement analysis of single-particle trajectories with localization error: Brownian motion in isotropic medium. Phys Rev E Stat Nonlinear Soft Matter Phys 82(4 Pt 1)

Mondal PP, Diaspro A (2014) Fundamentals of fluorescence microscopy: exploring life with light. Springer, Dordrecht; New York. xv, 218 pages

Monici M (2005) Cell and tissue autofluorescence research and diagnostic applications. Biotechnol Annu Rev 11:227

Nicolson GL (2013) Update of the 1972 singer-Nicolson fluid-mosaic model of membrane structure, vol 1(1). Discoveries, Craiova

Oheim M et al (2019) Calibrating evanescent-wave penetration depths for biological TIRF microscopy. Biophys J 117(5):795–809

Panchuk-Voloshina N et al (1999) Alexa dyes, a series of new fluorescent dyes that yield exceptionally bright, photostable conjugates. J Histochem Cytochem 47(9):1179–1188

Pedelacq JD et al (2006) Engineering and characterization of a superfolder green fluorescent protein. Nat Biotechnol 24(1):79–88

Pierce KL et al (2001) Classical and new roles of beta-arrestins in the regulation of G-protein-coupled receptors. Nat Rev Neurosci 2(10)

Pitkeathly WT et al (2012) Auto-align—multi-modality fluorescence microscopy image co-registration. Traffic 13(2):204–217

Plant LD et al (2014) Individual IKs channels at the surface of mammalian cells contain two KCNE1 accessory subunits. Proc Natl Acad Sci USA 111(14):E1438–E1446

Prasher DC et al (1992) Primary structure of the Aequorea victoria green-fluorescent protein. Gene 111(2):229–233

R. E (1911) Brownian movement and molecular reality. Nature 86(2160)

Rao TC, Nawara TJ, Mattheyses AL (2022) Live-cell Total internal reflection fluorescence (TIRF) microscopy to investigate protein internalization dynamics. Methods Mol Biol 2438:45–58

Rashed R (1990) A pioneer in anaclastics: Ibn Sahl on burning mirrors and lenses. Isis 81(3):464–491

Rosenbaum DM, Rasmussen SG, Kobilka BK (2009) The structure and function of G-protein-coupled receptors. Nature 459(7245)

Rust MJ, Bates M, Zhuang X (2006) Sub-diffraction-limit imaging by stochastic optical reconstruction microscopy (STORM). Nat Methods 3(10):793–795

Saffman PG, Delbrück M (1975) Brownian motion in biological membranes. Proc Natl Acad Sci USA 72(8)

Seidel S et al (2023) Improvement of HEK293 cell growth by adapting hydrodynamic stress and predicting cell aggregate size distribution. Bioengineering (Basel) 10(4)

Shaner NC et al (2004) Improved monomeric red, orange and yellow fluorescent proteins derived from Discosoma sp. red fluorescent protein. Nat Biotechnol 22(12):1567–1572

Shaw PJ, Rawlins DJ (1991) The point-spread function of a confocal microscope: its measurement and use in deconvolution of 3-D data. J Microsc 163(2):151–165

Shen H et al (2017) Single particle tracking: from Theory to biophysical applications. Chem Rev 117(11)

Simon F et al (2024) A guide to single-particle tracking. Nat Rev Method Primers 4(1)

Simson R, Sheets ED, Jacobson K (1995) Detection of temporary lateral confinement of membrane proteins using single-particle tracking analysis. Biophysical J 69(3)

Singer SJ, Nicolson GL (1972) The fluid mosaic model of the structure of cell membranes. Science, NY 175(423):720

Sivankutty S et al (2020) Supercritical angle fluorescence for enhanced axial sectioning in STED microscopy. Methods 174:20–26

Smith DJ (2008) Ultimate resolution in the electron microscope? Mater Today 11:30–38

Stepanenko OV et al (2008) Fluorescent proteins as biomarkers and biosensors: throwing color lights on molecular and cellular processes. Curr Protein Pept Sci 9(4):338–369

Stolpner N, Dickinson DJ (2022) Single-cell single-molecule pull-down (sc-SiMPull) for detection of protein complexes from embryonic lysates. Methods Mol Biol 2438:59–81

Sungkaworn, T., et al., Single-molecule imaging reveals receptor–G protein interactions at cell surface hot spots. Nature 2017 550:7677

Tam J, Merino D (2015) Stochastic optical reconstruction microscopy (STORM) in comparison with stimulated emission depletion (STED) and other imaging methods. J Neurochem 135(4):643–658

Trautwein W et al (1958) Zum Mechanismus der Membranwirkung des Acetylcholin an der Herzmuskelfaser. Pflüger's Archiv für die gesamte Physiologie des Menschen und der Tiere 266(3):324

Tsien RY (1998) The green fluorescent protein. Annu Rev Biochem 67:509–544

Udenfriend S (1995) Development of the spectrophotofluorometer and its commercialization. Protein Sci 4(3):542–551

Ulbrich MH, Isacoff EY (2007) Subunit counting in membrane-bound proteins. Nat Methods 4(4):319–321

Ulfhake B et al (1992) In: Conn PM (ed) 7—Confocal fluorescence microscopy in three-dimensional analysis of axon terminal distribution, neuronal connectivity, and colocalization of messenger molecules in nervous tissue: computerized analysis, in methods in neurosciences. Academic Press, pp 94–128

Valentine MT et al (2001) Investigating the microenvironments of inhomogeneous soft materials with multiple particle tracking. Phys Rev E Stat Nonlinear Soft Matter Phys 64(6 Pt 1)

Valli J et al (2021) Seeing beyond the limit: a guide to choosing the right super-resolution microscopy technique. J Biol Chem 297(1):100791

Webb DJ, Brown CM (2013) Epi-fluorescence microscopy. Methods Mol Biol 931:29–59

Wilson KM et al (1996) Single particle tracking of cell-surface HLA-DR molecules using R-phycoerythrin labeled monoclonal antibodies and fluorescence digital imaging. J Cell Sci 109(Pt 8):8

Yamamura H, Suzuki Y, Imaizumi Y (2015) New light on ion channel imaging by total internal reflection fluorescence (TIRF) microscopy. J Pharmacol Sci 128(1):1–7

York AG et al (2013) Instant super-resolution imaging in live cells and embryos via analog image processing. Nat Methods 10(11):1122–1126

Zeghal M et al (2023) Profiling of basal and ligand-dependent GPCR activities by means of a polyvalent cell-based high-throughput platform. Nat Commun 14(1)

Zhao Q et al (2023) Nanobead-based single-molecule pull-down for single cells. Heliyon 9(11):e22306

Zou F, Bai L (2019) Using time-lapse fluorescence microscopy to study gene regulation. Methods 159–160:138–145

A Historical Outline of Fluorescence Spectroscopy in Ion Channel Research

6

Rikard Blunck

Abstract

Fluorescence spectroscopy has always played an essential role in biomedical research as a tool to isolate the proteins of interest from their background in a natural environment, and recently, fluorescence spectroscopy has provided dynamical information that can be precisely linked to the ever-increasing wealth of atomic-scale structures. In this chapter, we will discuss the theory underlying the use of fluorescence as a noninvasive tool, review the role it played in deciphering some of the essential mechanisms in ion channels, and outline the latest developments that point to where the field is heading.

Keywords

Fluorescence spectroscopy · Förster resonance energy transfer · Noncanonical amino acids · Ion channels

R. Blunck (✉)
Department of Physics, Université de Montréal, Montréal, Québec, Canada

Department of Pharmacology and Physiology, Université de Montréal, Montréal, Québec, Canada

CIRCA—Centre interdisciplinaire de recherche sur le cerveau et l'apprentissage, Université de Montréal, Montréal, Québec, Canada
e-mail: rikard.blunck@umontreal.ca

6.1 Introduction

Fluorescence has always played an essential role in research of living matter because it allows optically targeting the subject of study in a complex environment and obtaining dynamic information with it. This subject might be an atom or a building block in a macromolecule and entire macromolecules or assemblies and aggregates thereof in a cell or tissue. Any modern high- and super-resolution imaging technique is using fluorescence in one form or another. In this chapter, we will explore how fluorescence spectroscopy has proven useful for structure-function studies.

This chapter does not reflect the entire body of work that has been carried out on ion channels using fluorescence spectroscopy but rather follows selected research directions that have been strongly influenced by the use of fluorescence techniques. We will thereby concentrate on the contributions of the fluorescence to these aspects and will not have space to include the many excellent articles in the field, on which the fluorescence studies are based and without which they would had never been possible. We will also offer an outlook on the current development of the use of fluorescence to study ion channels.

A. Rosenhouse-Dantsker (ed.), *Decoding Ion Channels Structure and Function*, Advances in Experimental Medicine and Biology 1497, https://doi.org/10.1007/978-3-032-07523-9_6

6.2 Theory of Fluorescence

While fluorescence had been observed historically, it was studied in detail for the first time by Sir George Gabriel Stokes, when he studied the spectral properties of light using quinine as a dispersive (Stokes 1852). He realized that the solution started glowing blue when it was moved beyond the violet into the near UV range.[1] This anecdote shows the occurrence of the most important property of fluorescence: It alters the wavelength of the incoming light before emitting, thus rendering the fluorophores visible over a background.

The shift toward longer wavelengths is named after its observer *Stoke's Shift* and is caused by a loss of vibrational energy. Each molecule features vibrational energy levels above the ground electronic orbitals. Since the energies of these vibrational modes are higher than the available thermal energy, only the ground vibrational state is occupied in the absence of any excitation. If the energy difference of the electronic states falls within the energy of photons with wavelengths in the visible range ($\lambda = 400 - 750\,\text{nm}$), light will be absorbed. The molecule is thereby typically excited to a higher vibrational state since their orbital spatial distribution coincides better with the spatial distribution of the ground state S_0 (Franck-Condon principle; Condon 1926). This vibrational energy rapidly relaxes to the ground vibrational state (Fig. 6.1a), leading to a loss of part of the energy of the exciting photon.

In most molecules, the energy of excitation to S_1 is dissipated to the surrounding as thermal energy. In the fluorophores, however, the excited state is long lived enough to directly return to the ground state, which typically occurs within nanoseconds, the *lifetime of the fluorophore*. The energy is, in this case, emitted in form of a photon; nevertheless, the probability to transit to a higher vibrational state is again increased, so that only part of the energy is emitted (Fig. 6.1a). The rest is dissipated as thermal energy. The *Stokes shift*, the shift in the wavelength between the absorbed and emitted photon, is thus due to the double relaxation of the vibrational states during the excitation-emission cycle.

6.2.1 Fluorescence Lifetime

It is instructive to describe the fluorescence cycle as a kinetic model, considering the occupancy of the excited S_1 state (Fig. 6.1b). The occupancy of S_1 increases by absorption of the excitation wavelength λ_{exc}. The number of absorbed photons will be proportional to the number of non-excited fluorophores and the excitation intensity.

$$\frac{dS_1{}^+}{dt} = k_{exc} \cdot I_{exc} \cdot S_0, \tag{6.1}$$

where k_{exc} is a function of the excitation spectrum and the absorption coefficient of the fluorophore.

The occupancy of the excited state S_1 decreases by non-radiative decay when energy is dissipated to the surrounding (k_{nr}) and by emission of photons (k_f). Both processes would be proportional to the number of excited states S_1, such that the total change in occupancy results in

$$\frac{dS_1}{dt} = k_{exc} \cdot I_{exc} \cdot S_0 - (k_{nr} + k_f) \cdot S_1. \tag{6.2}$$

The steady-state fluorescence intensity $I_{f,\infty}$ is then the fraction of the generated excited states $S_{1,\infty}$ that will emit a photon.

$$I_{f,\infty} = k_f \cdot S_{1,\infty} = \frac{k_f \cdot (k_{exc} \cdot I_{exc})}{k_{nr} + k_f + k_{exc} \cdot I_{exc}}. \tag{6.3}$$

In most use cases, the rate of excitation is much smaller than the deexcitation rate ($k_{exc} I_{exc} \ll k_{nr} + k_f$), such that one can simplify Eq. (6.3) to

[1] Quinine has maxima of excitation and emission at 349 nm and 461 nm, respectively.

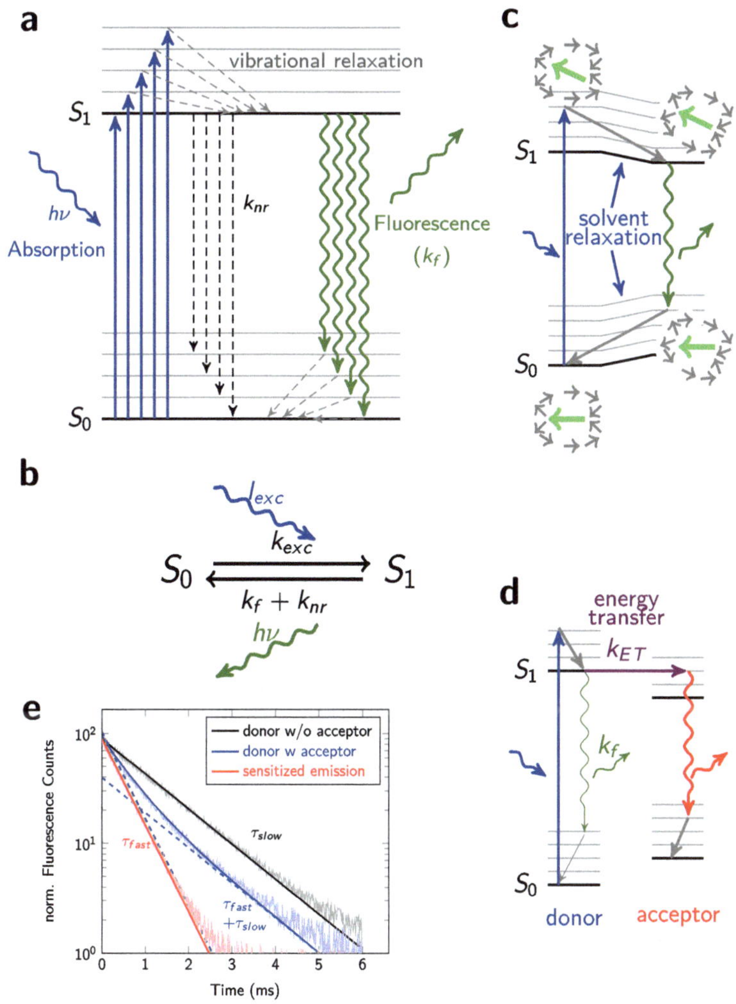

Fig. 6.1 (**a**) Jablonski diagram of the fluorescence process. Photons ($h\nu$) are absorbed and excite the fluorophore from the S_0 to the S_1 orbital with a higher vibrational energy level (*blue*). Vibrational energy is dissipated (*gray*). The fluorophore returns to the ground state via non-radiative decay (k_{nr}, *black*) or by emitting a photon (k_f,*green*), again to a higher vibrational energy level, which gets dissipated rapidly. (**b**) Kinetic Markov model to describe excitation and emission of the fluorophores shown in (**a**). (**c**) Principle of solvent relaxation. Due to the misalignment of the solvent's dipoles, the initial energy levels after excitation and emission are increased, leading to a longer emission wavelength (lower energy photons). Green and gray arrows indicate the dipole moments of the fluorophore and solvent, respectively. (**d**) Principle of Förster resonance energy transfer. If within a short distance ($\approx 2 \cdot R_0$) of the donor fluorophore an acceptor fluorophore with suitable excitation spectrum is found, the excited state can directly be transferred to the acceptor. (**e**) Orthogonality of exponential decays in lifetime measurements. The donor alone will decay with the intrinsic lifetime of the fluorophore (τ_{slow}, *black*). An acceptor with an intrinsically faster lifetime, to which the energy is transferred, will decay with the time constant of energy transfer (τ_{fast}, *red*). The donor, in the presence of acceptor, will also decay with the lifetime of the energy transfer; however, any unpaired donor will decay with the intrinsic lifetime. In the *blue* trace, one can clearly distinguish both components (LRET measurements reproduced from Faure et al. (2012)).

$$
\begin{aligned}
I_f &= \frac{k_f \cdot (k_{exc} \cdot I_{exc})}{k_{nr} + k_f + k_{exc} \cdot I_{exc}} \\
&\approx \frac{k_f}{k_{nr} + k_f} \cdot (k_{exc} \cdot I_{exc}) \\
&= QY \cdot (k_{exc} \cdot I_{exc}),
\end{aligned}
$$

with QY the *quantum yield*, i.e., the fraction of deexcitations that will emit a photon. If the fluorescence is excited with a pulsed light source, a specific number of fluorophores become excited ($S_1(t = 0)$), and the excitation intensity becomes zero after the pulse ($I_{exc}(t > 0) \equiv 0$). Therefore, fluorescence decay is observed

$$S_1(t) = S_1(0) \cdot e^{-t/\tau}, \tag{6.4}$$

with the lifetime of the fluorophore τ

$$\tau = \frac{1}{k_f + k_{nr}}.$$

6.2.2 Dependence of Fluorescence on Environment

All the processes mentioned above are intrinsic properties of the fluorophores. These properties are altered by the physical and chemical environment, such as the polarity of the solvent, interacting molecules, or presence of electric fields, and these effects can be exploited to probe the movement and environment of fluorescence probes attached to macromolecules.

Solvent Relaxation and Electrochromic Effect Solvent relaxation and the electrochromic effect both lead to a shift of the emission spectrum due to the interaction with electric fields. In the case of solvent relaxation, these are the dipole moments of the solvent; since fluorophores absorb electromagnetic radiation, they have an intrinsic dipole moment that alters during the transition from the ground to the excited state. If a polar solvent surrounds the fluorophore, the solvent's own intrinsic dipole moments will align with the fluorophore's as that would be the energetically most favorable state (Fig. 6.1c). Upon excitation of the fluorophore, its altered dipole moment is no longer aligned with the solvent's ones. Accordingly, the state has higher energy. After some time, the solvent realigns, and the energy of the excited state decreases. This lower energy is lost for the emitted photon in addition to the vibrational relaxation. When the fluorophore returns to the ground S_0 state, again, the dipoles are misaligned, increasing the energy of all S_0 states. Also this part of the energy is lost for the emitted photon, and the overall energy difference decreases, leading to a longer wavelength (Fig. 6.1c). Such a shift in the emission spectrum can be used to probe water accessibility of labeled positions in the protein (see also Sect. 6.3.2).

In the case of the *electrochromic effect*, the energy levels are altered by the interaction with an external electric field (*Stark effect*); in the case of ion channels this electric field is given by the membrane potential. The electric field causes a splitting of the energy levels such that the energy gap between S_0 and S_1 is altered, leading to a shift in the emission spectrum. Because this is an electronic effect intrinsic to the molecule, it occurs faster than any alteration of the membrane potential itself, thus truthfully following the dynamics of potential changes. Certain fluorophores, like ANEPPS,[2] are well known to sense membrane potential changes.

Quenching (Electron Transfer) Quenching—or *electron transfer*—is a process where the interaction between the excited fluorophore with a molecule (the quencher) allows an additional pathway (k_q) to the ground state S_0. Accordingly, the steady-state fluorescence intensity (Eq. (6.3)) decreases to

[2] di-N-ANEPPS is the electrochromic dye ANEP (Amino-Naphthyl-Ethenyl-Pyridinium) linked to propanesulfonate (PS) on one end and, on the other end, two carbon chains of length N (di-4 or di-8), which anchor it into the membrane.

$$\begin{aligned} I_f &= k_f \cdot S_1 \\ &= \frac{k_f \cdot (k_{exc} \cdot I_{exc})}{k_q + k_{nr} + k_f + k_{exc} \cdot I_{exc}} \\ &\approx QY_q \cdot k_{exc} I_{exc} \end{aligned} \tag{6.5}$$

with the reduced quantum yield

$$QY_q = \frac{k_f}{k_q + k_{nr} + k_f}. \tag{6.6}$$

In case the Van der Waals radii overlap, the excited state (the exciton) can be transferred from the fluorophore to the quencher. The excited electron in the quencher then, similar to non-radiative decay, returns to the ground state by non-radiative decay. Since the Van der Waals radii are the size of the atoms, very small movements that lower the overlap between the electron wave functions will immediately abolish the quenching. The quencher can be a molecule in solution, in which case the degree of quenching (for ensembles of fluorophores) is limited by quencher diffusion and concentration that will determine the number of collisions between quencher and fluorophore. More likely, it is a residue of the surrounding protein which can act as acceptor. In particular tryptophane has a high quenching efficiency (Marme et al. 2003). If the quenching amino acid comes close to the fluorophore in one conformation of the protein, entering this state will be associated with a decrease in fluorescence intensity.

Förster Resonance Energy Transfer (FRET) FRET is a near-field dipole-dipole interaction between a fluorophore and a suitable acceptor (Fig. 6.1d). Similar to electron transfer, also here an additional pathway is added to leave the S_1 state, resulting in a reduced quantum yield

$$QY_{ET} = \frac{k_f}{k_{ET} + k_{nr} + k_f}, \tag{6.7}$$

and—important for lifetime FRET measurements—also a reduced lifetime

$$\tau_{ET} = \frac{1}{k_{ET} + k_{nr} + k_f}. \tag{6.8}$$

The transition rate for energy transfer k_{ET} is a function of the overlap between the donor emission spectrum and the acceptor absorption spectrum, the relative orientation of donor and acceptor, as well as the sixth power of the distance between donor and acceptor normalized to a specific distance R_0, characteristic for the donor-acceptor pair. The acceptor does not need to be a fluorophore but only requires to absorb in the wavelength range of the excitation spectrum of the donor. With the distance dependence, FRET is not only suitable to detect conformational changes but can also be used as a measure for atomic distances within the macromolecule with Ångström resolution. The uncertainty of the measured distances arises from length and flexibility of the linker, attaching the probe to the macromolecule, and the relative orientation of donor and acceptor. The latter can be circumvented by using lanthanides as donors (*LRET*) or transition metals as acceptors (*tmFRET*). As (colored) atoms, they do not have a continuous dipole moment such that the orientation factor between donor and acceptor can be omitted.

In both cases, the quenching and FRET, the fluorescence intensity is the superposition of the signals of channels in both conformational states dependent on the relative occupancies. This will distort the distances to be determined. In this case, it is advantageous to use the fluorescence lifetimes. According to Eq. (6.8), each conformation will have a distinct lifetime with an exponential decay. Since exponentials are orthogonal to one another, this allows to directly determine the population occupancies and the exact distances for each population (Fig. 6.1e).

6.2.3 Labeling Ion Channels with Fluorophores

One of the most essential tasks when using fluorescence spectroscopy to study macromolecules is to specifically label them. While a certain fraction of unspecific labeling is tolerable if one focusses on dynamics, it will reduce the signal-to-noise ratio. In case one aims to measure FRET efficiencies, distribution of populations, or lifetimes of the fluorophores, it is essential that background fluorescence is kept to a minimum. This becomes increasingly difficult with more native systems. For purified ion channels, the proteins can be chemically labeled after isolation and excess label removed by affinity chromatography or gel filtration before reconstitution into proteoliposomes (see, e.g., Blunck et al. 2008), as routinely done for EPR experiments (e.g., Perozo et al. 1998). In more complex systems, specific labeling becomes essential.

Chemical Labeling Dependent on the use case, it might also be important to target a specific position within the ion channel; while for tracking expression patterns and trafficking, any labeling, including fluorescently tagged antibodies or toxins suffice, any structural and dynamic information can only be obtained if the labels are attached specifically at the position of interest. The most common way to achieve this is to use chemical labeling techniques, in particular thiol-reactive chemistry. Compounds such as maleimide, iodoacetamide, and methanethiosulfonate react with the thiol group of any accessible cysteines (Fig. 6.2a). To direct a label to a specific position within the protein, a cysteine is introduced at the position in question by site-directed mutagenesis. While chemical labeling is straightforward to accomplish, it has the limitation that only externally accessible sites can be labeled. In addition, any endogenous (accessible) cysteine, not part of a disulfide bridge, would also be labeled. In reconstitution systems where ion channels are typically overexpressed (Xenopus oocytes or transfected mammalian cell lines), this background labeling plays a minor role but will be problematic at many native expression levels.

Fluorescent Proteins Both limitations can be overcome with genetically encoded tags. The most commonly used are the many variations of the fluorescent proteins (Prasher et al. 1992; Heim et al. 1994; Kong et al. 2020). The excellent properties of GFP come with the drawback of a relatively large size of ≈ 27 kDa. The chromophore itself, which is formed by cyclization of three sequential amino acids (Ser-Tyr-Gly), is similarly sized as the typical synthetic organic fluorophore (Fig. 6.2b) but needs to be protected by a beta barrel. Not only the large size impedes targeting within the ion channel, but also the beta barrel protects it against modulation by the immediate environment. Therefore fluorescent proteins are not well suited for tracking dynamics or intramolecular distances. However, if space is available and the distance changes are sufficiently large, it can be used to follow dynamics in a FRET assay. This has been done for the cytosolic domain of the BK_{Ca} channel (Miranda et al. 2013) and in the detection of sodium channel activation kinetics (Ataka and Pieribone 2002).

In general, however, fluorescent proteins are better suited for tracking of ion channels to detect expression patterns and trafficking or to detect interaction partners either within the ion channel or between the ion channel and its auxiliary proteins (Biskup et al. 2004; Harley et al. 2021). FRET pairs of fluorescent proteins typically have a useful working distance of ≈ 50 Å, which is the perfect distance for detecting binding of two distinct units to a biological assembly. Intramolecular distances have been determined in TRPV1, for instance, (De-la Rosa et al. 2013).

Fluorescent proteins have been particularly powerful to determine stoichiometries. With certain limitations, this can be done using FRET assays (Zheng and Zagotta 2004), but single subunit counting offers a more precise alternative, where the sudden photophysical destruction of any fluorophore is exploited to directly count the number of fluorophores present in the complex (Fig. 6.2c) (Ulbrich and Isacoff 2007; McGuire et al. 2012; Blunck 2021). The photodestruction, or -bleaching, of the fluorescent protein can be seen on the single molecule level as a stepwise intensity decrease. This was initially used to determine stoichiometries in NMDA receptors (Ulbrich and Isacoff 2007) but rapidly became the method of choice to determine stoichiometry of ion channels. A high labeling efficiency is essential to the success of the counting, and

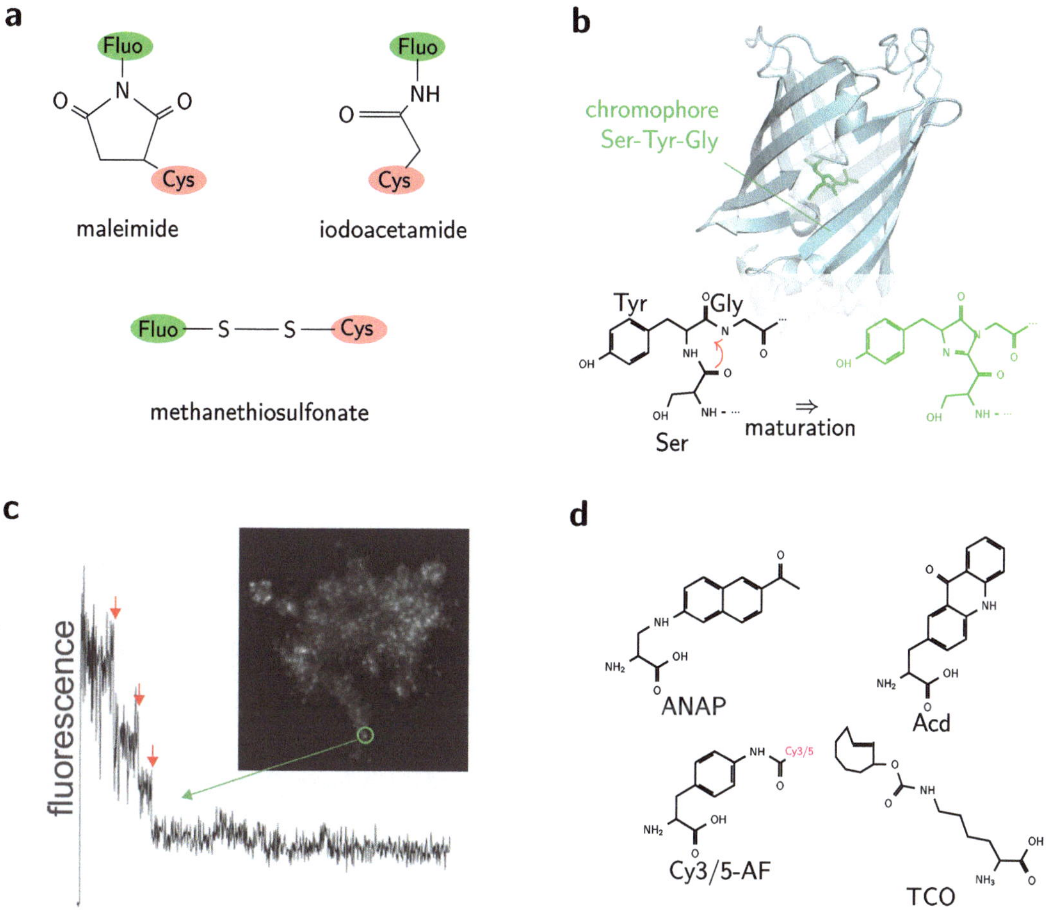

Fig. 6.2 (**a**) Typical linker compounds linked to the ion channels at the thiol group of the cysteine (Fluo = fluorophore; Cys = ion channel). (**b**) Structure of GFP and the chemical structure of its chromophore. (**c**) Structure of the fncAA Anap: *acetylnaphthalen-ylamino-aminopropanoic acid* (Lee et al. 2009), Cy3/5-AF: Cy3/5-para-amino-L-phenylalanine (Leisle et al. 2016), Acd: acridonylalanine (Jones et al. 2021), and the ccAA TCO: trans-cyclooctene (Tyagi and Lemke 2013). (**d**) Photobleaching in single subunit counting: Single fluorophores will photobleach in a sudden, distinct step (red arrows). GluK2 receptors were tagged with monomeric eGFP and expressed at low level in HEK293T cells. The single receptors will contain four subunits, such that following the fluorescence intensity for a single spot will lead to 1–4 photobleaching events. For details see (Blunck 2021)

fluorescent proteins, as genetically encoded, are technically present in 100% of all ion channels. However, the chromophore contained in the fluorescent proteins requires a maturation step (Fig. 6.2b), since the initial amino acid residues that form the chromoform are not intrinsically fluorescent. This *maturation* of the fluorescent proteins is not complete and has been reported as between 50 and 90% dependent on conditions.

Many variants of fluorescent proteins have been generated to be monitors of specific environmental parameters (see below).

Noncanonical Amino Acids Noncanonical amino acids can also be genetically encoded while still maintaining a small size comparable to chemical labels. Because of this property they are perfect probes for following conformational changes within the ion channel. During translation a fluorescent noncanonical amino acid (fncAA) is introduced into the protein of interest by nonsense suppression: A tRNA amino-acylated with the fncAA is provided for one of the stop-codons such that the ribosome will introduce the fncAA at this position (Diercks et al. 2021). There are three principal methods to introduce an fncAA into the ion channels. The first method is to aminoacylate synthetic tRNA with the desired synthesized fncAA in vitro and to inject these directly into the cell (Leisle et al. 2016). This method has least restrictions in terms of the nature and size of the ncAA, as long as the ion channel still folds correctly. It was used to intercalate the large Cy-dyes into ion channels (Leisle et al. 2016). However, in vitro aminoacylation and synthesis of the ncAA is technically very demanding.

One can let the natural system, the *tRNA-synthetase*, fulfill this job by evolving it to recognize specifically only the nonsense tRNA and the fncAA. While the evolution step is also time-consuming, evolved systems are readily available. To achieve expression, the modified tRNA-synthetase and tRNA-pair have to be cotransfected (Chatterjee et al. 2013)—or injected in case of Xenopus oocytes (Kalstrup and Blunck 2013). Anap (acetylnaphthalen-ylamino-aminopropanoic acid) (Lee et al. 2009; Chatterjee et al. 2013; Kalstrup and Blunck 2013) and Azd (acridonylalanine, Fig. 6.2c) (Jones et al. 2021) (Fig. 6.2c) are examples for available fncAAs with modified tRNA-synthetase. Anap was the first one to be used for measuring dynamics of Kv channels and has since be used for many channels (for details see below) (Kalstrup and Blunck 2013). Anap has a lower quantum yield than a synthetic organic fluorophore but therefore a size only slightly larger than a tryptophane. One can achieve expression levels similar to wildtype with Anap introduced into the protein (Kalstrup and Blunck 2017), and there are few restrictions as to where it can be placed inside the ion channel, similar to a tryptophane scan (but see Kalstrup and Blunck 2015).

To combine the high quantum yield of the organic fluorophores with the genetic targeting of the ncAA, one can use *click-chemistry noncanonical amino acids (ccAA)*. The ccAAs are introduced identically to Anap but are not themselves fluorescent. They are labeled similar to the cysteines in thiol-reactive labeling, but the reaction is more specific, and there is little to no background labeling, as the ccAA only exists in the target protein. One example is TCO (trans-cyclooctene) reactive to tetrazine-linked fluorophores via strain-promoted inverse electron-demand Diels-Alder cycloaddition (SPIEDAC, Fig. 6.2d; Tyagi and Lemke 2013). Many excellent synthetic organic fluorophores are available as tetrazines. The advantage of ccAA over Anap—other than the higher quantum yield—is that one can specifically label surface proteins at a specific time and track retrograde trafficking. Nevertheless, this can also be seen as a limitation, as only externally accessible sites can be studied—a limitation inherited from chemical labeling.

Other, larger genetically encoded tags, such as HALO or SNAP tags, are also available. While these are larger in size, selectivity among the tags enables bio-orthogonal labeling of different proteins within one system (see Sect. 6.6) (Jung et al. 2019).

Environmental Labels The above descriptions all referred to labeling of the ion channel itself. However, often, the function of the ion channel or its relative movement can be detected in its environment; be it through movement relative to the membrane, through changes of local concentrations or the membrane potential. Widely applied are the calcium-sensitive dyes that alter their fluorescence properties dependent on the calcium concentration. These were particularly useful for detecting activation of calcium channels (Shuai and Parker 2005) or nonselective cation channels (e.g., TRP channels Grandl et al. 2010). A great variety of environmental sensors based on fluorescent proteins were designed (for review see Gest et al. 2024), including those interesting for the study of ion channels such as protons, calcium, and potassium (Reifenrath and Boles 2018; Shen et al. 2019). In addition, sensors were developed for probing the membrane potential (Chanda et al. 2005; St-Pierre et al. 2014) and mechanical stress in the membrane (Meng et al. 2008). These probes either use FRET pairs that alter their distance based on the stimulus or lose their fluorescent properties in absence of the stimulus. To detect movements relative to the membrane normal, FRET is very powerful since the transmembrane distance ranges between 38 and 50 Å so that movement normal to the membrane increases the relative distance between the donor-acceptor pair with one of them anchored in the membrane (Chanda et al. 2005; Groulx et al. 2010).

6.3 Studying Ion Channel Structure and Function

6.3.1 Movements of the Voltage-Sensing Domain

Establishing structure-function relations is particularly applicable for ion channels, or more generally electrogenic transmembrane proteins, because their function is readily accessible with electrophysiological methods. Combination of fluorescence spectroscopy with various electrophysiological methods, including two-electrode voltage-clamp (Mannuzzu et al. 1996), cut-open oocyte voltage-clamp (Cha and Bezanilla 1997), patch-clamp (Zheng and Zagotta 2000), and planar lipid bilayers (McGuire and Blunck 2022; Groulx et al. 2010; Borisenko et al. 2003), allows to simultaneously obtain information on structural rearrangements via fluorescence and on function via electrophysiology (*voltage-clamp fluorometry, VCF*, Fig. 6.3a).

The first, and best-known, examples are the applications to the movement of the voltage-sensing domain of voltage-gated ion channels (Mannuzzu et al. 1996; Cha and Bezanilla 1997). In their seminal work, Hodgkin and Huxley (Abbott 2006) postulated the existence of four gating particles to enable conductance of potassium ions, which were directly proven by the measurements of gating currents by Armstrong and Bezanilla (1973). Due to the strong net charge of the S4 helix in the voltage-sensing domain, it was suspected to move toward the extracellular face of the plasma membrane. This movement was shown with the first VCF measurements of the Shaker Kv channel (Cha and Bezanilla 1997; Mannuzzu et al. 1996). Using thiol-reactive chemistry (see Sect. 6.2.3) organic fluorophores such as fluorescein- and tetramethylrhodamine (TMR)-maleimide were attached to a cysteine at the N-terminus of the S4 helix, which is accessible from the extracellular space. The fluorescence intensity decreased exponentially upon depolarization of the membrane (Fig. 6.3a) following the kinetics of a single voltage sensor. Accordingly, the fluorescence signal occurs at potentials more polarized than the opening of the channel; the fluorescence signals followed directly the gating currents.

Initially, the signal only conveys the information that the labeled position undergoes a relative movement. The exponential changes are due to the kinetics and reflect the time constant of transitioning from the resting to the activated state (see below). However, the direction of the movement can be analyzed if the quencher is identified. In the Shaker Kv channel, the movement of the voltage sensor is translated into a fluorescence signal mainly by quenching with a tryptophane in the S5 (W454) (Priest et al. 2021). One can exploit this quenching by tryptophane to directly probe the distance and direction of the movement. A detailed study on this was carried out in the large-conductance calcium-gated Kv channel BK_{Ca} by introducing the tryptophane at various positions throughout the channel's surface and using variable linker length (Pantazis and Olcese 2012; Pantazis et al. 2018, 2010). This way, it could be established that during activation of the BK_{Ca} voltage-sensing domain, the positively charged S4 moves away from S0, S1, and S2, whereas S2 and S1 approach each other. The degree of movement can be gauged using DEPET (*distance dependence of electron transfer*). Adding just a two-carbon linker between maleimide and fluorophore amplified the signal between S4 and S0 since the fluorophore is now within reach of the tryptophane, but the S1–S4 signal was attenuated since the fluorophore is quenched in both conformations (Fig. 6.3b) (Pantazis et al. 2018).

Quenching only occurs if the Van der Waals radii of the molecules overlap, which means that a small movement of only 1–2 Å suffices to

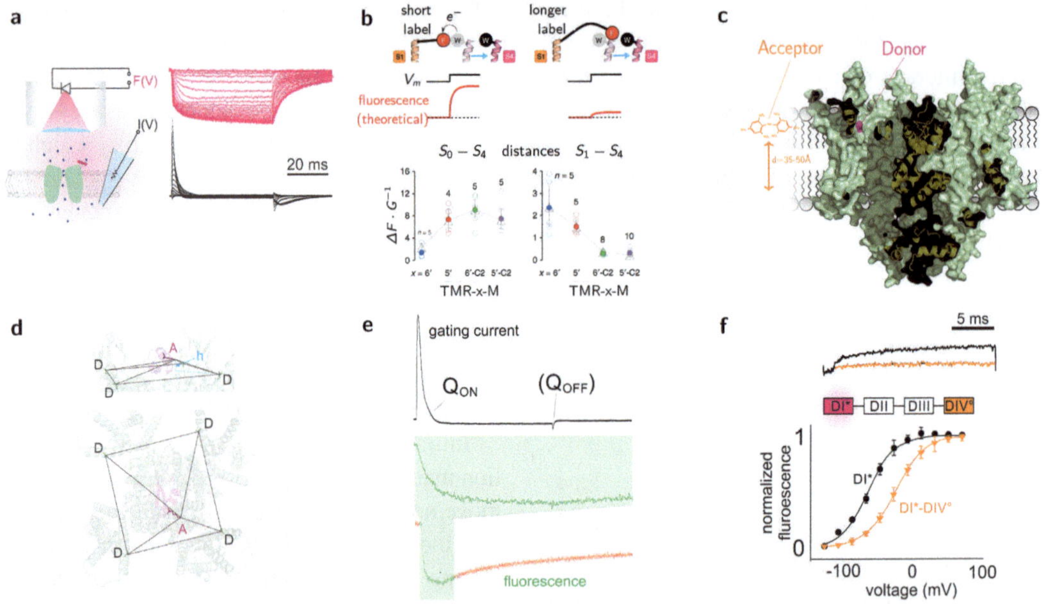

Fig. 6.3 (**a**) Example of a voltage-clamp fluorometry (VCF) experiment. Shaker-W434F (nonconductive) Kv was labeled at the N-terminus of the S4 helix with TMR. Gating currents (black) and fluorescence changes (magenta) are measured simultaneously as a response to a series of depolarizing pulses. (**b**) DEPET: Dequenching (intensity increase) occurs upon relative distancing between fluorophore and quencher. If the linker is too long, the signal gradually disappears. Dependence on the linker length of quenching of the fluorophore tetramethylrhodamine (TMR) by tryptophane on S0 (*left*) and S1 (*right*, figure from Pantazis et al. 2018). (**c**) Dipicrylamine (DPA) partitions into the membrane and distributes between the outer and inner leaflet dependent on the membrane potential. For a given position of DPA, movement of the donor (attached to the channel) perpendicular to the membrane will alter the distance and therefore the FRET signal. (**d**) 3D nanopositioning by triangulation (Hyde et al. 2012). If symmetric positions are labeled with the (lanthanide) donor and a single acceptor attached off-center (which is not on the central axis), the four different distances can be used to calculate the exact position including the height (h) relative to the acceptor (Hyde et al. 2012). (**e**) Detection of slow gating movements using fluorescence (Batulan et al. 2010): In the Shaker Kv-E395A mutant, the off-gating currents Q_{OFF} during deactivation apparently disappeared due to the slow return of the VSD. In the fluorescence signal (green/red), the slow return is clearly visible (Batulan et al. 2010). (**f**) VCF measurements in Nav channels. Shown are the voltage sensor movements of the S4 of domain I labeled with TMR (DI*). The mutation is inserted in domain IV (DIV°) and shifts the voltage dependence (figure adapted from Chanda et al. 2004)

quench or unquench the fluorophore, and these and earlier measurements in Shaker Kv (Cha et al. 1999) using LRET (see 6.2.2) indicated that the movements of the S4 were limited to a few Ångströms. This led to the question as to how ≈ 4.3 elementary charges per voltage sensor were moved through the electric field without a large movement normal to the membrane. To investigate this question, the electric field strength within the voltage sensor was measured. For any atom, the electronic states will split if the atom is entered into an electric field (*Stark effect*). For a fluorophore, this splitting leads to a shift in the emission spectrum proportional to the strength of the electric field (*electrochromic effect*, see Sect. 6.2.2). Attaching thiol-reactive ANEPPS dyes deep into the voltage-sensing domain of the Shaker Kv channel revealed that the electric field was concentrated on a small area such that small movements would move charges through most of the potential difference (Asamoah et al. 2003). This led to the proposal of water-filled crevices that carry the electric field into the voltage sensor and focus it on a small hydrophobic region (now called the *charge transfer center* (Tao et al. 2010)). These crevices were shown to exist experimentally (Starace and Bezanilla 2004; Tombola et al. 2007; Zhao and Blunck 2016) and are found in the atomic-scale structure of Kv channels (Tao et al. 2010).

Despite the earlier results suggesting a limited movement, a large S4 movement along the membrane normal was suggested based on the KvAP crystal structure (Jiang et al. 2003). This led to the challenge to detect movements perpendicular to the membrane rather than relative movements within the protein. One method to detect large movements normal to the membrane surface is to use energy transfer relative to an acceptor anchored in the membrane. With approximately 50 Å, movement through the membrane—such as observed for dipicrylamine (DPA)—is of perfect scale for FRET detection (Chanda et al. 2005; 2005; Groulx et al. 2010). DPA is an amphipathic compound that distributes between the leaflets of the membrane dependent on the membrane potential. If the DPA is located at the outer leaflet, a movement normal to the membrane toward the outer leaflet would lead to a reduction in the fluorescence intensity, since the distance between donor (the fluorophore attached to the protein) and acceptor (in the membrane) decreases (Fig. 6.3c). With this method (Chanda et al. 2005), as well as LRET distance measurements using a fluorescently labeled scorpion toxin bound to the pore as acceptor for a donor on S4 (Posson et al. 2005), large movements of the S4 through the membrane could be excluded.

The LRET method was further refined by triangulation (Fig. 6.3d). Lanthanides or transition metals as donor or acceptor, respectively, eliminate the uncertainty in the orientation factor between donor and acceptor and therefore allow exact determination of the relative distance between donor and acceptor. Using a triangulation method among several donors and a single acceptor, the exact coordinates in space relative to a fixed point can be determined. To this end, Hyde et al. (2012) labeled the scorpion toxin agitoxin in a way that the acceptor fluorophore is positioned off-center once attached to the Kv channel (Fig. 6.3d). As a consequence, it will have four different distances to the four symmetrically placed donors attached to the top of the S4 helix. Combination with molecular dynamics simulations of the entire system allowed fitting the luminescent decay as a function of the coordinates of the acceptors. The authors found that the S4 moves normal to the membrane by only a few Ångströms, and this movement occurs mainly during the relaxation—not deactivation—of the voltage sensor (Hyde et al. 2012).

Considering the many voltage sensor movements that have been observed using VCF but that largely follow the gating currents prompts the question as to what the advantage is of the classical VCF measurements over electrophysiological gating current measurements to detect voltage sensor movement. Gating currents are caused by the transient movement of the voltage sensor, and the amplitude is given by

$$I_{gating} = \frac{dQ_{gating}}{dt}, \tag{6.9}$$

i.e., it depends on the speed with which the charges are displaced. The number of charges is fixed, and each charge can only move once; as a consequence, very slow or non-synchronized displacements (see Sect. 6.3.2) are not detectable with electrophysiology. The fluorescence signal, on the other hand, is defined by the occupancy of the two conformations independent of the speed of the dynamics. Assuming a transition between state A and state B, which each can be assigned a fluorescence intensity F_A and F_B, respectively, the observed fluorescence intensity as a function of time $F(t)$ will be

$$F(t) = F_A \cdot A(t) + F_B \cdot B(t)$$

with $\Delta F = F_B - F_A$,

$$\begin{aligned} F(t) &= F_A \cdot (1 - B(t)) + (F_A + \Delta F) \cdot B(t) \\ &= F_A + \Delta F \cdot B(t) = F_B - \Delta F \cdot A(t). \end{aligned} \tag{6.10}$$

$F(t)$ follows exactly $A(t)$ and remains at an altered value proportional to the relative number of channels in state $A(t)$ (or $B(t)$) even when A and

B no longer change (with $A(t)+B(t) = 1$).[3] This is very well demonstrated in a mutant in the S4-S5 linker of the Shaker Kv channel, E395A (Batulan et al. 2010). E395A exhibited gating currents during activation, which were absent during deactivation due to its slow rate. Batulan et al. (2010) followed the slow movement of the voltage sensor with fluorescence, where the slow decay is clearly visible (Fig. 6.3e).

Another advantage of fluorescence over electrophysiology measurements becomes apparent in the analysis of relative movements among the monomers of the oligomeric channels. Ion channels are typically comprised of two to five monomers as either homo- or heteromers. For instance, voltage-gated proton channels (Hv) form dimers, and, on the other extreme, nicotinic acetylcholine receptors in the neuromuscular junctions form pentamers, which contain four different subunits. The electrical signal will provide the integrated response of the ensemble of monomers. With fluorescence, one can specifically follow the conformational change of a single monomer and observe how it is affected by the others. This technique was used to study cooperativity among the voltage sensors of the Shaker Kv channel (Pathak et al. 2005). It had been demonstrated that introducing the ILT mutations[4] led to separation of the initial independent charge movement (Q_1) from the final cooperative charge movement (Q_2) (Ledwell and Aldrich 1999; Smith-Maxwell et al. 1998). Labeling an ILT-monomer in the presence of three wildtype monomers or vice versa revealed that the labeled monomer did not move independently with its intrinsic kinetics but followed that of a cooperative movement in the second phase of the charge movement, indicating a strong energetic coupling (Pathak et al. 2005). A similar approach has been used to tackle the voltage sensor movements of voltage-gated sodium (Nav) and calcium (Cav) channels.

Role of Voltage Sensors in Voltage-Gated Sodium and Calcium Channels Nav and Cav channels differ from most other channels in that they are monomeric, where each monomer comprises four domains that mimic the monomers of the other hexahelical cation channels. Referring back to Hodgkin and Huxley (1952), the sodium channels were postulated to have three activating gating particles and one inactivating one. Since it had been established that the S4s of the four voltage sensors correspond to the gating particles in Kv channels, it is natural to assume that the S4s in the Nav and Cav channels adopt the same role. Each domain in the Nav and Cav channels has a distinct sequence, and the question remained as to which domain is responsible for activation and which for inactivation, since this is not immediately obvious from the primary sequence or the atomic-scale structures. While ionic currents report the conductance state of the central pore and gating currents report the ensemble charge movement of all four voltage sensors, fluorescence has the immense advantage that a single voltage sensor can be labeled.

In Nav channels, it was shown by electrophysiological studies that all four voltage sensors contribute to both activation and inactivation but that, in contrast to the three others, the S4 of the fourth domain is mainly involved in inactivation (Groome et al. 2000; Kontis et al. 1997; Kontis and Goldin 1997; Chahine et al. 1994). This was confirmed with tracking the movement of the Nav S4s with fluorescence quenching (Chanda and Bezanilla 2002). While it would be consistent with the model of Hodgkin and Huxley if one voltage sensor would inactivate the channel, the action could not be cleanly divided among the four sensors because the voltage sensors are energetically coupled, in particular between domain

[3] The total fluorescence is a function of the number of available fluorophores, which slowly decreases due to photobleaching. However, the relative fluorescence change $\Delta F/F_A$ remains constant since both values decrease proportionally.

[4] ILT: Shaker-Kv with the mutations V369**I**, I372**L**, and S376**T** from the Shaw-Kv channel (Smith-Maxwell et al. 1998).

1 and domain 4, which was shown by optically following movement of one domain while manipulating another one by site-directed mutagenesis (Fig. 6.3f) (Chanda et al. 2004).

In the voltage-gated calcium channels, the situation is similar. There were differences found among the Cavs; in *Cav1.2*, domain IV activates at very polarized potentials and does not seem to contribute to pore opening. The majority of the energy for pore opening is provided by domains II and III, which coincide with gating charge movement. Domain I triggers final opening and contributes the remaining energy (Pantazis et al. 2014).

In *Cav1.1*, the S4s of the domains are similarly staggered in voltage dependence and speed, but, here, domain I is not only linked to pore opening but also provides most of the energy. The much faster acting domains II-IV are structurally linked to the ryanodine receptors in the sarcoplasmic reticulum (Rios and Brum 1987; Bower et al. 2012). Therefore, the electrically silent fast activation of domains II-IV at more polarized potentials explains the triggering of Ca^{2+} release prior to Cav1.1 activation.

The movement of the S4 has been confirmed in many voltage-gated ion channels (Na_V (Chanda et al. 2004), Ca_V (Savalli et al. 2006), NaChBac (Blunck et al. 2004), EAG (Bannister et al. 2005), HCN (Bruening-Wright and Larsson 2007), Herg (Smith and Yellen 2002), and KCNQ (Kim et al. 2017)), including those that do not contain a pore domain (iVSD (Zhao and Blunck 2016), Hv (Gonzalez et al. 2010)) and those that are not channels at all (ciVSP (Murata and Okamura 2007)[5]). While these channels have a voltage sensor movement in common, the important differences occur, however, on the cytosolic surface of the channels where the different coupling mechanisms act upon the pore domain.

[5] *ciVSP*—voltage sensor containing phosphatase of *Ciona intestinalis*. ciVSP comprises a typical voltage-sensing domain linked to a cytosolic phosphatase domain.

Infobox 6.1 Lifetimes in the Frequency Domain

In frequency domain fluorescence lifetime measurements, fluorescence is excited with excitation light amplitude-modulated at different frequencies. Because of the low-pass filtering of the signal due to the delay between excitation and emission (the lifetime of the fluorophore τ), the emission amplitude decreases with increasing excitation frequency ω (Fig. 6.4a).

The excitation intensity I is sinusoidal of a specific frequency ω

$$I(t) = I_0 \, cos(\omega t) \tag{6.11}$$

or in complex form

$$I_c(t) = I_0 \cdot e^{i\omega t}. \tag{6.12}$$

The pulse response of the fluorescence is

$$F_p(t) = F_0 \cdot e^{-t/\tau}. \tag{6.13}$$

Convoluting both responses yields the response to a continuous sinusoidal excitation, whereby F_0 is replaced by the excitation intensity

$$F_c(t) = \int_0^t I_0 \cdot e^{i\omega t'} \cdot e^{-(t-t')/\tau} \, dt' \tag{6.14}$$

$$= \frac{I_0 \, \tau}{1+\omega^2\tau^2}(1 - i\omega t)(e^{i\omega t} - 1). \tag{6.15}$$

The steady-state real component is the fluorescence intensity $F(t)$

$$F(t) = \frac{I_0 \, \tau}{\sqrt{1+\omega^2\tau^2}} \cdot cos(\omega t + atan(\omega t)). \tag{6.16}$$

6.3.2 Movement on the Cytosolic Surface

The modulating domains, whether voltage sensors or ligand binding domains, are associated with the pore at the cytosolic surface. This raises the question as to how the energy is transferred to the pore and what is the conformation of the pore in the different states. The question can be addressed with various methods, including accessibility studies, where the efficiency of binding to the inner pore region as a function of the state of the ion channel is monitored by current blockage (Del Camino et al. 2005; Del Camino and Yellen 2001; Posson et al. 2013). While the accessibility studies can provide information on the state of the gate in different conformations, they are ultimately governed by the diffusion of the inhibitors. Consequently, we do not obtain any information on the dynamics of the opening. Opening can be visualized with fluorescence labeling on the cytosolic surface. As outlined in Sect. 6.2.3, chemical labeling in an intact cell is difficult but feasible for isolated proteins. Accordingly, the first fluorescence studies on gate opening were performed on isolated proteins.

Pore Opening of 2-TM Potassium Channels The prokaryotic proton-gated K^+ channel KcsA acts as a model system for the pore region of all K^+ channels and has become famous as the first channel, of which the crystal structure has been determined (Doyle et al. 1998). It rapidly inactivates after opening, triggered by a jump in the proton concentration. We know now that the external selectivity filter and the lower bundle crossing are linked to trigger the inactivation (Cordero-Morales et al. 2006; Cuello et al. 2010). One method to determine the state of the inner gate in the inactivated channel was to fluorescently label the bundle crossing. KcsA was expressed in bacteria and purified. In its isolated form, cysteines, introduced at the position of interest, can be labeled and any excess label removed by gel filtration or metal affinity chromatography (Blunck et al. 2006; Faure et al. 2012; Blunck et al. 2008). KcsA was subsequently reinserted into proteoliposomes formed from synthetic lipids, i.e., devoid of any background fluorescence.

To provide information on the relative occupancies of the conformations, fluorescence lifetime measurements need to be carried out since the exponential decays can be distinguished in the overall signal. KcsA, TMR-labeled at the bundle crossing, was subjected to frequency domain lifetime measurements (see Infobox 1). Because KcsA is ligand-gated, there was no need to control the membrane potential, which allowed these measurements to be done in a commercially available fluorescence spectrometer (Blunck et al. 2006). The measurements showed that in the non-inactivating mutant KcsA-E71A and the inactivating wildtype, the gating of the bundle crossing is unaltered, indicating that the lower gate controls activation but the upper selectivity filter controls inactivation (Fig. 6.4a) (Blunck et al. 2006; Cordero-Morales et al. 2006).

Electrophysiological recordings of KcsA in planar lipid bilayer had shown a certain extent of single channel events with lower single channel conductance, and it was not clear how many subunits had to be in the closed state to prevent ion conduction. To monitor the relative movement of the KcsA monomers, labeled KcsA channels were inserted into supported lipid bilayers formed on a glass substrate. While these measurements could not be correlated directly with the single channel current, it allowed to calculate the coupling energy among the single subunits and showed that pore opening occurs in a single cooperative step (Blunck et al. 2008).

Also in the *K_{IR} channels*, the opening of the helical bundle crossing was studied. The K_{IR} channel family comprises 15 members, and they are found throughout the body. In neurons and cardiomyocytes, they adjust the membrane potential and regulate pacing. They also regulate pancreatic insulin secretion and potassium homeostasis in the kidney (Driggers et al. 2023). To follow the movements, the K_{IR} channels were labeled at the "sliding helix," the helix attached to the first TM helix corresponding to the S4-S5 linker in voltage-gated cation channels. This helix is suggested to convey PIP_2 regulation to the gate (Kuo et al. 2003; Lee et al. 2013). The movement

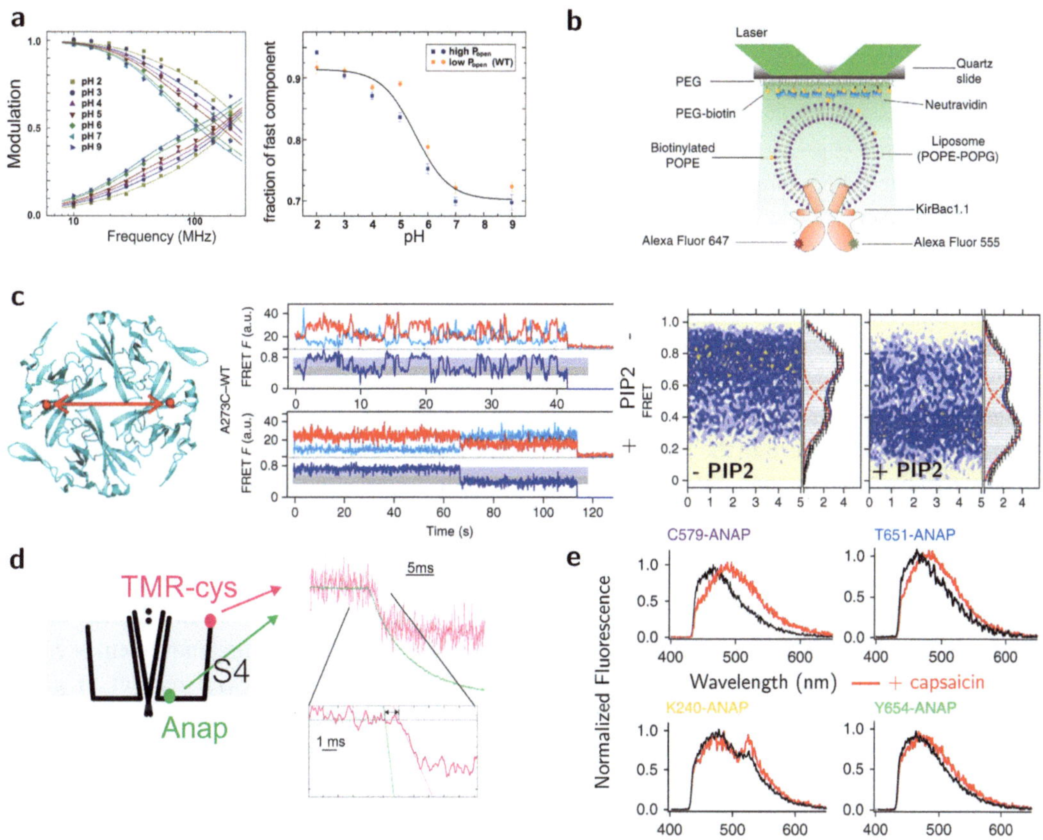

Fig. 6.4 (**a**) Frequency domain fluorescence lifetime measurements of KcsA. The fluorescence showed a fast and a slow time constant. The population shifted from the slow to the fast component with increasing proton concentration, leading to a modulation at higher frequencies. The shift coincided with the opening of the channel determined by Rubidium flux assay (figure from Blunck et al. 2006). (**b**) Attachment of proteoliposomes on quartz coverslip for single molecule fluorescence. Kir channels are reconstituted into proteoliposomes containing biotinylated POPE. The coverslips are PEG-neutravidin-coated (figure from Wang et al. 2016). (**c**) Single molecule FRET results for Kir labeled at the indicated position (*red*). Donor (*cyan*), acceptor (*red*), and FRET (*blue*) traces are shown in absence (*top, and left histogram*) and presence (*bottom and right*) of PIP2 (figure from Wang et al. 2016). (**d**) Dual bio-orthogonal labeling with Anap (*green*) at the S4-S5 linker and using thiol-reactive tetramethylrhodamine (TMR) maleimide (*magenta*) at the N-terminus of the S4. Both signals allow to determine relative timing (*right*) (figure partially from Kalstrup and Blunck 2018). (**e**) Accessibility study using Anap. TRPV1 channels were labeled with Anap at the indicated positions and the emission spectrum determined in the absence (*black*) and presence (*red*) of capsaicin. If the water accessibility altered, the spectrum shifted to the right (see also Sect. 6.2.2) (figure from Yang et al. 2018)

of the helices was then monitored using single molecule FRET (*smFRET*, Fig. 6.4b) (Wang et al. 2016).

In homotetrameric ion channels (incl. K^+ and TRP channels), obtaining distances from FRET measurements is often complicated by the mixtures of distances, which lead to a convolution of distances (see Infobox 2). These problems can be circumvented by constructing dimers with a single binding site. Liu et al. (2001) showed that in such constructs the identical monomers insert in opposite positions, leading to a single FRET (and EPR) distance. Using smFRET in proteoliposomes attached to a glass substrate (Fig. 6.4b), the sliding helix of K_{IR} adopted three distinct conformations with a distance change of $\approx$5Å from the central pore (Wang et al. 2016). The more proximal position was stabilized in the presence of PIP_2 in accordance with electrophysiological current measurements (Fig. 6.4b) (Wang et al. 2016).

The Role of the S4-S5 Linker For the voltage-gated ion channels, it had been shown that the energetic coupling between voltage sensor and pore domain is mediated by the coupling of the covalent link between the voltage-sensing and the pore domain (Lu et al. 2001; Labro et al. 2012; Haddad and Blunck 2011; Lu et al. 2002; Long et al. 2005). It remained unclear how the electrical energy, generated in the voltage sensor, evolves dynamically through the channel. The closed state of the Kv channels has long eluded structural determination and only recently was captured in a "dynamic" frozen state immediately after depolarization (Mandala and MacKinnon 2022, 2024). Previously, research focused on narrowing down on the closed state by experimentally measuring structural constraints that could be used in molecular dynamics simulations of the closed state (Campos et al. 2007; Faure et al. 2012; Vargas et al. 2011). In this context, lanthanide-based resonance energy transfer (*LRET*) was a perfect method to directly measure intramolecular distances in purified proteins.

Voltage sensors enter the activated state in the absence of an electric field, i.e., you require energy to push them into the "resting" state. Since in many configurations, including crystals or nanodiscs, no membrane potential is possible, one has to manipulate the electric field in a different way. This can, for instance, be achieved by altering the headgroup charges of the phospholipid membrane (Faure et al. 2014; Schmidt et al. 2006; Liin et al. 2016; Combs et al. 2013).

The particular influence of the headgroup charge to Kv gating is likely due to interaction with the S4 arginine (Faure et al. 2014; Borjesson and Elinder 2011). Using this influence, distances along the S4-S5 linker were determined in the open (inactivated, POPE:POPS:POPC mixture) and resting (DOTAP) state in concatemers generating dimers of dimers with each one donor and one acceptor (Faure et al. 2012). As donor, a lanthanide chelate (Tb-chelate) was used to prevent uncertainty with the relative orientation of donor and acceptor (see Sect. 6.2.3). It was established that the S4-S5 linker and the lower S6 only move a small distance of about 4 Å and that conduction is blocked by an isoleucine in the pathway. This was later confirmed in the cryoEM structures of TRPV1 (Cao et al. 2013) and Kv (Mandala and MacKinnon 2024). The measurements also suggested that the S4-S5 linker did not move as a unit but rather generates a "tension" during gating.

So how can the dynamics of this region be accessed? This was made possible with the fluorescent noncanonical amino acid Anap (Kalstrup and Blunck 2013, 2017, 2018), since it can be targeted to intracellular positions. Attaching Anap to the bundle crossing at the C-terminus of the S6 helix revealed that the pore undergoes two movements rather than a single opening step. The first movement occurred at voltages depolarized with respect to gating charge movement, whereas the second movement concurred with current development, indicating that the pore adopts a "pre-open" state before transitioning into opening. While we wrote movement here, it is possible that instead the environment alters, since these measurements are quenching measurements. For instance, dewetting of the pore has been seen in molecular dynamics simulations in the closed pore (Faure et al. 2012; Jensen et al. 2012) and might contribute to the second movement.

To follow the evolution of the movement along the S4-S5 linker that seems to "store" the activation energy, Anap was introduced along the S4-S5 linker, and its kinetics relative to charge movement determined (Kalstrup and Blunck 2018). This was possible because chemical labeling and insertion of fncAAs are bio-orthogonal, the temporal correlation of two positions can be observed simultaneously, TMR following gating charge movement and Anap following movement of the S4-S5 linker all the while the ionic current monitoring the state of the pore (Fig. 6.4d). The authors noted two observations: *(i)* In the Shaker Kv channel, the S4-S5 linker has a temporal delay with respect to charge movement that becomes progressively

Infobox 6.2 Förster Resonance Energy Transfer (FRET) in Multimers

When four identical binding sites are available, stoichiometric labeling will occur. Considering non-labeled sites and photobleached fluorophores, each site is either occupied by a donor and an acceptor or remains empty. Each combination will lead to a different lifetime[6] $\tau = k_f/(k_f + k_{nr} + k_{ET,app})$. Let us consider cases with a single donor first:

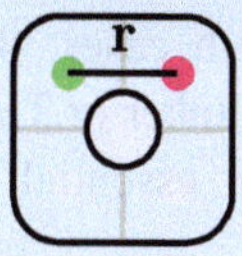

τ_1

$k_{ET,app} = k_{ET}(r)$

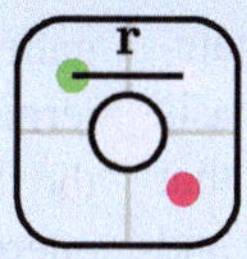

τ_2

$k_{ET}(\sqrt{2} \cdot r)$

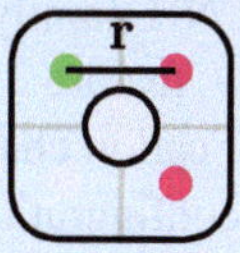

τ_3

$k_{ET}(\sqrt{2} \cdot r) + k_{ET}(\cdot r)$

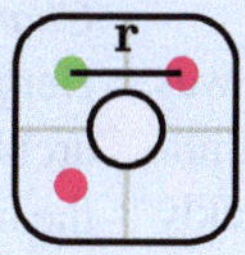

τ_4

$2 \cdot k_{ET}(r)$

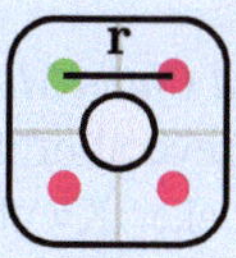

τ_5

$k_{ET}(\sqrt{2} \cdot r) + 2\, k_{ET}(r)$

If more than one donor is present, each donor will generate its own fluorescence decay with the time constants above:

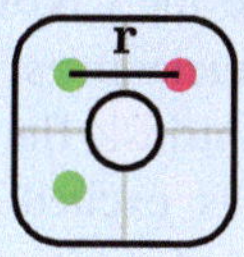

$\tau_1 : \tau_2$

$1 : 1$

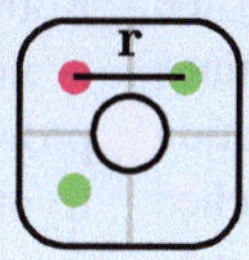

τ_1

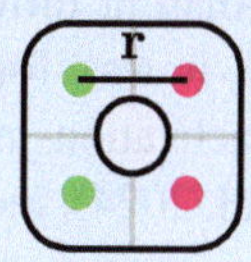

τ_3

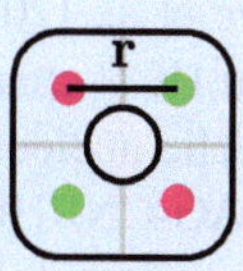

τ_1

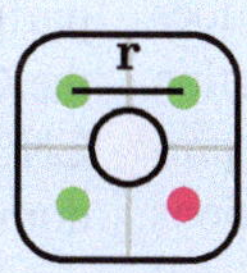

$\tau_1 : \tau_2$

$2 : 1$

While the above time constants would be difficult to separate, the superposition can be exactly calculated—and fitted—given that the stoichiometry is known. This can be achieved by labeling all positions with an identical donor and binding the acceptor off-center (e.g., on a toxin; *nanopositioning* (Hyde et al. 2012)).

longer from N- to C-terminus, and *(ii)* this delay disappears in the presence of the W434F mutation that renders the channel immediately C-type inactivated (Fig. 6.4d). This is in accordance with molecular dynamics simulations and structural information in KcsA (Li et al. 2018).

A very similar movement has been suggested for the capsaicin and heat-activated TRPV1 channels (Caterina et al. 1997). Capsaicin binds close to the S4-S5 linker in a binding pocket near the S4

[6] Any configuration where not at least one donor-acceptor pair exists is ignored.

(Liao et al. 2013). Yang et al. (2018) used a combination of simulations and fluorescence spectroscopy methods to follow the evolution of conformational changes through the channel. They used the shift of the Anap emission spectrum when exposed to water versus hydrophobic environments to probe water accessibility in a state-dependent manner and used the results as basis for the simulations (Fig. 6.4e). They found that the movement is initiated at the S4-S5 linker and transfers to the selectivity filter via the S6 bundle crossing (Yang et al. 2018).

The Ball and Chain Mechanism Early on, it was shown that several of the Kv channels, most notably the Shaker Kv channel, inactivate quickly by a peptide at the N-terminus (Zagotta et al. 1990). Following the movement by fluorescence remained elusive until the introduction of fncAAs. Fluorescent proteins were too large and inhibited inactivation (Blunck et al. 2004). Introducing Anap at different positions along the N-terminus, it was found that the N-terminus is sensing activation in addition to inactivation (Pandey et al. 2024), indicating that it is attached to the transmembrane domain in the resting state and released by activation. To narrow down on the resting position, FRET was used with the lanthanide terbium as a donor. Since terbium does not have a set dipole moment, the orientation factor can be determined accurately, which provided better accuracy in the distance calculations (LRET Selvin 2002). The fncAA Anap is not suitable as an acceptor for terbium emission; to pair terbium with an organic fluorophore without losing the advantage of genetic encoding, the cytosolic side was labeled using a click-chemistry ncAA, trans-cyclooct-L-lysine (*TCO*) (Tyagi and Lemke 2013) that was labeled with Cy3-tetrazine postexpression (Pandey et al. 2024). It was found that the N-terminus is located near the S4-S5 linker in resting position. The location was further narrowed down on by using transition metal FRET (tmFRET, see below).

6.3.3 Ligand Binding

Many of the hexahelical cation channels have a cytosolic ligand binding domain. These either modulate the voltage-dependent gating—as in case of BK_{Ca} channels—or are the dominant gating mechanism. In both cases, the helix C-terminal to the S6, the C-linker or TRP-box, and the S4-S5 linker play an essential role in the energetic coupling.

The Role of N- and C-termini of the KCNH Channel Family The KCNH family members, Herg, EAG, and ELK, feature the typical architecture of hexahelical cation channels but in the non-domain-swapped configuration (Whicher and MacKinnon 2016; Wang and MacKinnon 2017). They comprise both an N-terminal Per-Arnt-Sim (PAS) domain and a C-terminal cyclic nucleotide binding homology domain (CNBHD) (Warmke and Ganetzky 1994; Morais-Cabral and Robertson 2015). Despite the CNBHD, KCNH is not gated by cyclic nucleotides, which was proven by exploiting the intrinsic fluorescence of tryptophan. Tryptophan is excited in the UV (maximum at 280 nm) and emits in the near UV (maximum at 355 nm). Since the use of intrinsic tryptophan fluorescence is possible for purified proteins, Brelidze et al. (2009) isolated the C-termini of EAG and HCN channels with a CNBHD and an intact cyclic nucleotide binding domain (CNBD). The EAG-CNBHD harbors a single tryptophan near the nucleotide binding site, which does not alter fluorescence upon nucleotide binding. Introducing an analogous tryptophan in the HCN-CNBD shows quenching of tryptophan fluorescence with typical binding affinity. The authors also used a fluorescent derivative of cyclic nucleotides. The first fluorescent derivative of cAMP, ϵ-cAMP, was developed as early as 1972 (Secrist et al. 1972). Here, NBD-labeled cyclic nucleotides (Kraemer et al. 2001) were employed, which show marked fluorescence only when bound (Fig. 6.5a). These measurements confirmed that the cyclic nucleotides only bind to the HCN-CNBD but not the CNBHDs (Brelidze et al. 2009).

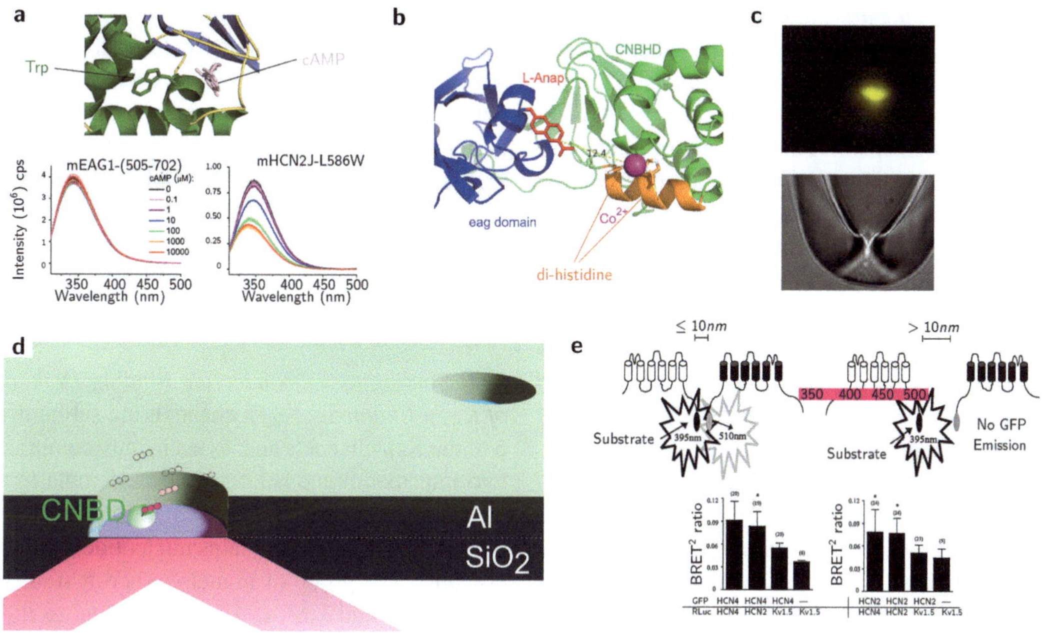

Fig. 6.5 (**a**) Intrinsic fluorescence of the CNB(H)D of EAG and HCN. cAMP in the binding site reduces fluorescence of HCN (figure adapted from Brelidze et al. 2009). (**b**) tmFRET in CNBHD of ELK channel: Two histidines (*orange*) introduced into an α-helix at positions i and $i+4$ coordinate a cobalt ion (*magenta*). The cobalt acts as acceptor from the Anap (*red*) in the EAG domain (figure adapted from Dai and Zagotta 2017). (**c**) Geometry of patch-clamp fluorometry: In this case, ELK channels are labeled with YFP (figure adapted from Dai and Zagotta 2017). (**d**) Zero-mode waveguides (ZMWs) are thin conducting coatings on a glass (fused silica) substrate. The coatings are perforated by nanopores smaller than half a wavelength of the light applied. Because of this constriction, light cannot pass through the pores but creates an evanescent excitation region above the glass surface. The volumes are so small that only a single ligand is present on average. (**e**) BRET assay in HCN channels. HCN2, HCN4, and Kv1.5 are each fused with RLUC or GFP. Energy transfer of all pairs is evaluated (figure from Whitaker et al. 2007)

The binding site in the CNBHDs of KCNH channels is occupied by intrinsic ligand formed by surrounding amino acid residues (Brelidze et al. 2012), which is required for the interaction between the N-terminal PAS domain and the CNBHD, as shown by FRET assays (Codding and Trudeau 2019) in full-length channels. The interaction site undergoes conformational changes, as evidenced by the state dependence of FRET between the CNBHD and a fluorescently labeled antibody against the interaction site (Harley et al. 2021; Dai and Zagotta 2017). The conformational change was further probed using transition metal FRET between the fncAA Anap introduced into the PAS domain of ELK and a cobalt ion in the CNBHD (Fig. 6.5b) (Dai and Zagotta 2017). As mentioned earlier, the acceptor of FRET does not need to be a fluorophore but merely needs to absorb in the wavelength range of the donor emission spectrum. Any colored compound would work; this includes cobalt (*magenta*) and nickel (*blue*). Since both ions bind to histidine—the underlying principle of metal affinity chromatography—and coordinate between two histidines in the appropriate distance, introducing a dihistidine motif in an α-helix at positions i and $i+4$ would lead to coordination of a single cobalt ion to which Anap can transfer (Fig. 6.5b). The specific distance R_0 is 12 Å, indicating that tmFRET is capable of measuring very small distance changes.

Ligand Binding to CNBDs in CNG and HCN Channels In contrast to the KCNH channels, the CNBDs in CNG and HCN channels do bind to cyclic nucleotides, which is a part of the activation mechanism. The first approach to use fluorescence to follow conformational changes in CNGA channels used thiol-reactive labeling to the CNBD that was quenched during binding (Zheng and Zagotta 2000). By switching the fluorescent tag to the ligands (see above), it was possible to quantify the binding affinities. This led to the idea of cooperativity among the binding sites (Biskup et al. 2007), which was later confirmed in HCN channels (Kusch et al. 2012).

In all studies on the CNBD, patch-clamp fluorometry with excised macro-patches was used since this gave access for the fluorescent ligands to the binding site (Fig. 6.5c) (Zheng and Zagotta 2000; Biskup et al. 2007; Kusch et al. 2012). This also meant that unbound fluorescent ligands were freely diffusing in solution. The challenge when using fluorescent derivatives of the ligands is to separate the "bulk" fluorescence, i.e., from those molecules in the external solution versus the bound fluorophores. In the excised patches, the membranes are clearly visible once fluorescence is bound, but the signal extends into the solution. Due to the form of the pipette, this background fluorescence has a specific shape. By adding a calibration dye into the bulk solution, this shape can be exactly reproduced and subtracted from the original (Fig. 6.5c) (Biskup et al. 2007).

Recently, time-resolved tmFRET lifetimes were used to determine the free energies of binding for the different compounds isolated CNBDs (see also Fig. 6.1e) (Eggan et al. 2024).

A different strategy was employed by Goldschen-Ohm et al. (2016). Using *zero-mode waveguides (ZMWs)*, they reduced the excitation volume around the binding site to such an extent that only bound fluorophores were present in the excitation volume whereas fluorophores in solution remain dark (Fig. 6.5d). The top layer of the ZMW contains pores in diameter smaller than half a wavelength. This prevents any light from passing through these openings (like a Faraday cage) but generates an exponentially decaying light field within the pore. Isolated and labeled CNBDs are bound to the surface of these ZMWs inside the pores and a fluorescent cAMP derivate added (Fig. 6.5d). A CNBD of HCN2 with all but one cysteine removed was isolated, and the cysteine was labeled with the organic fluorophore DyLight547. The labeled CNBD was attached to the surface of the ZMW. Binding to the CNBD results in a single molecule FRET signal, which allows for a detailed study of the binding kinetics (Goldschen-Ohm et al. 2016).

ATP Binding to the Nucleotide Binding Domain of K_{ATP} Channels K_{ATP} channels are subfamily 6 of the K_{IR} channels and, as such, only comprise two transmembrane helices. They assemble to a larger biological assembly with sulfonylurea receptor (SUR), which contains nucleotide binding domains. The group of Francis Ashcroft studied the binding and kinetics of ATP binding to this binding site using the fncAA Anap (Puljung et al. 2019; Usher et al. 2020). In patch-clamp fluorometry measurements, they performed FRET from fluorescently labeled ATP to Anap bound to the nucleotide binding domain and found that a single ATP bound is sufficient to evoke currents (Usher et al. 2020). They also established that while magnesium is required for K_{ATP} activation, ATP will still bind in magnesium-free conditions (Puljung et al. 2019).

6.4 Assembly of Ion Channel Complexes

Just like K_{ATP}, many ion channels do not comprise a single subunit or homooligomers, but they assemble as heteromers or recruit a number of auxiliary subunits. This raises questions about which proteins interact with each other, in which stoichiometry these proteins assemble and what the driving force for assembly is.

6.4.1 Bioluminescence Resonance Energy Transfer (BRET)

The first question can be answered by a method coined *Bioluminescence Resonance Energy*

Transfer (BRET) (Charest et al. 2005; Bouvier et al. 2007). BRET was originally developed to monitor the activation of G-protein-coupled receptors. BRET follows the photophysics of Förster resonance energy transfer with the modulation that—rather than using a fluorescent protein as a donor—BRET uses a bioluminescent protein, the luciferase of *Renilla reinformis*, a coral that luminesces upon movement. *Renilla* is well known to generate luminescence at the beach in the dark when the waves disturb them because they express both the luciferase (*RLuc*) and green fluorescent protein. BRET is therefore a natural phenomenon in *Renilla*. The luciferase does not require external excitation but converts chemical energy (in form of ATP) into excitation and eventually to emitted photons. The excited states can be energy transferred to fluorescent proteins in a distance-dependent manner. The distances are longer than what is typically used for intramolecular distance measurements, which makes it a perfect probe for protein assembly rather than conformational changes.

BRET has been, for instance, applied to verify whether the hyperpolarization-activated cyclic nucleotide-modulated (HCN) channel isoforms 2 and 4 assemble in heterotetramers, which was not clear since the electrophysiological differences were too subtle. Whitaker et al. (2007) fused RLuc and GFP each to one of the interaction partners. BRET signals upon addition of RLuc substrate for both the homo- and heterotetramers. A low energy transfer was also observed between HCN and Kv1.5 subunits, since mere proximity due to overexpression will still generate some energy transfer (Fig. 6.5e).

6.4.2 Single Subunit Counting

It is very challenging to use FRET assays for determining stoichiometry of biological assemblies because the stoichiometry might not be fixed and photobleaching and nonmatured fluorescent proteins might alter the ratios. Single subunit counting, in contrast, uses the photobleaching events to determine the number of fluorophores and thereby the number of labeled subunits. Photobleaching is a chemical destruction of the fluorophore itself that renders it permanently non-fluorescent (Turro et al. 2010). Fluorescently tagging the subunit in question and observing the photobleaching events allows to count the distinct steps of each photobleaching event (Fig. 6.2d) (Ulbrich and Isacoff 2007; Blunck 2021). Fluorescent labeling for single subunit counting can be achieved by fusion with the monomeric version of the fluorescent proteins (McGuire et al. 2012), chemical labeling (Thompson et al. 2011), or insertion of fncAAs (Leisle et al. 2016; Pantoja et al. 2009). Each method will lead to different labeling efficiencies and photobleaching before the experiment would also lower the number of occupied or fluorescent subunits. Therefore, one will obtain a binomial distribution that can be deconvoluted into the distribution of stoichiometries that are present in the populations (Moller et al. 2020). The single subunit counting method has been employed to a number of different ion channels, including CNGA (Ulbrich and Isacoff 2007), KCNE (Lussier et al. 2019; Nakajo et al. 2010), glutamate receptors (Hastie et al. 2013), silent Kv subunits (Moller et al. 2020; Pisupati et al. 2018), voltage-gated proton channels (Tombola et al. 2008), and acid sensing channels (Yu et al. 2012).

6.4.3 Clustering of Ion Channels

In addition to the formation of biological assemblies, ion channels also form clusters, which variably alter their behavior. Clusters were first described for Kv channels that bind to post-synaptic density (PSD). In the axon initial segment, super-resolution imaging using STORM revealed that the β spectrin is organized in a very distinct architecture that recruits Nav channels in a very precise arrangement (Leterrier et al. 2015). A similar result has been observed for Kv2 and Cav2 channels in neurons; both assemble together in "macro-clusters" in the interface of the plasma membrane (PM) and endoplasmic reticulum (ER) (Vierra et al. 2019). This clustering of Cav and Kv

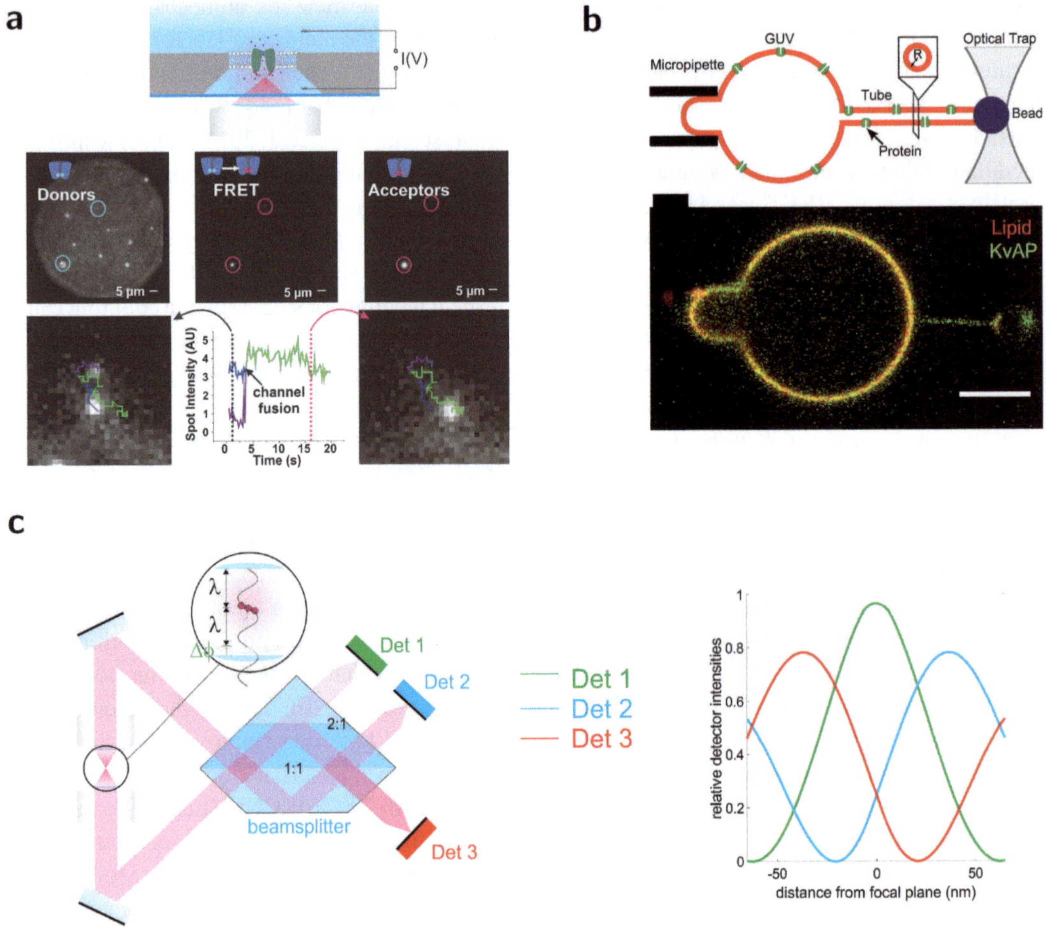

Fig. 6.6 (**a**) In a horizontal bilayer architecture, single channel fluorescence can be recorded (McGuire and Blunck 2022). (*center row*) fusing donor- and acceptor-labeled KcsA leads to FRET signals from the KcsA clusters. (*bottom row*) Tracking single KcsA channel diffusion shows fusion of two channels, which continue to diffuse as a cluster. (**b**) Sorting of KvAP into high curvature membranes. Fluorescently tagged KvAPs are reconstituted into giant unilamellar vesicles (GUVs). Small tubes are pulled out of the GUV by pulling out an enclosed bead with optical tweezers. The strength of the pull determines the radius of the tube, which is measured using confocal imaging (scale bar = 5 μm, figure adapted from Aimon et al. 2011). (**c**) iPALM: The subject is observed with two objectives from the top and bottom. The two beams pass through a dual beam splitter and are superposed in such a way that three interference patterns are generated with the original phase difference ϕ (*detector 1*) and with an added phase difference of $\phi \pm 120°$ (*detectors 2*/*3*). The ratio of the intensities (*right*) provides the exact height of the subject

channels in the PM-ER interface can be observed with confocal fluorescence imaging, which indicate that the macro-clusters are very located but larger than the diffraction limit ($\approx$300 nm). This left open the question as to what the underlying mechanism is for the ion channels to assemble.

Assembly can be followed using fluorescent single particle tracking. Single particle tracking was first done with pore forming toxins using planar lipid bilayers or droplet interface membranes (Borisenko et al. 2003; Thompson et al. 2011). The K^+ channel KcsA was also studied by imaging single ion channels in planar lipid bilayers while measuring the ionic current (Fig. 6.6a) (McGuire and Blunck 2022). KcsA function is very dependent on the lipid composition (LeMasurier et al. 2001; McGuire and Blunck 2022). When following clustering of KcsA, it was found that it occurs within minutes after incorporation into planar lipid bilayers (Fig. 6.6a), and the cluster size increases with time (McGuire and Blunck 2022). This occurs in lipids, in which KcsA is active, as well as those in which KcsA is inactive. In both cases, cooperative opening occurs after the clustering, indicating that energetic coupling among the channels occurs mediated by the lipid matrix. Analysis of the lipid composition indicates that KcsA prefers a lipid environment with a negative curvature. As the channels assemble in negative curvature regions, they also induce a negative curvature in their environment, which triggers self-assembly of the clusters (Fig. 6.6a) (McGuire and Blunck 2022). The energetic load on KcsA that prevents opening is shared among the clustered channels.

A similar result has been observed for KvAP and Piezo1 channels (Aimon et al. 2014; Yang et al. 2022). KvAP was purified, labeled, and reconstituted into giant unilamellar vesicles (GUVs). The GUVs were held in place and small membrane tubes pulled out of them using optical tweezers (Fig. 6.6b) (Aimon et al. 2014). KvAP was enriched in the curved regions of the membrane, indicating their preference for a specific curvature.

For *Piezo1*, the effect was even stronger. Piezo channels alter their angle in the open and closed state so, similar to KcsA, they prefer a certain curvature. However, in Piezo, the bending stress in the membrane is the actual gating mechanism in contrast to KcsA, where the effect is more passive. GFP-Piezo1 was tested directly in mammalian cells or liposomes, where Piezo1 left strongly convex curvatures and enriched in concave curvatures (Yang et al. 2022), in line with their geometry (Guo and MacKinnon 2017).

6.5 Nanopositioning

With cryoEM, the 3D structure of all atoms can be visualized directly, but of course with a much smaller wavelength and without access to dynamics. For optical methods, the diffraction limit falls in the range of 200–300 nm. We already discussed the triangulation of the position of the S4 voltage sensors above, which allows much higher precision. However, these are intramolecular distances. How can such a resolution be achieved in an imaging technique, where the relative distances of all ion channels in the field of view need to be assessed? With super-resolution imaging a resolution of 20–50 nm can be achieved in the x-y plane with less resolution normal to the focal plane. With *interferometric photoactivated localization microscopy (iPALM)* also the third dimension can be determined with high accuracy (Shtengel et al. 2009). This is achieved by observing single (or few) fluorescent molecules with two objectives from the top and bottom. The two beams will have a slight difference in length of double the distance of the fluorophore from the focal plane (Fig. 6.6c, *insert*). Small distances are typically measured by interferometry, i.e., superposition of two waves with a small phase difference. A displacement from zero to half a wavelength (250 nm equivalent to a phase difference of π) will lead to destructive interference, and the intensity disappears. In iPALM, in addition the beams are split into three and then superposed in a way that the three interference patterns comprise an additional phase shift of—120°, 0°, and +120°, allowing a more accurate z-resolution of $\approx$10 nm.

iPALM has been used to measure the movement of Piezo normal to the membrane (Mulhall et al. 2023). Fluorophores were attached to the extreme ends of the three blades of Piezo using the click-chemistry ncAA TCO (Fig. 6.2d) labeled with AlexaFluor647-tetrazine. The movement in all three dimensions was then imaged using iPALM. Piezo is particularly susceptible to the iPALM measurement because of its large size ($\approx$50 nm) and large movements, which are required to amplify the mechanical deformations of the membrane (Mulhall et al. 2023).

6.6 Visualizing Ion Channel Movement In Situ

In recent years, an increased interest has developed to measure biological assemblies and conformational changes in a more native environment. This has been made possible by the development of gene editing methods to efficiently manipulate the ion channels in situ.

Endogenous Voltage Sensor Activity Probe (EVAP) In the endogenous voltage sensor activity probe (EVAP), the authors made use of the high selectivity of toxins to voltage-gated ion channels, in this case Kv2 (Thapa et al. 2021). By labeling the tarantula toxin GxTX-594 with a 2-photon probe, Kv2 channels were selectively labeled in CA-1 hippocampal neurons in brain slices, allowing to detect activity of Kv2 with low temporal resolution in the 1s range (Thapa et al. 2021). The temporal resolution was reduced because the fluorescence did not directly track the voltage sensor movement, but rather the binding of the toxin to the channel, which led to a fluorescence change. Repeated or prolonged stimulation of Kv2 led to increased binding monitoring activity.

Noncanonical Amino Acid Labeling in Live Neurons The click-chemistry and fluorescent ncAA are an optimal method to genetically encode small tags into the protein of interest, and it is orthogonal to other labeling methods (Arsic et al. 2022). Bessa-Neto et al. (2021) used the ccAA TCO (Fig. 6.2d) and introduced an amber stop codon into the transmembrane AMPA receptor associated protein (TARP). The TCOs were labeled with Cy3-tetrazine postexpression. TARPs were co-expressed with AMPA receptors containing a SNAP-tag in their N-terminal ligand binding domain, to which the fluorophore, here Alexafluor-488, was bound so that two bio-orthogonal labels could be attached. The authors showed FRET between the TARP and AMPAR in dissociated primary culture and organotypic brain slices (Bessa-Neto et al. 2021).

Single Molecule Calcium Imaging in iPSC-Derived Neurons and Organoids One of the most challenging problems is to genetically encode a labeling method without disturbing the endogenous expression; in fact, the aim is to use the endogenous expression system to preserve the natural expression patterns. This can be achieved by using gene editing (CRISPR/CAS, Doudna and Charpentier 2014). In the Piezo1-HaloTag hiPSC technique, a halo-tag has been inserted in human iPSCs, and the iPSCs were subsequently differentiated to neural stem cells (Bertaccini et al. 2024). The halo-tag is very versatile as it allows linking a variety of different probes to the protein, including a calcium-sensitive fluorophore BAPTA. Using total internal reflection fluorescence (TIRF) microscopy, the authors succeeded to resolve the calcium flow through single Piezo1 channels optically with a temporal resolution in the millisecond range (Fig. 6.7a)

In addition to the single channel calcium studies in neurons, the iPSCs can also be differentiated to organoids of 50–200 μm diameter that can be studied using light sheet microscopy to identify the location of the Piezo1 channels (Fig. 6.7b) (Bertaccini et al. 2024).

6.7 High-Throughput Evaluation of Ion Channel Function

The most significant advantage of optical versus electrical measurements is the noninvasiveness (apart from the labeling) and the spatial resolution that allows to highly parallelize measurements throughout the field of view. In the last two decades high-throughput (HTP) electrophysiological methods have become available and play an essential role in the pharmaceutical industry (Rogers et al. 2024). However, also fluorescence has become an important tool to quickly scan ion channel function, either instead of or in parallel with the electrophysiological methods.

Fluorescence-based HTP methods typically do not directly detect the function of the ion channel but their influence on environmental factors such as membrane potential and internal proton or calcium concentrations. A calcium-based method was used initially to scan for TRPV channel locations that are linked to temperature sensitivity (Grandl et al. 2008). The assay was built using a fluorescence plate reader for 384-well plates. Since TRPV channels are calcium conductive, the cells were loaded with a calcium dye such that

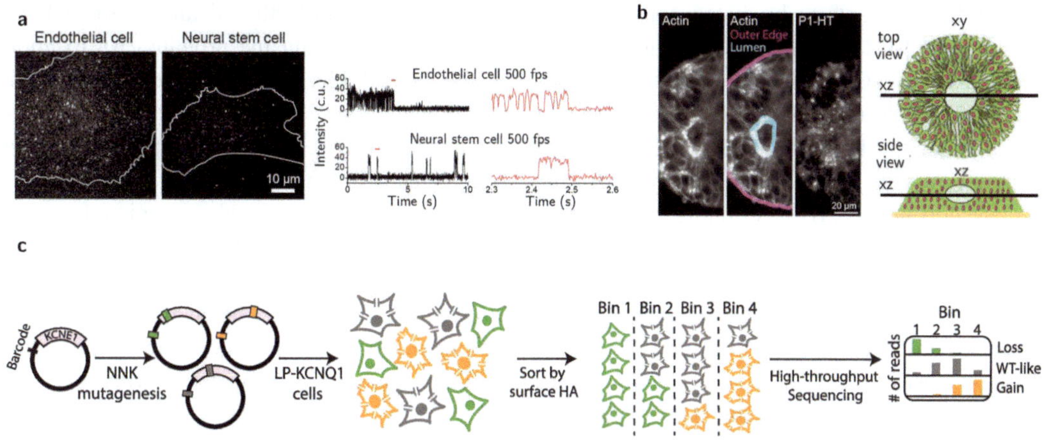

Fig. 6.7 (**a**) Calcium imaging through single piezo channels in endothelial and neural stem cells (figure adapted from Bertaccini et al. 2024). (**b**) Expression of gene-edited Piezo1-HaloTag in iPSC-derived neuronal organoids (figure adapted from Bertaccini et al. 2024). (**c**) Principle of multiplexed assays of variant effect MAVE (Glazer et al. 2019)

activation would lead to a fluorescence signal. As a stimulus, in this case, heat was required. Therefore, the plate reader was equipped with a Peltier element controlling the temperature of a plate and 384 pins reaching into the wells. The small volume of the wells allows for a precise and rapid temperature control. The authors identified five mutations all located at the outer loop between the filter and the S6 (Grandl et al. 2008).

Especially for excitatory cells with spontaneous activity—neurons and cardiomyocytes—voltage sensitive dyes can report changes in activity upon drug addition. We discussed the voltage-sensitive dyes (e.g., ANEP-PS) and fluorescent proteins above (see Sect. 6.2.3). These have been employed in cardiomyocytes, primary neurons, and hiPSC-derived cardiomyocytes.

The above examples were screening for mutants (TRPV) and active compounds (cardiomyocytes). While compounds can be simply added to the sample robotically, it is more complex to screen a large number of mutants unless each mutant is cultured separately. In a 500-amino acid ion channel, there exist 10000 single point mutations (including nonsense) and exponentially more if multiple mutations are considered. For this purpose *multiplexed assays of variant effects (MAVEs)* have been developed. This method attempts to tackle two challenges that one encounters when scanning a high number of variants: first, to generate the mutants and, second, to assess them efficiently. To address the first problem, each residue can be replaced using sequence saturation mutagenesis to obtain a population of all possible single point mutations (Wong et al. 2004), and a stable cell line of this population can be generated where each mutant is assigned a specific DNA barcode (Fig. 6.7c) (Glazer et al. 2019; Muhammad et al. 2024; Ma et al. 2024). Expression level can then be analyzed using fluorescently tagged antibodies against the construct. To more efficiently determine the effect of the entire population, the analysis is undertaken using fluorescence-assisted cell sorting (FACS), where the cells are classified dependent on their expression level. Each class then undergoes deep sequencing to determine the frequency of each barcode (mutant).

However, it does not suffice to monitor only expression levels: While many genetic variants linked to disease are causing trafficking defects, many others alter function. One possibility is to scan all mutants using automated patch-clamp (Glazer et al. 2019; Ma et al. 2024); however, if a fluorescence signal can be used, such as the calcium signaling for TRPV, the same multiplexed analysis using cell sorting can be employed. Muhammad et al. (2024) co-expressed the KCNE1 variants to be tested not only with wildtype KCNQ1 but also with a variant KCNQ1-S140G. The S140G replacement shifted the activation threshold of KCNQ1 close to the resting potential of the HEK293T cells, in which they were expressed. KCNE1 additionally shifts the voltage dependence. By positioning the threshold so close to the resting potential, any gain- or loss-of-function would lead to increased or decreased potassium flux at resting potential, respectively. Over several days of culturing, the cells would develop lower fitness and propagate slower. By repeating the FACS analysis and deep sequencing after 8 days, the effect of the genetic variant could be classified by assessing survival rates (Muhammad et al. 2024).

6.8 Conclusions and Outlook

In recent decades, driven by developments in the field of electron microscopy, the availability of structural information has increased dramatically. At the same time, genetic screening of patients has provided a wealth of information on the intramolecular changes that require a causal chain to understand how the diseases develop. These developments require methods to link structural information to the kinetics observed in function. In particular, for ion channels, the kinetics of function is easily obtained via electrophysiological studies, while fluorescence allows to follow structural dynamics.

Currently, the research focuses on linking molecular mechanisms to cellular and tissue functions, as only then causality for disease development can be established beyond doubt. We are now at a turning point in this endeavor since many of the puzzle pieces have come to maturity. As outlined in this chapter, recent developments in genome editing, protein engineering, genetic code expansion, and imaging techniques provide the tools to study the expression and dynamics of ion channels with fluorescence spectroscopic methods in complex systems. Protocols are available to develop stem cells with specific genetic background, and gene editing methods allow manipulation of the genes of interest without further disturbing the balance of expression among the proteins. An example is the tagged Piezo channels in the organoids mentioned earlier. We are no longer confined to reconstitution systems, so that biophysical studies of ion channels can be carried out in situ, directly determining the effect on physiological function. With these current developments, fluorescence spectroscopy will play a key role not only in the study of the molecular mechanisms but also in optimizing personalized medicine by providing tools for rapid testing of genetic variants.

Acknowledgments This chapter is dedicated to the memory of Sergei Noskov, whom we miss in the ion channel and biophysics community.

References

Abbott GW (2006) Molecular mechanisms of cardiac voltage-gated potassium channelopathies. Curr Pharm Des 12(28):3631–44

Aimon S, Manzi J, Schmidt D, Poveda Larrosa JA, Bassereau P, Toombes GE (2011) Functional reconstitution of a voltage-gated potassium channel in giant unilamellar vesicles. PLoS One 6(10):e25529. https://doi.org/10.1371/journal.pone.0025529

Aimon S, Callan-Jones A, Berthaud A, Pinot M, Toombes GS, Bassereau P (2014) Membrane shape modulates transmembrane protein distribution. Dev Cell 28(2):212–218. https://doi.org/10.1016/j.devcel.2013.12.012

Armstrong CM, Bezanilla F (1973) Currents related to movement of the gating particles of the sodium channels. Nature 242(5398):459–461

Arsic A, Hagemann C, Stajkovic N, Schubert T, Nikic-Spiegel I (2022) Minimal genetically encoded tags for fluorescent protein labeling in living neurons. Nat Commun 13(1):314. https://doi.org/10.1038/s41467-022-27956-y

Asamoah OK, Wuskell JP, Loew LM, Bezanilla F (2003) A fluorometric approach to local electric field measurements in a voltage-gated ion channel. Neuron 37(1):85–97

Ataka K, Pieribone VA (2002) A genetically targetable fluorescent probe of channel gating with rapid kinetics. Biophys J 82(1):509–516. https://doi.org/10.1016/S0006-3495(02)75415-5

Bannister JP, Chanda B, Bezanilla F, Papazian DM (2005) Optical detection of rate-determining ion-modulated conformational changes of the ether-a-go-go K+ channel voltage sensor. Proc Natl Acad Sci USA 102(51):18718–18723

Batulan Z, Haddad GA, Blunck R (2010) An intersubunit interaction between S4-S5 linker and S6 is responsible for the slow off-gating component in Shaker K+ channels. J Biol Chem 285(18):14005–19. https://doi.org/10.1074/jbc.M109.097717

Bertaccini GA, Casanellas I, Evans EL, Nourse JL, Dickinson GD, Liu G, et al (2024) Visualizing PIEZO1 localization and activity in hiPSC-Derived single cells and organoids with HaloTag technology. bioRxiv. https://doi.org/10.1101/2023.12.22.573117

Bessa-Neto D, Beliu G, Kuhlemann A, Pecoraro V, Doose S, Retailleau N, et al (2021) Bioorthogonal labeling of transmembrane proteins with non-canonical amino acids unveils masked epitopes in live neurons. Nat Commun 12(1):6715. https://doi.org/10.1038/s41467-021-27025-w

Biskup C, Zimmer T, Benndorf K (2004) FRET between cardiac Na+ channel subunits measured with a confocal microscope and a streak camera. Nat Biotechnol 22(2):220–4. https://doi.org/10.1038/nbt935

Biskup C, Kusch J, Schulz E, Nache V, Schwede F, Lehmann F, et al (2007) Relating ligand binding to activation gating in CNGA2 channels. Nature 446(7134):440–3. https://doi.org/10.1038/nature05596

Blunck R, Starace D, Correa A, Bezanilla F (2004) Detecting conformational changes of the prokaryotic sodium channel NaChBac in mammalian cells using site-directed fluorescence techniques. Biophys J 86(1):167A

Blunck R, Cordero-Morales JF, Cuello LG, Perozo E, Bezanilla F (2006) Detection of the opening of the bundle crossing in KcsA with fluorescence lifetime spectroscopy reveals the existence of two gates for ion conduction. J Gen Physiol 128(5):569–81. https://doi.org/10.1085/jgp.200609638

Blunck R, McGuire H, Hyde HC, Bezanilla F (2008) Fluorescence detection of the movement of single KcsA subunits reveals cooperativity. Proc Natl Acad Sci USA 105(51):20263–20268. https://doi.org/10.1073/pnas.0807056106

Blunck R (2021) Determining stoichiometry of ion channel complexes using single subunit counting. Methods Enzymol 653:377–404. https://doi.org/10.1016/bs.mie.2021.02.017

Borisenko V, Lougheed T, Hesse J, Fureder-Kitzmuller E, Fertig N, Behrends JC, et al (2003) Simultaneous optical and electrical recording of single gramicidin channels. Biophys J 84(1):612–22. https://doi.org/10.1016/S0006-3495(03)74881-4

Borjesson SI, Elinder F (2011) An electrostatic potassium channel opener targeting the final voltage sensor transition. J Gen Physiol 137(6):563–77. https://doi.org/10.1085/jgp.201110599

Bouvier M, Heveker N, Jockers R, Marullo S, Milligan G (2007) BRET analysis of GPCR oligomerization: newer does not mean better. Nat Methods 4(1):3–4; author reply 4. https://doi.org/10.1038/nmeth0107-3

Bower NI, de la Serrana DG, Cole NJ, Hollway GE, Lee HT, Assinder S, et al (2012) Stac3 is required for myotube formation and myogenic differentiation in vertebrate skeletal muscle. J Biol Chem 287(52):43936–49. https://doi.org/10.1074/jbc.M112.361311

Brelidze TI, Carlson AE, Zagotta WN (2009) Absence of direct cyclic nucleotide modulation of mEAG1 and hERG1 channels revealed with fluorescence and electrophysiological methods. J Biol Chem 284(41):27989–97. https://doi.org/10.1074/jbc.M109.016337

Brelidze TI, Carlson AE, Sankaran B, Zagotta WN (2012) Structure of the carboxy-terminal region of a KCNH channel. Nature 481(7382):530–3. https://doi.org/10.1038/nature10735

Bruening-Wright A, Larsson HP (2007) Slow conformational changes of the voltage sensor during the mode shift in hyperpolarization-activated cyclic-nucleotide-gated channels. J Neurosci. 27(2):270–8. https://doi.org/27/2/270[pii]10.1523/JNEUROSCI.3801-06.2007

Campos FV, Chanda B, Roux B, Bezanilla F (2007) Two atomic constraints unambiguously position the S4 segment relative to S1 and S2 segments in the closed state of Shaker K channel. Proc Natl Acad Sci USA 104(19):7904–7909

Cao E, Liao M, Cheng Y, Julius D (2013) TRPV1 structures in distinct conformations reveal activation mechanisms. Nature 504(7478):113–8. https://doi.org/10.1038/nature12823

Caterina MJ, Schumacher MA, Tominaga M, Rosen TA, Levine JD, Julius D (1997) The capsaicin receptor: a heat-activated ion channel in the pain pathway. Nature 389(6653):816–24. https://doi.org/10.1038/39807

Cha A, Bezanilla F (1997) Characterizing voltage-dependent conformational changes in the Shaker K+ channel with fluorescence. Neuron 19(5): 1127–1140

Cha A, Snyder GE, Selvin PR, Bezanilla F (1999) Atomic scale movement of the voltage-sensing region in a potassium channel measured via spectroscopy. Nature 402(6763):809–813

Chahine M, Bennett PB, George JAL, Horn R (1994) Functional expression and properties of the human skeletal muscle sodium channel. Pflugers Arch 427(1-2):136–142

Chanda B, Bezanilla F (2002) Tracking voltage-dependent conformational changes in skeletal muscle sodium channel during activation. JGenPhysiol 120(5):629–645

Chanda B, Asamoah OK, Bezanilla F (2004) Coupling interactions between voltage sensors of the sodium channel as revealed by site-specific measurements. JGenPhysiol 123(3):217–230

Chanda B, Asamoah OK, Blunck R, Roux B, Bezanilla F (2005) Gating charge displacement in voltage-gated ion channels involves limited transmembrane movement. Nature 436(7052):852–856. https://doi.org/10.1038/nature03888

Chanda B, Blunck R, Faria LC, Schweizer FE, Mody I, Bezanilla F (2005) A hybrid approach to measuring electrical activity in genetically specified neurons. NatNeurosci 8(11):1619–1626. https://doi.org/10.1038/nn1558

Charest PG, Terrillon S, Bouvier M (2005) Monitoring agonist-promoted conformational changes of beta-arrestin in living cells by intramolecular BRET.

EMBO Rep 6(4):334–40. https://doi.org/10.1038/sj.embor.7400373
Chatterjee A, Guo J, Lee HS, Schultz PG (2013) A genetically encoded fluorescent probe in mammalian cells. J Am Chem Soc 135(34):12540–3. https://doi.org/10.1021/ja4059553
Codding SJ, Trudeau MC (2019) The hERG potassium channel intrinsic ligand regulates N- and C-terminal interactions and channel closure. J Gen Physiol 151(4):478–488. https://doi.org/10.1085/jgp.201812129
Combs DJ, Shin HG, Xu Y, Ramu Y, Lu Z (2013) Tuning voltage-gated channel activity and cellular excitability with a sphingomyelinase. J Gen Physiol 142(4):367–380. https://doi.org/10.1085/jgp.201310986
Condon E (1926) A theory of intensity distribution in band systems. Phys Rev 28(6):1182–1201. https://doi.org/10.1103/PhysRev.28.1182
Cordero-Morales JF, Cuello LG, Zhao Y, Jogini V, Cortes DM, Roux B, et al (2006) Molecular determinants of gating at the potassium-channel selectivity filter. NatStructMolBiol 13(4):311–318
Cordero-Morales JF, Cuello LG, Perozo E (2006) Voltage-dependent gating at the KcsA selectivity filter. NatStructMolBiol 13(4):319–322
Cuello LG, Jogini V, Cortes DM, Perozo E (2010) Structural mechanism of C-type inactivation in K(+) channels. Nature 466(7303):203–8. https://doi.org/nature09153[pii]10.1038/nature09153
Dai G, Zagotta WN (2017) Molecular mechanism of voltage-dependent potentiation of KCNH potassium channels. eLife 6:e26355. https://doi.org/10.7554/eLife.26355
De-la Rosa V, Rangel-Yescas GE, Ladrón-de Guevara E, Rosenbaum T, Islas LD (2013) Coarse architecture of the transient receptor potential vanilloid 1 (TRPV1) ion channel determined by fluorescence resonance energy transfer. J Biol Chem 288(41):29506–17. https://doi.org/10.1074/jbc.M113.479618
Del Camino D, Yellen G (2001) Tight steric closure at the intracellular activation gate of a voltage-gated K(+) channel. Neuron 32(4):649–656
Del Camino D, Kanevsky M, Yellen G (2005) Status of the intracellular gate in the activated-not-open state of shaker K+ channels. JGenPhysiol 126(5):419–428
Diercks CS, Dik DA, Schultz PG (2021) Adding new chemistries to the central dogma of molecular biology. Chem 7(11):2883–2895. https://doi.org/10.1016/j.chempr.2021.09.014
Doudna JA, Charpentier E (2014) Genome editing. The new frontier of genome engineering with CRISPR-Cas9. Science 346(6213):1258096. https://doi.org/10.1126/science.1258096
Doyle DA, Morais CJ, Pfuetzner RA, Kuo A, Gulbis JM, Cohen SL, et al (1998) The structure of the potassium channel: molecular basis of K+ conduction and selectivity. Science 280(5360):69–77
Driggers C, Sung MW, Shyng SL (2023) 9. In: Zheng J, Trudeau MC (eds) Inward Rectifier Potassium Channels, vol. 2. CRC Press, Boca Raton https://doi.org/10.1201/9781003096276
Eggan P, Gordon SE, Zagotta WN (2024) Ligand-coupled conformational changes in a cyclic nucleotide-gated ion channel revealed by time-resolved transition metal ion FRET. Elife 13. https://doi.org/10.7554/eLife.99854
Faure E, Starek G, McGuire H, Berneche S, Blunck R (2012) A Limited 4 angstrom radial displacement of the S4-S5 linker is sufficient for internal gate closing in Kv channels. J Biol Chem 287(47):40091–40098. https://doi.org/10.1074/jbc.M112.415497
Faure E, Thompson C, Blunck R (2014) Do lipids show state-dependent affinity to the voltage-gated potassium channel KvAP?. JBC 289(23):16452–61. https://doi.org/10.1074/jbc.M113.537134
Gest AMM, Sahan AZ, Zhong Y, Lin W, Mehta S, Zhang J (2024) Molecular spies in action: genetically encoded fluorescent biosensors light up cellular signals. Chem Rev 124(22):12573–12660. https://doi.org/10.1021/acs.chemrev.4c00293
Glazer AM, Wada Y, Li B, Muhammad A, Kalash OR, O'Neill MJ, et al (2019) High-throughput reclassification of *SCN5A* variants. bioRxiv, p 858175. https://doi.org/10.1101/858175
Goldschen-Ohm MP, Klenchin VA, White DS, Cowgill JB, Cui Q, Goldsmith RH, et al (2016) Structure and dynamics underlying elementary ligand binding events in human pacemaking channels. Elife 5. https://doi.org/10.7554/eLife.20797
Gonzalez C, Koch HP, Drum BM, Larsson HP (2010) Strong cooperativity between subunits in voltage-gated proton channels. Nat Struct Mol Biol 17(1):51–6. https://doi.org/10.1038/nsmb.1739
Grandl J, Hu H, Bandell M, Bursulaya B, Schmidt M, Petrus M, et al (2008) Pore region of TRPV3 ion channel is specifically required for heat activation. Nat Neurosci 11(9):1007–13
Grandl J, Kim SE, Uzzell V, Bursulaya B, Petrus M, Bandell M, et al (2010) Temperature-induced opening of TRPV1 ion channel is stabilized by the pore domain. Nat Neurosci 13(6):708–14. https://doi.org/nn.2552[pii]10.1038/nn.2552
Groome JR, Fujimoto E, Ruben PC (2000) The delay in recovery from fast inactivation in skeletal muscle sodium channels is deactivation. Cell Mol Neurobiol 20(4):521–7. https://doi.org/10.1023/a:1007040731407
Groulx N, Juteau M, Blunck R (2010) Rapid topology probing using fluorescence spectroscopy in planar lipid bilayer: the pore-forming mechanism of the toxin Cry1Aa of Bacillus thuringiensis. J Gen Physiol 136(5):497–513. https://doi.org/jgp.200910347[pii]10.1085/jgp.200910347
Guo YR, MacKinnon R (2017) Structure-based membrane dome mechanism for Piezo mechanosensitivity. Elife 6. https://doi.org/10.7554/eLife.33660
Haddad GA, Blunck R (2011) Mode shift of the voltage sensors in Shaker K+ channels is caused by energetic coupling to the pore domain. J Gen Physiol 137(5):455–72. https://doi.org/10.1085/jgp.201010573

Harley CA, Bernardo-Seisdedos G, Stevens-Sostre WA, Jones DK, Azevedo MM, Sampaio P, et al (2021) Conformation-sensitive antibody reveals an altered cytosolic PAS/CNBh assembly during hERG channel gating. Proc Natl Acad Sci USA 118(44). https://doi.org/10.1073/pnas.2108796118

Hastie P, Ulbrich MH, Wang HL, Arant RJ, Lau AG, Zhang Z, et al (2013) AMPA receptor/TARP stoichiometry visualized by single-molecule subunit counting. Proc Natl Acad Sci USA 110(13):5163–8. https://doi.org/10.1073/pnas.1218765110

Heim R, Prasher DC, Tsien RY (1994) Wavelength mutations and posttranslational autoxidation of green fluorescent protein. Proc Natl Acad Sci USA 91(26):12501–4

Hodgkin AL, Huxley AF (1952) A quantitative description of membrane current and its application to conduction and excitation in nerve. JPhysiol 117(4):500–544

Hyde HC, Sandtner W, Vargas E, Dagcan AT, Robertson JL, Roux B, et al (2012) Nano-positioning system for structural analysis of functional homomeric proteins in multiple conformations. Structure 20(10):1629–40. https://doi.org/10.1016/j.str.2012.08.022

Jensen MO, Jogini V, Borhani DW, Leffler AE, Dror RO, Shaw DE (2012) Mechanism of voltage gating in potassium channels. Science 336(6078):229–33. https://doi.org/10.1126/science.1216533

Jiang Y, Lee A, Chen J, Ruta V, Cadene M, Chait BT, et al (2003) X-ray structure of a voltage-dependent K+ channel. Nature 423(6935):33–41

Jones CM, Robkis DM, Blizzard RJ, Munari M, Venkatesh Y, Mihaila TS, et al (2021) Genetic encoding of a highly photostable, long lifetime fluorescent amino acid for imaging in mammalian cells. Chem Sci 12(36):11955–11964. https://doi.org/10.1039/d1sc01914g

Jung KH, Kim SF, Liu Y, Zhang X (2019) A Fluorogenic AggTag Method Based on Halo- and SNAP-Tags to Simultaneously Detect Aggregation of Two-Proteins in Live Cells. Chembiochem 20(8):1078–1087. https://doi.org/10.1002/cbic.201800782

Kalstrup T, Blunck R (2013) Dynamics of internal pore opening in K-V channels probed by a fluorescent unnatural amino acid. Proc Natl Acad Sci USA 110(20):8272–8277. https://doi.org/10.1073/pnas.1220398110

Kalstrup T, Blunck R (2015) Reinitiation at non-canonical start codons leads to leak expression when incorporating unnatural amino acids. Sci Rep 5:11866. https://doi.org/10.1038/srep11866

Kalstrup T, Blunck R (2017) Voltage-clamp fluorometry in Xenopus oocytes using fluorescent unnatural amino acids. J Vis Exp (123):55598. https://doi.org/10.3791/55598

Kalstrup T, Blunck R (2018) S4-S5 linker movement during activation and inactivation in voltage-gated K(+) channels. Proc Natl Acad Sci USA 115(29):E6751–E6759. https://doi.org/10.1073/pnas.1719105115

Kim RY, Pless SA, Kurata HT (2017) PIP2 mediates functional coupling and pharmacology of neuronal KCNQ channels. Proc Natl Acad Sci 114(45):E9702–E9711. https://doi.org/doi:10.1073/pnas.1705802114

Kong J, Wang Y, Qi W, Huang M, Su R, He Z (2020) Green fluorescent protein inspired fluorophores. Adv Colloid Interf Sci 285:102286. https://doi.org/10.1016/j.cis.2020.102286

Kontis KJ, Goldin AL (1997) Sodium channel inactivation is altered by substitution of voltage sensor positive charges. J Gen Physiol 110(4):403–13. https://doi.org/10.1085/jgp.110.4.403

Kontis KJ, Rounaghi A, Goldin AL (1997) Sodium channel activation gating is affected by substitutions of voltage sensor positive charges in all four domains. J Gen Physiol 110(4):391–401. https://doi.org/10.1085/jgp.110.4.391

Kraemer A, Rehmann HR, Cool RH, Theiss C, de Rooij J, Bos JL, et al (2001) Dynamic interaction of cAMP with the rap guanine-nucleotide exchange factor epac11 1Edited by A. R. Fersht. J Mol Biol 306(5):1167–1177. https://doi.org/10.1006/jmbi.2001.4444

Kuo A, Gulbis JM, Antcliff JF, Rahman T, Lowe ED, Zimmer J, et al (2003) Crystal structure of the potassium channel KirBac1.1 in the closed state. Science 300(5627):1922–1926. https://doi.org/doi:10.1126/science.1085028

Kusch J, Thon S, Schulz E, Biskup C, Nache V, Zimmer T, et al (2012) How subunits cooperate in cAMP-induced activation of homotetrameric HCN2 channels. Nat Chem Biol 8(2):162–9. https://doi.org/10.1038/nchembio.747

Labro AJ, Lacroix JJ, Villalba-Galea CA, Snyders DJ, Bezanilla F (2012) Molecular mechanism for depolarization-induced modulation of Kv channel closure. J Gen Physiol 140(5):481–93. https://doi.org/10.1085/jgp.201210817

Ledwell JL, Aldrich RW (1999) Mutations in the S4 region isolate the final voltage-dependent cooperative step in potassium channel activation. JGenPhysiol 113(3):389–414

Lee HS, Guo J, Lemke EA, Dimla RD, Schultz PG (2009) Genetic incorporation of a small, environmentally sensitive, fluorescent probe into proteins in Saccharomyces cerevisiae. J Am Chem Soc 131(36):12921–3. https://doi.org/10.1021/ja904896s

Lee SJ, Wang S, Borschel W, Heyman S, Gyore J, Nichols CG (2013) Secondary anionic phospholipid binding site and gating mechanism in Kir2.1 inward rectifier channels. Nat Commun 4(1):2786. https://doi.org/10.1038/ncomms3786

Leisle L, Chadda R, Lueck JD, Infield DT, Galpin JD, Krishnamani V, et al (2016) Cellular encoding of Cy dyes for single-molecule imaging. Elife 5. https://doi.org/10.7554/eLife.19088

LeMasurier M, Heginbotham L, Miller C (2001) KcsA: it's a potassium channel. JGenPhysiol 118(3):303–314

Leterrier C, Potier J, Caillol G, Debarnot C, RuedaăBoroni F, Dargent B (2015) Nanoscale architecture of the axon initial segment reveals an organized and robust scaffold. Cell Rep 13(12):2781–2793. https://doi.org/10.1016/j.celrep.2015.11.051

Li J, Ostmeyer J, Cuello LG, Perozo E, Roux B (2018) Rapid constriction of the selectivity filter underlies C-type inactivation in the KcsA potassium channel. J Gen Physiol 150(10):1408–1420. https://doi.org/10.1085/jgp.201812082

Liao M, Cao E, Julius D, Cheng Y (2013) Structure of the TRPV1 ion channel determined by electron cryo-microscopy. Nature 504(7478):107–12. https://doi.org/10.1038/nature12822

Liin SI, Karlsson U, Bentzen BH, Schmitt N, Elinder F (2016) Polyunsaturated fatty acids are potent openers of human M-channels expressed in *Xenopus laevis* oocytes. Acta Physiol (Oxf) 218(1):28–37. https://doi.org/10.1111/apha.12663

Liu YS, Sompornpisut P, Perozo E (2001) Structure of the KcsA channel intracellular gate in the open state. NatStructBiol 8(10):883–887

Long SB, Campbell EB, Mackinnon R (2005) Voltage sensor of Kv1.2: structural basis of electromechanical coupling. Science 309(5736):903–8. https://doi.org/10.1126/science.1116270

Lu Z, Klem AM, Ramu Y (2001) Ion conduction pore is conserved among potassium channels. Nature 413(6858):809–813

Lu Z, Klem AM, Ramu Y (2002) Coupling between voltage sensors and activation gate in voltage-gated K+ channels. JGenPhysiol 120(5):663–676

Lussier Y, Furst O, Fortea E, Leclerc M, Priolo D, Moeller L, et al (2019) Disease-linked mutations alter the stoichiometries of HCN-KCNE2 complexes. Sci Rep 9. https://doi.org/10.1038/s41598-019-45592-3

Ma JG, O'Neill MJ, Richardson E, Thomson KL, Ingles J, Muhammad A, et al (2024) Multisite validation of a functional assay to adjudicate SCN5A Brugada syndrome-associated variants. Circ Genom Precis Med 17(4):e004569. https://doi.org/10.1161/CIRCGEN.124.004569

Mandala VS, MacKinnon R (2022) Voltage-sensor movements in the Eag Kv channel under an applied electric field. Proc Natl Acad Sci USA 119(46):e2214151119. https://doi.org/10.1073/pnas.2214151119

Mandala VS, MacKinnon R (2024) Electric field-induced pore constriction in the human Kv2.1 channel. bioRxiv, p 2024.12.18.629208. https://doi.org/10.1101/2024.12.18.629208

Mannuzzu LM, Moronne MM, Isacoff EY (1996) Direct physical measure of conformational rearrangement underlying potassium channel gating. Science 271(5246):213–216

Marme N, Knemeyer JP, Sauer M, Wolfrum J (2003) Inter- and intramolecular fluorescence quenching of organic dyes by tryptophan. BioconjugChem 14(6):1133–1139

McGuire H, Blunck R (2022) Studying KcsA channel clustering using single channel voltage-clamp fluorescence imaging. Front Physiol 13:863375. https://doi.org/10.3389/fphys.2022.863375

McGuire H, Aurousseau M, Bowie D, Blunck R (2012) An automated method to study oligomerization of single membrane-bound proteins using fluorescence imaging. Biophys J 102(3):114A–115A

McGuire H, Aurousseau MR, Bowie D, Blunck R (2012) Automating single subunit counting of membrane proteins in mammalian cells. JBC 287(43):35912–21. https://doi.org/10.1074/jbc.M112.402057

Meng F, Suchyna TM, Sachs F (2008) A fluorescence energy transfer-based mechanical stress sensor for specific proteins in situ. FEBS J 275(12):3072–3087. https://doi.org/10.1111/j.1742-4658.2008.06461.x

Miranda P, Contreras JE, Plested AJ, Sigworth FJ, Holmgren M, Giraldez T (2013) State-dependent FRET reports calcium- and voltage-dependent gating-ring motions in BK channels. Proc Natl Acad Sci USA. https://doi.org/10.1073/pnas.1219611110

Moller L, Regnier G, Labro AJ, Blunck R, Snyders DJ (2020) Determining the correct stoichiometry of Kv2.1/Kv6.4 heterotetramers, functional in multiple stoichiometrical configurations. Proc Natl Acad Sci USA 117(17):9365–9376. https://doi.org/10.1073/pnas.1916166117

Morais-Cabral JH, Robertson GA (2015) The enigmatic cytoplasmic regions of KCNH channels. J Mol Biol. 427(1):67–76. https://doi.org/10.1016/j.jmb.2014.08.008

Muhammad A, Calandranis ME, Li B, Yang T, Blackwell DJ, Harvey ML, et al (2024) High-throughput functional mapping of variants in an arrhythmia gene, KCNE1, reveals novel biology. Genome Med 16(1):73. https://doi.org/10.1186/s13073-024-01340-5

Mulhall EM, Gharpure A, Lee RM, Dubin AE, Aaron JS, Marshall KL, et al (2023) Direct observation of the conformational states of PIEZO1. Nature 620(7976):1117–1125. https://doi.org/10.1038/s41586-023-06427-4

Murata Y, Okamura Y (2007) Depolarization activates the phosphoinositide phosphatase Ci-VSP, as detected in Xenopus oocytes coexpressing sensors of PIP2. J Physiol 583(Pt 3):875–89. https://doi.org/10.1113/jphysiol.2007.134775

Nakajo K, Ulbrich MH, Kubo Y, Isacoff EY (2010) Stoichiometry of the KCNQ1 - KCNE1 ion channel complex. Proc Natl Acad Sci USA 107(44):18862–7. https://doi.org/10.1073/pnas.1010354107

Pandey R, Kalstrup T, Blunck R (2024) From resting state to inactivation: the dynamic journey of the inactivation peptide in Shaker Kv channels. bioRXiv. https://doi.org/10.1101/2024.07.10.602816

Pantazis A, Olcese R (2012) Relative transmembrane segment rearrangements during BK channel activation resolved by structurally assigned fluorophore-quencher pairing. J Gen Physiol 140(2):207–18. https://doi.org/10.1085/jgp.201210807

Pantazis A, Gudzenko V, Savalli N, Sigg D, Olcese R (2010) Operation of the voltage sensor of a human voltage- and Ca2+-activated K+ channel. Proc Natl Acad Sci USA 107(9):4459–64. https://doi.org/0911959107[pii]10.1073/pnas.0911959107

Pantazis A, Savalli N, Sigg D, Neely A, Olcese R (2014) Functional heterogeneity of the four voltage sensors of a human L-type calcium channel. Proc Natl Acad Sci USA 111(51):18381–6. https://doi.org/10.1073/pnas.1411127112

Pantazis A, Westerberg K, Althoff T, Abramson J, Olcese R (2018) Harnessing photoinduced electron transfer to optically determine protein sub-nanoscale atomic distances. Nat Commun . 9(1):4738. https://doi.org/10.1038/s41467-018-07218-6

Pantoja R, Rodriguez EA, Dibas MI, Dougherty DA, Lester HA (2009) Single-molecule imaging of a fluorescent unnatural amino acid incorporated into nicotinic receptors. Biophys J 96(1):226–37. https://doi.org/10.1016/j.bpj.2008.09.034

Pathak M, Kurtz L, Tombola F, Isacoff E (2005) The cooperative voltage sensor motion that gates a potassium channel. JGenPhysiol 125(1):57–69

Perozo E, Cortes DM, Cuello LG (1998) Three-dimensional architecture and gating mechanism of a K+ channel studied by EPR spectroscopy. NatStructBiol 5(6):459–469

Pisupati A, Mickolajczyk KJ, Horton W, van Rossum DB, Anishkin A, Chintapalli SV, et al (2018) The S6 gate in regulatory Kv6 subunits restricts heteromeric K(+) channel stoichiometry. J Gen Physiol 150(12):1702–1721. https://doi.org/10.1085/jgp.201812121

Posson DJ, Ge P, Miller C, Bezanilla F, Selvin PR (2005) Small vertical movement of a K+ channel voltage sensor measured with luminescence energy transfer. Nature 436(7052):848–851

Posson DJ, McCoy JG, Nimigean CM (2013) The voltage-dependent gate in MthK potassium channels is located at the selectivity filter. Nat Struct Mol Biol 20(2):159–66. https://doi.org/10.1038/nsmb.2473

Prasher DC, Eckenrode VK, Ward WW, Prendergast FG, Cormier MJ (1992) Primary structure of the *Aequorea victoria* green-fluorescent protein. Gene 111(2): 229–33

Priest MF, Lee EEL, Bezanilla F (2021) Tracking the movement of discrete gating charges in a voltage-gated potassium channel eLife 10:e58148. https://doi.org/10.7554/eLife.58148

Puljung M, Vedovato N, Usher S, Ashcroft F (2019) Activation mechanism of ATP-sensitive K+ channels explored with real-time nucleotide binding. eLife 8:e41103. https://doi.org/10.7554/eLife.41103

Reifenrath M, Boles E (2018) A superfolder variant of pH-sensitive pHluorin for in vivo pH measurements in the endoplasmic reticulum. Sci Rep 8(1):11985. https://doi.org/10.1038/s41598-018-30367-z

Rios E, Brum G (1987) Involvement of dihydropyridine receptors in excitation—contraction coupling in skeletal muscle. Nature 325(6106):717–720. https://doi.org/10.1038/325717a0

Rogers M, Obergrussberger A, Kondratskyi A, Fertig N (2024) Using automated patch clamp electrophysiology platforms in ion channel drug discovery: an industry perspective. Expert Opin Drug Discov 19(5):523–535. https://doi.org/10.1080/17460441.2024.2329104

Savalli N, Kondratiev A, Toro L, Olcese R (2006) Voltage-dependent conformational changes in human Ca(2+)- and voltage-activated K(+) channel, revealed by voltage-clamp fluorometry. ProcNatlAcadSciUSA 103(33):12619–12624

Schmidt D, Jiang QX, MacKinnon R (2006) Phospholipids and the origin of cationic gating charges in voltage sensors. Nature 444(7120):775–779

Secrist JA, Barrio JR, Leonard NJ (1972) A fluorescent modification of adenosine triphosphate with activity in enzyme systems: 1,N6-Ethenoadenosine Triphosphate. Science 175(4022):646–647. https://doi.org/doi:10.1126/science.175.4022.646

Selvin PR (2002) Principles and biophysical applications of lanthanide-based probes. Annu Rev Biophys Biomol Struct. 31:275–302. https://doi.org/10.1146/annurev.biophys.31.101101.140927

Shen Y, Wu SY, Rancic V, Aggarwal A, Qian Y, Miyashita SI, et al (2019) Genetically encoded fluorescent indicators for imaging intracellular potassium ion concentration. Commun Biol 2:18. https://doi.org/10.1038/s42003-018-0269-2

Shtengel G, Galbraith JA, Galbraith CG, Lippincott-Schwartz J, Gillette JM, Manley S, et al (2009) Interferometric fluorescent super-resolution microscopy resolves 3D cellular ultrastructure. Proc Natl Acad Sci 106(9):3125–3130. https://doi.org/doi:10.1073/pnas.0813131106

Shuai J, Parker I (2005) Optical single-channel recording by imaging Ca2+ flux through individual ion channels: theoretical considerations and limits to resolution. Cell Calcium 37(4):283–99. https://doi.org/10.1016/j.ceca.2004.10.008

Smith PL, Yellen G (2002) Fast and slow voltage sensor movements in HERG potassium channels. JGenPhysiol 119(3):275–293

Smith-Maxwell CJ, Ledwell JL, Aldrich RW (1998) Uncharged S4 residues and cooperativity in voltage-dependent potassium channel activation. JGenPhysiol 111(3):421–439

St-Pierre F, Marshall JD, Yang Y, Gong Y, Schnitzer MJ, Lin MZ (2014) High-fidelity optical reporting of neuronal electrical activity with an ultrafast fluorescent voltage sensor. Nat Neurosci 17(6):884–9. https://doi.org/10.1038/nn.3709

Starace DM, Bezanilla F (2004) A proton pore in a potassium channel voltage sensor reveals a focused electric field. Nature 427(6974):548–553

Stokes GG (1852) On the change of refrangibility of light. Philos Trans R Soc Lond 142:463–562. https://doi.org/doi:10.1098/rstl.1852.0022

Tao X, Lee A, Limapichat W, Dougherty DA, MacKinnon R (2010) A gating charge transfer center in voltage sensors. Science 328(5974):67–73. https://doi.org/328/5974/67[pii]10.1126/science.1185954

Thapa P, Stewart R, Sepela RJ, Vivas O, Parajuli LK, Lillya M, et al (2021) EVAP: a two-photon imaging tool to study conformational changes in endogenous Kv2 channels in live tissues. J Gen Physiol 153(11). https://doi.org/10.1085/jgp.202012858

Thompson JR, Cronin B, Bayley H, Wallace MI (2011) Rapid assembly of a multimeric membrane protein pore. Biophys J 101(11):2679–83. https://doi.org/10.1016/j.bpj.2011.09.054

Tombola F, Pathak MM, Gorostiza P, Isacoff EY (2007) The twisted ion-permeation pathway of a resting voltage-sensing domain. Nature 445(7127):546–549

Tombola F, Ulbrich MH, Isacoff EY (2008) The voltage-gated proton channel Hv1 has two pores, each controlled by one voltage sensor. Neuron 58(4):546–556

Turro NJ, Ramamurthy V, Scaiano JC (2010) Modern molecular photochemistry of organic molecules. University Science Books, Sausalito. http://bvbm2.bib-bvb.de:8993/F?func=service&doc_library=BVB01&doc_number=017293662&line_number=0001&func_code=DB_RECORDS&service_type=MEDIAInhaltsverzeichnis.

Tyagi S, Lemke EA (2013) Genetically encoded click chemistry for single-molecule FRET of proteins. Methods Cell Biol 113:169–87. https://doi.org/10.1016/B978-0-12-407239-8.00009-4

Ulbrich MH, Isacoff EY (2007) Subunit counting in membrane-bound proteins. NatMethods 4(4): 319–321

Usher SG, Ashcroft FM, Puljung MC (2020) Nucleotide inhibition of the pancreatic ATP-sensitive K+ channel explored with patch-clamp fluorometry. eLife 9:e52775. https://doi.org/10.7554/eLife.52775

Vargas E, Bezanilla F, Roux B (2011) In search of a consensus model of the resting state of a voltage-sensing domain. Neuron 72(5):713–20. https://doi.org/10.1016/j.neuron.2011.09.024

Vierra NC, Kirmiz M, van der List D, Santana LF, Trimmer JS (2019) Kv2.1 mediates spatial and functional coupling of L-type calcium channels and ryanodine receptors in mammalian neurons. eLife 8:e49953. https://doi.org/10.7554/eLife.49953

Wang WW, MacKinnon R (2017) Cryo-EM Structure of the Open Human Ether-a-go-go-Related K+ Channel hERG. Cell 169(3):422–430. https://doi.org/10.1016/j.cell.2017.03.048

Wang S, Vafabakhsh R, Borschel WF, Ha T, Nichols CG (2016) Structural dynamics of potassium-channel gating revealed by single-molecule FRET. Nat Struct Mol Biol 23(1):31–36. https://doi.org/10.1038/nsmb.3138

Warmke JW, Ganetzky B (1994) A family of potassium channel genes related to eag in Drosophila and mammals. Proc Natl Acad Sci USA 91(8):3438–42

Whicher JR, MacKinnon R (2016) Structure of the voltage-gated K+ channel Eag1 reveals an alternative voltage sensing mechanism. Science 353(6300):664–669. https://doi.org/10.1126/science.aaf8070

Whitaker GM, Angoli D, Nazzari H, Shigemoto R, Accili EA (2007) HCN2 and HCN4 isoforms self-assemble and co-assemble with equal preference to form functional pacemaker channels. J Biol Chem 282(31):22900–9. https://doi.org/10.1074/jbc.M610978200

Wong TS, Tee KL, Hauer B, Schwaneberg U (2004) Sequence saturation mutagenesis (SeSaM): a novel method for directed evolution. Nucleic Acids Res 32(3):e26. https://doi.org/10.1093/nar/gnh028

Yang F, Xiao X, Lee BH, Vu S, Yang W, Yarov-Yarovoy V, et al (2018) The conformational wave in capsaicin activation of transient receptor potential vanilloid 1 ion channel. Nat Commun 9(1):2879. https://doi.org/10.1038/s41467-018-05339-6.

Yang S, Miao X, Arnold S, Li B, Ly AT, Wang H, et al (2022) Membrane curvature governs the distribution of Piezo1 in live cells. Nat Commun 13(1):7467. https://doi.org/10.1038/s41467-022-35034-6

Yu Y, Ulbrich MH, Li MH, Dobbins S, Zhang WK, Tong L, et al (2012) Molecular mechanism of the assembly of an acid-sensing receptor ion channel complex. Nat Commun 3:1252. https://doi.org/10.1038/ncomms2257

Zagotta WN, Hoshi T, Aldrich RW (1990) Restoration of inactivation in mutants of Shaker potassium channels by a peptide derived from ShB. Science 250(4980):568–571

Zhao J, Blunck R (2016) The isolated voltage sensing domain of the Shaker potassium channel forms a voltage-gated cation channel. Elife 5. https://doi.org/10.7554/eLife.18130

Zheng J, Zagotta WN (2000) Gating rearrangements in cyclic nucleotide-gated channels revealed by patch-clamp fluorometry. Neuron 28(2):369–374

Zheng J, Zagotta WN (2004) Stoichiometry and assembly of olfactory cyclic nucleotide-gated channels. Neuron 42(3):411–21

Part II

Insights into Ion Channel Structure and Function

Molecular Determinants of Voltage-Dependent Gating in Kv11.1 Channels

7

Diana V. Hunter and Tom W. Claydon

Abstract

Kv11.1 channels (also known as hERG channels) are voltage-gated K^+ ion channels that are critical mediators of repolarization in numerous cell types. Their role is best studied in the heart, where they underlie the rapid delayed rectifier current, I_{Kr}, which supports rhythmic cardiac electrical activity. The importance of this role is highlighted by the association of Kv11.1 channel dysfunction with arrhythmias and sudden death. The critical role of Kv11.1 channels is dependent on their unusual biophysical properties, which produce unique gating behaviors that enable control of membrane repolarization and cellular excitability. In this chapter, we discuss the role of Kv11.1 channels in cellular physiology and how this depends on unique structural features and biophysical function. We then discover what can be learned about Kv11.1 channel gating from studying genetic causes of channel dysfunction. Finally, we consider more recent efforts investigating pharmacological modifiers of Kv11.1 channel gating that may lead to restoration of function and consequently protection against cardiac arrhythmogenicity.

D. V. Hunter · T. W. Claydon (✉)
Department of Biomedical Physiology & Kinesiology, Simon Fraser University, Burnaby, BC, Canada
e-mail: thomas_claydon@sfu.ca

Keywords

K^+ channel · Voltage-gated channel · hERG · Kv11.1 · Gating · Biophysics

7.1 Introduction

The human *KCNH2* gene encodes the α subunit of the voltage-gated K^+ (Kv) channel, Kv11.1. Kv11.1 channels are frequently referred to as *human-ether-à-go-go-related gene* (hERG) because of their relation to the *Drosophila ether-a-go-go* (eag) gene that induced leg shaking movements when mutated (Warmke and Ganetsky 1994). Kv11.1 channels are expressed in numerous tissues, including brain, GI tract, and uterus, and have important roles in cell proliferation and cancer progression. However, their role is most widely studied in the heart. Following the discovery (Noble and Tsien 1969a, b; Sanguinetti and Jurkiewicz 1990) that two major K^+ conductance, now known as the slow (I_{Ks}) and rapid (I_{Kr}) delayed rectifier currents, contribute to cardiac repolarization, the gene associated with I_{Kr} was identified as coding for

A. Rosenhouse-Dantsker (ed.), *Decoding Ion Channels Structure and Function*, Advances in Experimental Medicine and Biology 1497, https://doi.org/10.1007/978-3-032-07523-9_7

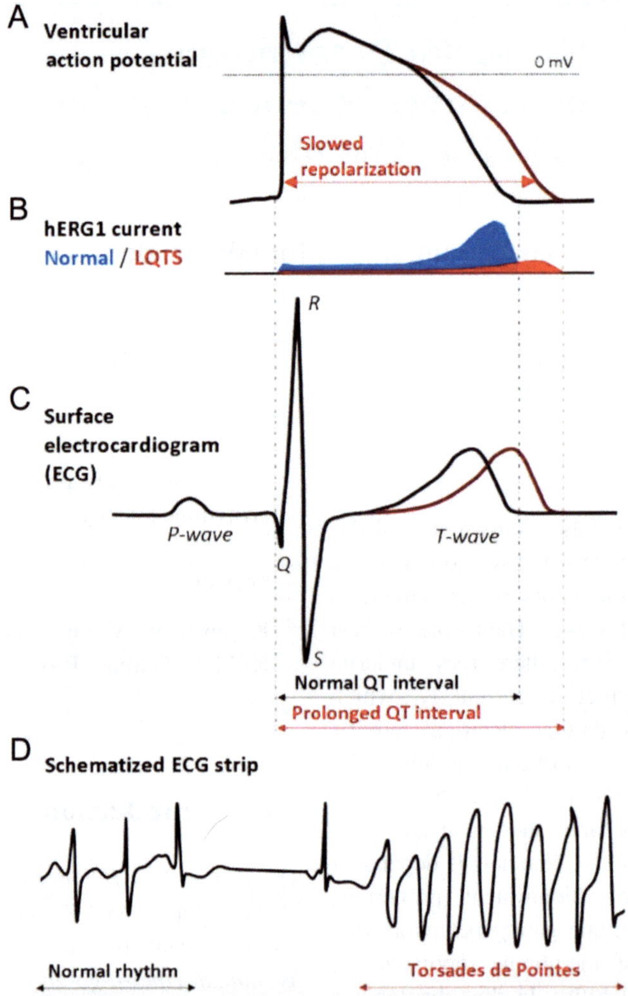

Fig. 7.1 Kv11.1 channels are critical for the maintenance of normal rhythm. (**a–c**), Loss of Kv11.1 (hERG1) channel function associated with Long QT Syndrome (LQTS) reduces the cardiac current, I_{Kr}, prolonging the action potential duration and QT interval. (**d**), Reduced repolarizing drive is associated with ventricular tachyarrhythmia, such as Torsades de Pointes (TdP). (Reproduced from Grilo et al. (2010))

the Kv11.1 channel (Sanguinetti et al. 1995; Trudeau et al. 1995). Kv11.1 channels are expressed broadly throughout the human atria, ventricle, sinoatrial node, and Purkinje fibers (Lees-Miller et al. 1997) and produce the I_{Kr} current critical for repolarization of the action potential and maintenance of cardiac rhythm (Fig. 7.1). The role of Kv11.1 channels and I_{Kr} repolarization is highlighted by genetic variants or drug interactions that cause loss of function or gain of function. Loss of Kv11.1 channel function reduces I_{Kr}, which delays repolarization, prolonging the action potential duration (APD). As a consequence, the heart rate-corrected QT interval (QT_c) is lengthened, leading to the electrical disorder Long QT Syndrome (LQTS), which predisposes to ventricular arrhythmia and sudden death. Conversely, gain of Kv11.1 channel function is associated with a shortened APD and QT interval, and arrhythmia associated with early ventricular repolarization (Fig. 7.1).

Since the first association of Kv11.1 with arrhythmias and sudden death (Curran et al. 1995), hundreds of genetic variants have been identified, and Kv11.1 channels have also been shown to be the major target for drug-induced arrhythmias occurring as unintended off-target action of a very wide range of drug compounds (Sanguinetti et al. 1995; Roden and Viswanathan 2005). Intensive efforts have enabled extensive screening to improve understanding of genotype-phenotype correlations (O'Neill et al. 2024; Ng et al. 2022; Kozek et al. 2020; Ma et al. 2023; Jiang et al. 2022), and requirements from regulatory bodies for cardiac safety screening of drugs and lead compounds have provided a wealth of data and insight. However, uncertainties with assigning pathogenic risk to variants of unknown significance and with accurate and efficient detection of arrhythmogenicity of novel compounds persist and still present significant hurdles. This has led to combinations of translationally relevant approaches, such as determination of high-resolution structures, generation of toxicological machine learning algorithms, and quests for targeted genetic or pharmacological modulation of Kv11.1 function. Such endeavors increasingly use human-induced pluripotent stem cells-derived cardiomyocytes (hiPSC-CMs) to model electrical dysfunction in beating human cardiomyocytes in vitro and/or with mathematical modeling of the consequences of Kv11.1 channel functional alterations in the context of the multiple relevant ion channels expressed in the heart, as is encapsulated in the Comprehensive in vitro Proarrhythmia Assay (CiPA) initiative.

7.2 Kv11.1 Channel Architecture

Like other Kv channels, Kv11.1 forms as a tetrameric assembly of α-subunits that creates a central ion-conducting pore. Each α-subunit comprises six α-helical transmembrane segments, S1-S6, and cytoplasmic N- and C-terminal regions that form functional domains (Fig. 7.2a and b). The S1-S4 transmembrane segments of each α-subunit form the voltage-sensing domain (VSD), with the S4 segment presenting six regularly spaced basic side-chain amino acids (Arg or Lys) along its length that enable it to sense transmembrane voltage changes and act as the primary voltage sensor. Translocation of the S4 segment charges in response to changes in the electric field provides the energetic force that initiates opening of the ion-conducting pore. Segments S1-S3 of the voltage-sensing domain possess critical negative amino acid side chains that serve as counter-charges to the S4 basic residues (Papazian et al. 1987). In resting and activated states of the voltage sensor, these counter-charges stabilize the S4 segment and define the energetic pathway for the movement of the S4 charges through the transmembrane electric field (Chanda and Bezanilla 2008; Catterall 2010). High-resolution structural determinations of Kv11.1 reveal the activated state of the voltage sensor (Wang and MacKinnon 2017). However, there are no structures to date that define the resting state of the Kv11.1 voltage sensor, so the dynamic rearrangements within the S1-S4 voltage-sensing unit that govern transitions of Kv11.1 channels from the closed to the open state are not well-defined.

The pore domain of the Kv11.1 channel α-subunit is formed by its S5 and S6 segments along with the intervening re-entrant pore helix and loop (P-loop). This pore unit is arranged in four-fold symmetry to form an ion conduction pathway across the membrane with a narrow ion selectivity filter at the extracellular end and a wider internal water-filled pore cavity (Fig. 7.2c) (Wang and MacKinnon 2017). The selectivity filter provides four distinct coordination sites for dehydrated K^+ ions that are formed by the backbone carbonyl oxygen atoms of residues lining the filter. High-resolution structural data show that the selectivity filter forms a highly conductive cylindrical shape with all four K^+ coordination sites occupied when the external K^+ concentration is high (e.g., 300 mM), and a non-conducting configuration with regions of dilatation and constriction with low external K^+ (e.g., 3 mM) (Fig. 7.2d) (Lau et al. 2024). These structures highlight how high-throughput ion conduction is achieved as a result of electrostatic repulsion between ions that are aligned in a single file across the transmembrane voltage gradient (Doyle et al. 1998). These

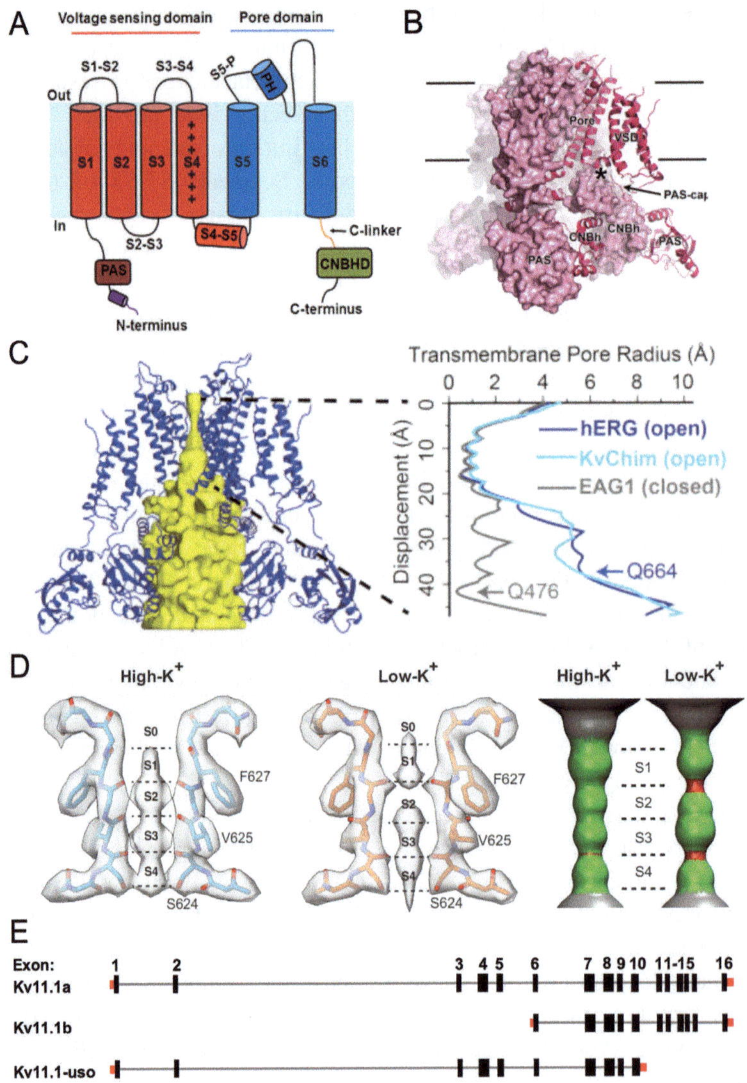

Fig. 7.2 Kv11.1 channel architecture. (**a**) cartoon membrane topology of a Kv11.1 channel α subunit. (Reproduced from Shi et al. (2020a)). (**b**) surface rendered representation of the cryo-EM structure of a Kv11.1 α subunit, highlighting critical functional domains. Voltage-sensing domain (VSD), cyclic nucleotide homology binding domain (cNBh). (Reproduced from Robertson and Morais Cabral (2019) with permission). (**c**) cryo-EM structure of the open Kv11.1 channel (left) and comparison of the permeation pathway diameter with that in the closed EAG channel (right). (Reproduced from Wang and Mackinnon (2017) with permission). (**d**) Cryo-EM structures of the Kv11.1 channel selectivity filter in high (conducting) and low (inactivated) extracellular K^+. (Reproduced from Lau et al. (2024)). (**e**) Exon map of alternate transcripts, Kv11.1b and Kv11.1-uso that are expressed in cardiac tissues in addition to the primary Kv11.1a gene product

atomic-level resolution structures reveal mechanisms underlying unique gating transitions in Kv11.1 channels, such as inactivation (see Sect. 7.3.4). The inner pore region below the selectivity filter is formed by the S6 transmembrane helices, which create a cavity lined by hydrophobic amino acid residue side chains. Atomic structures indicate that a single hydrated K^+ ion coordinates in the center of the cavity with the base of the pore helices contributing to localized negative electrostatic charge (Wang and MacKinnon 2017).

At the intracellular face of the membrane, the convergence of the S6 transmembrane helices forms an activation gate that controls ion access between the cytoplasm and channel pore (Fig. 7.2a). In canonical Kv channels, the flexible S4-S5 linker mechanically transduces voltage sensor energetics to open and close the intracellular activation gate of the pore domain. This electromechanical transduction is facilitated by a domain-swapped structure, in which the voltage sensor and S4-S5 linker of one subunit interacts with the pore domain of the adjacent subunit. Early mechanistic studies suggested a similar mechanism of electromechanical coupling in Kv11.1 channels (Tristani-Firouzi et al. 2002; Ferrer et al. 2006; Van Slyke et al. 2010; Cheng and Claydon 2012; Ng et al. 2012; Hull et al. 2014). However, Kv11.1 channels have a very short S4-S5 linker and lack a PXP motif that in other Kv channels contributes to both the formation of the activation gate aperture (del Camino et al. 2000; del Camino and Yellen 2001) and to the mechanical transduction of voltage sensor movement to the pore (Long et al. 2005; Lu et al. 2002). In the absence of the PXP motif, the intracellular activation gate of Kv11.1 channels is formed at residue Q664 (Wang and MacKinnon 2017; Wynia-Smith et al. 2008; Thouta et al. 2014) and is coupled with voltage sensor gating via an alternate pathway (Whicher and MacKinnon 2016; Thouta et al. 2014; Lorinczi et al. 2015; Malak et al. 2017, 2019) that involves cytoplasmic domains formed by the N- and C-termini (Adaixo et al. 2013).

Kv11.1 channels contain an N-terminal Per-Arnt-Sim (PAS) domain, known to mediate protein-protein interactions (Henry and Crosson 2011; Möglich et al. 2009; McIntosh et al. 2010), and a C-terminal cyclic nucleotide binding homology domain (CNBHD) with an intrinsic ligand that precludes nucleotide binding (Brelidze et al. 2009, 2012, 2013; Marques-Carvalho et al. 2012) (Fig. 7.2b). These N- and C-terminal domains interact with one another and manipulations of this assembly robustly modify Kv11.1 pore opening and closing, indicating that they are critical parts of the gating machinery (Adaixo et al. 2013).

The full-length α subunit of the Kv11.1 channel is the primary transcript from the *KCNH2* gene. However, other transcripts are prevalent in cardiac tissues and contribute to I_{Kr} and repolarization. As a result, the channel encoded by the primary transcript is termed K_v11.1a (or hERG1a). An alternately spliced Kv11.1b (or hERG1b) gene product that is broadly expressed in cardiac tissues (Jones et al. 2004) is identical to Kv11.1a from exon 6 onward, but lacks the first 5 exons that code for the majority of the N-terminal region, including the PAS domain. Co-assemblies of Kv11.1a with Kv11.1b display modified channel kinetics and drug binding (Lees-Miller et al. 1997; Jones et al. 2004, 2014; McNally et al. 2017; Tschirhart and Zhang 2020; Johnson et al. 2022; London et al. 1997; Phartiyal et al. 2007, 2008; Larsen et al. 2008; Sale et al. 2008; Larsen and Olesen 2010; Abi-Gerges et al. 2011), and there is also evidence for co-dependence of these transcripts. For example, Kv11.1a expression facilitates the membrane trafficking of Kv11.1b subunits (Phartiyal et al. 2008; McNally et al. 2017), a Kv11.1b-specific loss-of-function variant produces LQTS (Jones et al. 2016), and a Kv11.1a variant increases the ratio of Kv11.1b in Kv11.1a/Kv11.1b multimers (Feng et al. 2021). Indeed, co-translational biogenesis of functionally related channels may reduce electrical heterogeneity and suppress arrhythmia generation (Jameson et al. 2023). Another isoform detected in cardiac tissue, Kv11.1-uso, arises from premature polyadenylation following exon 9 (Fig. 7.2e). As a result of this premature truncation, Kv11.1-uso lacks much of the C-terminus and does not traffic to the cell membrane. Co-assembly of Kv11.1-uso with Kv11.1a subunits reduces membrane expression by 20% (Kupershmidt et al. 1998; Gong et al. 2014). The majority of studies involving Kv11.1 to date focus solely on Kv11.1a without much consideration of the relative impact of alternate transcripts.

7.3 Kv11.1 Channel Gating

At the cardiac cell resting membrane potential, Kv11.1 channels are typically closed. Membrane depolarization initiates voltage-dependent opening of channels, a process termed 'activation,' that is reversed with membrane repolarization, which closes channels, a process termed 'deactivation.' Kv11.1 channels also display a process termed 'inactivation' that involves the transition of channels into a non-conducting state during membrane depolarization. Upon membrane repolarization, channels rapidly recover from the inactivated state, producing a highly characteristic hooked tail current (Fig. 7.3a–c).

The unusual gating properties of Kv11.1 channels underlie their role in cellular repolarization. Relative to canonical Kv channels, Kv11.1 channels activate and deactivate slowly, but inactivate and recover from inactivation rapidly and with a uniquely strong dependence upon membrane voltage (Sanguinetti et al. 1995; Trudeau et al. 1995) (Fig. 7.3a–c). As a result of these behaviors, Kv11.1 channels open upon membrane depolarization but transition rapidly into a stable inactivated state, resulting in low occupancy of the open conducting channel and little outward current. This essentially stores repolarization potential until the end of the action potential. Upon membrane repolarization, Kv11.1 channels rapidly recover from inactivation back into the open conducting state, enabling conduction of K^+. Since Kv11.1 channels are slow to close upon the membrane repolarization, the open conducting state is long-lived, producing a robust repolarizing drive that facilitates termination of the action potential. In the heart, this functionality produces the resurgent current, I_{Kr}, that is critical for cardiac repolarization, as well as a protective latent repolarizing force long into the diastolic period that resists premature depolarizations (Lu et al. 2001) (Fig. 7.3d). In this way, each depolarization primes Kv11.1 channels, but holds them in reserve, releasing them into function to shape repolarization dependent on the repolarization need, and to protect against premature initiation of the next action potential.

Figure 7.4a shows an example behavior of single Kv11.1 channels. Consistent with the gating phenotype described above, Kv11.1 channel open probability is low during depolarization and increases markedly upon membrane repolarization. The single-channel conductance of Kv11.1 was reported to be 1–2 pS with flickering openings that indicated one (Kiehn et al. 1996) or two (Zou et al. 1997) open states with mean open times of 3 ms and 12 ms and mean closed times of 1 ms and 15–26 ms (Kiehn et al. 1996; Zou et al. 1997). Such single-channel recordings enable scrutiny of channel gating behaviors and their range of operation, including increased channel conductance with elevated external K^+ (Kiehn et al. 1996, 1999), impact of inactivation removal (Zou et al. 1997), and accelerated channel deactivation with acidic pH (Van Slyke et al. 2012). Mathematical representations of channel gating schemes that are based on available microscopic (single channel) and macroscopic (whole cell) data have arisen as powerful tools to describe and predict Kv11.1 channel gating behaviors. A number of linear Markov state models derived from the original Hodgkin and Huxley formulations (Hodgkin and Huxley 1952) have been used to describe Kv11.1 gating. These models introduce varying numbers of discrete closed, open, and inactivated channel states that channels transition between with variable kinetics and voltage dependence. One of the earliest models describes linear transitions between three closed states, one open state, and one inactivated state (Fig. 7.4b) (Wang et al. 1997). This model recapitulates the salient features of Kv11.1 channel gating behaviors, such as the weakening of the voltage dependence of activation kinetics at strongly depolarized membrane voltages (see Sect. 7.3.1), and preferential access to the inactivated state via the open state (see Sect. 7.3.4). Indeed, the Wang et al. model fared well in cross-model comparisons (Bett et al. 2011), and a modified version with two open states and co-operative transitions between subunits has been adapted to incorporate measurements of gating charge movement (Zhang et al. 2004; Shi et al. 2019). More recent cross-comparison of available models highlights the uncertainties and challenges

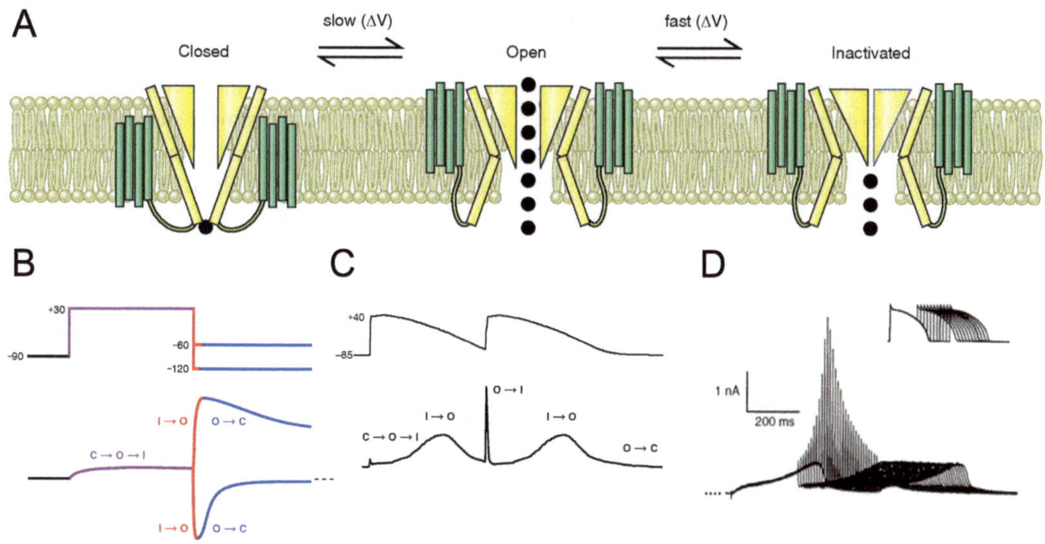

Fig. 7.3 Overview of Kv11.1 channel gating. (**a**) cartoon representation of the unusual gating processes in Kv11.1 channels. (**b**, **c**) representations of Kv11.1 channel currents in response to square wave (B) or action potential (B) voltage waveforms highlight the unique timing of Kv11.1 channel currents. (Adapted from Vandenberg et al. (2012) with permission) (**d**) demonstration of protective Kv11.1 channel currents conducted in response to premature depolarizations. (Reproduced from Lu et al. (2001) with permission)

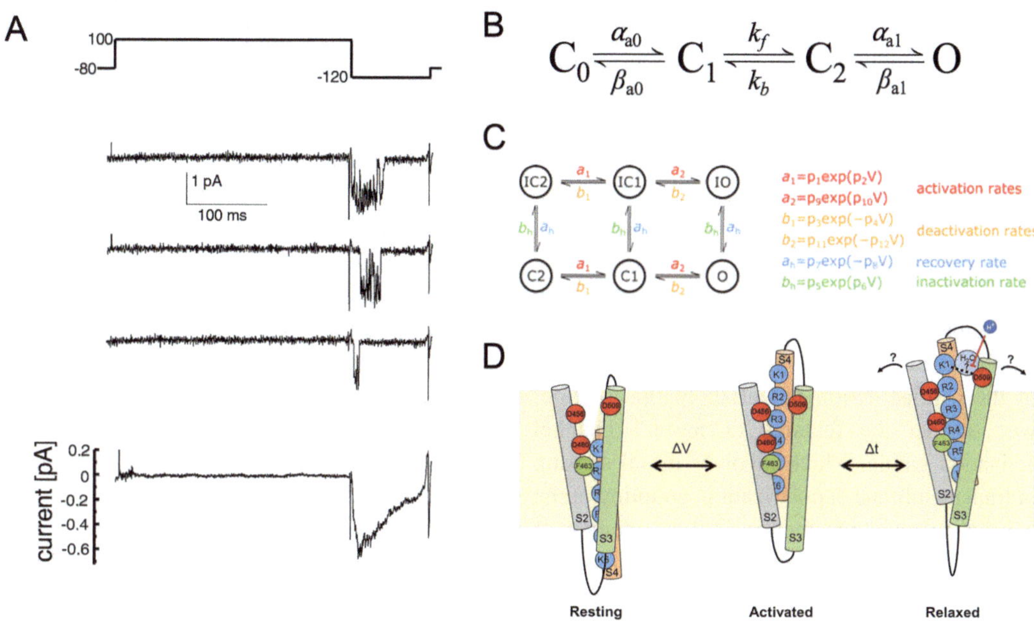

Fig. 7.4 Kv11.1 channel gating mechanisms. (**a**) recordings of single Kv11.1 channel activity. (Adapted from Kiehn et al. (1999) with permission). (**b**, **c**) kinetic schemes used to describe Kv11.1 channel gating pathways. (Redrawn from Wang et al. (1997) (b), and reproduced from Kemp et al. (2021) (c)). (**d**) cartoon representation of voltage-sensing unit interactions that regulate activation of Kv11.1 channels. (Reproduced from Shi et al. (2020a))

of modeling Kv11.1 kinetic behavior (Beattie et al. 2013) and proposes simplified cell-specific gating schemes built on information-rich data sets recorded at physiological temperature (Fig. 7.4c) (Lei et al. 2019a, b; Kemp et al. 2021). These models have significant potential to describe and predict the effects of genetic channel variants and consequences of drug interactions with Kv11.1 channels.

The voltage-dependent occupancy of open, closed, and inactivated channel states and the kinetics of transitions between them dynamically regulate Kv11.1 channel availability and determine action potential duration, restitution, and the effective refractory period, which are critical mediators of cellular excitability. Below, we discuss existing knowledge related to each gating step (see Cheng and Claydon (2012) and Vandenberg et al. (2012) for detailed reviews).

7.3.1 Kv11.1 Channel Activation

Kv11.1 channels and I_{Kr} measured in cardiac myocytes activate with a half-maximal open probability of −15 mV (Sanguinetti et al. 1995; Trudeau et al. 1995; Shibasaki 1987; Liu et al. 1996) and with a time constant of 125 ms at +30 mV (Wang et al. 1997). This activation is an order of magnitude slower than is observed in canonical Kv channels. The combination of slow activation and rapid inactivation creates stored repolarization which is released toward the end of the cardiac action potential. In this section, we consider the features of Kv11.1 channel activation gating, which involves initial sensing of transmembrane depolarization, coupling of the consequent dynamic motions to the opening of the intracellular pore gate, and ultimately ion conduction. These will be considered in the sequence they occur.

7.3.1.1 Voltage Sensing in Kv11.1 Channels

Hodgkin and Huxley (1952) proposed that the voltage-dependent conductance of K^+ and Na^+ across the squid axon membrane was dictated by charged particles within the membrane electric field. This idea was later demonstrated as a general concept of voltage-gated channels by the measurement of gating currents that were associated with the movement of gating charges across the cell membrane prior to channel opening (Armstrong and Bezanilla 1973; Gilly and Armstrong 1980). Subsequently, the S4 transmembrane segment, with its conserved sequence of regularly spaced positively charged amino acid side-chain residues, was identified as the primary voltage sensor in voltage-gated channels (Papazian et al. 1987; Noda et al. 1984). High-resolution three-dimensional crystal and cryo-EM structures of Kv channels show, with atomic-level detail, how the voltage-sensing unit (S1-S4) is arranged at the periphery of the pore region (S5-S6) (Wang and MacKinnon 2017; Long et al. 2005). Structures of the Kv11.1 voltage sensor domain depict wide internal and external aqueous cavities formed by membrane invaginations that dip toward a hydrophobic barrier (Wang and MacKinnon 2017; Khan et al. 2021). Functional data suggest that this hydrophobic plug acts as a gating charge transfer center that catalyzes the movement of S4 gating charges across the membrane between internal and external aqueous cavities (Tao et al. 2010; Cheng et al. 2013). Similar to canonical voltage-gated channels, the S4 segment of the Kv11.1 channel contains regularly spaced basic amino acid side chains that act as gating charges. Interestingly, the Kv11.1 S4 segment contains six positive charges (K1, R2-R5, K6), one fewer than found in the canonical *Shaker* Kv channel. When these charges are sequentially mutated, the outer three (not four as in *Shaker* channels), K1, R2, and R3 of Kv11.1 channels are found to contribute to gating currents that initiate channel activation (Subbiah et al. 2004; Piper et al. 2005). This is consistent with the translocation of these charges across the hydrophobic barrier from the internal to the external aqueous cavity when measured by the state dependence of chemical modification of the voltage sensor side chains (Elliott et al. 2009). While the outer four positive charges of the voltage sensor exhibit state-dependent accessibility in *Shaker* channels, only K1, R2, and R3 move to an externally accessible location upon activation of Kv11.1 channels (Zhang et al. 2004; Elliott et al. 2009). These functional studies are in excellent agreement with cryo-EM structures, which show K1, R2, and R3 above the gating charge transfer center and R4, K5, and K6 below (Wang and MacKinnon 2017). As such, estimates of the number of gating charges moved during channel activation are lower (6–8 e_0) in Kv11.1 channels (Zhang et al. 2004) than in *Shaker* channels (12–14 e_0) (Aggarwal and MacKinnon 1996; Seoh et al. 1996; Chanda and Bezanilla 2008; Catterall 2010). This has led to the suggestion that charge movement may be more restricted in Kv11.1 channels.

The activated state of the Kv11.1 channel is stable in the absence of an electric field (i.e., at 0 mV) and as such all atomic structures resolved to date show an open pore with voltage sensors in the activated position (Wang and MacKinnon 2017). The activated voltage sensor configuration is remarkable in that it is loosely packed relative to the *Shaker*-like channel voltage sensor with more limited lipid interactions that likely contribute to the altered energetics (Subbiah et al. 2004; Kudaibergenova et al. 2022). Following initial structural determinations, the Kv11.1 channel activated voltage sensor configuration was subsequently refined to account for a variety of interactions shown to be functionally important (Khan et al. 2021); however, there are currently no structures of the resting state of the voltage sensor, and an understanding of its dynamics must be interpreted from functional evidence. The arrangement of S4 voltage sensor charges is unusual in the Kv11.1 channel, with a lysine residue as the basic charge at both extremes (i.e., K1 and K6 are residues K525 and K538), while *Shaker*-type channels have an arginine at the outer end of S4. Mutations of either K1 or K6 shift the voltage dependence of Kv11.1 channel activation to occur at more hyperpolarized membrane voltages (Zhang et al. 2004, 2005; Subbiah et al. 2004, 2005; Piper et al. 2005; Liu et al. 2003; Subbotina et al. 2010), suggesting that the presence of K1 and K6 stabilizes the closed state of Kv11.1 channels. On the other hand, neutralization or charge reversal of the arginine residues (R2 and R3, and to a lesser degree, R4 and R5) causes only moderate shifts of activation in the depolarizing direction and also affects the rates of activation and deactivation (Zhang et al. 2004, 2005; Subbiah et al. 2004, 2005; Piper et al. 2005), suggesting that the middle S4 charges stabilize the open state or facilitate transitions into the open state. These S4 basic charges engage in functional interactions with acidic sidechain counter-charges (D411 in S1, D456, D460, D466 in S2, D501, D509 in S3) and a critical phenylalanine residue (F463 in S2) that forms a hydrophobic plug separating internal and external aqueous crevices, which has been described as a gating charge transfer center (Cheng and Claydon 2012; Tao et al. 2010; Cheng et al. 2013; Long et al. 2007). The narrowness of the gap separating internal and external crevices is highlighted by the observation that mutations that interrupt the interaction between K1 and F463 (Cheng et al. 2013), produce an alternate ion pathway across the membrane via the voltage sensor, known as an 'omega' current (Kudaibergenova et al. 2022; Starace et al. 1997; Starace and Bezanilla 2004). These studies help create a functional understanding of the voltage sensor movements from the resting to the activated configuration that underlies Kv11.1 channel opening (Fig. 7.4d) that involves an initial resting state of the voltage sensor that is stabilized at the outer and inner ends by interactions involving K1 and K6, respectively. Upon membrane depolarization, K1 and K6 interactions are broken, and sequential interactions between R2, R3, and likely R4 with acidic counter-charges form as the S4 segment is extruded. This sequential coupling of S4 charges with interacting partners orchestrates the shuttling of gating charges across the membrane electric field and governs the kinetics of Kv11.1 channel activation gating.

In the canonical Kv channel, *Shaker*, each of the four voltage sensor subunits activates independently and very rapidly (Cha and Bezanilla 1997, 1998; Zagotta et al. 1994; Bezanilla et al. 1994; Loots and Isacoff 1998; Kanevsky and Aldrich 1999) with a final rate-limiting concerted step associated with opening of the channel pore (Fedida and Hesketh 2001; Bezanilla 2000; Ledwell and Aldrich 1999). In contrast, gating current recordings and voltage clamp fluorimetry reports of voltage sensor dynamics imply slow rate-limiting voltage sensor movement in the Kv11.1 channel activation pathway (Van Slyke et al. 2010; Piper et al. 2003, 2005; Smith and Yellen 2002; Thouta et al. 2017). Gating currents recorded from Kv11.1 channels show that the majority of gating charge movement upon membrane depolarization is slow (τ 50 ms), though there are initial rapid components of charge movement (Piper et al. 2003; Wang et al. 2013; Goodchild and Fedida 2014). The voltage dependence of the majority of voltage sensor charge movement in Kv11.1 channels is left-shifted from channel opening, which is consistent with voltage sensor movement energetically preceding channel opening and suggests that slow charge movement limits the opening of the channel pore (Piper et al. 2003). The faster elements of charge movement and fluorescence change from the S4 may reflect transitions between closed states early in the activation pathway (Piper et al. 2003; Es-Salah-Lamoureux et al. 2010). Consistent with these reports of gating charge movement, kinetic simulations of ionic currents predict a voltage-independent, rate-limiting step in the activation pathway of Kv11.1 channels. This was introduced into simulations to account for the observation that the time constant of activation kinetics saturates at strongly depolarized potentials, becoming voltage-independent and rate-limiting (Wang et al. 1997; Liu et al. 1996; Subbiah et al. 2004). This is reflected in the kinetic scheme in Fig. 7.4b, which shows three closed states preceding a final open state with a voltage-independent transition between the second and third closed states, immediately before the voltage-dependent final opening step (Wang et al. 1997). A subsequent model built to incorporate the biphasic nature of Kv11.1 gating current measurements (Piper et al. 2003) predicts that each subunit independently undergoes two sequential voltage-dependent rate-limiting transitions before concerted movement through the two final closed states and into the open state. This expanded model simulates both ionic and gating current measurements and provides a quantitative view of the activation pathway of Kv11.1 channels.

7.3.1.2 Coupling Voltage Sensor Movement to Kv11.1 Channel Pore Opening

Atomic resolution structures show that the intracellular pore gate in Kv11.1 channels is formed by the convergence of the inner S6 segment helices at residue Q664 with gate opening that is facilitated by S6 helix flexibility afforded by a glycine hinge at G648 (Wang and MacKinnon 2017; Jiang et al. 2002). These structural data are consistent with prior functional data describing Q664 as the location of the gate (Wynia-Smith et al. 2008; Thouta et al. 2014). In canonical Kv channels, voltage sensor displacement electromechanically couples to opening of the intracellular pore gate via a lever-like action that is transduced via the S4-S5 linker, which connects the two functional units (Long et al. 2007). Early reports suggested a similar role for the amphipathic S4-S5 linker in Kv11.1 channels. Consistent with a role for the S4-S5 linker in electromechanical coupling in Kv11.1 channels, mutations of G546 greatly reduced the constraints on voltage sensor movements and facilitated pore opening (Van Slyke et al. 2010). In addition, mutation of D540 enabled channels to re-open at hyperpolarized membrane voltages, and this was dependent on the S6 pore gate region residue, R665 (Tristani-Firouzi et al. 2002; Ferrer et al. 2006; Sanguinetti and Xu 1999; Mitcheson et al. 2000). Moreover, cross-linking D540 with the L666 pore residue also immobilized a portion of the voltage sensor gating charge (Ferrer et al. 2006). Cryo-EM structures of the Kv11.1 channel, however, show an organization that suggests a different mechanism of voltage-sensing coupling with pore gating (Wang and MacKinnon 2017). These structures depict a loosely packed S4 and a very short S4-S5 linker. The contacts of the S4-S5 linker with the lower S6 segment are less robust than in canonical Kv structures, and the pore region of Kv11.1 channels lacks the conserved PXP motif important for engaging the lower S6 segment with the S4-S5 linker. Furthermore, the Kv11.1 tetramer is 'non-domain-swapped,' i.e., the S4-S5 linker of one subunit interacts with the lower S6 of the same subunit, rather than the neighboring subunit. This would be expected to alter the trajectory of mechanical force raising questions as to how voltage sensor motions might open the Kv11.1 channel pore. These structural observations amplify functional studies showing that channels split within the S4-S5 linker assemble and activate in a manner similar to wild-type channels (Lorinczi et al. 2015). As a result, studies have looked elsewhere to understand the molecular determinants of the coupling between voltage sensor and pore, and found that the mechanism likely involves complex integration of several cytoplasmic domains.

Comparison of the cryo-EM structure of the open Kv11.1 channel with the closed EAG1 channel (Whicher and MacKinnon 2016; Wang and MacKinnon 2017) reveals that channel opening involves displacement of the pore lining S6 helices below a glycine hinge (G648) that opens an 6 Å inner pore aperture. These structures also show critical association of the N-terminal PAS-Cap region (residues 1–25) with the S4-S5 linker and the C-linker, a flexible region that connects the CNBHD to the intracellular S6 pore region (Fig. 7.2b and c). Functional evidence also describes the PAS-Cap as a signal integrator that mediates closed-to-open transitions in Kv11.1 channels (Ng et al. 2012; de la Pena et al. 2011; Wang et al. 1998). Together, these observations suggest that upon depolarization, voltage sensor movement engages the PAS-Cap tail with the C-linker region, causing it to rotate, which loosens S6 and opens the pore gate (Whicher and MacKinnon 2016; Wang and MacKinnon 2017). There is evidence that this mechanism may involve the N- and C-terminus more broadly, since deletion of proximal N-terminus regions (residues 138–373 or 362–372 including the KIKER motif) promotes activation (Viloria et al. 2000; Gomez-Varela et al. 2002; Alonso-Ron et al. 2008; Saenen et al. 2006). Moreover, the N-terminal PAS domain and the C-terminal CNBHD are well-documented to regulate Kv11.1 channel function (Gianulis et al. 2013; Haitin et al. 2013; Gustina and Trudeau 2011). Rather than simply positioning the PAS-cap at the interface between the S4-S5 linker and C-linker, recent evidence suggests that dynamic allosteric modulation of PAS-CNBHD interactions regulate gating (Harley et al. 2021). These observations suggest that in addition to barriers or constraints to outward S4 gating charge movement, complex integrated signaling via cytoplasmic domains must be engaged before Kv11.1 channels can open.

7.3.2 Kv11.1 Channel Deactivation

Slow deactivation gating is a hallmark feature of Kv11.1 channels that enables robust resurgent repolarizing current during the cardiac action potential (Sanguinetti et al. 1995; Smith et al. 1996). As mentioned above, the observation that the closed-open equilibrium is inherently biased toward the open state (open probability is near maximal at 0 mV) indicates that energetic work must be applied to close Kv11.1 channels. It follows then that regulation of deactivation is likely a critical mediator of Kv11.1 channel function. The mechanistic basis for the slow closing of Kv11.1 channels appears more complex than in *Shaker* channels. Measurements of Kv11.1 gating currents demonstrate that gating charge return upon repolarization of the membrane is slow, suggesting that slow voltage sensor return may be a contributing factor (Piper et al. 2003; Thouta et al. 2017); however, an extensive range of studies has demonstrated that many domains of the Kv11.1 channel exert profound influence on the closing process. This has led to the suggestion of the presence of a 'master-switch,' where a variety of perturbations act through a common mechanism to control the slow rate of closing (Liu et al. 2003). It is also interesting to note that the overwhelming majority of mutations and manipulations that have been studied accelerate Kv11.1 deactivation gating, suggesting that multiple interdependent elements are required for the slow deactivation gating process that is so critical for the generation of the resurgent current.

Early studies seeking to understand the mechanistic basis of the rapid inactivation kinetics observed in Kv11.1 channels discovered that deletion of most of the N-terminus, including the globular PAS domain (from residues 2–354 or 2–373), accelerated deactivation, suggesting that N-terminal structures stabilized the open state of the channel (Spector et al. 1996; Schonherr and Heinemann 1996). Subsequent studies homed in on a key role for the PAS-Cap region (residues 1–25) (Wang et al. 1998; Ng et al. 2011), which contains an α helical region (residues 13–25) and an unstructured tail region (Wang and MacKinnon 2017; Muskett et al. 2011; Li et al. 2010; Gayen et al. 2011). Indeed, slow deactivation in N-terminally deleted (Δ2–354) fast deactivating Kv11.1 channels can be partially reconstituted by exogenous application of a PAS-Cap-like peptide (residues 1–16) (Wang et al. 2000) and fully reconstituted by application of a complete globular PAS domain fragment (Gianulis et al. 2013; Morais Cabral et al. 1998; Gustina and Trudeau 2011, 2009, 2012, 2013). The physiological significance of PAS domain regulation of closing is readily demonstrated by co-expression of Kv11.1a with Kv11.1b subunits. Kv11.1b has a truncated N-terminus, lacking both the globular PAS domain and PAS-Cap, and readily associates with Kv11.1a (Lees-Miller et al. 1997; London et al. 1997). In these Kv11.1a/Kv11.1b heterotetramers, with a reduced complement of PAS domains, deactivation kinetics are accelerated, and remarkably, the closing rate can be titrated by the relative expression of the Kv11.1b subunit (Sale et al. 2008; McNally et al. 2017; Trudeau et al. 2011). These observations suggest that dynamic globular PAS domain and in particular PAS-Cap interactions regulate the stability of open and closed states of the Kv11.1 channel.

PAS domain regulation of Kv11.1 channel deactivation is in concert with the C-terminal CNBHD. Initial studies showed that deletion of the C-terminus accelerated deactivation of Kv11.1 channels, and that further acceleration occurred upon deletion of the N-terminal globular PAS domain (Aydar and Palmer 2001). Similarly, application of an exogenous PAS domain peptide did not restore slow deactivation in N-terminally truncated Kv11.1 channels that lacked the CNBHD (Gustina and Trudeau 2011). Tight structural association of a PAS-CNBHD assembly is demonstrated in cryo-EM structures (Wang and MacKinnon 2017), confirming prior functional and structural data that mapped interaction surfaces (Muskett et al. 2011; Morais Cabral et al. 1998; Al-Owais et al. 2009). Cryo-EM structures also reveal that the globular PAS domain of one subunit interacts with the CNBHD of the neighboring subunit (Wang and MacKinnon 2017) in a manner that is reminiscent of the domain-swapped coupling of the voltage sensor and pore domains in canonical Kv channels. The binding pocket for cyclic nucleotides within the CNBHD forms part of the interaction surface of the PAS-CNBHD assembly. However, this site is not a functional coordination site for cyclic nucleotides; rather, it is occupied by an 'intrinsic ligand' (Brelidze et al. 2012, 2013; Marques-Carvalho et al. 2012; Zhao et al. 2017) that mediates PAS-CNBHD interactions (Zhao et al. 2017; Codding and Trudeau 2019).

There is extensive evidence from mutation-based and structural studies for a role of the S4-S5 linker in regulating Kv11.1 channel deactivation gating transitions. Modification of G546C with a bulky N-ethylmaleimide (NEM) adjunct accelerated deactivation to the same extent as deletion of the globular PAS domain, suggesting that both manipulations affect a common mechanism (Wang et al. 1998). Subsequently, other mutations at G546 were shown to alter voltage sensor return during deactivation, suggesting that flexibility of the S4-S5 linker at this point critically regulates voltage sensor gating and pore closing in Kv11.1 channels (Van Slyke et al. 2010). Other studies showed that mutation of Y542 prevents reconstitution of slow deactivation by exogenous application of a globular PAS domain peptide in N-terminally truncated channels (Fernandez-Trillo et al. 2011). Moreover, cysteines introduced at V3C in the PAS-Cap tail and Y542C in the S4-S5 linker are in close enough proximity to cross-link (de la Pena et al. 2011). Other mutation-based evidence has amplified the observation that mutations within the S4-S5 linker disrupt the same mechanism impacted by the removal of the globular PAS domain (Alonso-Ron et al. 2008). Indeed, NMR structures suggest a direct interaction between the S4-S5 linker and PAS domain

(Li et al. 2010). Based on the complex gating-modifying effects observed from an alanine substitution scan of the S4-S5 linker, Ng et al. proposed that the S4-S5 linker serves as an integrator that mediates transitions between closed and open Kv11.1 channel states (Ng et al. 2012). This idea is borne out by cryo-EM studies of open Kv11.1 where the PAS-Cap tail is tucked into a pocket at the interface of the S4-S5 linker and the C-linker (Wang and MacKinnon 2017). Most recently, molecular dynamics simulations of gating effects induced by mutations within this interface demonstrate a critical role for the hydrophobicity of the nexus formed by the amphipathic PAS-Cap tail α-helix, globular PAS domain, and CNBHD in sustaining PAS-Cap tip interactions with the S4-S5 linker (Stevens-Sostre et al. 2024). Taken together, these findings define a critical nexus at the intracellular membrane surface that couples S4 voltage sensor movements with the pore gate via the S4-S5 linker and PAS-CNBHD assembly controlling slow deactivation in Kv11.1 channels.

7.3.3 Kv11.1 Channel Relaxation

The activation and deactivation processes of Kv11.1 channels display marked energetic hysteresis (see Shi et al., 2020a for an in-depth review). That is to say, the voltage dependence of channel activation and deactivation is separated on the voltage axis, suggesting that an altered energetic landscape is experienced during deactivation (Fig. 7.5a–d). Such hysteresis is common in a variety of voltage-gated channels (Hull et al. 2014; Shi et al. 2019; Lacroix et al. 2011; Haddad and Blunck 2011; Labro et al. 2012, 2015; Priest et al. 2013; Corbin-Leftwich et al. 2016; Tan et al. 2012; Goodchild et al. 2015; Dierich et al. 2018; Elinder et al. 2006; Piper et al. 2003; Bruening-Wright and Larsson 2007; Thouta et al. 2017; Bezanilla et al. 1982; Kuzmenkin et al. 2004; Brum and Rios 1987; Brum et al. 1988; Shirokov et al. 1992; Tilegenova et al. 2017), and early investigations suggested that transitions to inactivated states underlie hysteretic behavior. Subsequent studies using isolated voltage sensor units showed that hysteresis is an intrinsic property of the voltage sensor (Villalba-Galea 2014; Zhao and Blunck 2016; Villalba-Galea et al. 2008) and is influenced by the mechanical load associated with interactions with the pore domain (Haddad and Blunck 2011; Batulan et al. 2010). Studies showed that hysteresis is caused by prolonged depolarization, during which the voltage sensor repositions from its activated state into a more stable 'relaxed' state by a process known as voltage sensor 'relaxation' (Lacroix et al. 2011; Labro et al. 2012; Priest et al. 2013). These observations indicate that channel deactivation is limited by the return of the voltage sensor from the stable relaxed configuration.

Voltage sensor relaxation can be observed in Kv11.1 channels as a progressive slowing of pore closure and gating charge return in response to depolarization steps of increasing duration (Fig. 7.5e and f) (Shi et al. 2019; Thouta et al. 2017; Goodchild and Fedida 2014; Goodchild et al. 2015). The slowing of deactivation shows a biphasic dependence on depolarization duration with one phase reflecting stabilization of the activated voltage sensor in response to opening of the pore gate, and a second much slower component that is kinetically dissociated from known gating steps and appears to arise intrinsically from the voltage sensor (Thouta et al. 2017). This voltage sensor relaxation likely contributes to the observed hysteresis of voltage-dependent Kv11.1 channel gating, by increasing the energetic barrier to voltage sensor return and slowing deactivation. As such, when voltage steps of physiological duration are used, Kv11.1 channel activation and deactivation show pronounced hysteresis, with the voltage dependence of open and closing separated by 65 mV (Thouta et al. 2017). The connection between relaxation and hysteresis is further demonstrated by the observation that mutation of a critical S3 segment acidic counter-charge, D509, destabilizes voltage sensor relaxation and accelerates deactivation kinetics greatly diminishing hysteresis (Shi et al. 2014, 2019). Similarly, external acidification reduces hysteresis and accelerates Kv11.1 channel deactivation by destabilizing the relaxed state

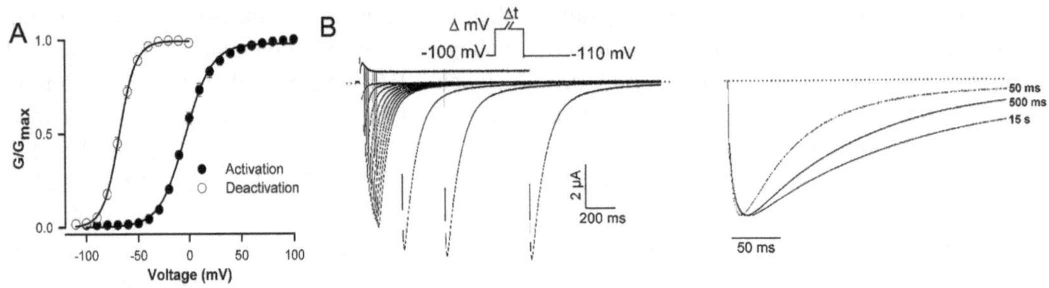

Fig. 7.5 Kv11.1 channel gating hysteresis. (**a**) physiologically relevant voltage step durations suggest a large energetic separation of activation and deactivation pathways. (**b**) increasing the duration of the activation step slows deactivation kinetics, suggesting an activation-induced stabilization of Kv11.1 channels in a relaxed configuration. (Adapted from Thouta et al. (2017) with permission)

of the voltage sensor (Shi et al. 2019) (Fig. 7.4d). Intriguingly, in wild-type Kv11.1 channels, long voltage step durations, which enable more complete capture of steady-state conditions, produce a discrepancy between the hysteresis of gating charge movement and the hysteresis of pore opening (40 mV vs 15 mV, respectively) (Thouta et al. 2017). This has led to the suggestion that pore closure is energetically more favorable than voltage sensor charge return, such that only one voltage sensor per channel tetramer is required to return to its resting configuration for the pore to close (Thouta et al. 2017). Taken together, these studies suggest that opening of the Kv11.1 channel pore stabilizes activated voltage sensors in a relaxed configuration, which increases the energetic barrier for voltage sensor return and channel closure, thus producing hysteresis. While electrostatic interactions involving D509 appear to be involved in the voltage sensor relaxation process, existing cryo-EM structures are not well resolved in this region. One proposed possibility is that water molecules may bridge H-bond or electrostatic interactions between D509 and S4 basic gating charges, stabilizing the relaxed state (Shi et al. 2019).

As described above, the N-terminal PAS domain of Kv11.1 channels is an integral regulator of deactivation gating that stabilizes the open state (Wang et al. 1998, 2000; Gustina and Trudeau 2011; Ng et al. 2011; Muskett et al. 2011; Morais Cabral et al. 1998). Interestingly, deletion of the PAS domain, or individual mutations within the PAS-Cap tail, accelerates the return of voltage sensor charge upon repolarization without reducing voltage sensor relaxation or hysteresis (Thouta et al. 2017; Tan et al. 2012; Jones 2022). These observations support the notion that relaxation is an intrinsic property of the voltage sensor and suggest that deactivation is modified at the interface of the voltage sensor and pore gate by the PAS domain independent of voltage sensor relaxation (Tan et al. 2012). High-resolution structures of these regions may help to visualize this mechanism. Future studies will help understand the relationship between dynamic changes in action potential duration and the magnitude of hysteresis. This will reveal the role of relaxation in the modulation of the timing and amplitude of repolarizing I_{Kr} current during beat-to-beat and rate-dependent changes in action potential duration.

7.3.4 Kv11.1 Channel Inactivation

From recordings of I_{Kr} in rabbit nodal tissue, Shibaski reported that the characteristic hooked tail current observed upon membrane repolarization arose as a result of rapid recovery from fast inactivation that occurred during the preceding depolarizing step (Shibasaki 1987). First measurements from cloned Kv11.1 channels recapitulated these behaviors, enabling initial descriptions of inactivation parameters, such as its unusually strong voltage dependence that correlated with I_{Kr} recordings (Sanguinetti et al. 1995). Although rapid, inactivation in Kv11.1 channels is not mediated by a ball-an-chain N-type mechanism where an N-terminal peptide occludes the intracellular pore. Rather, Kv11.1 channel inactivation involves loss of conduction at the outer mouth of the pore, indicating a P/C-type mechanism (Spector et al. 1996; Schonherr and Heinemann 1996). These studies showed that, like P/C-type inactivation, Kv11.1 channel inactivation is sensitive to external application of TEA, to K^+ occupancy of the selectivity filter, and to mutations introduced in the outer mouth of the pore. This suggested that inactivation in Kv11.1 channels involves a localized collapse of the selectivity filter that limits ion conduction (Smith et al. 1996). However, unlike P/C-type inactivation in canonical Kv channels, inactivation in Kv11.1 channels exhibits very rapid and strongly voltage-dependent kinetics (Smith et al. 1996; Spector et al. 1996) resulting in the characteristic rectification properties that produce the restricted outward current during depolarizing voltage steps and the critical resurgent cardiac repolarizing current. As such, the unusual mechanism of inactivation in Kv11.1 channels has been the subject of intense study (e.g., see Vandenberg et al. (2012) for a comprehensive review).

Studies in canonical K^+ channels, *Shaker* and *KcsA*, aimed at understanding the molecular mechanisms of P/C-type inactivation revealed broadly similar processes, but each has its own unique twist. Both involve hydrogen bond interactions behind the selectivity filter that maintain its configuration, but the mechanisms by which ion coordination is impeded appear divergent (Fig. 7.6) (Lau et al. 2024). In *KcsA* channels, interaction of E71 and D80 via a water molecule forms spring-like tension on the selectivity filter, stabilizing the conducting state. Inactivation involves rotation of the backbone carbonyl residues away from the central axis of the pore and entry of additional stabilizing water molecules behind the selectivity filter, resulting in an hourglass-like non-conducting structure (Doyle et al. 1998; Cordero-Morales et al. 2007; Cuello et al. 2010; Zhou et al. 2001; Ostmeyer et al. 2013). In *Shaker* and *Shaker*-like channels, interactions involving F434 and D447 stabilize the conducting selectivity filter, and inactivation involves dilatation of the outer part of the filter that radiates conformational change and reduces permeant ion coordination (Tan et al. 2022; Reddi et al. 2022). These findings are consistent with long-standing observations that inactivation in *Shaker* channels is particularly sensitive to mutation of T449 in the outer pore (Lopez-Barneo et al. 1993). Early mutagenesis studies in Kv11.1 channels showed a similar critical dependence of inactivation on S631, the equivalent residue to *Shaker* T449, as well as S620, the equivalent residue to E71 in KcsA (Smith et al. 1996; Schonherr and Heinemann 1996; Herzberg et al. 1998; Ficker et al. 1998). Very recently, Lau et al. demonstrated that the mechanism of inactivation in Kv11.1 channels involves an analogous, but distinct mechanism of ion conduction collapse (Lau et al. 2024). Comparing cryo-EM structures solved in high and low extracellular K^+, which prevent and encourage inactivation, respectively, they showed that inactivation was associated with a critical rotation of V625 away from the central axis of the selectivity filter that is stabilized by interaction with S620. The stabilized inactivated configuration produces an hourglass-like constricted shape with reduced ion occupancy (Fig. 7.2d). Molecular dynamics simulations suggest that the energetic barrier to the rotation of V625 is low such that rapid flipping of V625 may underlie the rapid kinetics of inactivation, which in Kv11.1 channels is two orders of magnitude faster than observed in *Shaker* channels (Lau et al. 2024). This mechanism is further supported by the observation that the structural differences between the Kv11.1 channel selectivity filter in high and low K^+ are broadly consistent with those between WT and S631A inactivation-deficient Kv11.1 channels (Wang and MacKinnon 2017). Comparison of this collection of structures has identified a potential role for upward displacement of the selectivity filter that enables flipping of V625 and stabilization with S620 (Lau et al. 2024).

While these structures provide plausible mechanisms underlying the unusually rapid inactivation process observed in Kv11.1 channels, little insight can be gleaned regarding its unique

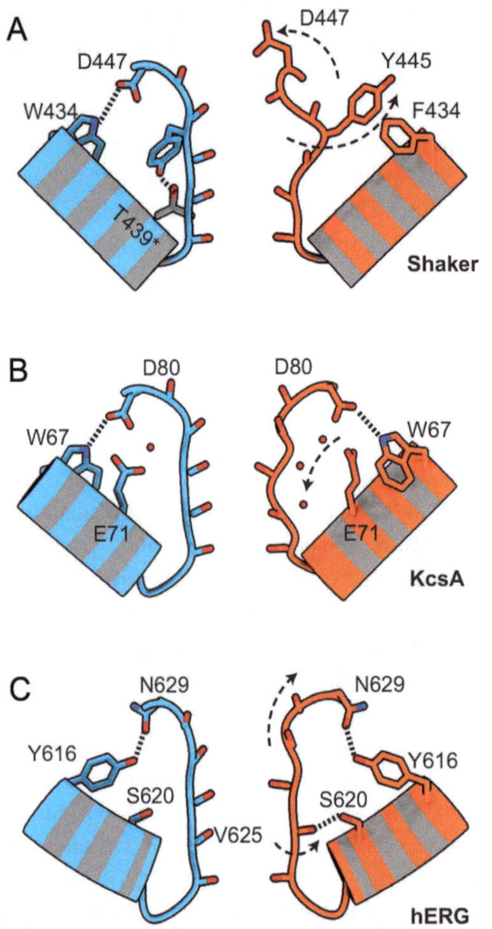

Fig. 7.6 Mechanism of Kv11.1 channel inactivation. Comparison of selectivity filter reconfigurations associated with inactivation in Kv11.1 channels (**c**) with those described in canonical K^+ channels, *Shaker* (**a**) and KcsA (**b**). (Reproduced from Lau et al. (2024))

voltage dependence. In both comparisons of WT and S631A structures (Wang and MacKinnon 2017) and low and high K^+ WT structures (Lau et al. 2024), there is no evidence of significant rearrangements in the voltage sensor domain. This is consistent with previous functional studies suggesting that the voltage dependence of inactivation is not derived from S4 voltage sensor basic charges (Zhang et al. 2004) and has distinct origins from that of activation (Piper et al. 2005). Interestingly, although mutations in the outer S5-P-loop have been shown to modify inactivation (Clarke et al. 2006) and a fluorescent reporter in this region appears to track events occurring on a fast timescale (Fougere et al. 2011), cryo-EM structures show limited motions of the S5-P-loop in relation to inactivation of the Kv11.1 selectivity filter (Lau et al. 2024). As such, the mechanistic basis for the unique voltage dependence of Kv11.1 channel inactivation remains unanswered, although one possibility posed is that the upward displacement of the selectivity filter noted above may derive a voltage dependence by altering the membrane electric field. This would be consistent with a previous 'Japanese puzzle box' model in which inactivation is initiated by K^+ efflux from the pore (Wang et al. 2011). This idea, however, remains to be tested.

7.4 Modification of Kv11.1 Gating by Genetic Variants

More than 1300 variants have been identified in Kv11.1 channels (Alameh et al. 2023). While the majority of missense mutations reduce the number of Kv11.1 channels expressed at the membrane as a result of protein misfolding or reduced protein stability in the membrane (Gong et al. 2006; Anderson et al. 2006, 2014; Ke et al. 2013), mutations can also alter a range of Kv11.1 channel gating properties. Moreover, following the rescue of the protein trafficking deficit, the majority of non-trafficking channel variants show altered gating properties (Perry et al. 2016). It is therefore worth considering the mechanisms by which variants affect Kv11.1 channel gating.

In a detailed study measuring the gating properties of 49 missense variants broadly selected from a LQTS2 registry, Perry et al. discovered that following rescue of trafficking, most of the variants affected gating (Perry et al. 2016). Some variants altered the voltage dependence of activation, producing a loss of function, while others shifted the voltage dependence of inactivation either toward more hyperpolarized voltages (suggesting a loss of function) or more depolarized voltages (suggesting a gain of function). Interestingly, many of the rescued variants exhibited accelerated deactivation kinetics, which

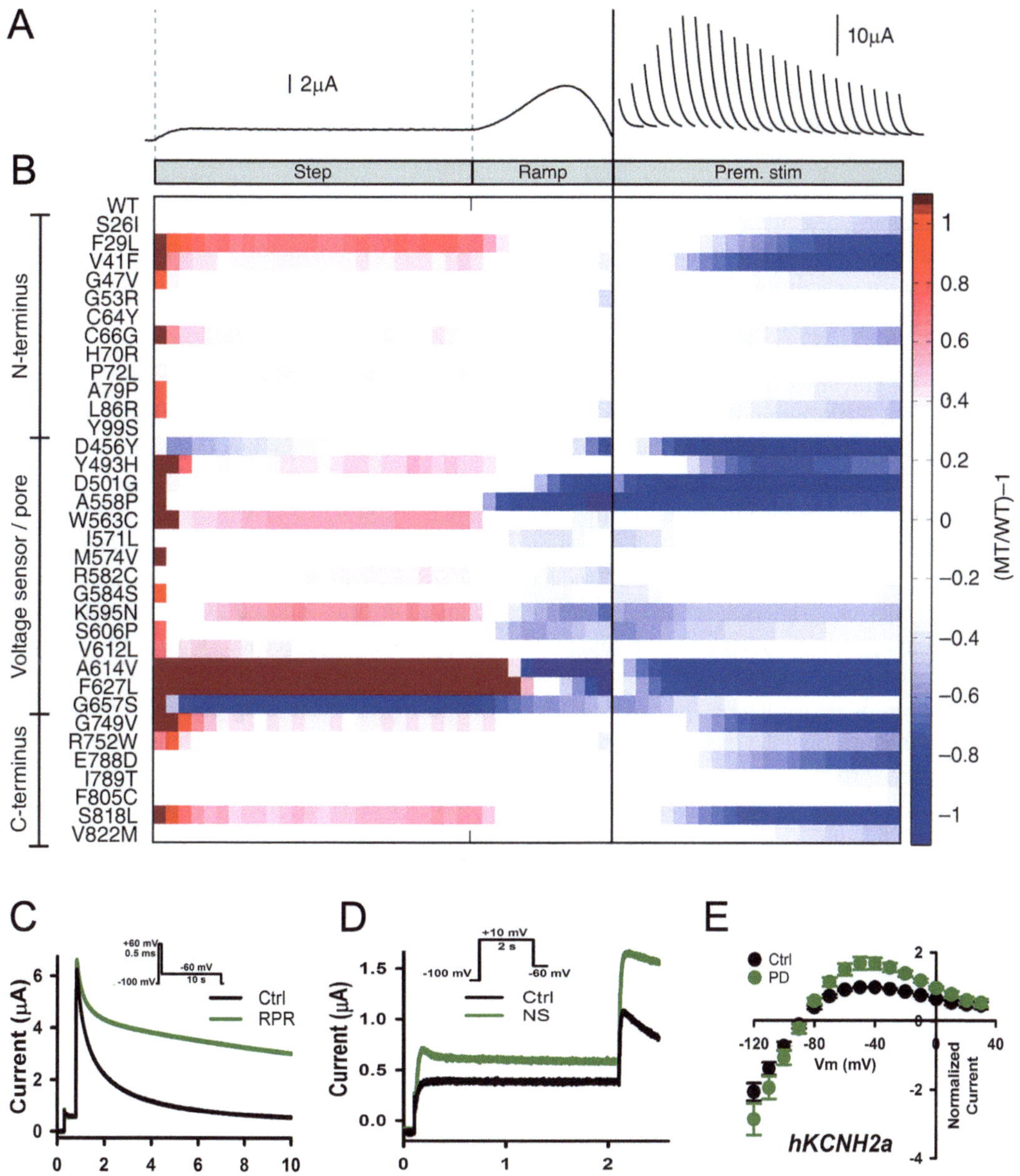

Fig. 7.7 Impact of genetic variants and activator compounds on Kv11.1 channel gating. (**a**, **b**) summary heatmap of gating effects in a large cohort of Kv11.1 variants highlights the diversity of gating deficits even following rescue of trafficking. (Reproduced from Perry et al. (2016) with permission). (**c–e**) examples of the gating-specific effects of different Kv11.1 activator compounds. (Adapted from Hull et al. (2019))

suggests a destabilized open state of the channel. Consistent with this, many variants presented reduced Kv11.1 channel current in response to premature membrane depolarizations. This study provides a simplified functional test, which elegantly captures the overall outcome of the variant-specific effects on different channel gating properties that can be used to understand functional consequences in the patient (Fig. 7.7).

Understanding the mechanism by which a variant modifies channel gating properties can be challenging. In the case of some well-studied variants, structural information is available to complement functional evidence. One example is the variant, R56Q, which primarily accelerates Kv11.1 channel deactivation kinetics, although the membrane protein expression is reduced somewhat (Foo et al. 2016) and the voltage dependence of inactivation is shifted slightly (Gustina and Trudeau 2009). Residue R56 is located in the globular PAS domain and its mutation to glutamine has been suggested to reduce PAS-CNBHD stability. This has led to the proposition that the R56Q mutation causes the PAS-Cap to dissociate from the Kv11.1 channel gating interface leading to destabilization of the open state and accelerated deactivation gating (Harley et al. 2021; Gustina and Trudeau 2009; Gianulis and Trudeau 2011). Other variant examples are the inactivation-reduced mutations S631A and S620T. As discussed above, high-resolution cryo-EM structures provide atomic-level information about the mechanism by which each variant produces an inactivation gating disturbance (Wang and MacKinnon 2017; Lau et al. 2024). In both cases, the variants affect ion occupancy of the selectivity filter by interfering with reorganizations that trigger collapse of ion conduction, with S631A altering rotation of F627, and S620T influencing the stability of the flipped configuration of V625. These examples may represent common mechanisms of gating disturbance upon which many variants converge and highlight the rich insight provided from coupling structural and functional studies.

7.5 Pharmacological Modification of Kv11.1 Gating

A rich diversity of pharmacological compounds interacts with Kv11.1 channels. Many compounds bind within the intracellular pore with an affinity that is unparalleled amongst other ion channels.

Cryo-EM structures of the Kv11.1 channel pore indicate novel features that may contribute to such diverse and high-affinity drug binding, such as internal extensions of the pore and a focused electrostatic charge (Wang and MacKinnon 2017). In combination with the presence of unique pore lining residues, Y652 and F656, as well as other important sites (T623, S624, V625, G648, V659), π-π and/or π-stacking electrostatic interactions stabilize drug coordination once the pore gate has opened (Mitcheson et al. 2000, 2005; Kamiya et al. 2006; Chen et al. 2002; Fernandez et al. 2004; Mitcheson 2008). As such, many drugs are open state blockers of Kv11.1 channels, where their binding requires channel activation and stabilizes non-conducting states. Drug block of Kv11.1 channels reduces resurgent repolarizing current during the action potential and delays cardiac repolarization, resulting in acquired or drug-induced LQTS (Fenichel et al. 2004). As a consequence, mandated screening of drug interactions with Kv11.1 channels is an essential part of the drug development process, and machine learning algorithms have become increasing useful in predicting the cardiac risk of Kv11.1 channel blocker compounds (Lee et al. 2021; DeMarco et al. 2021; Yang et al. 2020; Wacker and Noskov 2018; Wacker et al. 2017).

A growing number of compounds have been discovered to activate or enhance Kv11.1 channel function, and these would be expected to increase current during cardiac repolarization. Their mechanisms of action are diverse and often overlapping, and this has been well-summarized recently (El Harchi and Brincourt 2022). The first Kv11.1 channel activator to be described increased channel function by slowing deactivation gating (Kang et al. 2005), and activators with this mechanism of action are classed as Type 1 (Fig. 7.7). Type 2 Kv11.1 channel activators inhibit channel inactivation, while other activator types increase channel open probability or shift the voltage dependence of activation gating (Fig. 7.7). The Type 1 activator, RPR260243, interacts with L553 and F557 in a pocket between S5 and S6 and binding slows deactivation gating, stabilizing the open state of the pore (Gardner and Sanguinetti 2015; Wu et al. 2015; Perry et al. 2007) without affecting voltage sensor movements (Abbruzzese et al. 2010). These observations suggest that this Type 1 activator slows deactivation kinetics by altering coupling at the hydrophobic nexus between the voltage sensor and the pore. Its effectiveness to selectively increase protective Kv11.1 currents in response to premature depolarizations has also been demonstrated (Kemp et al. 2021; Venkateshappa et al. 2023; Shi et al. 2020b). The Type 2 Kv11.1 channel activator, NS1643, induces a depolarizing shift of the voltage dependence of inactivation, resulting in a marked increase in channel current (Casis et al. 2006). Molecular studies suggest several possible interaction sites for NS1643 (Durdagi et al. 2012; Guo et al. 2015; Xu et al. 2008; Grunnet et al. 2011), including a site that involves N629 at the outer mouth of the pore. Recent insight into the role of N629 interactions in stabilizing the conducting configuration of the Kv11.1 selectivity filter (Lau et al. 2024) would suggest that this interaction site may be critical for modulation of inactivation gating by this Type 2 activator. The effectiveness of NS1643 to enhance Kv11.1 channel current has been explored in cardiac tissues and in silico studies (Colman et al. 2017; Hansen et al. 2006; Peitersen et al. 2008) with these and other studies suggesting that over-correction of Kv11.1 channel function leading to abbreviated repolarization represents a significant risk of Type 2 activator compounds (Perry et al. 2020). The activator, PD118057, interacts at a hydrophobic site in the lower S6 created by residues F619, L622, and L646 where it may stabilize the conducting state of the selectivity filter (Perry et al. 2009). It remains to be seen whether rational design of small-molecule Kv11.1 channel activators with desirable gating effects will offer safe therapeutic potential.

7.6 Summary and Outlook

Since their first cloning 30 years ago, Kv11.1 channels have presented a uniquely interesting and challenging area of study. Kv11.1 channel gating properties are complex and have proved challenging to measure accurately. Despite this, much has been revealed through diligent application of biochemical, electrophysiological, structural, and in silico modeling approaches. These studies have provided a tantalizing view of the diversity of Kv channel function and structure. Improved clinical identification of variant types, functional characterization of variant and drug-bound channel behavior, and rigorous mathematical modeling have laid the foundation for predictive indices of Kv11.1 function in a wide range of physiological and pathophysiological conditions. These exciting advances have created a unique opportunity to explore the benefits of personalized approaches for the correction of Kv11.1 channel dysfunction in humans.

References

Abbruzzese J, Sachse FB, Tristani-Firouzi M, Sanguinetti MC (2010) Modification of hERG1 channel gating by Cd^{2+}. J Gen Physiol 136(1540–7748 (Electronic)):203–224

Abi-Gerges N, Holkham H, Jones EMC, Pollard CE, Valentin JP, Robertson GA (2011) HERG subunit composition determines differential drug sensitivity. Br J Pharmacol 164(2 B):419–432

Adaixo R, Harley CA, Castro-Rodrigues AF, Morais-Cabral JH (2013) Structural properties of PAS domains from the KCNH potassium channels. PLoSOne 8(1932–6203 (Electronic)):e59265

Aggarwal SK, MacKinnon R (1996) Contribution of the S4 segment to gating charge in the Shaker K^+ channel. Neuron 16(0896–6273 SB-M):1169–1177

Alameh M, Oliveira-Mendes BR, Kyndt F, Rivron J, Denjoy I, Lesage F, et al (2023) A need for exhaustive and standardized characterization of ion channels activity. The case of KV11.1. Front Physiol [Internet] [cited 2024 Dec 10];14. Available from: https://pubmed.ncbi.nlm.nih.gov/36860515/

Alonso-Ron C, de la Pena P, Miranda P, Dominguez P, Barros F (2008) Thermodynamic and kinetic properties of amino-terminal and S4-S5 loop HERG channel mutants under steady-state conditions. Biophys J 94(1542–0086 (Electronic)):3893–3911

Al-Owais M, Bracey K, Wray D (2009) Role of intracellular domains in the function of the herg potassium channel. Eur Biophys J [Internet] [cited 2019 Jun 18];38(5):569–576. Available from: http://link.springer.com/10.1007/s00249-009-0408-2

Anderson CL, Delisle BP, Anson BD, Kilby JA, Will ML, Tester DJ, et al (2006) Most LQT2 mutations reduce Kv11.1 (hERG) current by a class

2 (trafficking-deficient) mechanism. Circulation [Internet] [cited 2024 Dec 5];113(3):365–373. Available from: https://www.ahajournals.org/doi/10.1161/CIRCULATIONAHA.105.570200

Anderson CL, Kuzmicki CE, Childs RR, Hintz CJ, Delisle BP, January CT (2014) Large-scale mutational analysis of Kv11.1 reveals molecular insights into type 2 long QT syndrome. Nat Commun [Internet] [cited 2024 Dec 5];5. Available from: https://pubmed.ncbi.nlm.nih.gov/25417810/

Armstrong CM, Bezanilla F (1973) Currents related to movement of the gating particles of the sodium channels. Nature [Internet] [cited 2024 Jun 13];242(5398):459–461. Available from: https://pubmed.ncbi.nlm.nih.gov/4700900/

Aydar E, Palmer C (2001) Functional characterization of the C-terminus of the human ether-à-go-go-related gene K(+) channel (HERG). J Physiol [Internet] [cited 2024 Dec 5];534(Pt 1):1–14. Available from: https://pubmed.ncbi.nlm.nih.gov/11432987/

Batulan Z, Haddad GA, Blunck R (2010) An intersubunit interaction between S4-S5 linker and S6 is responsible for the slow off-gating component in Shaker K+ channels. J Biol Chem 285(1083-351X (Electronic)):14005–14019

Beattie KA, Luscombe C, Williams G, Munoz-Muriedas J, Gavaghan DJ, Cui Y et al (2013) Evaluation of an in silico cardiac safety assay: using ion channel screening data to predict QT interval changes in the rabbit ventricular wedge. J Pharmacol Toxicol Methods 68(1):88–96

Bett GCL, Zhou Q, Rasmusson RL (2011) Models of HERG gating. Biophys J [Internet] [cited 2024 Dec 5];101(3):631–642. Available from: https://pubmed.ncbi.nlm.nih.gov/21806931/

Bezanilla F (2000) The voltage sensor in voltage-dependent ion channels. Physiol Rev 80(0031–9333 SB-IM):555–592

Bezanilla F, Taylor RE, Fernández JM (1982) Distribution and kinetics of membrane dielectric polarization. 1. Long-term inactivation of gating currents. J Gen Physiol 79(1):21–40

Bezanilla F, Perozo E, Stefani E (1994) Gating of Shaker K^+ channels: II. The components of gating currents and a model of channel activation. Biophys J 66(0006–3495 (Print)):1011–1021

Brelidze TI, Carlson AE, Zagotta WN (2009) Absence of direct cyclic nucleotide modulation of mEAG1 and hERG1 channels revealed with fluorescence and electrophysiological methods. J Biol Chem [Internet] [cited 2024 Dec 5];284(41):27989–27997. Available from: https://pubmed.ncbi.nlm.nih.gov/19671703/

Brelidze TI, Carlson AE, Sankaran B, Zagotta WN (2012) Structure of the carboxy-terminal region of a KCNH channel. Nature 481(7382):530–533

Brelidze TI, Gianulis EC, DiMaio F, Trudeau MC, Zagotta WN (2013) Structure of the C-terminal region of an ERG channel and functional implications. Proc Natl Acad Sci U S A 110(1091–6490 (Electronic)):11648–11653

Bruening-Wright A, Larsson HP (2007) Slow conformational changes of the voltage sensor during the mode shift in hyperpolarization-activated cyclic-nucleotide-gated channels. J Neurosci 27(1529–2401 (Electronic)):270–278

Brum G, Rios E (1987) Intramembrane charge movement in frog skeletal muscle fibres. Properties of charge 2. J Physiol 387(1):489–517

Brum G, Ríos E, Stéfani E (1988) Effects of extracellular calcium on calcium movements of excitation-contraction coupling in frog skeletal muscle fibres. J Physiol 398(1):441–473

Casis O, Olesen SP, Sanguinetti MC (2006) Mechanism of action of a novel human ether-a-go-go-related gene channel activator. Mol Pharmacol 69(0026–895X (Print)):658–665

Catterall WA (2010) Ion channel voltage sensors: Structure, function, and pathophysiology. Neuron 67:915–928

Cha A, Bezanilla F (1997) Characterizing voltage-dependent conformational changes in the Shaker K^+ channel with fluorescence. Neuron 19(0896–6273 SB-M):1127–1140

Cha A, Bezanilla F (1998) Structural implications of fluorescence quenching in the Shaker K^+ channel. J Gen Physiol 112(0022-1295 (Print)):391–408

Chanda B, Bezanilla F (2008) A common pathway for charge transport through voltage-sensing domains. Neuron 57:345–351

Chen J, Seebohm G, Sanguinetti MC (2002) Position of aromatic residues in the S6 domain, not inactivation, dictates cisapride sensitivity of HERG and eag potassium channels. Proc Natl Acad Sci U S A 99(0027–8424 (Print)):12461–12466

Cheng YM, Claydon TW (2012) Voltage-dependent gating of hERG potassium channels. Front Pharmacol 3 MAY

Cheng YM, Hull CM, Niven CM, Qi J, Allard CR, Claydon TW (2013) Functional interactions of voltage sensor charges with an S2 hydrophobic plug in hERG channels. J Gen Physiol 142(3):289

Clarke CE, Hill AP, Zhao J, Kondo M, Subbiah RN, Campbell TJ et al (2006) Effect of S5P alpha-helix charge mutants on inactivation of hERG K^+ channels. J Physiol 573(0022–3751 (Print)):291–304

Codding SJ, Trudeau MC (2019) The hERG potassium channel intrinsic ligand regulates N- and C-terminal interactions and channel closure. J Gen Physiol 151(4):478–488

Colman MA, Perez Alday EA, Holden AV, Benson AP (2017) Trigger vs. substrate: multi-dimensional modulation of QT-prolongation associated arrhythmic dynamics by a hERG channel activator. Front Physiol 8(OCT):757

Corbin-Leftwich A, Mossadeq SM, Ha J, Ruchala I, Le AH, Villalba-Galea CA (2016) Retigabine holds KV7 channels open and stabilizes the resting potential. J Gen Physiol 147(1540-7748 (Electronic)):229–241

Cordero-Morales JF, Jogini V, Lewis A, Vasquez V, Cortes DM, Roux B et al (2007) Molecular driving forces determining potassium channel slow inactivation. Nat

Struct Mol Biol 14(1545–9985 (Electronic)):1062–1069

Cuello LG, Jogini V, Cortes DM, Perozo E (2010) Structural mechanism of C-type inactivation in K(+) channels. Nature 466(1476–4687 (Electronic)):203–208

Curran ME, Splawski I, Timothy KW, Vincent GM, Green ED, Keating MT (1995) A molecular basis for cardiac arrhythmia: HERG mutations cause long QT syndrome. Cell 80(0092–8674 (Print)):795–803

de la Pena P, Alonso-Ron C, Machin A, Fernandez-Trillo J, Carretero L, Dominguez P et al (2011) Demonstration of physical proximity between the N terminus and the S4-S5 linker of the human ether-a-go-go-related gene (hERG) potassium channel. J Biol Chem 286(1083-351X (Electronic)):19065–19075

del Camino D, Yellen G (2001) Tight steric closure at the intracellular activation gate of a voltage-gated K+ channel. Neuron 32(0896–6273):649–656

del Camino D, Holmgren M, Liu Y, Yellen G (2000) Blocker protection in the pore of a voltage-gated K^+ channel and its structural implications. Nature 403:321–325

DeMarco KR, Yang PC, Singh V, Furutani K, Dawson JRD, Jeng MT, et al (2021) Molecular determinants of pro-arrhythmia proclivity of d- and l-sotalol via a multi-scale modeling pipeline. J Mol Cell Cardiol [Internet] [cited 2024 Dec 5];158:163–177. Available from: https://pubmed.ncbi.nlm.nih.gov/34062207/

Dierich M, Evers S, Wilke BU, Leitner MG (2018) Inverse modulation of neuronal Kv 12.1 and kv 11.1 channels by 4-aminopyridine and NS1643. Front Mol Neurosci 11(January):1–13

Doyle DD, Cabral JM, Pfuetzner RA, Kuo A, Gulbis JM, Cohen SL et al (1998) The structure of the potassium channel: molecular basis of K^+ conduction and selectivity. Science (80-) 280:69–77

Durdagi S, Guo J, Lees-Miller JP, Noskov SY, Duff HJ (2012) Structure-guided topographic mapping and mutagenesis to elucidate binding sites for the human ether-a-go-go-related gene 1 potassium channel (KCNH2) activator NS1643. J Pharmacol Exp Ther 342(1521–0103 (Electronic)):441–452

El Harchi A, Brincourt O (2022) Pharmacological activation of the hERG K+ channel for the management of the long QT syndrome: a review. J Arrhythmia 38(4):554–569

Elinder F, Männikkö R, Pandey S, Larsson HP (2006) Mode shifts in the voltage gating of the mouse and human HCN2 and HCN4 channels. J Physiol 575(2):417–431

Elliott DJ, Dondas NY, Munsey TS, Sivaprasadarao A (2009) Movement of the S4 segment in the hERG potassium channel during membrane depolarization. Mol Membr Biol 26(1464–5203 (Electronic)):435–447

Es-Salah-Lamoureux Z, Fougere R, Xiong PY, Robertson GA, Fedida D (2010) Fluorescence-tracking of activation gating in human ERG channels reveals rapid S4 movement and slow pore opening. PLoS One 5(1932–6203 (Electronic)):e10876

Fedida D, Hesketh JC (2001) Gating of voltage-dependent potassium channels. Prog Biophys Mol Biol 75(0079–6107):165–199

Feng L, Zhang J, Lee CH, Kim G, Liu F, Petersen AJ, et al (2021) Long QT syndrome KCNH2 variant induces hERG1a/1b subunit imbalance in patient-specific induced pluripotent stem cell-derived cardiomyocytes. Circ Arrhythmia Electrophysiol [Internet] [cited 2023 Feb 25];14(4):E009343. Available from: https://www.ahajournals.org/doi/abs/10.1161/CIRCEP.120.009343

Fenichel RR, Malik M, Antzelevitch C, Sanguinetti M, Roden DM, Priori SG, et al (2004) Drug-induced torsades de pointes and implications for drug development. J Cardiovasc Electrophysiol [Internet] [cited 2019 Jun 17];15(4):475–495. Available from: http://www.ncbi.nlm.nih.gov/pubmed/15090000

Fernandez D, Ghanta A, Kauffman GW, Sanguinetti MC (2004) Physicochemical features of the HERG channel drug binding site. J Biol Chem 279(0021–9258 (Print)):10120–10127

Fernandez-Trillo J, Barros F, Machin A, Carretero L, Dominguez P, de la Pena P (2011) Molecular determinants of interactions between the N-terminal domain and the transmembrane core that modulate hERG K+ channel gating. PLoS One 6(1932–6203 (Electronic)):e24674

Ferrer T, Rupp J, Piper DR, Tristani-Firouzi M (2006) The S4-S5 linker directly couples voltage sensor movement to the activation gate in the human ether-a'-go-go-related gene (hERG) K^+ channel. J Biol Chem 281(0021–9258 (Print)):12858–12864

Ficker E, Jarolimek W, Kiehn J, Baumann A, Brown AM (1998) Molecular determinants of dofetilide block of HERG K^+ channels. Circ Res 82(0009–7330 (Print)):386–395

Foo B, Williamson B, Young JC, Lukacs G, Shrier A (2016) hERG quality control and the long QT syndrome. J Physiol [Internet] [cited 2024 Dec 5];594(9):2469. Available from: https://pmc.ncbi.nlm.nih.gov/articles/PMC4850197/

Fougere RR, Es-Salah-Lamoureux Z, Rezazadeh S, Eldstrom J, Fedida D (2011) Functional characterization of the LQT2-causing mutation R582C and the associated voltage-dependent fluorescence signal. Heart Rhythm [Internet] [cited 2024 Dec 5];8(8):1273–80. Available from: https://pubmed.ncbi.nlm.nih.gov/21376840/

Gardner A, Sanguinetti MC (2015) C-linker accounts for differential sensitivity of ERG1 and ERG2 K^+ channels to RPR260243-induced slow deactivation. Mol Pharmacol 88(1):19–28

Gayen S, Li Q, Chen AS, Nguyen THT, Huang Q, Hill J, et al (2011) An NMR study of the N-terminal domain of wild-type hERG and a T65P trafficking deficient hERG mutant. Proteins [Internet] [cited 2024 Dec 5];79(8):2557–2565. Available from: https://pubmed.ncbi.nlm.nih.gov/21661061/

Gianulis EC, Trudeau MC (2011) Rescue of aberrant gating by a genetically encoded PAS (Per-Arnt-Sim) domain in several long QT syndrome mutant human ether-á-go-go-related gene potassium channels. J Biol Chem 286(25):22160–22169

Gianulis EC, Liu Q, Trudeau MC (2013) Direct interaction of eag domains and cyclic nucleotide-binding homology domains regulate deactivation gating in hERG channels. J Gen Physiol 142(4):351–366

Gilly WF, Armstrong CM (1980) Gating current and potassium channels in the giant axon of the squid. Biophys J [Internet] [cited 2024 Dec 5];29(3):485–492. Available from: https://pubmed.ncbi.nlm.nih.gov/6271271/

Gomez-Varela D, de la Pena P, Garcia J, Giraldez T, Barros F (2002) Influence of amino-terminal structures on kinetic transitions between several closed and open states in human erg K+ channels. J Membr Biol 187(0022–2631 (Print)):117–133

Gong Q, Jones MA, Zhou Z (2006) Mechanisms of pharmacological rescue of trafficking-defective hERG mutant channels in human long QT syndrome. J Biol Chem [Internet] [cited 2024 Dec 5];281(7):4069–4074. Available from: http://www.jbc.org/article/S0021925819483877/fulltext

Gong Q, Stump MR, Zhou Z (2014) Upregulation of functional Kv11.1 isoform expression by inhibition of intronic polyadenylation with antisense morpholino oligonucleotides. J Mol Cell Cardiol [Internet] [cited 2019 Oct 9];76:26–32. Available from: http://www.ncbi.nlm.nih.gov/pubmed/25128783

Goodchild SJ, Fedida D (2014) Gating charge movement precedes ionic current activation in hERG channels. Channels 8(1):84–89

Goodchild SJ, Macdonald LC, Fedida D (2015) Sequence of gating charge movement and pore gating in HERG activation and deactivation pathways. Biophys J 108(1542–0086 (Electronic)):1435–1447

Grilo LS, Carrupt PA, Abriel H (2010) Stereoselective inhibition of the hERG1 potassium channel. Front Pharmacol 1 NOV:7878

Grunnet M, Abbruzzese J, Sachse FB, Sanguinetti MC (2011) Molecular determinants of human ether-à-go-go-related gene 1 (hERG1) K^{+} channel activation by NS1643. Mol Pharmacol [Internet] [cited 2024 Dec 5];79(1):1. Available from: https://pmc.ncbi.nlm.nih.gov/articles/PMC3014275/

Guo J, Cheng YM, Lees-Miller JP, Perissinotti LL, Claydon TW, Hull CM et al (2015) NS1643 interacts around L529 of hERG to alter voltage sensor movement on the path to activation. Biophys J 108(6):1400

Gustina AS, Trudeau MC (2009) A recombinant N-terminal domain fully restores deactivation gating in N-truncated and long QT syndrome mutant hERG potassium channels. Proc Natl Acad Sci USA 106(1091–6490 (Electronic)):13082–13087

Gustina AS, Trudeau MC (2011) hERG potassium channel gating is mediated by N- and C-terminal region interactions. J Gen Physiol 137(1540-7748 (Electronic)):315–325

Gustina AS, Trudeau MC (2012) HERG potassium channel regulation by the N-terminal eag domain. Cell Signal 24(1873–3913 (Electronic)):1592–1598

Gustina AS, Trudeau MC (2013) The eag domain regulates hERG channel inactivation gating via a direct interaction. J Gen Physiol 141(2):229–241

Haddad GA, Blunck R (2011) Mode shift of the voltage sensors in Shaker K+ channels is caused by energetic coupling to the pore domain. J Gen Physiol 137(1540-7748 (Electronic)):455–472

Haitin Y, Carlson AE, Zagotta WN (2013) The structural mechanism of KCNH-channel regulation by the eag domain. Nature 501(1476–4687 (Electronic)):444–448

Hansen RS, Diness TG, Christ T, Demnitz J, Ravens U, Olesen SP et al (2006) Activation of human ether-a-go-go-related gene potassium channels by the diphenylurea 1,3-bis-(2-hydroxy-5-trifluoromethyl-phenyl)-urea (NS1643). Mol Pharmacol 69(0026-895X (Print)):266–277

Harley CA, Bernardo-Seisdedos G, Stevens-Sostre WA, Jones DK, Azevedo MM, Sampaio P, et al (2021) Conformation-sensitive antibody reveals an altered cytosolic PAS/CNBh assembly during hERG channel gating. Proc Natl Acad Sci U S A [Internet] [cited 2024 Dec 5];118(44). Available from: https://pubmed.ncbi.nlm.nih.gov/34716268/

Henry JT, Crosson S (2011) Ligand-binding PAS domains in a genomic, cellular, and structural context. Annu Rev Microbiol 65(1545–3251 (Electronic)):261–286

Herzberg IM, Trudeau MC, Robertson GA (1998) Transfer of rapid inactivation and sensitivity to the class III antiarrhythmic drug E-4031 from HERG to M-eag channels. J Physiol 511(0022–3751 (Print)):3–14

Hodgkin AL, Huxley AF (1952) A quantitative description of membrane current and its application to conduction and excitation in nerve. J Physiol [Internet] [cited 2024 Jun 13];117(4):500. Available from: https://www.ncbi.nlm.nih.gov/pmc/articles/PMC1392413/

Hull CM, Sokolov S, Van Slyke AC, Claydon TW (2014) Regional flexibility in the S4–S5 linker regulates hERG channel closed-state stabilization. Pflugers Arch Eur J Physiol 466(10):1911

Hull CM, Genge CE, Hobbs Y, Rayani K, Lin E, Gunawan M, et al (2019) Investigating the utility of adult zebrafish ex vivo whole hearts to pharmacologically screen hERG channel activator compounds. Am J Physiol Integr Comp Physiol [Internet] [cited 2020 Jan 8];317(6):R921–R931. Available from: https://www.physiology.org/doi/10.1152/ajpregu.00190.2019

Jameson MB, Ríos-Pérez EB, Liu F, Eichel CA, Robertson GA (2023) Pairwise biosynthesis of ion channels stabilizes excitability and mitigates arrhythmias. Proc Natl Acad Sci U S A [Internet] [cited 2024 Jun 17];120(42):e2305295120. Available from: https://www.pnas.org/doi/abs/10.1073/pnas.2305295120

Jiang Y, Lee A, Chen J, Cadene M, Chait BT, MacKinnon R (2002) The open pore conformation of potassium channels. Nature 417(0028–0836 SB-IM):523–526

Jiang C, Richardson E, Farr J, Hill AP, Ullah R, Kroncke BM et al (2022) A calibrated functional patch-clamp assay to enhance clinical variant interpretation in KCNH2-related long QT syndrome. Am J Hum Genet 109(7):1199–1207

Johnson AA, Crawford TR, Trudeau MC (2022) The N-linker region of hERG1a upregulates hERG1b potassium channels. J Biol Chem [Internet] [cited 2023 Feb 25];298(9):102233. Available from: http://www.jbc.org/article/S0021925822006755/fulltext

Jones DK (2022) Hysteretic hERG channel gating current recorded at physiological temperature. Sci Reports 2022 121 [Internet] [cited 2024 Jun 13];12(1):1–13. Available from: https://www.nature.com/articles/s41598-022-10003-7

Jones EM, Roti Roti EC, Wang J, Delfosse SA, Robertson GA (2004) Cardiac IKr channels minimally comprise hERG 1a and 1b subunits. J Biol Chem 279(0021–9258 (Print)):44690–44694

Jones DK, Liu F, Vaidyanathan R, Eckhardt LL, Trudeau MC, Robertson GA (2014) hERG 1b is critical for human cardiac repolarization. Proc Natl Acad Sci [Internet] [cited 2019 Oct 9];111(50):18073–18077. Available from: http://www.ncbi.nlm.nih.gov/pubmed/25453103

Jones DK, Liu F, Dombrowski N, Joshi S, Robertson GA (2016) Dominant negative consequences of a hERG 1b-specific mutation associated with intrauterine fetal death. Prog Biophys Mol Biol [Internet] [cited 2019 Oct 9];120(1–3):67–76. Available from: http://www.ncbi.nlm.nih.gov/pubmed/26772437

Kamiya K, Niwa R, Mitcheson JS, Sanguinetti MC (2006) Molecular determinants of HERG channel block. Mol Pharmacol 69(0026–895X (Print)):1709–1716

Kanevsky M, Aldrich RW (1999) Determinants of voltage-dependent gating and open-state stability in the S5 segment of Shaker potassium channels. J Gen Physiol 114(0022–1295 SB-IM):215–242

Kang J, Chen XL, Wang H, Ji J, Cheng H, Incardona J et al (2005) Discovery of a small molecule activator of the human ether-a-go-go-related gene (HERG) cardiac K+ channel. Mol Pharmacol 67(0026-895X (Print)):827–836

Ke Y, Ng CA, Hunter MJ, Mann SA, Heide J, Hill AP, et al (2013) Trafficking defects in PAS domain mutant Kv11.1 channels: roles of reduced domain stability and altered domain–domain interactions. Biochem J [Internet] [cited 2024 Dec 5];454(1):69–77. Available from: /biochemj/article/454/1/69/46291/Trafficking-defects-in-PAS-domain-mutant-Kv11-1

Kemp JM, Whittaker DG, Venkateshappa R, Pang Z, Johal R, Sergeev V, et al (2021) Electrophysiological characterization of the herg r56q lqts variant and targeted rescue by the activator rpr260243. J Gen Physiol [Internet] [cited 2023 Feb 25];153(10). Available from: https://doi.org/10.1085/jgp.202112923

Khan HM, Guo J, Duff HJ, Tieleman DP, Noskov SY (2021) Refinement of a cryo-EM structure of hERG: Bridging structure and function. Biophys J [Internet] [cited 2024 Dec 5];120(4):738–748. Available from: http://www.cell.com/article/S0006349521000485/fulltext

Kiehn J, Lacerda AE, Wible B, Brown AM (1996) Molecular physiology and pharmacology of HERG. Single-channel currents and block by dofetilide. Circulation 94(0009–7322 (Print)):2572–2579

Kiehn J, Lacerda AE, Brown AM (1999) Pathways of HERG inactivation. Am J Physiol [Internet] [cited 2024 Dec 5];277(1). Available from: https://pubmed.ncbi.nlm.nih.gov/10409198/

Kozek KA, Glazer AM, Ng CA, Blackwell D, Egly CL, Vanags LR, et al (2020) High-throughput discovery of trafficking-deficient variants in the cardiac potassium channel KV11.1. Heart Rhythm [Internet] [cited 2024 Jun 17];17(12):2180–2189. Available from: http://www.heartrhythmjournal.com/article/S1547527120305427/fulltext

Kudaibergenova M, Guo J, Khan HM, Lees-Miller J, Mousaei M, Miranda W, et al. (2022) The voltage-sensing domain of a hERG1 mutant is a cation-selective channel. Biophys J [Internet] [cited 2024 Dec 5];121(23):4585. Available from: https://pmc.ncbi.nlm.nih.gov/articles/PMC9748372/

Kupershmidt S, Snyders DJ, Raes A, Roden DM (1998) A K+ channel splice variant common in human heart lacks a C-terminal domain required for expression of rapidly activating delayed rectifier current. J Biol Chem [Internet] [cited 2019 Oct 9];273(42):27231–27235. Available from: http://www.ncbi.nlm.nih.gov/pubmed/9765245

Kuzmenkin A, Bezanilla F, Correa AM (2004) Gating of the bacterial sodium channel, NaChBac. J Gen Physiol 124(4):349–356

Labro AJ, Lacroix JJ, Villalba-Galea CA, Snyders DJ, Bezanilla F (2012) Molecular mechanism for depolarization-induced modulation of Kv channel closure. J Gen Physiol 140(1540–7748 (Electronic)):481–493

Labro AJ, Priest MF, Lacroix JJ, Snyders DJ, Bezanilla F (2015) Kv3.1 uses a timely resurgent K(+) current to secure action potential repolarization. Nat Commun 6(2041-1723 (Electronic)):10173

Lacroix JJ, Labro AJ, Bezanilla F (2011) Properties of deactivation gating currents in Shaker channels. Biophys J 100(1542–0086 (Electronic)):L28–L30

Larsen AP, Olesen SP (2010) Differential expression of hERG1 channel isoforms reproduces properties of native IKr and modulates cardiac action potential characteristics. PLoS One [Internet] [cited 2023 Feb 25];5(2):e9021. Available from: https://journals.plos.org/plosone/article?id=10.1371/journal.pone.0009021

Larsen AP, Olesen SP, Grunnet M, Jespersen T (2008) Characterization of hERG1a and hERG1b potassium channels – a possible role for hERG1b in the IKr current. Pflugers Arch Eur J Physiol 456(6):1137–1148

Lau CHY, Flood E, Hunter MJ, Williams-Noonan BJ, Corbett KM, Ng CA, et al (2024) Potassium dependent structural changes in the selectivity filter of HERG potassium channels. Nat Commun [Internet] [cited 2024 Dec 5];15(1):7470. Available from: https://pmc.ncbi.nlm.nih.gov/articles/PMC11362469/

Ledwell JL, Aldrich RW (1999) Mutations in the S4 region isolate the final voltage-dependent cooperative step in

potassium channel activation. J Gen Physiol 113(0022–1295 (Print)):389–414
Lee KH, Fant AD, Guo J, Guan A, Jung J, Kudaibergenova M, et al (2021) Toward reducing hERG affinities for DAT inhibitors with a combined machine learning and molecular modeling approach. J Chem Inf Model [Internet] [cited 2024 Dec 5];61(9):4266–4279. Available from: https://pubmed.ncbi.nlm.nih.gov/34420294/
Lees-Miller JP, Kondo C, Wang L, Duff HJ (1997) Electrophysiological characterization of an alternatively processed ERG K+ channel in mouse and human hearts. Circ Res [Internet] [cited 2023 Feb 25];81(5):719–726. Available from: https://www.ahajournals.org/doi/abs/10.1161/01.RES.81.5.719
Lei CL, Clerx M, Gavaghan DJ, Polonchuk L, Mirams GR, Wang K (2019a) Rapid characterization of hERG channel kinetics I: using an automated high-throughput system. Biophys J [Internet] [cited 2023 Feb 25];117(12):2438–2454. Available from: https://pubmed.ncbi.nlm.nih.gov/31447109/
Lei CL, Clerx M, Beattie KA, Melgari D, Hancox JC, Gavaghan DJ, et al (2019b) Rapid characterization of hERG channel kinetics II: temperature dependence. Biophys J [Internet] [cited 2023 Feb 25];117(12):2455. Available from: /pmc/articles/PMC6990152/
Li Q, Gayen S, Chen AS, Huang Q, Raida M, Kang C (2010) NMR solution structure of the N-terminal domain of hERG and its interaction with the S4-S5 linker. Biochem Biophys Res Commun 403(1090–2104 (Electronic)):126–132
Liu S, Rasmusson RL, Campbell DL, Wang S, Strauss HC (1996) Activation and inactivation kinetics of an E-4031-sensitive current from single ferret atrial myocytes. Biophys J [Internet] [cited 2024 Dec 5];70(6):2704–2715. Available from: https://pubmed.ncbi.nlm.nih.gov/8744308/
Liu J, Zhang M, Jiang M, Tseng GN (2003) Negative charges in the transmembrane domains of the HERG K channel are involved in the activation- and deactivation-gating processes. J Gen Physiol 121(0022–1295 (Print)):599–614
London B, Trudeau MC, Newton KP, Beyer AK, Copeland NG, Gilbert DJ, et al (1997) Two isoforms of the mouse ether-a-go-go–related gene coassemble to form channels with properties similar to the rapidly activating component of the cardiac delayed rectifier K^+ current. Circ Res [Internet] [cited 2023 Feb 25];81(5):870–878. Available from: https://www.ahajournals.org/doi/abs/10.1161/01.RES.81.5.870
Long SB, Campbell EB, MacKinnon R (2005) Voltage sensor of Kv1.2: structural basis of electromechanical coupling. Science (80-) 309(1095–9203 (Electronic)):903–908
Long SB, Tao X, Campbell EB, MacKinnon R (2007) Atomic structure of a voltage-dependent K+ channel in a lipid membrane-like environment. Nature 450(1476–4687 (Electronic)):376–382
Loots EA, Isacoff EY (1998) Protein rearrangements underlying slow inactivation of the Shaker K^+ channel. J Gen Physiol 112:377–389
Lopez-Barneo J, Hoshi T, Heinemann SH, Aldrich RW (1993) Effects of external cations and mutations in the pore region on C-type inactivation of Shaker potassium channels. Recept Channels 1:61–71
Lorinczi E, Gomez-Posada JC, de la Pena P, Tomczak AP, Fernandez-Trillo J, Leipscher U et al (2015) Voltage-dependent gating of KCNH potassium channels lacking a covalent link between voltage-sensing and pore domains. Nat Commun 6(2041–1723 (Electronic)):6672
Lu Y, Mahaut-Smith MP, Varghese A, Huang CL, Kemp PR, Vandenberg JI (2001) Effects of premature stimulation on HERG K(+) channels. J Physiol 537(0022–3751):843–851
Lu Z, Klem AM, Ramu Y (2002) Coupling between voltage sensors and activation gate in voltage-gated K^+ channels. J Gen Physiol 120(0022–1295):663–676
Ma JG, Vandenberg JI, Ng C-A (2023) Development of automated patch clamp assays to overcome the burden of variants of uncertain significance in inheritable arrhythmia syndromes. Front Physiol [Internet] [cited 2024 Dec 5];14:1294741. Available from: http://www.ncbi.nlm.nih.gov/pubmed/38089476
Malak OA, Es-Salah-Lamoureux Z, Loussouarn G (2017) hERG S4-S5 linker acts as a voltage-dependent ligand that binds to the activation gate and locks it in a closed state. Sci Rep 7(1):1–12
Malak OA, Gluhov GS, Grizel AV, Kudryashova KS, Sokolova OS, Loussouarn G (2019) Voltage-dependent activation in EAG channels follows a ligand-receptor rather than a mechanical-lever mechanism. J Biol Chem 294(16):6506–6521
Marques-Carvalho MJ, Sahoo N, Muskett FW, Vieira-Pires RS, Gabant G, Cadene M et al (2012) Structural, biochemical, and functional characterization of the cyclic nucleotide binding homology domain from the mouse EAG1 potassium channel. J Mol Biol 423(1):34–46
McIntosh BE, Hogenesch JB, Bradfield CA (2010) Mammalian Per-Arnt-Sim proteins in environmental adaptation. Annu Rev Physiol [Internet] [cited 2024 Dec 5];72:625–645. Available from: https://pubmed.ncbi.nlm.nih.gov/20148691/
McNally BA, Pendon ZD, Trudeau MC (2017) hERG1a and hERG1b potassium channel subunits directly interact and preferentially form heteromeric channels. J Biol Chem [Internet] [cited 2023 Feb 25];292(52):21548–21557. Available from: https://pubmed.ncbi.nlm.nih.gov/29089383/
Mitcheson JS (2008) hERG potassium channels and the structural basis of drug-induced arrhythmias. Chem Res Toxicol 21(1520–5010 (Electronic)):1005–1010
Mitcheson JS, Chen J, Sanguinetti MC (2000) Trapping of a methanesulfonanilide by closure of the HERG potassium channel activation gate. J Gen Physiol 115(0022–1295 (Print)):229–240
Mitcheson J, Perry M, Stansfeld P, Sanguinetti MC, Witchel H, Hancox J (2005) Structural determinants for high-affinity block of hERG potassium channels. Novartis Found Symp [Internet] [cited

2024 Dec 5];266:136–150. Available from: https://pubmed.ncbi.nlm.nih.gov/16050266/

Möglich A, Ayers RA, Moffat K (2009) Structure and signaling mechanism of Per-ARNT-Sim domains. Structure [Internet] [cited 2024 Dec 5];17(10):1282–1294. Available from: https://pubmed.ncbi.nlm.nih.gov/19836329/

Morais Cabral JH, Lee A, Cohen SL, Chait BT, Li M, MacKinnon R (1998) Crystal structure and functional analysis of the HERG potassium channel N terminus: a eukaryotic PAS domain. Cell 95(0092–8674 (Print)):649–655

Muskett FW, Thouta S, Thomson SJ, Bowen A, Stansfeld PJ, Mitcheson JS (2011) Mechanistic insight into human ether-a-go-go-related gene (hERG) K+ channel deactivation gating from the solution structure of the EAG domain. J Biol Chem 286(1083-351X (Electronic)):6184–6191

Ng CA, Hunter MJ, Perry MD, Mobli M, Ke Y, Kuchel PW et al (2011) The N-terminal tail of hERG contains an amphipathic alpha-helix that regulates channel deactivation. PLoS One 6(1932–6203 (Electronic)):e16191

Ng CA, Perry MD, Tan PS, Hill AP, Kuchel PW, Vandenberg JI (2012) The S4-S5 linker acts as a signal integrator for HERG K+ channel activation and deactivation gating. PLoS One 7(1932–6203 (Electronic)):e31640

Ng CA, Ullah R, Farr J, Hill AP, Kozek KA, Vanags LR, et al (2022) A massively parallel assay accurately discriminates between functionally normal and abnormal variants in a hotspot domain of KCNH2. Am J Hum Genet [Internet] [cited 2024 Jun 17];109(7):1208–1216. Available from: https://pubmed.ncbi.nlm.nih.gov/35688148/

Noble D, Tsien RW (1969a) Outward membrane currents activated in the plateau range of potentials in cardiac Purkinje fibres. J Physiol [Internet] [cited 2024 Dec 5];200(1):205–231. Available from: https://pubmed.ncbi.nlm.nih.gov/5761944/

Noble D, Tsien RW (1969b) Reconstruction of the repolarization process in cardiac Purkinje fibres based on voltage clamp measurements of membrane current. J Physiol [Internet] [cited 2024 Dec 5];200(1):233. Available from: https://pmc.ncbi.nlm.nih.gov/articles/PMC1350426/

Noda M, Shimizu S, Tanabe T, Takai T, Kayano T, Ikeda T et al (1984) Primary structure of Electrophorus electricus sodium channel deduced from cDNA sequence. Nature 312(0028–0836 SB-IM):121–127

O'Neill MJ, Ng C-A, Aizawa T, Sala L, Bains S, Winbo A, et al (2024) Multiplexed assays of variant effect and automated patch clamping improve KCNH2-LQTS variant classification and cardiac event risk stratification. Circulation [Internet] [cited 2024 Dec 5];150(23):1869–1881. Available from: http://www.ncbi.nlm.nih.gov/pubmed/39315434

Ostmeyer J, Chakrapani S, Pan AC, Perozo E, Roux B (2013) Recovery from slow inactivation in K+ channels is controlled by water molecules. Nature [Internet] [cited 2024 Dec 5];501(7465):121–124. Available from: https://pubmed.ncbi.nlm.nih.gov/23892782/

Papazian DM, Schwarz TL, Tempel BL, Jan YN, Jan LY (1987) Cloning of genomic and complementary DNA from Shaker, a putative potassium channel gene from Drosophila. Science (80-) 237(0036–8075 SB-IM):749–753

Peitersen T, Grunnet M, Benson AP, Holden AV, Holstein-Rathlou NH, Olesen SP (2008) Computational analysis of the effects of the hERG channel opener NS1643 in a human ventricular cell model. Heart Rhythm [Internet] [cited 2024 Dec 5];5(5):734–741. Available from: https://pubmed.ncbi.nlm.nih.gov/18452879/

Perry M, Sachse FB, Sanguinetti MC (2007) Structural basis of action for a human ether-a-go-go-related gene 1 potassium channel activator. Proc Natl Acad Sci 104(34):13827–13832

Perry M, Sachse FB, Abbruzzese J, Sanguinetti MC (2009) PD-118057 contacts the pore helix of hERG1 channels to attenuate inactivation and enhance K+ conductance. Proc Natl Acad Sci U S A [Internet] [cited 2024 Dec 5];106(47):20075. Available from: https://pmc.ncbi.nlm.nih.gov/articles/PMC2785294/

Perry MD, Ng CA, Phan K, David E, Steer K, Hunter MJ et al (2016) Rescue of protein expression defects may not be enough to abolish the pro-arrhythmic phenotype of long QT type 2 mutations. J Physiol 594(1469–7793 (Electronic)):4031–4049

Perry MD, Ng CA, Mangala MM, Ng TYM, Hines AD, Liang W, et al (2020) Pharmacological activation of IKr in models of long QT type 2 risks overcorrection of repolarization. Cardiovasc Res [Internet] [cited 2023 Feb 25];116(8):1434–1445. Available from: https://pubmed.ncbi.nlm.nih.gov/31628797/

Phartiyal P, Jones EMC, Robertson GA (2007) Heteromeric assembly of human ether-à-go-go-related gene (hERG) 1a/1b channels occurs cotranslationally via N-terminal interactions. J Biol Chem 282(13):9874–9882

Phartiyal P, Sale H, Jones EMC, Robertson GA (2008) Endoplasmic reticulum retention and rescue by heteromeric assembly regulate human ERG 1a/1b surface channel composition. J Biol Chem [Internet] [cited 2023 Feb 25];283(7):3702–3207. Available from: http://www.jbc.org/article/S0021925820696788/fulltext

Piper DR, Varghese A, Sanguinetti MC, Tristani-Firouzi M (2003) Gating currents associated with intramembrane charge displacement in hERG potassium channels. Proc Natl Acad Sci USA 100(0027–8424 (Print)):10534–10539

Piper DR, Hinz WA, Tallurri CK, Sanguinetti MC, Tristani-Firouzi M (2005) Regional specificity of human ether-a'-go-go-related gene channel activation and inactivation gating. J Biol Chem 280(0021–9258 (Print)):7206–7217

Priest MF, Lacroix JJ, Villalba-Galea CA, Bezanilla F (2013) S3-S4 linker length modulates the relaxed state of a voltage-gated potassium channel. Biophys J 105(1542–0086 (Electronic)):2312–2322

Reddi R, Matulef K, Riederer EA, Whorton MR, Valiyaveetil FI (2022) Structural basis for C-type inactivation in a Shaker family voltage-gated K+ channel.

Sci Adv [Internet] [cited 2024 Dec 5];8(16). Available from: https://pubmed.ncbi.nlm.nih.gov/35452285/

Robertson GA, Morais-Cabral JH (2019) hERG function in light of structure. Biophys J [Internet] [cited 2024 Dec 6];118(4):790. Available from: https://pmc.ncbi.nlm.nih.gov/articles/PMC7036721/

Roden DM, Viswanathan PC (2005) Genetics of acquired long QT syndrome. J Clin Invest [Internet] [cited 2024 Dec 5];115(8):2025–2032. Available from: https://pubmed.ncbi.nlm.nih.gov/16075043/

Saenen JB, Labro AJ, Raes A, Snyders DJ (2006) Modulation of HERG gating by a charge cluster in the N-terminal proximal domain. Biophys J [Internet] [cited 2024 Dec 5];91(12):4381–4391. Available from: https://pubmed.ncbi.nlm.nih.gov/16997865/

Sale H, Wang J, O'Hara TJ, Tester DJ, Phartiyal P, He J-Q et al (2008) Physiological properties of hERG 1a/1b heteromeric currents and a hERG 1b-specific mutation associated with long QT syndrome. Circ Res 103(7):e81–e95

Sanguinetti MC, Jurkiewicz NK (1990) Two components of cardiac delayed rectifier K^+ current. Differential sensitivity to block by class III anti arrhythmic agents. J Gen Physiol 96(0022–1295 SB-IM):195–215

Sanguinetti MC, Xu QP (1999) Mutations of the S4-S5 linker alter activation properties of HERG potassium channels expressed in *Xenopus* oocytes. J Physiol 514(3):667–675

Sanguinetti MC, Jiang C, Curran ME, Keating MT (1995) A mechanistic link between an inherited and an acquird cardiac arrthytmia: HERG encodes the IKr potassium channel. Cell 81(2):299–307

Schonherr R, Heinemann SH (1996) Molecular determinants for activation and inactivation of HERG, a human inward rectifier potassium channel. J Physiol 493(0022–3751) Print:635–642

Seoh SA, Sigg D, Papazian DM, Bezanilla F (1996) Voltage-sensing residues in the S2 and S4 segments of the Shaker K^+ channel. Neuron 16(0896–6273 SB-IM):1159–1167

Shi YP, Cheng YM, Van Slyke AC, Claydon TW (2014) External protons destabilize the activated voltage sensor in hERG channels. Eur Biophys J 43(2–3):59–69

Shi YP, Thouta S, Cheng YM, Claydon TW (2019) Extracellular protons accelerate hERG channel deactivation by destabilizing voltage sensor relaxation. J Gen Physiol 151(2):231

Shi YP, Thouta S, Claydon TW (2020a). Modulation of hERG K^+ channel deactivation by voltage sensor relaxation. Front Pharmacol Feb 28:11:139. https://doi.org/10.3389/fphar.2020.00139.

Shi YP, Pang Z, Venkateshappa R, Gunawan M, Kemp J, Truong E, et al (2020b) The hERG channel activator, RPR260243, enhances protective IKr current early in the refractory period reducing arrhythmogenicity in zebrafish hearts. Am J Physiol – Hear Circ Physiol [Internet] [cited 2023 Feb 25];319(2):H251–H261. Available from: https://journals.physiology.org/doi/10.1152/ajpheart.00038.2020

Shibasaki T (1987) Conductance and kinetics of delayed rectifier potassium channels in nodal cells of the rabbit heart. J Physiol [Internet] [cited 2024 Dec 5];387(1):227–250. Available from: https://pubmed.ncbi.nlm.nih.gov/2443680/

Shirokov R, Levis R, Shirokova N, Rios E (1992) Two classes of gating current from L-type Ca channels in Guinea pig ventricular myocytes. J Gen Physiol 99(0022-1295 (Print)):863–895

Smith PL, Yellen G (2002) Fast and slow voltage sensor movements in hERG potassium channels. J Gen Physiol 119(0022-1295 (Print)):275–293

Smith PL, Baukrowitz T, Yellen G (1996) The inward rectification mechanism of the hERG cardiac potassium channel. Nature 379:833–836

Spector PS, Curran ME, Zou A, Keating MT, Sanguinetti MC (1996) Fast inactivation causes rectification of the IKr channel. J Gen Physiol 107(0022-1295 (Print)):611–619

Starace DM, Bezanilla F (2004) A proton pore in a potassium channel voltage sensor reveals a focused electric field. Nature 427(1476–4687):548–553

Starace DM, Stefani E, Bezanilla F (1997) Voltage-dependent proton transport by the voltage sensor of the Shaker K^+ channel. Neuron 19(0896–6273 SB-IM):1319–1327

Stevens-Sostre WA, Flores-Aldama L, Bustos D, Li J, Morais-Cabral JH, Delemotte L et al (2024) An intracellular hydrophobic nexus critical for hERG1 channel slow deactivation. Biophys J 123(14):2024–2037

Subbiah RN, Clarke CE, Smith DJ, Zhao J, Campbell TJ, Vandenberg JI (2004) Molecular basis of slow activation of the human ether-a-go-go related gene potassium channel. J Physiol 558(0022–3751 (Print)):417–431

Subbiah RN, Kondo M, Campbell TJ, Vandenberg JI (2005) Tryptophan scanning mutagenesis of the HERG K+ channel: the S4 domain is loosely packed and likely to be lipid exposed. J Physiol 569(0022–3751 (Print)):367–379

Subbotina J, Yarov-Yarovoy V, Lees-Miller J, Durdagi S, Guo J, Duff HJ, et al (2010) Structural refinement of the hERG1 pore and voltage-sensing domains with ROSETTA-membrane and molecular dynamics simulations. Proteins [Internet] [cited 2024 Dec 5];78(14):2922–2934. Available from: https://pubmed.ncbi.nlm.nih.gov/20740484/

Tan PS, Perry MD, Ng CA, Vandenberg JI, Hill AP (2012) Voltage-sensing domain mode shift is coupled to the activation gate by the N-terminal tail of hERG channels. J Gen Physiol 140(1540–7748 (Electronic)):293–306

Tan XF, Bae C, Stix R, Fernández-Mariño AI, Huffer K, Chang TH, et al (2022) Structure of the Shaker Kv channel and mechanism of slow C-type inactivation. Sci Adv [Internet] [cited 2024 Dec 5];8(11). Available from: https://pubmed.ncbi.nlm.nih.gov/35302848/

Tao X, Lee A, Limapichat W, Dougherty DA, MacKinnon R (2010) A gating charge transfer center in voltage sensors. Science (80-) 328(1095–9203 (Electronic)):67–73

Thouta S, Sokolov S, Abe Y, Clark SJ, Cheng YM, Claydon TW (2014) Proline scan of the hERG channel S6 helix reveals the location of the intracellular pore gate. Biophys J 106(5):1057

Thouta S, Hull CM, Shi YP, Sergeev V, Young J, Cheng YM et al (2017) Stabilization of the activated hERG channel voltage sensor by depolarization involves the S4-S5 linker. Biophys J 112(1542–0086 (Electronic)):300–312

Tilegenova C, Cortes DM, Cuello LG (2017) Hysteresis of KcsA potassium channel's activation- deactivation gating is caused by structural changes at the channel's selectivity filter. Proc Natl Acad Sci USA 114(12):3234–3239

Tristani-Firouzi M, Chen J, Sanguinetti MC (2002) Interactions between S4-S5 linker and S6 transmembrane domain modulate gating of HERG K^+ channels. J Biol Chem 277(21):18994–19000

Trudeau MC, Warmke JW, Ganetzky B, Robertson GA (1995) HERG, a human inward rectifier in the voltage-gated potassium channel family. Science (80-) 269(0036–8075 (Print)):92–95

Trudeau MC, Leung LM, Roti ER, Robertson GA (2011) hERG1a N-terminal eag domain–containing polypeptides regulate homomeric hERG1b and heteromeric hERG1a/hERG1b channels: a possible mechanism for long QT syndrome. J Gen Physiol 138(6):581–592

Tschirhart JN, Zhang S (2020) Fentanyl-induced block of hERG channels is exacerbated by hypoxia, hypokalemia, alkalosis, and the presence of hERG1b. Mol Pharmacol [Internet] [cited 2023 Feb 25];98(4):508–517. Available from: https://molpharm.aspetjournals.org/content/early/2020/04/22/mol.119.119271

Van Slyke AC, Rezazadeh S, Snopkowski M, Shi P, Allard CR, Claydon TW (2010) Mutations within the S4-S5 linker alter voltage sensor constraints in hERG K^+ channels. Biophys J 99(9)

Van Slyke AC, Cheng YM, Mafi P, Allard CR, Hull CM, Shi YP et al (2012) Proton block of the pore underlies the inhibition of hERG cardiac K+ channels during acidosis. Am J Physiol Cell Physiol 302(1522–1563 (Electronic)):C1797–C1806

Vandenberg JI, Perry MD, Perrin MJ, Mann SA, Ke Y, Hill AP (2012) hERG K(+) channels: structure, function, and clinical significance. Physiol Rev 92(1522–1210 (Electronic)):1393–1478

Venkateshappa R, Hunter D V., Muralidharan P, Nagalingam RS, Huen G, Faizi S, et al (2023) Targeted activation of human ether-à-go-go-related gene channels rescues electrical instability induced by the R56Q+/− long QT syndrome variant. Cardiovasc Res [Internet] [cited 2024 Jun 17];119(15):2522–2535. Available from: https://pubmed.ncbi.nlm.nih.gov/37739930/

Villalba-Galea CA (2014) Hv1 proton channel opening is preceded by a voltage-independent transition. Biophys J 107(7):1564–1572

Villalba-Galea CA, Sandtner W, Starace DM, Bezanilla F (2008) S4-based voltage sensors have three major conformations. Proc Natl Acad Sci USA 105(1091–6490 (Electronic)):17600–17607

Viloria CG, Barros F, Giraldez T, Gomez-Varela D, de la Pena P (2000) Differential effects of amino-terminal distal and proximal domains in the regulation of human erg K(+) channel gating. Biophys J 79(0006–3495 (Print)):231–246

Wacker S, Noskov SY (2018) Performance of machine learning algorithms for qualitative and quantitative prediction drug blockade of hERG1 channel. Comput Toxicol (Amsterdam, Netherlands) [Internet] [cited 2024 Dec 5];6:55–63. Available from: https://pubmed.ncbi.nlm.nih.gov/29806042/

Wacker S, Noskov SY, Perissinotti LL (2017) Computational models for understanding of structure, function and pharmacology of the cardiac potassium channel Kv11.1 (hERG). Curr Top Med Chem [Internet] [cited 2024 Dec 5];17(23):2681–2702. Available from: https://pubmed.ncbi.nlm.nih.gov/28413954/

Wang W, MacKinnon R (2017) Cryo-EM structure of the open human ether-à-go-go-related K+ channel hERG. Cell [Internet] [cited 2024 Jun 17];169(3):422–430.e10. Available from: https://pubmed.ncbi.nlm.nih.gov/28431243/

Wang S, Liu S, Morales MJ, Strauss HC, Rasmusson RL (1997) A quantitative analysis of the activation and inactivation kinetics of HERG expressed in Xenopus oocytes. J Physiol 502(0022–3751 (Print)):45–60

Wang J, Trudeau MC, Zappia AM, Robertson GA (1998) Regulation of deactivation by an amino terminal domain in human ether-à-go-go-related gene potassium channels. J Gen Physiol 112(5):637–647

Wang J, Myers CD, Robertson GA (2000) Dynamic control of deactivation gating by a soluble amino-terminal domain in HERG K(+) channels. J Gen Physiol 115(0022–1295 (Print)):749–758

Wang DT, Hill AP, Mann SA, Tan PS, Vandenberg JI (2011) Mapping the sequence of conformational changes underlying selectivity filter gating in the K(v)11.1 potassium channel. Nat Struct Mol Biol 18(1545–9985 (Electronic)):35–41

Wang Z, Dou Y, Goodchild SJ, Es-Salah-Lamoureux Z, Fedida D (2013) Components of gating charge movement and S4 voltage-sensor exposure during activation of hERG channels. J Gen Physiol 141(1540–7748 (Electronic)):431–443

Warmke JW, Ganetsky B (1994) A family of potassium channel genes related to *eag* in Drosophila and mammals. Proc Natl Acad Sci USA 91:3438–3442

Whicher JR, MacKinnon R (2016) Structure of the voltage-gated K(+) channel Eag1 reveals an alternative voltage sensing mechanism. Science (80-) 353(1095–9203 (Electronic)):664–669

Wu W, Gardner A, Sanguinetti MC (2015) Concatenated hERG1 tetramers reveal stoichiometry of altered channel gating by RPR-260243. Mol Pharmacol 87(3):401–409

Wynia-Smith SL, Gillian-Daniel AL, Satyshur KA, Robertson GA (2008) hERG gating microdomains defined by S6 mutagenesis and molecular modeling. J Gen Physiol 132(1540–7748 (Electronic)):507–520

Xu X, Recanatini M, Roberti M, Tseng GN (2008) Probing the binding sites and mechanisms of action of two human ether-a-go-go-related gene channel activators, 1,3-bis-(2-hydroxy-5-trifluoromethyl-phenyl)-urea (NS1643) and 2-[2-(3,4-dichloro-phenyl)-2,3-dihydro-1H-isoindol-5-ylamino]-nicotinic acid (PD3). Mol Pharmacol 73(1521–0111 (Electronic)):1709–1721

Yang PC, Demarco KR, Aghasafari P, Jeng MT, Dawson JRD, Bekker S, et al (2020) A computational pipeline to predict cardiotoxicity: from the atom to the rhythm. Circ Res [Internet] [cited 2024 Dec 5];126(8):947–964. Available from: https://pubmed.ncbi.nlm.nih.gov/32091972/

Zagotta WN, Hoshi T, Dittman J, Aldrich RW (1994) Shaker potassium channel gating. II: transitions in the activation pathway. J Gen Physiol 103:279–319

Zhang M, Liu J, Tseng G-N (2004) Gating charges in the activation and inactivation processes of the hERG channel. J Gen Physiol 124(6):703–718

Zhang M, Liu J, Jiang M, Wu DM, Sonawane K, Guy HR et al (2005) Interactions between charged residues in the transmembrane segments of the voltage-sensing domain in the hERG channel. J Membr Biol 207(0022–2631 (Print)):169–181

Zhao J, Blunck R (2016) The isolated voltage sensing domain of the Shaker potassium channel forms a voltage-gated cation channel. elife 5(OCTOBER):1–18

Zhao Y, Goldschen-Ohm MP, Morais-Cabral JH, Chanda B, Robertson GA (2017) The intrinsically liganded cyclic nucleotide-binding homology domain promotes KCNH channel activation. J Gen Physiol 149(2):249–260

Zhou Y, Morais-Cabral JH, Kaufman A, MacKinnon R (2001) Chemistry of ion coordination and hydration revealed by a K^+ channel-Fab complex at 2.0 A resolution. Nature 414(0028–0836 SB-IM):43–48

Zou A, Curran ME, Keating MT, Sanguinetti MC (1997) Single HERG delayed rectifier K+ channels expressed in Xenopus oocytes. Am J Physiol [Internet] [cited 2024 Dec 5];272(3 Pt 2). Available from: https://pubmed.ncbi.nlm.nih.gov/9087606/

Structure–Function Relationships of K_{2P} Channels

8

Julian A. Schreiber and Guiscard Seebohm

Abstract

In the 1950s, Hodgkin and Huxley predicted the existence of ion channels that produce a continuous potassium (K^+) background current, maintaining the resting membrane potential. In the mid-1990s, these leak channels were identified and named two-pore domain potassium channels (K_{2P}) based on their unique monomer structure. In a very short period, 15 different genes encoding for K_{2P} channel subunits were identified within the human genome that form functional homo as well as heterodimeric K^+ channels. Based on their functional properties, homodimeric K_{2P} channels were subdivided into the six different subfamilies TWIK, TREK, TASK, TALK, THIK, and TRESK. Intensive research uncovered the critical contribution of K_{2P} channels to basal physiological processes in all kinds of different tissues. Since 2012, functional studies have been aligned with structural examinations via crystallography and cryo-electron microscopy. The combination of functional and structural research generates a deeper understanding for physiological as well as pathophysiological processes supporting the development of specific ion channel modulators. Hereafter, we will recapitulate the current knowledge from a structural and functional perspective, focusing on the different domains of K_{2P} ion channels as well as their contribution to certain physiological processes.

Keywords

K_{2P} channels · Ion channels · Potassium · Electrophysiology · Structural biology

8.1 The Discovery and Common Structure of K_{2P} Channels

The family of two pore domain potassium (K_{2P}) channels clearly distinguishes from voltage-gated (K_v) or ligand-gated (K_{ir}) potassium channels due to their dimeric character as well as their two-pore domain containing monomers (Taura et al. 2021). Their discovery and characterization represent a significant advancement in our understanding of ion channel biology and underline the importance of background conductance for physiological and

J. A. Schreiber (✉)
University of Münster, Institute of Pharmaceutical and Medicinal Chemistry, Münster, Germany
e-mail: j.schreiber@uni-muenster.de

G. Seebohm
University Hospital Münster, Institute for Genetics of Heart Diseases (IfGH), Department of Cardiovascular Medicine, Münster, Germany
e-mail: guiscard.seebohm@ukmuenster.de

A. Rosenhouse-Dantsker (ed.), *Decoding Ion Channels Structure and Function*, Advances in Experimental Medicine and Biology 1497, https://doi.org/10.1007/978-3-032-07523-9_8

pathophysiological processes (Lee et al. 2021). The journey of K_{2P} channel discovery began in the late twentieth century with the identification of TOK1 from *Saccharomyces cerevisiae* and the finding that potassium channels were not limited to a single pore-forming domain within one monomer (Ketchum et al. 1995). Instead of having 6 or 2 transmembrane domains and one pore-forming domain like previously observed in K_v or K_{ir} channel monomers, TOK1 monomers possess eight transmembrane domains with two pore-forming domains (Houdinet et al. 2023). Furthermore, TOK1 channels assemble as dimers, while K_v and K_{ir} channels assemble as tetramers (Ketchum et al. 1995; Houdinet et al. 2023). Shortly after the discovery of TOK1, a K^+ channel named ORK1 (open rectifier K^+) was cloned from *Drosophila melanogaster* in 1996, encompassing a similar two-pore domain architecture within one monomer (Goldstein et al. 1996). Compared to TOK1, only four transmembrane helices (M1–M4) with two pore-forming domains between M1 / M2 and M3 / M4 were predicted (Goldstein et al. 1996). In the very same year, the first human representative of K_{2P} channels, TWIK-1 (Tandem of P domains in a weak inward rectifying K^+ channel; $K_{2P}1.1$), was described, which showed high structural similarity with ORK1 and defined the overall structural features of K_{2P} channel monomers (Fig. 8.1a) (Lesage et al. 1996a). Besides the four transmembrane helices (M1–M4), K_{2P} channel monomers are characterized by two extracellular helices (E1 and E2) forming the cap domain, an archway above the central ion pore (Fig. 8.1a, b) (Pope et al. 2020). Together with the two pore-forming domains consisting of one pore helix (P1 and P2) and one selectivity filter-forming loop (SF1 and SF2), this extracellular domain facilitates the dimerization process (Pope et al. 2020). Last, most K_{2P} channel monomers possess long N- and/or C-terminal domains that bear binding sites for many different regulatory proteins, shaping the channel properties and enabling channel activity regulation by different physiological stimuli (Zuzarte et al. 2009; Kamuene et al. 2021; Rajan et al. 2005; Blin et al. 2014; Renigunta et al. 2014; Koh et al. 2001; Honoré 2007; García et al. 2021). Under physiological conditions, the activity of K_{2P} channels is regulated almost voltage independently, providing substantial background potassium conductance, which was already predicted and ratified by Hodgkin and Huxley in the 1950s (Fig. 8.1c) (Kamuene et al. 2021; Hodgkin and Huxley 1952). Consequently, K_{2P} channels play an important role in all kinds of different tissues for the maintenance of background K^+ conductance (Lee et al. 2021). Within a few years after $K_{2P}1.1$ discovery, more and more K_{2P} channels were discovered, and by 2007, the human genome was found to encode a total of 15 different K_{2P} channels named $K_{2P}1.1$ to $K_{2P}18.1$, that can be subdivided into six different subfamilies: TWIK, TREK (TWIK-related), TALK (TWIK-related alkaline pH activated), TASK (TWIK-related acid sensitive), THIK (tandem pore domain halothane-inhibited), and TRESK (TWIK-related spinal cord; Fig. 8.1d) (Mathie and Veale 2007; Goldstein et al. 2005). Although all structurally characterized members share a similar overall structure, each family is attributed with different properties and modulatory pathways including various stimuli (pH, mechanical stretch, temperature, etc.) showcasing their versatile roles in physiology (Patel et al. 1999). The physiological relevance of K_{2P} channels began to be unraveled through experimental studies. K_{2P} channels play a crucial role in regulating cellular excitability and maintaining physiological homeostasis by allowing largely voltage-independent potassium leak currents (Taura et al. 2021). Consequently, researchers discovered that these channels play critical roles in various biological processes, including neuronal excitability, cardiac rhythm regulation, breathing rhythm, and vascular tone control (Lee et al. 2021).

8.2 The Subfamilies of K_{2P} Channels: TWIK, TREK, and TASK

The six subfamilies of the K_{2P} channel family encompass 15 members that form homodimeric

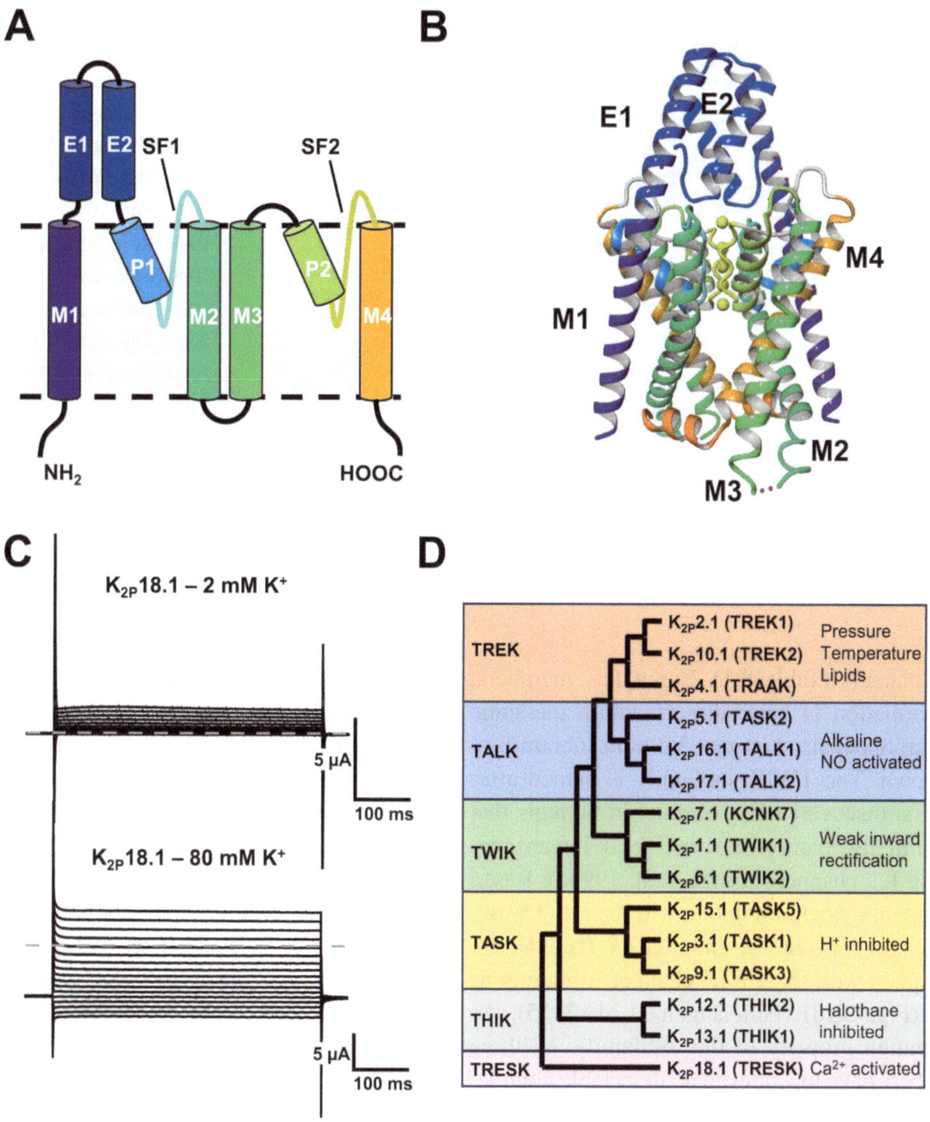

Fig. 8.1 Characteristics of K_{2P} channel family. (**a**) Schematic depiction of K_{2P} channel monomer including the four transmembrane helices M1–M4, the extracellular cap forming helices E1 and E2 as well as the pore-forming domain consisting of two pore helices (P1, P2) and two selectivity filter-forming loops (SF1, SF2). (**b**) Crystal structure of human $K_{2P}1.1$ ion channel (PDB 3UKM (Miller and Long 2012)). Color code for structural elements is defined in **a**. K^+ ions within the selectivity filter are colored yellow. (**c**) Representative current traces from two-electrode voltage clamp recordings of human $K_{2P}18.1$ channel expressed in *Xenopus laevis* oocytes in the presence of different extracellular K^+ concentrations. (**d**) Phylogenetic tree of human K_{2P} protein sequences including subfamily names, trivial names, and typical subfamily characteristics referring to Natale et al. (2021) and Schreiber et al. (2022a)

Table 8.1 Nomenclature of K_{2P} channels including IUPHAR nomenclature, common and alternative names as well as year of first appearance in literature

IUPHAR (Goldstein et al. 2005)	Common name	Other names	Discovery
$K_{2P}1.1$	TWIK-1	hOHO	1996 (Lesage et al. 1996a)
$K_{2P}2.1$	TREK-1	TPKC1	1996 (Fink et al. 1996)
$K_{2P}3.1$	TASK-1	TBAK1, OAT1	1997 (Duprat 1997)
$K_{2P}4.1$	TRAAK	KT4	1998 (Fink 1998)
$K_{2P}5.1$		TASK-2	1998 (Reyes et al. 1998)
$K_{2P}6.1$	TWIK-2	TOSS	1999 (Chavez et al. 1999)
$K_{2P}7.1$		Kcnk8	1999 (Salinas et al. 1999)
$K_{2P}9.1$	TASK-3		2000 (Rajan et al. 2000)
$K_{2P}10.1$	TREK-2		2000 (Bang et al. 2000)
$K_{2P}12.1$	THIK-2		2001 (Rajan et al. 2001)
$K_{2P}13.1$	THIK-1		2001 (Rajan et al. 2001)
$K_{2P}15.1$	TASK-5	KCNK11, KCNK14	2001 (Kim and Gnatenco 2001; Ashmole et al. 2001)
$K_{2P}16.1$	TALK-1		2001 (Girard et al. 2001)
$K_{2P}17.1$	TALK-2	TASK-4	2001 (Decher et al. 2001)
$K_{2P}18.1$	TRESK	$K_{2P}19.1$	2003 (Sano et al. 2003)

ion channels (Table 8.1). For a few members, the formation of heterodimers within the same subfamily but also between different subfamilies, is known. The TWIK subfamily is named after the first discovered K_{2P} channel in humans that was initially characterized as a weak inward rectifying K^+ channel (Lesage et al. 1996a). Based on protein sequence homology to $K_{2P}1.1$ (TWIK-1), $K_{2P}6.1$ (TWIK-2), and $K_{2P}7.1$ (kcnk8) are associated with the TWIK channel subfamily as well (Fig. 8.1d) (Feliciangeli et al. 2015). As a common property of this subfamily, all three members have low or even no detectable conductivity in heterologous or native expression systems (Goldstein et al. 1997; Chavez et al. 1999; Salinas et al. 1999). High amounts of mRNA are found in the heart and brain, while lower amounts are found in placenta, lung, liver, kidney, and pancreas (Lesage et al. 1996a). Furthermore, heterodimerization was observed in primary cultures of rat neurons with other K_{2P} channel subunits like $K_{2P}2.1$ (TREK-1), $K_{2P}3.1$ (TASK-1), and $K_{2P}9.1$ (TASK-3) (Mi Hwang et al. 2014; Plant et al. 2012). $K_{2P}1.1$ channel activity was found to be inhibited via intracellular and extracellular acidification, which also changes ion selectivity of the channel, increasing conductance of Na^+ ions (Lesage et al. 1996a; Turney et al. 2022; Chatelain et al. 2024). Similar to other inward rectifiers, rectification of outward currents at positive membrane potentials is associated with a partial block of conductance via intracellular Mg^{2+} (Lesage et al. 1996a). The reasons for low $K_{2P}1.1$ conductance in different expression systems are associated with various processes including regulation of surface expression via SUMOlation and rapid endocytosis as well as regulation of ion conductance via selectivity filter instability and hydrophobic dewetting, which is caused by the inner channel vestibule creating an energetic barrier for K^+ permeation (Rajan et al. 2005; Plant et al. 2010; Feliciangeli et al. 2010; Nematian-Ardestani et al. 2020). Modification of $K_{2P}1.1$ at lysine 274 with a small ubiquitin-like modifier (SUMO) protein results in reversible silencing of $K_{2P}1.1$ conductance (Rajan et al. 2005). Additionally, $K_{2P}1.1$ is associated with rapid dynamin-dependent endocytosis, which is transmitted via interactions with Ile293 and Ile294 (Feliciangeli et al. 2010).

Compared to $K_{2P}1.1$, much less is known about the other two members of the TWIK subfamily. $K_{2P}6.1$ and $K_{2P}7.1$ show higher sequence homology to each other than to $K_{2P}1.1$ (Fig. 8.1d). Both channels are expressed in peripheral tissues like the eye, lung, and stomach (Salinas et al. 1999). Interestingly, based on homology and syntenic location within the genome, kcnk8 isolated from mice was identified as the human homologue of $K_{2P}7.1$ and subsequently renamed (Bockenhauer et al. 2000). Consequently, no protein named $K_{2P}8.1$ exists so far. $K_{2P}6.1$ expression in vascular smooth muscle cells is associated with the regulation of vascular tone by regulation of the membrane potential (Lloyd et al. 2011). In line with these findings, $K_{2P}6.1$-deficient mice suffer from vascular dysfunction as well as systemic and pulmonary hypertension (Lloyd et al. 2011; Kitagawa et al. 2019). Contrary, knockout of *KCNK7* in mice resulted in animals that were indistinguishable compared to wild-type mice, leading to the conclusion that $K_{2P}7.1$ plays either a minor role in physiological processes or that $K_{2P}7.1$ function might be compensated by other K_{2P} channels (Yost et al. 2008). Interestingly, $K_{2P}7.1$ shows an unusual sequence motif within the SF2 that contributes to the formation of the selectivity filter (Salinas et al. 1999). While all other K_{2P}-forming subunits stick to a classical K^+ selectivity filter motive G-F/L-G in SF2, one of the glycines in $K_{2P}7.1$ is replaced with an unflexible and charged glutamate (G-L-E), emphasizing the role of a silent K_{2P} channel subunit (Salinas et al. 1999).

The TREK (TWIK-related K^+) channel subfamily comprises the channels $K_{2P}2.1$ (TREK-1), $K_{2P}4.1$ (TRAAK; TWIK-related arachidonic acid-stimulated), and $K_{2P}10.1$ (TREK-2) (Fink et al. 1996; Fink 1998; Bang et al. 2000). While $K_{2P}4.1$ is more exclusively expressed in the central nervous system, $K_{2P}2.1$ and $K_{2P}10.1$ are also found in peripheral tissues like the heart, lung, pancreas, or smooth muscles (Medhurst et al. 2001). All three members respond to a wide range of different physiological stimuli including membrane tension, heat, changes in intracellular pH as well as membrane lipids like Phosphatidylinositol-4,5-bisphosphate (PIP_2) rendering this subfamily the most versatile regulated (Riel et al. 2022; Schewe et al. 2016; Sorum et al. 2024). Although strong associations with neuronal processes and high expression rates in the central nervous system are claimed for all TREK channels, knockout mice show only mildly altered neuronal phenotypes, that are associated with alterations of pain and temperature sensations (Mirkovic et al. 2012; Pereira et al. 2014). More pronounced changes were observed in $K_{2P}2.1$ knockout mice, which showed higher sensitivity to ischemia and epilepsy as well as changes in neuronal inflammation and neurodegeneration (Heurteaux et al. 2004; Bittner et al. 2013). Furthermore, cardiac alterations like hypertrophy but also reduced fibroblast proliferation and migration, were observed in global $K_{2P}2.1$ knockout mice (Abraham et al. 2018). Increased sensitivity to heat and mechanically induced pain was observed in $K_{2P}10.1$ knockout mice as well (Pereira et al. 2014). Previous studies already identified $K_{2P}10.1$ together with $K_{2P}18.1$ (TRESK) as responsible for the major K^+ background component in nociceptive neurons of the dorsal root ganglion, regulating cell excitability (Kang and Kim 2006). In accordance with these results, the $K_{2P}4.1$ function is associated with the regulation of neuronal excitability in some hippocampal neurons (Viatchenko-Karpinski et al. 2018; de la Peña et al. 2012). The common name of $K_{2P}4.1$, TRAAK, is derived from the stimulatory effect of unsaturated fatty acids like arachidonic acid, which plays an important role not only in pathophysiological conditions like seizures, inflammation, or brain ischemia but also in physiological synaptic transmission (Fink 1998). Recent studies show that the expression of $K_{2P}4.1$ together with $K_{2P}2.1$ is not only restricted to nodes of Ranvier but also found at the axon initial segment modulating neuronal excitability in the hippocampus and cortex (Luque-Fernández et al. 2024).

Compared to the TREK channel subfamily, members of the TASK (acid-sensitive K^+) channel subfamily including $K_{2P}3.1$ (TASK-2), $K_{2P}9.1$ (TASK-3), and $K_{2P}15.1$ (KCNK11, KCNK14, TASK-5) are more ubiquitously expressed (Kim and Gnatenco 2001; Rajan et al. 2000; Ashmole et al. 2001; Duprat 1997). As indicated by the subfamily name, the common feature of these channels is the inhibition via extracellular acidification (Kim and Gnatenco

2001; Rajan et al. 2000; Duprat 1997). The mechanism behind this inhibition is related to a histidine residue, that is found near the SF1 section on the extracellular side (Rajan et al. 2000; Czirják and Enyedi 2002). Interestingly, this histidine is also present in $K_{2P}1.1$, that shows sensitivity to increased extracellular H^+ concentrations as well (Chatelain et al. 2024). $K_{2P}3.1$ and $K_{2P}9.1$ are expressed in different tissues like the pancreas, brain, lung, heart, and kidney, where they also form functional heterodimers (Czirják and Enyedi 2002; Bayliss and Barrett 2008; Kim et al. 2009). Therefore, it is not surprising that different phenotypes in knockout mice as well as associations with different diseases, can be observed (Bayliss and Barrett 2008). Knockout of $K_{2P}3.1$ and $K_{2P}9.1$ in mice induces respiratory changes including abnormal breathing under hypoxic conditions and pulmonary hypertension (Buehler et al. 2017; Trapp et al. 2008). Cardiac phenotypes of $K_{2P}3.1$ knockout are associated with prolonged QT intervals and atrial fibrillation (Wiedmann et al. 2022; Decher et al. 2011). Furthermore, both TASK channels contribute to cardiac remodeling under elevated pressure load (Duan et al. 2020). In humans, several *KCNK3* mutations were identified as possible disease-causing genes of heritable pulmonary arterial hypertension, while mutations in the *KCNK9* gene are associated with the Birk-Barel syndrome, which is characterized by mental retardation, hypotonia, and hyperactivity (Barel et al. 2008; Cunningham et al. 2019). Furthermore, both channels might contribute to the development of sleep apnea by affecting the lung function (Song et al. 2024). Last, $K_{2P}3.1$ and $K_{2P}9.1$ are identified as potential targets for autoimmune diseases as well as cancer due to their pro-oncogenic potential (Rinné et al. 2024). In contrast to the broad characterization of $K_{2P}3.1$ and $K_{2P}9.1$, much less is known about $K_{2P}15.1$ due to its nonfunctional or silent character (Kim and Gnatenco 2001; Ashmole et al. 2001). Homodimeric $K_{2P}15.1$ channels remain silent due to their inefficient transport to the plasma membrane (Rinné et al. 2024). However, very recently functional $K_{2P}15.1$ heterodimers with $K_{2P}3.1$ and $K_{2P}9.1$ were described, which are transported to the plasma membrane and contribute to physiological and pathophysiological conditions (Rinné et al. 2024). These findings open up a new point of view since $K_{2P}15.1$ shows a similar expression pattern to other TASK channels and is dysregulated in some cancer types as well, underscoring the potential as a drug target (Ashmole et al. 2001).

8.3 The Subfamilies of K_{2P} Channels: TALK, THIK, and TRESK

The TALK (alkaline pH-activated K^+) channel subfamily encompassing $K_{2P}5.1$ (TASK-2), $K_{2P}16.1$ (TALK-1), and $K_{2P}17.1$ (TALK-2, TASK-4) is characterized by the common features of activation via extracellular as well as intracellular alkalization, which are based on two separate mechanisms (Girard et al. 2001; Decher et al. 2001; Reyes et al. 1998; Niemeyer et al. 2010). As recognizable by their trivial names, $K_{2P}5.1$ and $K_{2P}17.1$ were initially described as new members of the TASK subfamily (Decher et al. 2001; Reyes et al. 1998). However, protein sequence comparisons revealed that these channels are more related to the TREK subfamily members than to $K_{2P}3.1$ or $K_{2P}9.1$ (Girard et al. 2001). Furthermore, TALK-related channels show, in contrast to TASK-related channels, only low activity at physiological pH values (Niemeyer et al. 2010). High expression rates of $K_{2P}5.1$ are found in human liver, lung, pancreas, intestines, and kidney, where it was initially isolated from (Reyes et al. 1998). Several studies underscored the relevance of $K_{2P}5.1$ function for different renal and intestinal processes as well as cell volume regulation (Warth et al. 2004; Barriere et al. 2003; L'Hoste et al. 2007; Julio-Kalajzić et al. 2018). Furthermore, missense variants of *KCNK5* are associated with Balkan endemic nephropathy (Reed et al. 2016). Less is known about the physiological relevance of $K_{2P}16.1$ and $K_{2P}17.1$. While $K_{2P}17.1$ is expressed in many different tissues like heart, placenta, liver, and pancreas, high expression rates of $K_{2P}16.1$ were only found in the pancreas, contributing to the physiological β cell function and insulin release via modulation of the Ca^{2+} handling within the endoplasmic reticulum (Girard et al. 2001; Vierra et al. 2017). In line with these findings, the $K_{2P}16.1$ polymorphism was associated with type 2 diabetes in humans (Vierra et al. 2015). Dysregulation of $K_{2P}17.1$ is associated with different cardiac disorders including arrhythmia, atrial fibrillation, and long QT syndrome (Chai et al. 2018; Friedrich et al. 2014). Moreover, $K_{2P}17.1$ was linked to ischemic stroke and cerebral hemorrhage, although the exact mechanisms are not fully understood (Domingues-Montanari et al. 2010; He et al. 2014).

The THIK (halothane-inhibited K^+) subfamily encompassing $K_{2P}12.1$ (THIK-2) and $K_{2P}13.1$ (THIK-1) was named after the volatile anesthetic halothane, which surprisingly inhibited $K_{2P}13.1$ currents instead of amplifying them, like it was previously observed for other K_{2P} channels like $K_{2P}2.1$ and $K_{2P}3.1$ (Girard et al. 2001; Rajan et al. 2001). $K_{2P}13.1$ shows robust surface expression and K^+ currents, that can be amplified via application of arachidonic acid as well (Rajan et al. 2001). Contrary to this, $K_{2P}12.1$ forms homodimers with low intrinsic channel activity (Renigunta et al. 2014; Bichet et al. 2015). The reason for this silent appearance is related to the intracellular accumulation due to an endoplasmic reticulum retention motif at the N-terminal domain (Renigunta et al. 2014; Bichet et al. 2015). On the other hand, functional heterodimers of $K_{2P}12.1$ and $K_{2P}13.1$ are described, leading to the hypothesis that $K_{2P}12.1$ might regulate surface expression and function of $K_{2P}13.1$, which is further supported by the similar expression patterns of both subunits (Lee et al. 2021; Blin et al. 2014). Pronounced expression rates of $K_{2P}12.1$ and $K_{2P}13.1$ are found in several brain regions, while $K_{2P}13.1$ was also found in cardiac tissues (Lee et al. 2021; Bürli et al. 2024; Staudacher et al. 2019). Interestingly, elevated expression rates of $K_{2P}13.1$ were found postmortem in microglia cells of Alzheimer patients, rendering inhibition of $K_{2P}13.1$ as a possible target for pharmacological intervention (Bürli et al. 2024). Furthermore, it was found that patients with arrhythmia or heart failure showed reduced atrial expression levels of $K_{2P}13.1$, rendering pharmacological modulation of the channel as a new option for these diseases as well (Staudacher et al. 2019). However, most contributions of $K_{2P}12.1$ and $K_{2P}13.1$ to physiological effects are not well understood.

The only remaining member of the TRESK (TWIK-related spinal cord K^+) subfamily is $K_{2P}18.1$ (TRESK), since $K_{2P}19.1$ was later identified as the mice homologue of the human $K_{2P}18.1$ channel (Sano et al. 2003; Kang et al. 2004; Schreiber et al. 2022a). Although it is assumed that the structure of TRESK is similar to previously characterized K_{2P} channels, the protein sequence of TRESK shares only around 20% identity with the other K_{2P} channels (Schreiber et al. 2022a). TRESK possesses several structural and functional properties that are unique within the K_{2P} channel family. First, in contrast to the other members, the N- and C-terminal domains of TRESK are short, encompassing less than 20 amino acids (Sano et al. 2003). Second, TRESK is the only member with an extended intracellular loop between the M2 and M3 helices, which encompasses more than 100 amino acids harboring binding sites for calcineurin, 14-3-3 proteins, protein kinase A, and tubulin (Sano et al. 2003; Schreiber et al. 2022a). Last, TRESK is the only member that is functionally controlled by rising intracellular Ca^{2+} levels (Czirják and Enyedi 2006). Functional studies as well as evaluations of patient-related mutations associate $K_{2P}18.1$ with several diseases including neuropathic pain, migraine, and epilepsy (Shkodra and Caraceni 2022; Liu et al. 2021; Yang et al. 2018; Pettingill et al. 2019; Imbrici et al. 2020; Huang et al. 2021). Recent studies also indicate a dominant role of $K_{2P}18.1$ in immune-related progenitor cells regulating the differentiation and proliferation of regulatory T cells in the thymus, which have anti-inflammatory effects (Ruck et al. 2022; Huntemann et al. 2022). Consequently, modulation of TRESK might be an interesting target for immune-related diseases as well as for add-on therapies in some cancer treatments (Ruck et al. 2022; Huntemann et al. 2022).

8.4 N- and C-Terminal Domains of K_{2P} Channels

The N- and C-terminal domains are key structural components of K_{2P} channels that are crucial for function, regulation, and interactions with other molecules, allowing the modulation of K^+ background currents in response to various physical and chemical stimuli. While all K_{2P} channels share a similar core structure for transmembrane and extracellular domains, significant differences in their N- and C-terminal domains are observed, enabling distinct functional specifics of K_{2P} channels. These specifics determine how individual K_{2P} channels react to different stimuli and what roles they play in various physiological processes.

The N-terminal domain in K_{2P} channels often impacts channel folding, cellular transport, and membrane localization (Zuzarte et al. 2009). In some K_{2P} channels, the N-terminal regions serve as hubs for regulatory proteins or signaling molecules that modulate the channel's activity (Zuzarte et al. 2009). Some K_{2P} channels have specific N-terminal motifs that enable them to respond to intracellular signals, such as G-proteins or phosphorylation events. An example is $K_{2P}3.1$ (TASK-1), where the N-terminal domain is important for phosphorylation-dependent regulation via various kinase-mediated signaling pathways (Zuzarte et al. 2009). For $K_{2P}3.1$ and $K_{2P}9.1$, the N-terminus plays an important role in membrane trafficking from the endoplasmic reticulum (ER) through interaction with 14-3-3 proteins, which are adapter proteins for a vast number of signaling proteins, allowing for channel cAMP-dependent protein kinase (PKA) phosphorylation, subsequently affecting channel opening and altering its activity in response to cellular stimuli (Zuzarte et al. 2009).

The C-terminal domain plays a crucial role in modulating channel activity in many K_{2P} channels as well (Harleton et al. 2013; Nagaraj et al. 2013; Seyler et al. 2012). It is also involved in interactions with other proteins, such as cytoskeletal structures, signaling proteins, or second messengers. Again, $K_{2P}3.1$ is a good example of multiple phosphorylation events: S293 of rat $K_{2P}3.1$ channel is positioned in the classical consensus sequence RRx(S/T), allowing for phosphorylation by PKA, which is essential for 14-3-3 binding to a dileucine motif (L263 and L264), forming a modulating protein complex (Mant et al. 2011; Kilisch et al. 2015; O'Kelly and Goldstein 2008). $K_{2P}3.1$ is further phosphorylated by Rho kinase at residues S336 and S393 and by Src kinase at Y419 leading to altered function and subcellular localization (Nagaraj et al. 2013; Seyler et al. 2012). Further studies identified $K_{2P}3.1$ T383 as a protein kinase C (PKC) ε-dependent phosphorylation site (Harleton et al. 2013).

A second well-studied example is the mechanosensitive channel $K_{2P}2.1$ (TREK-1). The C-terminal domain of $K_{2P}2.1$ contains binding sites for phospholipids like arachidonic acid or PIP_2 that activate the channel responding to membrane deformations or temperature changes (Schmidtpeter et al. 2023). $K_{2P}2.1$ activity is also regulated through interactions with PKA and PKC, which phosphorylate the C-terminal domain, inducing channel inhibition (García et al. 2021; Schmidtpeter et al. 2023). Regulation of $K_{2P}2.1$ channels by PKA-dependent phosphorylation of S333 causes closure of channels (Patel et al. 1999). Further, murine $K_{2P}2.1$ is activated by cGMP-dependent phosphorylation at S351, which was verified by mutational studies (S351A) at the PKG consensus sequence (Koh et al. 2001). A small degree of current inhibition occurred when cells with the S351A mutant channels were exposed to 8-Br-cGMP, suggesting that cGMP-dependent mechanisms (i.e., either activation of PKG or cross-over activation of PKA) may result in phosphorylation of S333. Application of 8-Br-cAMP to cells with the mutant murine $K_{2P}2.1$ channels marked and sustained suppression of currents (Koh et al. 2001). Thus, the major regulatory effects of cGMP-dependent and cAMP-dependent mechanisms occur via different sites of phosphorylation. Similar to $K_{2P}2.1$, the other members of the TREK subfamily ($K_{2P}4.1$ and $K_{2P}10.1$) respond to C-terminal mediated stimuli like mechanical and thermal stimuli as well (Ávalos Prado et al. 2022).

In comparison to other K_{2P} channels, the C- and N-terminal domains of the $K_{2P}18.1$

channel are much shorter (Schreiber et al. 2022a). However, the intracellular domains of $K_{2P}18.1$ interact with intracellular modifiers as well and strongly alter the ion channel activity (Schreiber et al. 2022a). Recently, regulation via the C-terminus of $K_{2P}18.1$ was described (Pergel et al. 2024). $K_{2P}18.1$ channels interact with Nedd4 family-interacting protein 1 (NDFIP1), which in turn contains a PPxY motif that couples the channel to the Nedd4 degradation system and thus regulates channel surface density (Pergel et al. 2024). Besides this C-terminal modulation, most regulatory proteins interact with $K_{2P}18.1$ channels via the unique intracellular domain, which is localized between the M2 and M3 helices, encompassing more than 100 amino acids (Schreiber et al. 2022a). This domain is responsible for the unique $K_{2P}18.1$ regulation via intracellular Ca^{2+} ions, which is transduced by calcium-dependent phosphatase calcineurin, that dephosphorylates human TRESK at S252, S262, S264, and S269, inducing channel activation (Schreiber et al. 2022a). Consequently, high intracellular Ca^{2+} levels induce channel activation. This calcium-dependent regulation sets TRESK apart from other K_{2P} channels (Schreiber et al. 2022a).

Summarizing the relevance of the intracellular N- and C-terminal domains of K_{2P} channels, these structures are crucial for channel function, regulation, and interaction with cellular signaling pathways. While the N-terminal domain is often important for phosphorylation-dependent regulation and cell transport, the C-terminal domain is typically responsible for modulation by mechanical, thermal, and chemical stimuli.

8.5 Structure and Function of M1–M4 Transmembrane Domains

K_{2P} channel monomers share a common structure encompassing four transmembrane helices (M1–M4), which form the backbone of the channel and play critical roles in ion conduction and channel gating (Miller and Long 2012; Natale et al. 2021). Together with the pore-forming domain, these helices are arranged in a specific pattern to form the ion channel pore (Fig. 8.2a) (Natale et al. 2021). The inner vestibule, which is localized at the cytosolic site beneath the selectivity filter, is formed by the M2 and M4 helices. From the cytosolic point of view, M2 and M4 helices are arranged in an alternating pattern with opposing M2/M4 helices from the different channel subunits (Miller and Long 2012). The outer transmembrane segment is formed by the M1 and M3 helices, which are also arranged with opposing M1/M3 helices from the different monomers (Miller and Long 2012). M1–M4 are not independent from each other but rather work in concerted motions to stabilize the open or closed states of K_{2P} channels, allowing for dynamic response to different physiological conditions. The interplay between the transmembrane helices as well as their interactions with the surrounding membrane shapes the gating properties as well as the channel activity via different mechanisms, leading to a highly versatile functionality in all K_{2P} channels (Lolicato et al. 2020).

The N-terminal M1 helix helps to anchor the protein within the membrane and contributes to the formation of the channels' outer surface, interacting with the lipid environment and impacting gating dynamics of the channel (Schmidtpeter et al. 2023). M1 spans the complete lipid bilayer in an almost perpendicular manner (Schmidtpeter et al. 2023). At the extracellular side, M1 is connected to E1, which continues with the helical structure in an angled fashion (Schmidtpeter et al. 2023). Compared to the other transmembrane helices, M1 is domain-swapped, leading to interactions with M2, P1, and M4 of the second channel-forming subunit (Natale et al. 2021). Direct interactions of M1 with the loop connecting M2 and SF1 affect the movement flexibility of the selectivity filter, subsequently influencing the so-called C-type gating of K_{2P} channels, which plays a major role in the regulation of K_{2P} channel conductance (Lolicato et al. 2020). However, C-type gating is modulated by all transmembrane helices including M2, M3, and M4 (Schmidtpeter et al. 2023; Lolicato et al. 2020; Lolicato et al. 2014).

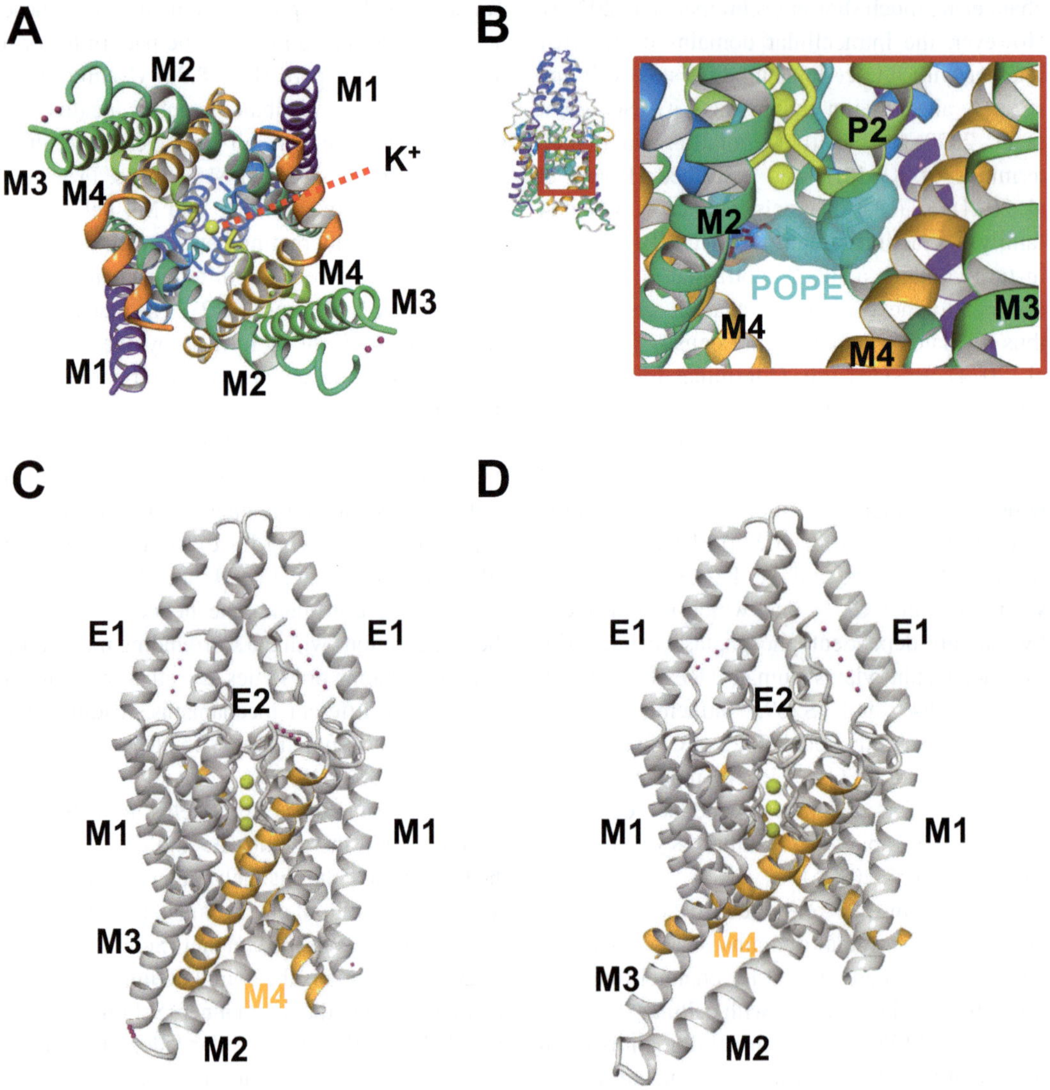

Fig. 8.2 Transmembrane segments and fenestration site. (**a**) Crystal structure of human $K_{2P}1.1$ ion channel from a cytosolic point of view (PDB 3UKM (Miller and Long 2012)). Color code for structural elements is defined in Fig. 8.1a. K^+ ions within the selectivity filter are colored yellow. The inner vestibule beneath the selectivity filter is formed by M2/M4 helices, while M1/M3 helices face the lipid bilayer. (**b**) Cryo-EM structure of $K_{2P}2.1$ zebrafish ortholog (PDB 8DE9 (Schmidtpeter et al. 2023)) with highlighted fenestration site (red box) occupied by 1-palmitoyl-2-oleoyl phosphatidylethanolamine (POPE, cyan). (**c, d**) Crystal structures of human $K_{2P}10.1$ ion channel in down (**c**; PDB 4XDJ (Dong et al. 2015)) and up (**d**; PDB 4BW5 (Dong et al. 2015)) conformation of the M4 helix

The M2 helix contributes to the inner pore lining and shapes the ion conductance pathway, contributing to the channel gating in different ways (Lolicato et al. 2014). Together with the M4 helix, M2 forms the inner cavity, which plays a vital role in determining ion selectivity by stabilizing K^+ ions as they pass through the pore (Lolicato et al. 2014; Aryal et al. 2014). Consistently, mutations in the M2 region alter channel gating properties, leading to changes in ion flux rate, ion selectivity, and open probability (Natale et al. 2021). The inner cavity harbors a conserved binding site that interacts with M2 and M4 residues and is targeted by different K_{2P} channel activators as well as inhibitors (Dong et al. 2015; Schewe et al. 2019). The formation of this site is characterized by rearrangements of the M2, M3, and M4 helices connecting the ion pathway to the membrane lipids from the surrounding bilayer (Dong et al. 2015). Therefore, this site is also called the fenestration side, enabling the direct modulation of the conduction pathway by certain membrane lipids (Fig. 8.2b) (Dong et al. 2015). The flexibility and the needed rearrangement of the inner transmembrane helices M2 and M4 are shown in several structural studies (Turney et al. 2022; Miller and Long 2012; Lolicato et al. 2014; Dong et al. 2015; Li et al. 2020). They can be induced by extracellular or intracellular stimuli, such as changes in pH, mechanical stretch, or temperature shifts (Turney et al. 2022; Chatelain et al. 2024; Sorum et al. 2024; Lolicato et al. 2014; Li et al. 2020; Brohawn et al. 2012). For members of the TREK channel subfamily (K_{2P}2.1, K_{2P}4.1, and K_{2P}10.1), a conserved GXG motif at M2 can be identified, which contributes to the mechanosensitive properties of this subfamily (Lolicato et al. 2014). Conformational changes of the M4 helix reduce the interaction surface with M2 from the neighboring subunit while it increases the interaction surface with M2 from the same subunit (Lolicato et al. 2014). Subsequently, the GXG motif induces a buckling of the helix, which is associated with a stabilization of the selectivity filter, inducing C-type dependent activation of TREK channels (Lolicato et al. 2014). Interestingly, this motif is also partially conserved in other K_{2P} channels, indicating a similar contribution of M2 to their gating as well (Natale et al. 2021).

The M3 helix is positioned in front of the second pore-forming domain, facing the outer side of the K_{2P} channels (Turney et al. 2022; Miller and Long 2012; Li et al. 2020). M3 contributes to K_{2P} channel gating by a conserved glutamate in the upper α-helix segment, which interacts via H bonds with the long SF2-M4 linker, optimizing the S1 and S2 K^+ binding sites of the selectivity filter for K^+ interactions to facilitate conduction and thus leading to ion channel activation (Lolicato et al. 2020). In K_{2P}2.1 (TREK-1), the M3 region has been shown to participate in mechano-gating, where mechanical forces modulate the opening and closing of the channel (Honoré 2007). Consistently, mutations in the M3 segment can alter gating properties including mechanosensation (Honoré 2007). As contributors to the fenestration side, lower M3 residues are identified as interacting partners for modulatory compounds and endogenous lipid molecules as well (Dong et al. 2015).

The last transmembrane α-helix is M4, which is positioned between the second pore-forming and the C-terminal domain (Dong et al. 2015). Several studies suggest that M4 is the key player of the transmembrane helices in terms of C-type gating modulation (Dong et al. 2015; Türkaydin et al. 2024; Schreiber et al. 2023; Rödström et al. 2020). Based on structural studies, M4 exists in different conformational states, that have a great impact on the conformational stability of the selectivity filter (Dong et al. 2015; Türkaydin et al. 2024). In reference to the central axis of the ion channel and the membrane, these states are named up and down (Dong et al. 2015). In the down state, M4 helix appears in an almost straight conformation and crosses the membrane at a nearly 45° angle (Fig. 8.2c) (Dong et al. 2015). Compared to this, the upstate is characterized by a bent helix conformation (enabled by a conserved glycine) moving the second half of the helix closer to the membrane (Fig. 8.2d) (Natale et al. 2021; Dong et al. 2015). This upward movement of M4 increases the contact surface with M2, subsequently closing the fenestration side (Dong et al. 2015). Initial studies denoted the up state as the active conformation, while the down state was associated with the less conductive or inhibited state (Dong et al. 2015). However, follow-up studies showed that this association is not absolute and regulation of K_{2P} channel activity by conformational changes of M4 is much more complex (Natale et al. 2021). For example, some gain-of-function mutations of K_{2P}2.1 and K_{2P}4.1 induce conformational changes in the M4 helix that are closer to the down than to the up conformation (Lolicato et al. 2014). Furthermore, different mechanisms are discussed that might be responsible for an increased channel activity in the up state compared to the down state (Natale et al. 2021). One explanation is the higher accessibility of the fenestration side in the down state, which increases the affinity of channel inhibitors including endogenous lipids (Riel et al. 2022; Dong et al. 2015). In line with these results, the structure of $K_{2P}2.1$ (PDB 8DE9) shows an additional electron density of a phospholipid entering the central ion channel pore via the fenestration side, indicating a lipid block mechanism (Schmidtpeter et al. 2023). Nevertheless, a general lipid block in the down state cannot be assumed, since many studies describe down states with preserved conductance (Schmidtpeter et al. 2023). Another hypothesis for some K_{2P} channels is that reorientation of M2 and M4 might change the hydrophobic surface of the inner cavity beneath the selectivity filter, enabling a hydrophobic gating mechanism (Aryal et al. 2014). However, it is not clear how these changes in hydrophobicity are linked to distinct M4 conformations (Aryal et al. 2014). Last, conformational changes of M4 are assumed to influence the stability of the selectivity filter, subsequently influencing the conductance process and therefore changing the channel activity (Türkaydin et al. 2024; Schreiber et al. 2023). In summary, many studies showed correlations between M4 mobility as well as conformational changes and the K_{2P} channel activity; however, molecular mechanisms clearly connecting distinct M4 states to channel activity states are missing. Further experimental data are needed to uncover the exact regulatory mechanisms.

8.6 Cap Domain and Extracellular Ion Pathway

Compared to other K^+ channels, K_{2P} channels possess an extended extracellular domain, which is formed by the assembly of two helices (E1 and E2) from each channel monomer, which are located between M1 and the first pore-forming domain (P1 and SF1) (Miller and Long 2012). The so-called cap domain shields the central ion channel pore, leading to a unique ion pathway at the extracellular side and reducing sensitivity to extracellular agents (Fig. 8.3a) (Brohawn et al. 2012). Furthermore, the cap domain serves important regulatory and functional roles, influencing channel assembly, and gating.

The assembly of the cap domain is facilitated by hydrophobic interactions at the inner leaflet with E2 helices forming the central cap domain (Goldstein et al. 2016). For most K_{2P} channels, cap domain dimerization is also assisted by a cysteine bridge at the top of the domain (Goldstein et al. 2016; Lesage et al. 1996b). However, this cysteine is not present in members of the TASK ($K_{2P}3.1$, $K_{2P}9.1$, and $K_{2P}15.1$) and THIK ($K_{2P}12.1$ and $K_{2P}13.1$) subfamily, which are able to form a functional cap domain as well (Goldstein et al. 2016). Furthermore, assembly of the cap domain within the channel dimer seems to be variable since domain-swapped as well as non-swapped structures are reported (Zúñiga and Zúñiga 2016). Molecular dynamics simulations indicate that at least for $K_{2P}4.1$, energetic differences between these two assembling modes exist, which show a preference for the domain-swapped cap conformation (Navarro-Retamal and Caballero 2018).

Independently from the assembling mode, localization of the cap domain restricts K^+ ions to cross the channel in a direct perpendicular manner (Miller and Long 2012). Under physiological conditions, K^+ ions exit K_{2P} channels through one of the two water-accessible extracellular ion pathways (EIP), which are formed laterally at the bottom of the cap domain (Miller and Long 2012). Because of this shielded ion pathway, most classical blockers for tetrameric K^+ channels are ineffective at K_{2P} channels (Brohawn et al. 2012). On the other hand, small inhibitors like Ruthenium Red can enter the EIP, subsequently inhibiting the ion permeation (Pope et al. 2020). The inhibitory effect is restricted to K_{2P} channels like $K_{2P}10.1$ or the $K_{2P}2.1$ mutant I110D, which possess an acidic residue at the bottom of the E2 helix immobilizing the positively charged compound directly above the selectivity filter, leading to obstruction of the EIP by Ruthenium Red (Fig. 8.3b) (Pope et al. 2020). While the negatively charged

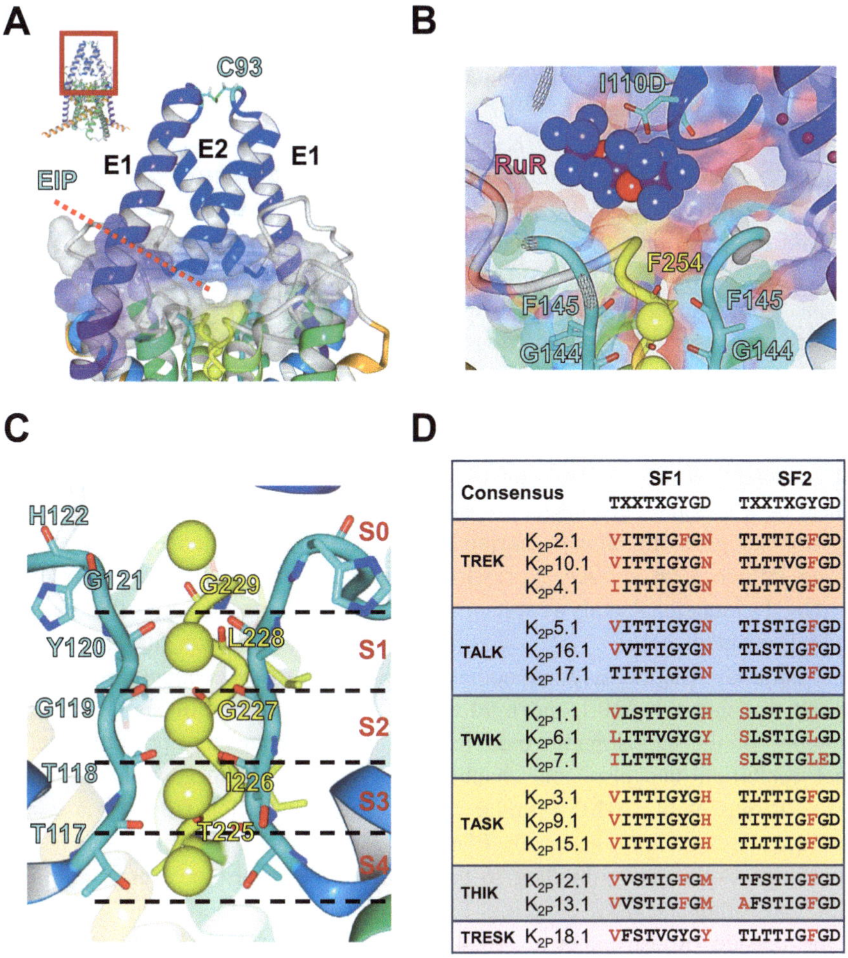

Consensus		SF1 TXXTXGYGD	SF2 TXXTXGYGD
TREK	$K_{2P}2.1$	VITTIGFGN	TLTTIGFGD
	$K_{2P}10.1$	VITTIGYGN	TLTTVGFGD
	$K_{2P}4.1$	IITTIGYGN	TLTTVGFGD
TALK	$K_{2P}5.1$	VITTIGYGN	TISTIGFGD
	$K_{2P}16.1$	VVTTIGYGN	TLSTIGFGD
	$K_{2P}17.1$	TITTIGYGN	TLSTVGFGD
TWIK	$K_{2P}1.1$	VLSTTGYGH	SLSTIGLGD
	$K_{2P}6.1$	LITTVGYGY	SLSTIGLGD
	$K_{2P}7.1$	ILTTTGYGH	SLSTIGLED
TASK	$K_{2P}3.1$	VITTIGYGH	TLTTIGFGD
	$K_{2P}9.1$	VITTIGYGH	TITTIGFGD
	$K_{2P}15.1$	VITTIGYGH	TLTTIGFGD
THIK	$K_{2P}12.1$	VVSTIGFGM	TFSTIGFGD
	$K_{2P}13.1$	VVSTIGFGM	AFSTIGFGD
TRESK	$K_{2P}18.1$	VFSTVGYGY	TLTTIGFGD

Fig. 8.3 Cap domain, extracellular ion pathway (EIP) and selectivity filter. (**a**) Illustration of the cap domain using Cryo-EM structure of $K_{2P}2.1$ zebrafish ortholog (PDB 8DE9 (Schmidtpeter et al 2023)). The cap domain is formed at the extracellular side by E1 and E2 helices. The molecular surface of lower cap domain residues is shown in gray. For all K_{2P} channels except for members of the TASK and THIK subfamilies dimerization of the cap via hydrophobic interactions is accompanied by a cysteine bridge at the top of the domain. Positioning directly above the selectivity filter induces lateral ion permeation through the extracellular ion pathways (EIP) at both sides of the cap domain. (**b**) Crystal structure of murine $K_{2P}2.1$ channel mutant I110D with bound Ruthenium red (RuR; PDB 6V3I (Pope et al. 2020)). RuR binds within the EIP (molecular surface) above the selectivity filter (SF1 cyan and SF2 yellow). Binding as well as inhibitory activity is facilitated by ionic interactions with aspartate residue I110D. (**c**) Depiction of selectivity filter from $K_{2P}1.1$ crystal structure (PDB 3UKM (Miller and Long 2012)). The selectivity filter is formed by carbonyl backbone oxygens (red) from SF1 (cyan) and SF2 (yellow) and encompasses four binding sites (S1–S4) for K^+ ions. (**d**) Protein sequences of SF1 and SF2 from $K_{2P}1.1$ to $K_{2P}18.1$ compared to the consensus motif for K^+ channels (Salinas et al. 1999; Natale et al. 2021; Heginbotham et al. 1994). Deviations are highlighted in red

residue at the bottom of E2 is responsible for Ruthenium Red sensitivity, other E2 residues as well as residues in spatial proximity are directly involved in the formation of the ion pathway (Goldstein et al. 2016; Luo et al. 2017). Therefore, it is not surprising that mutations within this region lead to strongly reduced conductivity (Goldstein et al. 2016).

The top of the cap domain is also known as a region for possible asparagine-linked glycosylation (N-glycosylation), which can influence surface expression of K_{2P} channels like $K_{2P}2.1$, $K_{2P}17.1$, and $K_{2P}18.1$ (Wiedmann et al. 2019a, b; Egenberger et al. 2010). In case of $K_{2P}3.1$, $K_{2P}9.1$, $K_{2P}13.1$, and $K_{2P}18.1$ protein sequences, an asparagine suitable for glycosylation is directly next to the cysteine, which assists the dimerization process (Goldstein et al. 2016). For $K_{2P}18.1$, N-glycosylation at this asparagine has a significant impact on surface expression (Egenberger et al. 2010). Mutations at this position reduced protein surface expression in a range from 20 to 80% depending on the protein sequence for human or mouse $K_{2P}18.1$ channel (Egenberger et al. 2010). However, the existence of a possible glycosylation site at the cap domain does not always correlate with a strong influence on surface expression as visualized by the mutational analysis at $K_{2P}3.1$ channels (Goldstein et al. 2016).

Recently, antibody fragments (nanobodies) were designed that target $K_{2P}10.1$ and bind to different regions of the cap domain, inducing ion channel inhibition but also ion channel activation (Rödström et al. 2024). Similar to Ruthenium red, the inhibitory effect of most nanobodies can be explained by binding to the cap domain and simultaneously clogging the EIP from the outside (Rödström et al. 2024). However, the mechanism of action for activating nanobodies (e.g. Nb-activator-67) needs to be further investigated (Rödström et al. 2024). Instead of EIP obstruction, these activating nanobodies might stabilize the EIP pathway or the selectivity filter by interactions with E1 and the SF2-M4 linker, which is known for its influence on K_{2P} channel gating (Rödström et al. 2024).

Obstruction of the EIP might also play a role in the gating modulation by extracellular acidification of some K_{2P} channels (Turney et al. 2022). Structural studies using $K_{2P}1.1$ show that extracellular acidification induces protonation of a histidine at the top of the selectivity filter inducing conformational changes that significantly lower the diameter of the EIP, preventing K^+ ion permeation (Turney et al. 2022). Interestingly, the histidine residue is conserved in other K_{2P} channels like $K_{2P}3.1$, $K_{2P}9.1$, and $K_{2P}18.1$. Further investigations are needed to verify a common mechanism of action for the observed EIP diameter obstruction (Turney et al. 2022).

8.7 Regulation of K^+ Conductance in K_{2P} Channels

Since K_{2P} channels belong to the K^+ channel superfamily, it is not surprising that they share several properties with other members like voltage-gated (K_v) or inward rectifying (K_{ir}) channels (Taura et al. 2021). On the other hand, K_{2P} channels possess unique structural features that support their differences in gating functionality (Miller and Long 2012; Lolicato et al. 2020). K_{2P} channels are sometimes named "leak" channels due to their ability to conduct K^+ ions under many physiological conditions giving the impression that they are unregulated pores selectively conducting K^+ ions (Cohen et al. 2009). However, this opinion is changing due to the vast number of physiological stimuli that can alter K_{2P} channel activity dramatically.

The core region for the ion conductance as well as for the K^+ ion specificity is the selectivity filter, which is also present in other K^+ ion channels (Fig. 8.3c) (Taura et al. 2021). The K_{2P} selectivity filter is formed by dimerization of two K_{2P} monomers at the SF1 and SF2 pore-forming loops (Miller and Long 2012). In contrast to this, other K^+ channels are built up as a tetramer in which each monomer contributes with only one pore-forming loop to the selectivity filter (Taura et al. 2021). The protein sequence within this region is highly conserved in most K^+ ion channels, generating a consensus motif (TXXTXGYGD) that is altered only by small variations (Heginbotham et al. 1994). The carbonyl oxygens of these peptide backbones create a narrow, four-sided pathway that perfectly fits dehydrated K^+ ions (Miller and Long 2012). Moreover, the spatial arrangement of the carbonyl oxygens mimics the specific hydration shell that surrounds K^+ ions in solution, allowing the permeation through the filter without generating a larger energetic barrier (Miller and Long 2012). The pathway diameter as well as the specific spatial arrangement of the hydration shell imitation are also responsible for the selectivity for K^+ ions over other physiological cations (Miller and Long 2012). For example, despite being smaller than K^+, Na^+ ions are not conducted under physiological conditions because they do not fit into the specific hydration shell geometry created by the selectivity filter strongly increasing the energy needed for ion permeation (Chatelain et al. 2024). However, some K_{2P} channels are able to conduct other ions like Tl^+ or Rb^+, which can be utilized to elucidate processes that are directly connected to the function of the selectivity filter (Schreiber et al. 2022b, 2023; Wright et al. 2019). Interestingly, selectivity for K^+ over Na^+ can be reduced by mutations at the top of the selectivity filter (Turney et al. 2022; O'Connell et al. 2005).

The arrangement of carbonyl oxygens forms four binding sites named S1–S4, which mimic the hydration shell of K^+ ions (Fig. 8.3c) (Lolicato et al. 2020; Zhou and MacKinnon 2003). To permeate through the selectivity filter, each dehydrated K^+ ion moves stepwise from one binding site to the next (Zhou and MacKinnon 2003; Mironenko et al. 2021). The general occupancy of the four K^+ binding sites depends on the physiological circumstances and is not fully understood so far (Lolicato et al. 2020). Robust and sustained conduction is assumed for a typical occupation of all four binding sites by K^+ generating strong electrostatic forces due to directly neighbored positive charges. Therefore, this hypothesis for ion permeation is known as hard or direct knock-on (Mironenko et al. 2021). However, the existence of a different binding site occupation under physiological conditions cannot be fully excluded (Mironenko et al. 2021). For example, an alternative hypothesis called soft knock-on assumes the presence of an alternating scheme of K^+ ions and water molecules permeating the selectivity filter (Mironenko et al. 2021). Recently, structures with occupations of less than four sites were described, which might be conductive as well (Lolicato et al. 2020). These results are in accordance with the observations that the selectivity filter can be fully depleted under certain conditions, indicating the necessity of different transition states with different occupation patterns (Schewe et al. 2016). Although further experimental data is needed, the direct knock-on model seems to be favorable for K_{2P} channels (Mironenko et al. 2021).

Despite the functional similarities for conduction with K_v and K_{ir}, K_{2P} channel selectivity filters have properties that distinguish them from other K^+ channels. For example, the protein sequence of the pore loops SF1 and SF2 differs slightly from the consensus motif previously defined for other K^+ channels (Fig. 8.3d) (Natale et al. 2021). Moreover, protein sequences of SF1 and SF2 are not identical in any K_{2P} monomer, creating asymmetric selectivity filters in all K_{2P} channels, while homotetrameric K_v or K_{ir} channels possess identical protein sequences in each selectivity filter loop (Lolicato et al. 2020). Compared to the selectivity filter consensus motif, SF1 of all K_{2P} channels except $K_{2P}17.1$ possesses a hydrophobic (valine, leucine) residue replacing the first threonine (Natale et al. 2021). Furthermore, the final aspartate residue in this motif, which is present in most K_v and K_{Ca} channels, is only present in the SF2 of K_{2P} channels, while in SF1 it is replaced with asparagine (N), histidine (H), tyrosine (Y), or methionine (M) (Taura et al. 2021; Natale et al. 2021). Further alterations are observed for the GYG motif, which is present in almost all K_v, $K_{Ca,}$ and K_{ir} channels (Taura et al. 2021). In SF2 of K_{2P} channels, this motif is replaced by GFG or GLG (Natale et al. 2021). In SF1, the GYG motif is preserved except for $K_{2P}2.1$, $K_{2P}12.1$, and $K_{2P}13.1$, which bear a GFG sequence as well (Natale et al. 2021).

The selectivity filter is not only important for the permeation process itself, it is also able to control and adapt the ion channel gating, a mechanism that controls the opening and closing of ion channels, determining when and how many ions can pass through a channel pore. Due to distinct conformational changes as well as movement cooperation of SF1 and SF2, the selectivity filter alters the permeation process, acting as the upper gate (Lolicato et al. 2020). This gating process is known from other K^+ channels and was previously termed C-type gating (Baukrowitz and Yellen 1996). Although the exact mechanism behind this gating process is still not fully understood, C-type gating is the main mechanism for activity modulation in K_{2P} channels (Lolicato et al. 2020). The conductance process through the selectivity filter is facilitated by electrostatic interactions between the π electrons of the oxygen atoms and the positive charge (Zhou and MacKinnon 2003). Due to the permeation of K^+ ions, attracting forces are altered over the complete conduction process, inducing backbone movements, subsequently disturbing the conformation that is needed for optimal K^+ coordination (Lolicato et al. 2020; Zhou and MacKinnon 2003). As a result, the permeation process transporting K^+ from one binding site to the next binding site can slow down or stop completely (Mironenko et al. 2021). Moreover, the induced fluctuations of the selectivity filter backbone can lead to loss of occupation in the upper K^+ binding sites S1 and S2 (Lolicato et al. 2020). In K_v channels, this process is also induced by long-lasting opening events, inducing inactivation due to selectivity filter fluctuations (Schreiber et al. 2022b). Recently, the induced movements in the asymmetric selectivity filter of $K_{2P}2.1$ were described as a combination of SF1 pinching and SF2 dilation in a concerted motion (Lolicato et al. 2020).

By stabilizing the selectivity filter, subsequently reducing these motions, the conductance process can be preserved, leading to increased permeation of K^+ ions (Lolicato et al. 2017). Therefore, influencing the selectivity filter fluctuations provides an opportunity to adapt the activity of K^+ channels to certain conditions (Lolicato et al. 2017). Indeed, C-type gating is heavily influenced by extrinsic conditions such as ion concentrations, ion type (e.g., K^+ vs. Rb^+), pH, or mutations in the selectivity filter sequence (Chatelain et al. 2024; Duprat 1997; Heginbotham et al. 1994; Pusch et al. 2000). In K_{2P} channels, even more modulatory pathways influencing C-type gating are known, rendering this mechanism as most crucial for controlling K_{2P} channel activity (Niemeyer et al. 2010; Dong et al. 2015; Piechotta et al. 2011). As previously mentioned, there is clear evidence that conformational orientation of the M4 helix impacts the conductance process through the selectivity filter (Natale et al. 2021; Türkaydin et al. 2024). However, the exact interactions between different conformational M4 states and the selectivity filter activity need further experimental data to specify the molecular mechanisms. Additionally, the extracellular SF1-M2 and SF2-M4 loops interfere with the selectivity filter movement, influencing the C-type gating as well (Natale et al. 2021). Compared to classical K_v channels, all K_{2P} channels except $K_{2P}18.1$ possess extended SF2-M4 loops enabling more pronounced interactions with the extracellular ion channel domains including the M3 glutamate network, which is crucial for gating modulation via extracellular stimuli (Natale et al. 2021; Lolicato et al. 2020). Finally, voltage changes influence C-type gating as well, although K_{2P} channels are not known as classical voltage-gated ion channels (Schewe et al. 2016). Under pronounced voltage changes, the selectivity filter can act as a check valve, which induces channel inactivation for inward-directed ion flow (Schewe et al. 2016).

Besides the upper gate, which is crucial for most K^+ channels, many K^+ channels possess a second gate at the intracellular side, which is called the lower or activation gate (Taura et al. 2021). In K_v channels, this gate is mainly controlled by changes in membrane potentials (Sun and MacKinnon 2020). Depolarization triggers a conformational change in the voltage sensor domain that induces ion channel opening by conformational changes of the lower S6 helices, subsequently allowing for ion permeation (Sun and MacKinnon 2020). Although most K_{2P} channels are dominantly controlled by C-type gating, recent studies suggest that some K_{2P} channels possess domains at the intracellular side that resemble a lower gate. So far, intracellular gate-like structures have been observed for $K_{2P}3.1$, $K_{2P}5.1$, $K_{2P}9.1$, and very recently $K_{2P}17.1$ (Li et al. 2020; Rödström et al. 2020; Neelsen et al. 2024). Interestingly, these gate-like structures are formed by reorientations of M4 helix residues, which emphasize the role of M4 for K_{2P} channel gating (Li et al. 2020; Rödström et al. 2020; Neelsen et al. 2024). In addition to these findings, the inner cavity formed by M2 and M4 residues might influence the conduction process as well. For $K_{2P}1.1$ channels, it was shown that the inner surface of this cavity is characterized by hydrophobic residues, inducing a dewetting of the region below the selectivity filter (Aryal et al. 2014). Consequently, full hydration of the inner cavity is restricted by the hydrophobic surface, creating an energetic barrier that slows down the ion permeation and therefore reduces the overall conductivity (Aryal et al. 2014).

In summary, K_{2P} channels are not classical "leak" channels that just provide a continuous K^+ outward current. As implicated by many different studies, they respond to a vast number of different stimuli, which variate K_{2P} channel activity over a broad spectrum by adapting C-type gating of the selectivity filter. However, other aspects like the presence of a second gate, the surface properties of the inner cavity as well as the obstruction of the extracellular ion pathway by reorientations within the cap domain influence the channel activity as well.

8.8 Physiological Modulation of K_{2P} Channels

K_{2P} channels play a crucial role in maintaining cellular excitability and overall physiological balance. In order to fulfill their functions in a plethora of physiological conditions K_{2P} channels are modulated by several cues like pH levels, membrane stretch, lipid environment, neurotransmitters and hormones, and the temperature (Lee et al. 2021; Natale et al. 2021). Indeed, these physiological functionalities may be intertwined as well. For example, K_{2P} channels, such as $K_{2P}2.1$ (TREK-1), $K_{2P}10.1$ (TREK-2), and $K_{2P}4.1$ (TRAAK), are activated by increases in temperature and are regulated by heat through temperature-sensitive mechanisms that modulate their gating and activity (Lamas et al. 2019). This heat sensitivity plays a critical role in physiological processes such as thermoregulation, pain perception, and neuronal excitability by adjusting cellular potassium conductance in response to temperature changes (Lamas et al. 2019). Thermal changes may directly influence these channels by promoting conformational changes or indirectly through lipid interactions in the surrounding membrane (Schmidtpeter et al. 2023; Lamas et al. 2019).

Another physiological stimulus that regulates several K_{2P} channels is the extracellular and intracellular H^+ concentration (pH). The K_{2P} channels of the TASK subfamily as well as $K_{2P}1.1$ (TWIK-1) and $K_{2P}16.1$ (TALK-1), exhibit distinct pH sensitivity that influences their gating and activity (Turney et al. 2022; Rajan et al. 2000; Duprat 1997). The TASK channels $K_{2P}3.1$ and $K_{2P}9.1$ are potently inhibited by acidic extracellular pH due to protonation of key histidine residues in their extracellular pore regions, which act as pH sensors stabilizing closed states under protonation (Rajan et al. 2000). The $K_{2P}1.1$ channels are less sensitive to pH changes but can show modulation under extreme pH conditions, typically mediated via residues in their extracellular domain or pore region (Chatelain et al. 2024). In contrast, alkaline extracellular pH activates $K_{2P}16.1$ channels via deprotonation of specific residues, which promotes channel opening (Girard et al. 2001). In general, the structural basis for the diverse pH sensitivities in these channels lies in the location and reactivity of critical amino acids in proximity to the selectivity filter and/or gating domains (Turney et al. 2022). These residues sense changes in pH by altering their charge, leading to conformational changes in the channel protein that modulate ion flow (Turney et al. 2022). These pH-dependent regulations enable these channels to play roles in maintaining cellular ion homeostasis and responding to physiological pH fluctuations.

Some K_{2P} channels can be modulated by specific ligands, such as neurotransmitters, membrane lipids including cholesterol, or other signaling molecules. Examples are $K_{2P}2.1$ (TREK-1), $K_{2P}4.1$ (TRAAK), and $K_{2P}10.1$ (TREK-2) channels, which are sensitive to various ligands, including polyunsaturated fatty acids (PUFAs), such as arachidonic acid, and phospholipids (Riel et al. 2022). Further, TREK channels are mechanosensitive and respond to mechanical forces such as membrane stretch or pressure (Sorum et al. 2024). These channels are activated when the lipid bilayer is deformed, leading to structural changes that shift channel conformations toward open states (Sorum et al. 2024). The mechanical forces may act directly on the channels through the surrounding membrane, where membrane tensions induce conformational alterations within transmembrane gating regions (Sorum et al. 2024). Additionally, interactions with specific lipids in the membrane, such as phosphatidylinositol phosphates, further modulate their mechanosensitivity (Sorum et al. 2024). Here, the interaction with the lipids may mediate activation by mechanical stretch and heat, making the lipids an integral part of the sensory transduction machinery. In line, specific lipid metabolites, like unsaturated fatty acids (e.g., arachidonic acid), have been shown to activate channels like $K_{2P}2.1$, leading to physiologically relevant hyperpolarization of the cell membrane (Riel et al. 2022). Several lipids may have individual functions on K_{2P} channels as well complicating the picture even further (Riel et al. 2022). Allosteric interactions of the M domains with the pore loop and the surrounding membrane environment determine the function of the pore domain. Thus, the composition of the lipid environment determines K_{2P} functions (Riel et al. 2022). For example, PIP_2 is a membrane phospholipid that plays a key role in regulating various ion channels, including the K_{2P} channels (Riel et al. 2022). PIP_2 is primarily located in the inner leaflet of the plasma membrane and is involved in numerous signaling pathways, including those that regulate K_{2P} ion channel activity (Riel et al. 2022). Not surprisingly, the direct interactions between PIP_2 and K_{2P} channels are crucial for modulating their gating and activity in response to physiological stimuli (Riel et al. 2022). Examples for PIP_2-regulated K_{2P}s are $K_{2P}2.1$, $K_{2P}3.1$, and $K_{2P}9.1$ (Riel et al. 2022). PIP_2 enhances the stability of the open states of these channels by binding to specific residues in the pore domain and cytoplasmic domains, particularly in the C-terminal region. This interaction allows PIP_2 to increase channel activity, thereby facilitating potassium ion efflux and stabilizing the resting membrane potential. In case of $K_{2P}3.1$ and $K_{2P}9.1$, PIP_2 binding mediates their sensitivity to changes in extracellular pH, whereas PIP_2 can modulate the proton sensitivity of these TASK channels, linking their gating to their acidic environments (Riel et al. 2022). PIP_2 binds to a conserved binding site within the cytoplasmic domain, allowing the channel to remain open even under conditions that would otherwise inhibit its activity (Chemin et al. 2005). PIP_2 binding to $K_{2P}2.1$ increases the channel's open probability and thus enhances its background K^+ current. Consistently, depletion of the PIP_2–pool reduces $K_{2P}2.1$ activity, while its replenishment restores normal channel function (Chemin et al. 2005).

Recently, the structural basis for the interplay of P1/M4 interface and the membrane has been revealed for the $K_{2P}13.1$ (THIK-1 and KCNK13) channel. A unique structure underlying $K_{2P}13.1$ function highlights the importance of the P1/M4 interface in control of K_{2P}s by lipids, as well as natural and synthetic agents (Roy-Chowdhury et al. 2024). Using cryo-electron microscopy (cryo-EM) structures of THIK-1 and mutant channels in lipid nanodiscs and detergent revealed a novel hydrophilic pore restriction and lipid cofactor site (Roy-Chowdhury et al. 2024).

8.9 Pharmacological Modulation of K_{2P} Channels

Paralleled with the understanding that K_{2P} channels are more than just leak channels, they also became interesting for drug development. Due to their contribution to specific physiological and pathophysiological processes, they were identified as possible targets for the treatment of neurological, cardiovascular, and immune-related diseases (Mathie et al. 2021). However, clear drug candidates are rare and often hampered by insufficient selectivity or activity, indicating that further investigations and structure optimizations are needed (Decher et al. 2021). Despite their small size of 70–80 kDa, K_{2P} channels offer several modulatory sides that can be addressed by different natural and synthetic compounds, which can either increase or decrease the ion channel activity to influence physiological function (Fig. 8.4) (Pope and Minor 2021).

On the extracellular side, the channel is shielded by the cap domain, hampering the direct inhibition of the central pore region

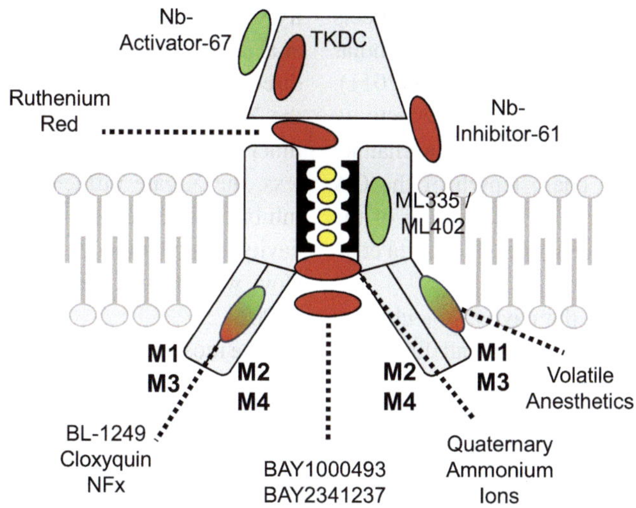

Fig. 8.4 Schematic depiction of binding sites for K_{2P} channel activators (green) and inhibitors (red)

(Brohawn et al. 2012). However, as previously mentioned, some compounds are able to block the extracellular ion pathway (EIP) by binding to the cap domain (e.g., Nb-inhibitor-61) or by entering the EIP clogging the ion pathway directly above the selectivity filter (e.g., Ruthenium red; Fig. 8.3b) (Pope et al. 2020; Rödström et al. 2024). Another mechanism of extracellular channel inhibition was described for members of the TREK channel subfamily (Luo et al. 2017). By binding to a small pocket between the E1 and E2 helix, the compound TKDC induces a separation of these helices, repositioning E2 of K_{2P}2.1, K_{2P}4.1, and K_{2P}10.1 in a way that E2 obstructs the conductive pathway leading to ion channel inhibition (Luo et al. 2017). Besides the mechanism of action, the authors could show that TKDC achieved antidepressant-like effects. These findings are in accordance with previously observed effects in K_{2P}2.1 knockout mice (Luo et al. 2017).

Most compounds influencing K_{2P} activity bind to the transmembrane or intracellular part of the ion channels, where they alter the accessibility of the central ion pore or modulate the stability of the selectivity filter, influencing C-type gating (Schewe et al. 2019; Schreiber et al. 2023; Lolicato et al. 2017). A good example for direct C-type gating modulation is given by the K_{2P}2.1 activators ML335 and ML402, that bind to a small pocket directly next to the selectivity filter (Lolicato et al. 2017). Binding to this pocket facilitates compound-assisted interactions between pore helix P1and the upper M4 segment, reducing selectivity filter fluctuations and therefore activating the ion channel (Lolicato et al. 2017). Both compounds are specific for K_{2P}2.1 and K_{2P}10.1, while K_{2P}4.1 is not sensitive to ML335 or ML402 (Lolicato et al. 2017). The reason for this subtype specificity is the presence of a cation-π interaction between the compound and a lysine side chain, which is not present in K_{2P}4.1 (Lolicato et al. 2017).

At the cytosolic side below the selectivity filter, the inner vestibule can harbor different binding sites for compounds that can activate or inhibit the channel activity (Schewe et al. 2016; Rödström et al. 2020; Piechotta et al. 2011; Proks et al. 2021). At the upper part of the cavity, quaternary ammonium (QA) ions can bind in close proximity to the S4 K^+ binding site (Piechotta et al. 2011). QA ions like tetraethyl ammonium (TEA) are useful pharmacological tools to study the characteristics of most K^+ channels (Piechotta et al. 2011). While K_{2P} channels are only poorly sensitive to applications of extracellular TEA, compounds with longer chain lengths like tetrapentyl ammonium (TPenA) can inhibit K_{2P} channels from the intracellular side with low micromolar to nanomolar IC_{50} values (Piechotta et al. 2011). Interestingly, differences in inhibitory activity can be observed depending on the K_{2P} channel subtype as well as the chain length of the QA ion, indicating differences in the shaping of the inner cavity for each K_{2P} channel (Piechotta et al. 2011).

On the lateral side of the cavity, K_{2P} channel activators as well as inhibitors can bind directly or in close proximity to the previously characterized fenestration side (Schewe et al. 2016; Schreiber et al. 2023; Proks et al. 2021). Together with the number of different modulators, the different modulatory mechanisms emphasize the importance of this modulatory site for the gating of K_{2P} channels (Natale et al. 2021). Depending on the binding mode of the compounds, they can stabilize a state with high or lower conductivity, leading to agonistic or antagonistic effects, which are predominantly related to modulation of the C-type gating (Schewe et al. 2019; Proks et al. 2021). Examples for agonists binding close or directly to the fenestration side are BL-1249, which activates members of the TREK channel subfamily, and the K_{2P}18.1-selective compounds cloxyquin and nitroxoline (Schewe et al. 2019; Schreiber et al. 2023). For inhibitory compounds, binding of norfluoxetine to the fenestration side was verified by crystallography (Dong et al. 2015). Moreover, many compounds including DCPIB, flufenamic acid, ML67–33, quinidine, or GI530159 are assumed to bind close to the fenestration side as well (Natale et al. 2021).

Subtype-specific modulation of K_{2P} channels is one of the major challenges, especially since many compounds targeting the fenestration side show modulatory effects at more than one member of the K_{2P} channel family (Decher et al. 2021). Higher specificity for K_{2P}3.1 and K_{2P}9.1 is achieved by ion channel inhibitors BAY1000493 and BAY2341237, which address the X-gate, a unique lower gate-like structure in these channels (Rödström et al. 2020). By crystallization of BAY1000493 together with K_{2P}3.1 as well as functional experiments it was shown that these inhibitors are bound within the center of the cavity obstructing the ion channel pore (Rödström et al. 2020). Interestingly, BAY1000493 does not directly interact with lower M4 residues that form the X-gate but more with the upper part of the inner cavity (Rödström et al. 2020). Nevertheless, the X-gate influences the activity of these inhibitors indirectly by trapping them in the inner cavity, which induces sustained inhibition with slow washout of the inhibitor (Rödström et al. 2020).

The most studied K_{2P} channel modulators with clinical relevance are volatile anesthetics (Mathie et al. 2021). Drugs such as halothane, isoflurane, or sevoflurane can induce increased activity for several K_{2P} channels like K_{2P}2.1, K_{2P}3.1, K_{2P}9.1, K_{2P}10.1, and K_{2P}18.1 (Schreiber et al. 2022a). The exact binding site for these drugs is still not fully discovered, but functional studies suggest that they might bind to a side at the outer channel surface formed by residues from M2, M3, and M4 (Wague et al. 2020). Although the EC_{50} values of volatile anesthetics at K_{2P} channels are in the high micromolar range, the amplification of hyperpolarizing potassium currents is still assumed as one important component of the anesthetic effect, inhibiting neuronal firing activity (Schreiber et al. 2022a; Mathie et al. 2021). Interestingly, not all K_{2P} channel currents get amplified by the presence of volatile anesthetics. For example, K_{2P}12.1 and K_{2P}13.1 are inhibited and not activated by halothane, giving this subfamily of K_{2P} channels the name "halothane-inhibited" (THIK) (Rajan et al. 2001).

Many rational opportunities to target K_{2P} channels therapeutically exist that are not limited to general anesthesia including neurological, cardiac as well as immune-related diseases (Mathie et al. 2021). For neurological diseases like epilepsy, neuropathic pain, or migraine, channel activators for the TREK and TRESK subfamilies are of particular interest, which can reduce neuronal hyperexcitability and therefore support the treatment of these diseases (Huntemann et al. 2022). Furthermore, pharmacological modulation of K_{2P} channels like K_{2P}3.1 might be a new strategy for the treatment of patients with atrial fibrillation or vascular hypertension (Wiedmann et al. 2021). Last, activation or inhibition of distinct K_{2P} channels is of particular interest for oncologic and autoimmune-related drug therapies (Lee et al. 2021). For example, recent studies identified $K_{2P}18.1$ as a regulatory switch for the proliferation and differentiation of regulatory T cells in the thymus, which are associated with immunosuppressive properties impacting the therapy of certain autoimmune-related diseases as well as distinct cancer types (Ruck et al. 2022).

8.10 Summary and Outlook

The family of K_{2P} channels is a fascinating and important member of the K^+ channel superfamily. Despite their genetic similarity with other K^+ channels, they possess several unique structural and functional features that clearly separate them from K_v, K_{Ca}, or K_{ir} channels. Typical structural characteristics of the K_{2P} channel family include their dimeric assembly, their cap domain, their asymmetric selectivity filter, as well as their four transmembrane helices per monomer. Due to their ability to conduct K^+ ions under many different physiological conditions K_{2P} channels have been characterized as leak potassium channels. However, this characterization is changing now rendering K_{2P} channels more as highly modulated ion channels, that adapt their activity to many different stimuli. While the overall structural appearance of K_{2P} channels is highly similar, the different members of the six different subfamilies clearly differ in their functionality as well as in their response to different stimuli, open up a broad spectrum to contribute to physiological processes in all kinds of tissues. The functional versatility of K_{2P} channels is even more boosted by their ability to form heterodimers with other K_{2P} channel-forming subunits. However, the physiological relevance of heterodimeric K_{2P} channels is poorly understood and needs to be extended by structural as well as functional analyses. Moreover, the understanding of gating modulation by interactions between the transmembrane helices and the selectivity filter is still not fully understood hampering the process of targeted compound development. New ion channel modulators could be used either for further research but also for treatment applications. Despite the vast number of potential applications for drug therapy, several challenges for specific drug development remain (Decher et al. 2021). The similarities in K_{2P} monomer protein sequence as well as the complexity of K_{2P} channel heterodimerization can complicate drug design, as selective targeting is essential for drug therapy. Additionally, the precise understanding of K_{2P} channel physiology and pathophysiology needs to be extended to avoid severe side effects induced by altered K_{2P} channel activity. Future studies should focus on elucidating the specific mechanisms by which pharmacological agents affect K_{2P} channels and how patients with specific diseases could benefit from this compound. The integration of high-throughput screening techniques, advanced molecular biology methods and high-resolution imaging may facilitate the discovery of new therapeutic agents with a precise pharmacological profile.

References

Abraham DM, Lee TE, Watson LJ et al (2018) The two-pore domain potassium channel TREK-1 mediates cardiac fibrosis and diastolic dysfunction. J Clin Invest 128:4843–4855

Aryal P, Abd-Wahab F, Bucci G, Sansom MSP, Tucker SJ (2014) A hydrophobic barrier deep within the inner pore of the TWIK-1 K2P potassium channel. Nat Commun 5:4377

Ashmole I, Goodwin P, Stanfield P (2001) TASK-5, a novel member of the tandem pore K + channel family. Pflügers Arch 442:828–833

Ávalos Prado P, Chassot A-A, Landra-Willm A, Sandoz G (2022) Regulation of two-pore-domain potassium TREK channels and their involvement in pain perception and migraine. Neurosci Lett 773:136494

Bang H, Kim Y, Kim D (2000) TREK-2, a new member of the mechanosensitive tandem-pore K+ channel family. J Biol Chem 275:17412–17419

Barel O, Shalev SA, Ofir R et al (2008) Maternally inherited Birk Barel mental retardation dysmorphism syndrome caused by a mutation in the genomically imprinted potassium channel KCNK9. Am J Hum Genet 83:193–199

Barriere H, Belfodil R, Rubera I, Tauc M, Lesage F, Poujeol C, Guy N, Barhanin J, Poujeol P (2003) Role of TASK2 potassium channels regarding volume regulation in primary cultures of mouse proximal tubules. J Gen Physiol 122:177–190

Baukrowitz T, Yellen G (1996) use-dependent blockers and exit rate of the last ion from the multi-ion pore of a K^+ channel. Science (1979) 271:653–656

Bayliss DA, Barrett PQ (2008) Emerging roles for two-pore-domain potassium channels and their potential therapeutic impact. Trends Pharmacol Sci 29:566–575

Bichet D, Blin S, Feliciangeli S, Chatelain FC, Bobak N, Lesage F (2015) Silent but not dumb: how cellular trafficking and pore gating modulate expression of TWIK1 and THIK2. Pflugers Arch 467:1121–1131

Bittner S, Ruck T, Schuhmann MK et al (2013) Endothelial TWIK-related potassium channel-1 (TREK1) regulates immune-cell trafficking into the CNS. Nat Med 19:1161–1165

Blin S, Chatelain FC, Feliciangeli S, Kang D, Lesage F, Bichet D (2014) Tandem pore domain halothane-inhibited K+channel subunits THIK1 and THIK2 assemble and form active channels. J Biol Chem 289:28202–28212

Bockenhauer D, Nimmakayalu MA, Ward DC, Goldstein SAN, Gallagher PG (2000) Genomic organization and chromosomal localization of the murine 2 P domain potassium channel gene Kcnk8: conservation of gene structure in 2 P domain potassium channels. Gene 261:365–372

Brohawn SG, del Mármol J, MacKinnon R (2012) Crystal structure of the human K2P TRAAK, a lipid- and mechano-sensitive K + ion channel. Science (1979) 335:436–441

Buehler PK, Bleiler D, Tegtmeier I, Heitzmann D, Both C, Georgieff M, Lesage F, Warth R, Thomas J (2017) Abnormal respiration under hyperoxia in TASK-1/3 potassium channel double knockout mice. Respir Physiol Neurobiol 244:17–25

Bürli RW, Doyle KJ, Dickson L et al (2024) Discovery of CVN293, a brain permeable KCNK13 (THIK-1) inhibitor suitable for clinical assessment. ACS Med Chem Lett 15:646–652

Chai S, Wan X, Ramirez-Navarro A, Tesar PJ, Kaufman ES, Ficker E, George AL, Deschênes I (2018) Physiological genomics identifies genetic modifiers of long QT syndrome type 2 severity. J Clin Invest 128:1043–1056

Chatelain FC, Gilbert N, Bichet D, Jauch A, Feliciangeli S, Lesage F, Bignucolo O (2024) Mechanistic basis of the dynamic response of TWIK1 ionic selectivity to pH. Nat Commun. https://doi.org/10.1038/s41467-024-48067-w

Chavez RA, Gray AT, Zhao BB, Kindler CH, Mazurek MJ, Mehta Y, Forsayeth JR, Yost CS (1999) TWIK-2, a new weak inward rectifying member of the tandem pore domain potassium channel family. J Biol Chem 274:7887–7892

Chemin J, Patel AJ, Duprat F, Lauritzen I, Lazdunski M, Honoré E (2005) A phospholipid sensor controls mechanogating of the K+ channel TREK-1. EMBO J 24:44–53

Cohen A, Ben-Abu Y, Zilberberg N (2009) Gating the pore of potassium leak channels. Eur Biophys J 39:61–73

Cunningham KP, Holden RG, Escribano-Subias PM, Cogolludo A, Veale EL, Mathie A (2019) Characterization and regulation of wild-type and mutant TASK-1 two pore domain potassium channels indicated in pulmonary arterial hypertension. J Physiol 597:1087–1101

Czirják G, Enyedi P (2002) Formation of functional heterodimers between the TASK-1 and TASK-3 two-pore domain potassium channel subunits. J Biol Chem 277:5426–5432

Czirják G, Enyedi P (2006) Targeting of calcineurin to an NFAT-like docking site is required for the calcium-dependent activation of the background K+ channel, TRESK. J Biol Chem 281:14677–14682

de la Peña E, Mälkiä A, Vara H, Caires R, Ballesta JJ, Belmonte C, Viana F (2012) The influence of cold temperature on cellular excitability of hippocampal networks. PLoS One 7:e52475

Decher N, Maier M, Dittrich W, Gassenhuber J, Brüggemann A, Busch AE, Steinmeyer K (2001) Characterization of TASK-4, a novel member of the pH-sensitive, two-pore domain potassium channel family. FEBS Lett 492:84–89

Decher N, Wemhöner K, Rinné S et al (2011) Knock-out of the potassium channel TASK-1 leads to a prolonged QT interval and a disturbed QRS complex. Cell Physiol Biochem 28:77–86

Decher N, Rinné S, Bedoya M, Gonzalez W, Kiper AK (2021) Molecular pharmacology of K2P potassium channels. Cell Physiol Biochem 55:87–107

Domingues-Montanari S, Fernández-Cadenas I, del Río-Espinola A et al (2010) KCNK17 genetic variants in ischemic stroke. Atherosclerosis 208:203–209

Dong YY, Pike ACW, Mackenzie A et al (2015) K2P channel gating mechanisms revealed by structures of TREK-2 and a complex with Prozac. Science (1979) 347:1256–1259

Duan W, Hicks J, Makara MA, Ilkayeva O, Abraham DM (2020) TASK-1 and TASK-3 channels modulate pressure overload-induced cardiac remodeling and dysfunction. Am J Phys Heart Circ Phys 318:H566–H580

Duprat F (1997) TASK, a human background K+ channel to sense external pH variations near physiological pH. EMBO J 16:5464–5471

Egenberger B, Polleichtner G, Wischmeyer E, Döring F (2010) N-linked glycosylation determines cell surface expression of two-pore-domain K+ channel TRESK. Biochem Biophys Res Commun 391:1262–1267

Feliciangeli S, Tardy MP, Sandoz G, Chatelain FC, Warth R, Barhanin J, Bendahhou S, Lesage F (2010) Potassium channel silencing by constitutive endocytosis and intracellular sequestration. J Biol Chem 285:4798–4805

Feliciangeli S, Chatelain FC, Bichet D, Lesage F (2015) The family of K_{2P} channels: salient structural and functional properties. J Physiol 593:2587–2603

Fink M (1998) A neuronal two P domain K+ channel stimulated by arachidonic acid and polyunsaturated fatty acids. EMBO J 17:3297–3308

Fink M, Duprat F, Lesage F, Reyes R, Romey G, Heurteaux C, Lazdunski M (1996) Cloning, functional expression and brain localization of a novel unconventional outward rectifier K+ channel. EMBO J 15:6854–6862

Friedrich C, Rinné S, Zumhagen S, Kiper AK, Silbernagel N, Netter MF, Stallmeyer B, Schulze-Bahr E, Decher N (2014) Gain-of-function mutation in <scp>TASK</scp> -4 channels and severe cardiac conduction disorder. EMBO Mol Med 6:937–951

García G, Méndez-Reséndiz KA, Oviedo N, Murbartián J (2021) PKC- and PKA-dependent phosphorylation modulates TREK-1 function in naïve and neuropathic rats. J Neurochem 157:2039–2054

Girard C, Duprat F, Terrenoire C, Tinel N, Fosset M, Romey G, Lazdunski M, Lesage F (2001) Genomic and functional characteristics of novel human pancreatic 2P domain K+ channels. Biochem Biophys Res Commun 282:249–256

Goldstein SAN, Price LA, Rosenthal DN, Pausch MH (1996) ORK1, a potassium-selective leak channel with two pore domains cloned from Drosophila melanogaster by expression in Saccharomyces cerevisiae

Goldstein SAN, Wang K-W, Ilan N, Pausch MH (1997) Sequence and function of the two P domain potassium channels: implications of an emerging superfamily. J Mol Med 76:13–20

Goldstein SAN, Bayliss DA, Kim D, Lesage F, Plant LD, Rajan S (2005) International Union of Pharmacology. LV. nomenclature and molecular relationships of two-P potassium channels. Pharmacol Rev 57: 527–540

Goldstein M, Rinné S, Kiper AK, Ramírez D, Netter MF, Bustos D, Ortiz-Bonnin B, González W, Decher N (2016) Functional mutagenesis screens reveal the 'cap structure' formation in disulfide-bridge free TASK channels. Sci Rep 6:19492

Harleton E, Besana A, Comas GM, Danilo P, Rosen TS, Argenziano M, Rosen MR, Robinson RB, Feinmark SJ (2013) Ability to induce atrial fibrillation in the Peri-operative period is associated with phosphorylation-dependent inhibition of TWIK protein-related acid-sensitive potassium channel 1 (TASK-1). J Biol Chem 288:2829–2838

He L, Ma Q, Wang Y, Liu X, Yuan Y, Zhang Y, Ou W, Liu L, Tan X, Wang X (2014) Association of Variants in KCNK17 gene with ischemic stroke and cerebral hemorrhage in a Chinese population. J Stroke Cerebrovasc Dis 23:2322–2327

Heginbotham L, Lu Z, Abramson T, MacKinnon R (1994) Mutations in the K+ channel signature sequence. Biophys J 66:1061–1067

Heurteaux C, Guy N, Laigle C et al (2004) TREK-1, a K+ channel involved in neuroprotection and general anesthesia. EMBO J 23:2684–2695

Hodgkin AL, Huxley AF (1952) A quantitative description of membrane current and its application to conduction and excitation in nerve. J Physiol 117:500–544

Honoré E (2007) The neuronal background K2P channels: focus on TREK1. Nat Rev Neurosci 8:251–261

Houdinet G, Guerrero-Galán C, Rose BD, Garcia K, Zimmermann SD (2023) Secrets of the fungus-specific potassium channel TOK family. Trends Microbiol 31:511–520

Huang W, Ke Y, Zhu J et al (2021) TRESK channel contributes to depolarization-induced shunting inhibition and modulates epileptic seizures. Cell Rep 36:109404

Huntemann N, Bittner S, Bock S, Meuth SG, Ruck T (2022) Mini-review: two brothers in crime—the interplay of TRESK and TREK in human diseases. Neurosci Lett 769:136376

Imbrici P, Nematian-Ardestani E, Hasan S, Pessia M, Tucker SJ, D'Adamo MC (2020) Altered functional properties of a missense variant in the TRESK K+ channel (KCNK18) associated with migraine and intellectual disability. Pflugers Arch 472:923–930

Julio-Kalajzić F, Villanueva S, Burgos J, Ojeda M, Cid LP, Jentsch TJ, Sepúlveda FV (2018) K $_{2P}$ TASK-2 and KCNQ1–KCNE3 K $^{+}$ channels are major players contributing to intestinal anion and fluid secretion. J Physiol 596:393–407

Kamuene JM, Xu Y, Plant LD (2021) The pharmacology of two-pore domain potassium channels. In: Pharmacology of potassium channels. pp 417–443

Kang D, Kim D (2006) TREK-2 (K2P10.1) and TRESK (K2P18.1) are major background K+ channels in dorsal root ganglion neurons. Am J Physiol Cell Physiol 291:C138–C146

Kang D, Mariash E, Kim D (2004) Functional expression of TRESK-2, a new member of the tandem-pore K + channel family. J Biol Chem 279:28063–28070

Ketchum KA, Joiner WJ, Sellers AJ, Kaczmarek LK, Goldstein SAN (1995) A new family of outwardly rectifying potassium channel proteins with two pore domains in tandem. Nature 376:690–695

Kilisch M, Lytovchenko O, Schwappach B, Renigunta V, Daut J (2015) The role of protein–protein interactions in the intracellular traffic of the potassium channels TASK-1 and TASK-3. Pflugers Arch 467:1105–1120

Kim D, Gnatenco C (2001) TASK-5, a new member of the tandem-pore K+ channel family. Biochem Biophys Res Commun 284:923–930

Kim D, Cavanaugh EJ, Kim I, Carroll JL (2009) Heteromeric TASK-1/TASK-3 is the major oxygen-sensitive background K $^{+}$ channel in rat carotid body glomus cells. J Physiol 587:2963–2975

Kitagawa MG, Reynolds JO, Durgan D, Rodney G, Karmouty-Quintana H, Bryan R, Pandit LM (2019) Twik-2 $^{-/-}$ mouse demonstrates pulmonary vascular heterogeneity in intracellular pathways for vasocontractility. Physiol Rep 7:e13950

Koh SD, Monaghan K, Sergeant GP, Ro S, Walker RL, Sanders KM, Horowitz B (2001) TREK-1 regulation by nitric oxide and cGMP-dependent protein kinase. J Biol Chem 276:44338–44346

L'Hoste S, Poet M, Duranton C, Belfodil R, Barriere H, Rubera I, Tauc M, Poujeol C, Barhanin J, Poujeol P (2007) Role of TASK2 in the control of apoptotic volume decrease in proximal kidney cells. J Biol Chem 282:36692–36703

Lamas JA, Rueda-Ruzafa L, Herrera-Pérez S (2019) Ion channels and Thermosensitivity: TRP, TREK, or Both? Int J Mol Sci. https://doi.org/10.3390/ijms20102371

Lee LM, Müntefering T, Budde T, Meuth SG, Ruck T (2021) Pathophysiological role of K2P channels in human diseases. Cell Physiol Biochem 55:65–86

Lesage F, Guillemare E, Fink M, Duprat F, Lazdunski M, Romey G, Barhanin J (1996a) TWIK-1, a ubiquitous human weakly inward rectifying K+ channel with a novel structure. EMBO J 15:1004–1011

Lesage F, Reyes R, Fink M, Duprat F, Guillemare E, Lazdunski M (1996b) Dimerization of TWIK-1 K+ channel subunits via a disulfide bridge. EMBO J 15:6400–6407

Li B, Rietmeijer RA, Brohawn SG (2020) Structural basis for pH gating of the two-pore domain K+ channel TASK2. Nature 586:457–462

Liu J-P, Jing H-B, Xi K, Zhang Z-X, Jin Z-R, Cai S-Q, Tian Y, Cai J, Xing G-G (2021) Contribution of TRESK two-pore domain potassium channel to bone cancer-induced spontaneous pain and evoked cutaneous pain in rats. Mol Pain 17:174480692110232

Lloyd EE, Crossland RF, Phillips SC, Marrelli SP, Reddy AK, Taffet GE, Hartley CJ, Bryan RM (2011) Disruption of K2P6.1 produces vascular dysfunction and hypertension in mice. Hypertension 58:672–678

Lolicato M, Riegelhaupt PM, Arrigoni C, Clark KA, Minor DL (2014) Transmembrane helix straightening and buckling underlies activation of mechanosensitive and thermosensitive K2P channels. Neuron 84:1198–1212

Lolicato M, Arrigoni C, Mori T, Sekioka Y, Bryant C, Clark KA, Minor DL (2017) K2P2.1 (TREK-1)-activator complexes reveal a cryptic selectivity filter binding site. Nature 547:364–368

Lolicato M, Natale AM, Abderemane-Ali F, Crottès D, Capponi S, Duman R, Wagner A, Rosenberg JM, Grabe M, Minor DL (2020) K2Pchannel C-type gating involves asymmetric selectivity filter order-disorder transitions. Sci Adv 6:1–14

Luo Q, Chen L, Cheng X et al (2017) An allosteric ligand-binding site in the extracellular cap of K2P channels. Nat Commun 8:378

Luque-Fernández V, Vanspauwen SK, Landra-Willm A, Arvedsen E, Besquent M, Sandoz G, Rasmussen HB (2024) An ankyrin G–binding motif mediates TRAAK periodic localization at axon initial segments of hippocampal pyramidal neurons. Proc Natl Acad Sci. https://doi.org/10.1073/pnas.2310120121

Mant A, Elliott D, Eyers PA, O'Kelly IM (2011) Protein kinase a is central for forward transport of two-pore domain potassium channels K2P3.1 and K2P9.1. J Biol Chem 286:14110–14119

Mathie A, Veale EL (2007) Therapeutic potential of neuronal two-pore domain potassium-channel modulators. Curr Opin Investig Drugs 8:555–562

Mathie A, Veale EL, Cunningham KP, Holden RG, Wright PD (2021) Two-pore domain potassium channels as drug targets: anesthesia and beyond. Annu Rev Pharmacol Toxicol 61:401–420

Medhurst AD, Rennie G, Chapman CG, Meadows H, Duckworth MD, Kelsell RE, Gloger II, Pangalos MN (2001) Distribution analysis of human two pore domain potassium channels in tissues of the central nervous system and periphery. Mol Brain Res 86:101–114

Mi Hwang E, Kim E, Yarishkin O et al (2014) A disulphide-linked heterodimer of TWIK-1 and TREK-1 mediates passive conductance in astrocytes. Nat Commun. https://doi.org/10.1038/ncomms4227

Miller AN, Long SB (2012) Crystal structure of the human two–pore domain potassium channel K2P1. Science (1979) 335:432–436

Mirkovic K, Palmersheim J, Lesage F, Wickman K (2012) Behavioral characterization of mice lacking Trek channels. Front Behav Neurosci. https://doi.org/10.3389/fnbeh.2012.00060

Mironenko A, Zachariae U, de Groot BL, Kopec W (2021) The persistent question of potassium channel permeation mechanisms. J Mol Biol 433:167002

Nagaraj C, Tang B, Balint Z et al (2013) Src tyrosine kinase is crucial for potassium channel function in human pulmonary arteries. Eur Respir J 41:85–95

Natale AM, Deal PE, Minor DL (2021) Structural insights into the mechanisms and pharmacology of K2P potassium channels. J Mol Biol 433:166995

Navarro-Retamal C, Caballero J (2018) Energetic differences between non-domain-swapped and domain-swapped chain connectivities in the K2P potassium channel TRAAK. RSC Adv 8:26610–26618

Neelsen LC, Riel EB, Rinné S et al (2024) Ion occupancy of the selectivity filter controls opening of a cytoplasmic gate in the K2P channel TALK-2. Nat Commun. https://doi.org/10.1038/s41467-024-51812-w

Nematian-Ardestani E, Abd-Wahab F, Chatelain FC, Sun H, Schewe M, Baukrowitz T, Tucker SJ (2020) Selectivity filter instability dominates the low intrinsic activity of the TWIK-1 K2P K+ channel. J Biol Chem 295:610–618

Niemeyer MI, Cid LP, Peña-Münzenmayer G, Sepúlveda FV (2010) Separate gating mechanisms mediate the regulation of K2P potassium channel TASK-2 by intra- and extracellular pH. J Biol Chem 285:16467–16475

O'Connell AD, Morton MJ, Sivaprasadarao A, Hunter M (2005) Selectivity and interactions of Ba $^{2+}$ and Cs $^{+}$ with wild-type and mutant TASK1 K $^{+}$ channels expressed in *Xenopus* oocytes. J Physiol 562:687–696

O'Kelly I, Goldstein SAN (2008) Forward transport of K $_{2P}$ 3.1: mediation by 14-3-3 and COPI, modulation by p11. Traffic 9:72–78

Schmidtpeter PAM, Petroff JT, Khajoueinejad L, Wague A, Frankfater C, Cheng WWL, Nimigean CM, Riegelhaupt PM (2023) Membrane phospholipids control gating of the mechanosensitive potassium leak channel TREK1. Nat Commun. https://doi.org/10.1038/s41467-023-36765-w

Patel AJ, Honoré E, Lesage F, Fink M, Romey G, Lazdunski M (1999) Inhalational anesthetics activate two-pore-domain background K+ channels. Nat Neurosci 2:422–426

Pereira V, Busserolles J, Christin M et al (2014) Role of the TREK2 potassium channel in cold and warm thermosensation and in pain perception. Pain 155:2534–2544

Pergel E, Tóth DJ, Baukál D, Veres I, Czirják G (2024) The ubiquitin ligase adaptor NDFIP1 interacts with TRESK and negatively regulates the background K+ current. Int J Mol Sci 25:8879

Pettingill P, Weir GA, Wei T et al (2019) A causal role for TRESK loss of function in migraine mechanisms. Brain 142:3852–3867

Piechotta PL, Rapedius M, Stansfeld PJ et al (2011) The pore structure and gating mechanism of K2P channels. EMBO J 30:3607–3619

Plant LD, Dementieva IS, Kollewe A, Olikara S, Marks JD, Goldstein SAN (2010) One SUMO is sufficient to silence the dimeric potassium channel K2P1. Proc Natl Acad Sci 107:10743–10748

Plant LD, Zuniga L, Araki D, Marks JD, Goldstein SAN (2012) SUMOylation silences heterodimeric TASK potassium channels containing K2P1 subunits in cerebellar granule neurons. Sci Signal. https://doi.org/10.1126/scisignal.2003431

Pope L, Minor DL (2021) The polysite pharmacology of TREK K2P channels, pp 51–65

Pope L, Lolicato M, Minor DL (2020) Polynuclear ruthenium amines inhibit K2P channels via a "finger in the dam" mechanism. Cell Chem Biol 27:511–524.e4

Proks P, Schewe M, Conrad LJ, Rao S, Rathje K, Rödström KEJ, Carpenter EP, Baukrowitz T, Tucker SJ (2021) Norfluoxetine inhibits TREK-2 K2P channels by multiple mechanisms including state-independent effects on the selectivity filter gate. J Gen Physiol 153:1–13

Pusch M, Bertorello L, Conti F (2000) Gating and Flickery block differentially affected by rubidium in homomeric KCNQ1 and heteromeric KCNQ1/KCNE1 potassium channels. Biophys J 78:211–226

Rajan S, Wischmeyer E, Xin Liu G, Preisig-Müller R, Daut J, Karschin A, Derst C (2000) TASK-3, a novel tandem pore domain acid-sensitive K+ channel. J Biol Chem 275:16650–16657

Rajan S, Wischmeyer E, Karschin C, Preisig-Müller R, Grzeschik K-H, Daut J, Karschin A, Derst C (2001) THIK-1 and THIK-2, a novel subfamily of tandem pore domain K+ channels. J Biol Chem 276:7302–7311

Rajan S, Plant LD, Rabin ML, Butler MH, Goldstein SAN (2005) Sumoylation silences the plasma membrane leak K+ channel K2P1. Cell 121:37–47

Reed AP, Bucci G, Abd-Wahab F, Tucker SJ (2016) Dominant-negative effect of a missense variant in the TASK-2 (KCNK5) K+ channel associated with Balkan endemic nephropathy. PLoS One 11:e0156456

Renigunta V, Zou X, Kling S, Schlichthörl G, Daut J (2014) Breaking the silence: functional expression of the two-pore-domain potassium channel THIK-2. Pflugers Arch 466:1735–1745

Reyes R, Duprat F, Lesage F, Fink M, Salinas M, Farman N, Lazdunski M (1998) Cloning and expression of a novel pH-sensitive two pore domain K+ channel from human kidney. J Biol Chem 273:30863–30869

Riel EB, Jürs BC, Cordeiro S, Musinszki M, Schewe M, Baukrowitz T (2022) The versatile regulation of K2P channels by polyanionic lipids of the phosphoinositide and fatty acid metabolism. J Gen Physiol. https://doi.org/10.1085/jgp.202112989

Rinné S, Schick F, Vowinkel K, Schütte S, Krasel C, Kauferstein S, Schäfer MK-H, Kiper AK, Müller T, Decher N (2024) Potassium channel TASK-5 forms functional heterodimers with TASK-1 and TASK-3 to break its silence. Nat Commun 15:7548

Rödström KEJ, Kiper AK, Zhang W et al (2020) A lower X-gate in TASK channels traps inhibitors within the vestibule. Nature 582:443–447

Rödström KEJ, Cloake A, Sörmann J et al (2024) Extracellular modulation of TREK-2 activity with nanobodies provides insight into the mechanisms of K2P channel regulation. Nat Commun 15:4173

Roy-Chowdhury S, Jang S, Abderemane-Ali F, Naughton F, Grabe M, Minor DL (2024) Structure of the human K2P13.1(THIK-1) channel reveals a novel hydrophilic pore restriction and lipid cofactor site. bioRxiv. https://doi.org/10.1101/2024.06.26.600491

Ruck T, Bock S, Pfeuffer S et al (2022) K2P18.1 translates T cell receptor signals into thymic regulatory T cell development. Cell Res 32:72–88

Salinas M, Reyes R, Lesage F, Fosset M, Heurteaux C, Romey G, Lazdunski M (1999) Cloning of a new mouse two-P domain channel subunit and a human homologue with a unique pore structure. J Biol Chem 274:11751–11760

Sano Y, Inamura K, Miyake A, Mochizuki S, Kitada C, Yokoi H, Nozawa K, Okada H, Matsushime H, Furuichi K (2003) A novel two-pore domain K+ channel, TRESK, is localized in the spinal cord. J Biol Chem 278:27406–27412

Schewe M, Nematian-Ardestani E, Sun H, Musinszki M, Cordeiro S, Bucci G, de Groot BL, Tucker SJ, Rapedius M, Baukrowitz T (2016) A non-canonical voltage-sensing mechanism controls gating in K2P K+ channels. Cell 164:937–949

Schewe M, Sun H, Mert Ü et al (2019) A pharmacological master key mechanism that unlocks the selectivity filter gate in K channels. Science (1979) 363:875–880

Schreiber JA, Düfer M, Seebohm G (2022a) The special one: architecture, physiology and pharmacology of the TRESK Channel. Cell Physiol Biochem 56:663–684

Schreiber JA, Möller M, Zaydman M et al (2022b) A benzodiazepine activator locks Kv7.1 channels open by electro-mechanical uncoupling. Commun Biol 5:301

Schreiber JA, Derksen A, Goerges G et al (2023) Cloxyquin activates hTRESK by allosteric modulation of the selectivity filter. Commun Biol 6:745

Seyler C, Duthil-Straub E, Zitron E, Gierten J, Scholz E, Fink R, Karle C, Becker R, Katus H, Thomas D (2012) TASK1 (K_{2P} 3.1) K^{+} channel inhibition by endothelin-1 is mediated through rho kinase-dependent phosphorylation. Br J Pharmacol 165:1467–1475

Shkodra M, Caraceni A (2022) Treatment of neuropathic pain directly due to cancer: an update. Cancers (Basel) 14:1992

Song H, Zhang Y, Ma Z-X, Yao Q, Ma X (2024) TWIK-related acid-sensitive potassium channels TASK-1 and TASK-3 may participate in the process of the coexistence of asthma and OSA. Sleep Breath 28:123–131

Sorum B, Docter T, Panico V, Rietmeijer RA, Brohawn SG (2024) Tension activation of mechanosensitive two-pore domain K+ channels TRAAK, TREK-1, and TREK-2. Nat Commun 15:3142

Staudacher I, Seehausen S, Illg C, Lugenbiel P, Schweizer PA, Katus HA, Thomas D (2019) Cardiac K2P13.1 (THIK-1) two-pore-domain K+ channels: pharmacological regulation and remodeling in atrial fibrillation. Prog Biophys Mol Biol 144:128–138

Sun J, MacKinnon R (2020) Structural basis of human KCNQ1 modulation and gating. Cell 180:340–347.e9

Taura J, Kircher DM, Gameiro-Ros I, Slesinger PA (2021) Comparison of K+ channel families. In: Pharmacology of potassium channels. pp 1–49

Trapp S, Isabel Aller M, Wisden W, Gourine AV (2008) A role for TASK-1 (KCNK3) channels in the chemosensory control of breathing. J Neurosci 28:8844–8850

Türkaydin B, Schewe M, Riel EB, Schulz F, Biedermann J, Baukrowitz T, Sun H (2024) Atomistic mechanism of coupling between cytosolic sensor domain and selectivity filter in TREK K2P channels. Nat Commun 15:4628

Turney TS, Li V, Brohawn SG (2022) Structural basis for pH-gating of the K+ channel TWIK1 at the selectivity filter. Nat Commun 13:3232

Viatchenko-Karpinski V, Ling J, Gu JG (2018) Characterization of temperature-sensitive leak K+ currents and expression of TRAAK, TREK-1, and TREK2 channels in dorsal root ganglion neurons of rats. Mol Brain 11:40

Vierra NC, Dadi PK, Jeong I, Dickerson M, Powell DR, Jacobson DA (2015) Type 2 diabetes–associated K+ channel TALK-1 modulates β-cell electrical excitability, second-phase insulin secretion, and glucose homeostasis. Diabetes 64:3818–3828

Vierra NC, Dadi PK, Milian SC, Dickerson MT, Jordan KL, Gilon P, Jacobson DA (2017) TALK-1 channels control β cell endoplasmic reticulum Ca^{2+} homeostasis. Sci Signal. https://doi.org/10.1126/scisignal.aan2883

Wague A, Joseph TT, Woll KA, Bu W, Vaidya KA, Bhanu NV, Garcia BA, Nimigean CM, Eckenhoff RG, Riegelhaupt PM (2020) Mechanistic insights into volatile anesthetic modulation of k2p channels. elife 9:1–23

Warth R, Barrière H, Meneton P et al (2004) Proximal renal tubular acidosis in TASK2 K^+ channel-deficient mice reveals a mechanism for stabilizing bicarbonate transport. Proc Natl Acad Sci 101:8215–8220

Wiedmann F, Schlund D, Voigt N, Ratte A, Kraft M, Katus HA, Schmidt C (2019a) N-glycosylation–dependent regulation of hK_{2P} 17.1 currents. Mol Biol Cell 30:1425–1436

Wiedmann F, Schlund D, Faustino F, Kraft M, Ratte A, Thomas D, Katus HA, Schmidt C (2019b) N-glycosylation of TREK-1/hK2P2.1 two-pore-domain potassium (K2P) channels. Int J Mol Sci 20:5193

Wiedmann F, Frey N, Schmidt C (2021) Two-pore-domain potassium (K2P-) channels: cardiac expression patterns and disease-specific remodelling processes. Cells 10:2914

Wiedmann F, Beyersdorf C, Zhou XB et al (2022) Treatment of atrial fibrillation with doxapram: TASK-1 potassium channel inhibition as a novel pharmacological strategy. Cardiovasc Res 118:1728–1741

Wright PD, McCoull D, Walsh Y, Large JM, Hadrys BW, Gaurilcikaite E, Byrom L, Veale EL, Jerman J, Mathie A (2019) Pranlukast is a novel small molecule activator of the two-pore domain potassium channel TREK2. Biochem Biophys Res Commun 520:35–40

Yang Y, Li S, Jin ZR et al (2018) Decreased abundance of TRESK two-pore domain potassium channels in sensory neurons underlies the pain associated with bone metastasis. Sci Signal. https://doi.org/10.1126/scisignal.aao5150

Yost CS, Oh I, Eger EI, Sonner JM (2008) Knockout of the gene encoding the K2P channel KCNK7 does not alter volatile anesthetic sensitivity. Behav Brain Res 193:192–196

Zhou Y, MacKinnon R (2003) The occupancy of ions in the K+ selectivity filter: charge balance and coupling of ion binding to a protein conformational change underlie high conduction rates. J Mol Biol 333:965–975

Zúñiga L, Zúñiga R (2016) Understanding the cap structure in K2P channels. Front Physiol. https://doi.org/10.3389/fphys.2016.00228

Zuzarte M, Heusser K, Renigunta V, Schlichthörl G, Rinné S, Wischmeyer E, Daut J, Schwappach B, Preisig-Müller R (2009) Intracellular traffic of the K+ channels TASK-1 and TASK-3: role of N- and C-terminal sorting signals and interaction with 14-3-3 proteins. J Physiol 587:929–952

9 Cryo-EM Brings the Mechanisms of Activation, Inactivation, and Inhibition of Ryanodine Receptors into Focus

Montserrat Samsó and Jack G. Wallace

Abstract

Ryanodine receptors (RyRs) are massive intracellular Ca^{2+} channels that generate large intracellular Ca^{2+} signals in myocytes, neurons, and other cell types. RyR channels, membrane proteins in the endoplasmic reticulum, are subject to complex regulation. Cryo-electron microscopy (cryo-EM) saw a major leap in resolution around 2016, when direct detectors combined with computational advances enabled near-atomic structures of proteins. Since then, unique cryo-EM structures of RyR in various states provide a structural framework for a wealth of existing functional data. These structures reveal an ultra-complex machinery assembled from diverse domains, including helical repeats, SPRY domains, EF hands, and zinc fingers, which form binding sites for multiple ligands. This chapter focuses on the conformational changes RyRs undergo in response to physiological effectors. Ca^{2+} and Mg^{2+} command activation, inhibition, and inactivation. CaM acts as a mobile subunit that changes domain connectivity according to Ca^{2+} concentration. FKBP props up the cytoplasmic shell and coordinates the four subunits. Single-point disease-causing mutations have a profound impact on long-range allostery, which, for RyR1, can be partially reversed by dantrolene, an FDA-approved drug. The integration of structural and functional knowledge provides a quintessential example of allostery, which is useful in the research of other proteins and offers valuable guidance to design new therapies.

Keywords

Ryanodine receptor · Ion channel · Muscle contraction · Allostery · Calcium · Magnesium

M. Samsó (✉) · J. G. Wallace
Department of Cellular, Molecular and Genetic Medicine, School of Medicine, Virginia Commonwealth University, Richmond, VA, USA
e-mail: msamso@vcu.edu

9.1 RyR Generates Fast and Large Amplitude Intracellular Ca^{2+} Signals of Special Importance in Muscle Cells

The primary function of RyR is to control the cytosolic concentration of Ca^{2+}, a major class of second messenger within the cell. With its high conductance, this channel plays a crucial

A. Rosenhouse-Dantsker (ed.), *Decoding Ion Channels Structure and Function*, Advances in Experimental Medicine and Biology 1497, https://doi.org/10.1007/978-3-032-07523-9_9

role in excitable cells such as myocytes and neurons. It controls the flow of Ca^{2+} from the endoplasmic reticulum (ER) into the cytoplasm, down a concentration gradient of 4 orders of magnitude. In skeletal and cardiac muscles, Ca^{2+} release from the ER is strictly linked to the plasma membrane depolarization wave physiologically observed during an action potential, in a process known as excitation-contraction (EC) coupling. To respond rapidly to the nerve impulse, RyR interacts directly, or indirectly, with a voltage-sensing membrane protein in the plasma membrane known as the dihydropyridine receptor (DHPR), an L-type Ca^{2+} channel. The communication between these proteins is supported by an intricate subcellular arrangement where the ER and the cellular membrane are in close proximity. From the cell membrane of muscle cells emerge long invaginations perpendicular to the long axis of the muscle fiber called transverse (T) tubules (Franzini-Armstrong and Porter 1964; Walker and Schrodt 1965); they form a tubular network that propagates the action potential to the interior of the cell. The ER (termed sarcoplasmic reticulum or SR in muscle cells) is located alongside the T-tubule, leaving an intervening cleft of 12 nm (Franzini-Armstrong et al. 1999) and creating a microenvironment with limited diffusion (Cully et al. 2018). The continuity of the T-tubule with the plasma membrane, and the closely apposed SR, can be appreciated in Fig. 9.1a. In cardiac muscle, SR cisternae expand around the T-tubule, forming dyad junctions (Fig. 9.1b), while peripheral junctions also form directly beneath the plasma membrane (Franzini-Armstrong et al. 1999; Scriven et al. 2013). In mammals, skeletal muscle has two SR terminal cisternae per TT, one on either side, forming the triad junction (Fig. 9.1c). Dyad and triad junctions deliver Ca^{2+} to the surrounding myofilaments and trigger muscle contraction. The sequence of events of EC coupling is as follows: (1) depolarization of the plasma membrane and T-tubule, (2) transient rise in cytoplasmic concentration of Ca^{2+}, and (3) muscle contraction (Ashley and Ridgway 1968). Ca^{2+} is pumped back into the SR against its gradient by the Ca^{2+} ATPase (aka SERCA), using the energy of ATP. Mitochondria and glycogen granules are located nearby to ensure adequate ATP production powering the Ca^{2+} ATPase and myosin ATPase. The subcellular architecture of the EC coupling machinery is repeated periodically at every sarcomere—the repeating unit of the muscle's contractile machinery (Flucher and Franzini-Armstrong 1996; Flucher et al. 1993).

Skeletal and cardiac muscle have specific morphological arrangements and dedicated protein isoforms to support specialized forms of EC coupling. RyR1 and RyR2 isoforms are located in skeletal and cardiac muscle, respectively, and interact with specific isoforms of the main subunit of the DHPR ($Ca_v1.1$ in skeletal muscle and $Ca_v1.2$ in cardiac muscle). The pairing of DHPR and RyR isoforms dictates how information is transmitted between the two. In heart, $Ca_V1.2$ opens in response to membrane depolarization, and the resulting Ca^{2+} entry, driven by its electrochemical gradient, opens the RyR2 channel. This leads to the subsequent release of Ca^{2+} stored within the lumen of the SR, amplifying the Ca^{2+} signal. This mechanism is referred to as the Ca^{2+}-induced Ca^{2+}-release (*CICR*) mechanism (Fabiato 1983). In skeletal muscle, the communication between DHPR and RyR1 is known as "*conformational coupling*" (Schneider and Chandler 1973; Rios and Pizarro 1991): despite that $Ca_V1.1$ also supports Ca^{2+} channel activity, Ca^{2+} influx through $Ca_V1.1$ is unnecessary for EC coupling in skeletal muscle (Dayal et al. 2017, Kobayashi et al. 2025). Instead, signal transmission takes place by physical interaction between the two proteins and additional involvement of Stac3 (Paolini et al. 2004a; Perni et al. 2017). The interaction, or lack thereof, between RyR and DHPR dictates different quaternary arrangements among RyRs, with few DHPRs controlling a cluster of around 15–100 RyR2s in heart (Hou et al. 2015; Scriven et al. 2010), and a ratio of 4 DHPRs interacting with every other RyR1 in skeletal muscle (Paolini et al. 2004b; Block et al. 1988). RyR3, the third isoform of the RyR family in mammals, also operates via the CICR mechanism. It is expressed

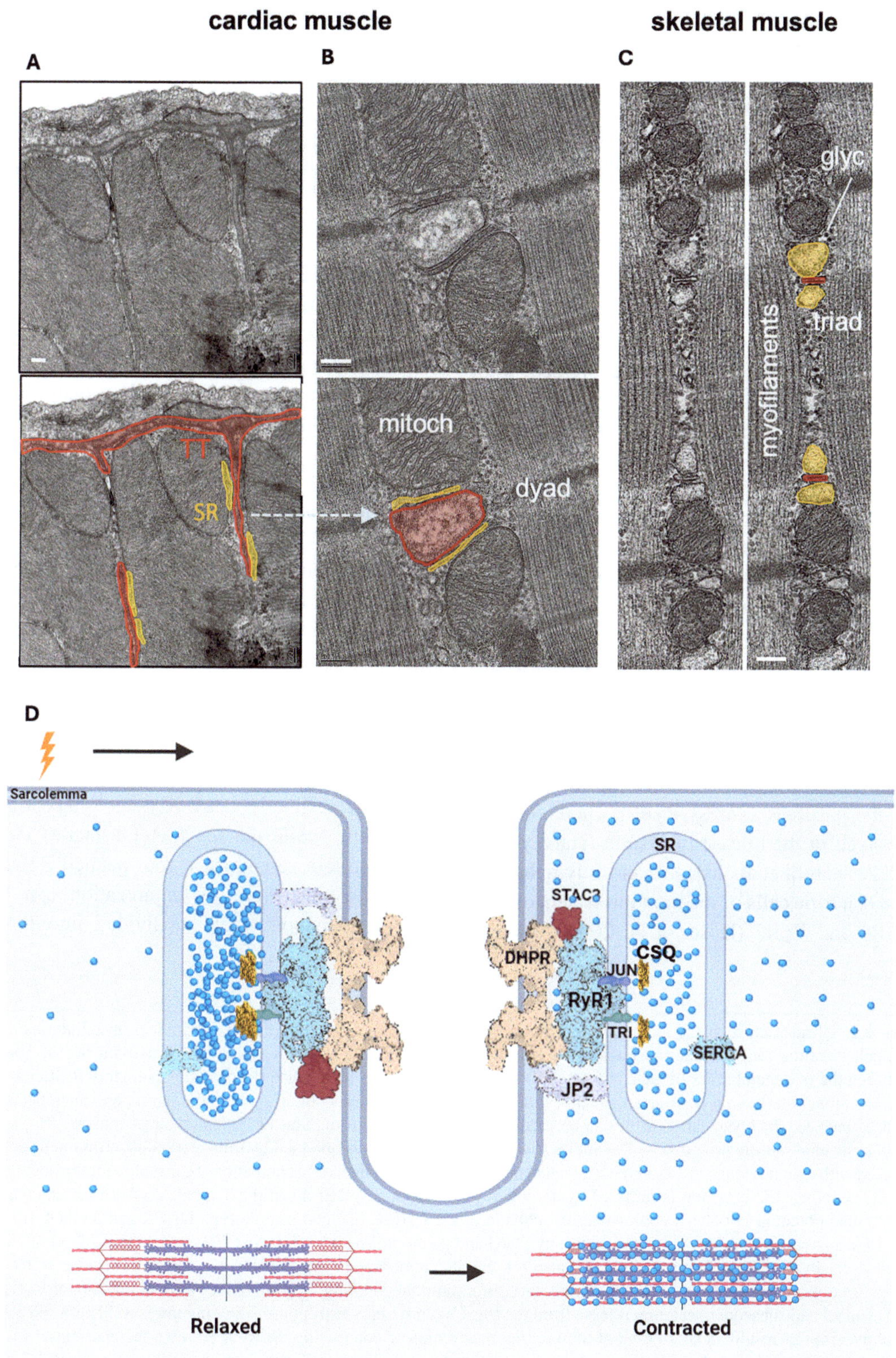

Fig. 9.1 (continued)

during the early development of skeletal muscles (Rossi and Sorrentino 2002) and, in adults, in the diaphragm, certain muscles in the neck and head, the extraocular muscle, and slow-twitch muscle, alongside RyR1 (Meissner 2017). Figure 9.1d illustrates the general arrangement of the main proteins involved in EC coupling and their relationship to the T-tubules and SR. Some components of the machinery will be described later in this chapter.

All three isoforms play significant roles in various other excitable tissues and cell types. In the brain, RyR2 is the most widely expressed, while RyR1 is expressed in specific brain regions such as the Purkinje cells of the cerebellum, the dentate gyrus of the hippocampus, the CA3 and CA1 cells of the Ammons horn, and in the olfactory bulb (Rossi and Sorrentino 2002; Giannini et al. 1995). Although RyR3 was first found in brain, and often known as the "brain isoform," it is expressed at low levels. This isoform is found in the granular cell layer of the cerebellum, the CA1 region of the Ammon's horn, the caudate/putamen nuclei, the mitral and granular cell layer of the olfactory bulb (Giannini et al. 1995), and appears to be the only isoform expressed in glial cells (Rossi and Sorrentino 2002). The beta cells of the pancreas express RyR2, where RyR2 is involved in the release of insulin (Harvey et al. 2022; Santulli et al. 2015). T and B lymphocytes and dendritic cells of the immune system express RyR1 and RyR2 (Hosoi et al. 2001; Vukcevic et al. 2008; O'Connell et al. 2002; Wolf et al. 2015). Smooth muscle in arteries, ureters, urinary bladder, and esophagus, among others, express RyR2 and RyR3, while cerebral arteries express all three isoforms (Ledbetter et al. 1994), and smooth muscle cells from other organs express different combinations of the three isoforms.

9.2 The 3D Structure of RyR Reveals an Intricate Molecular Machinery

RyR, a homotetramer with more than 5000 residues per subunit and a molecular weight of 2.2 MDa, is the largest known ion channel and one of the largest proteins in the body; it can be readily discerned in thin sections of muscle (Franzini-Armstrong and Nunzi 1983). The first 3D structure was determined by single-particle cryo-EM (cryo-EM) in 1989 (Wagenknecht et al. 1989). The cytoplasmic domain forms a characteristic square prism of 29x29x12 nm^3, while the transmembrane domain (TMD) has a footprint of 9x9 nm^2 within the membrane; the fourfold symmetry reflects its homotetrameric organization. The side view distinctly resembles the shape of a mushroom, while the top (cytoplasmatic) view of the protein resembles a four-petaled flower. From the side view, its organization can be better understood when subdivided into three

Fig. 9.1 Ultrastructure of the EC coupling machinery. (**a**) The plasma membrane emits periodical invaginations into the cell, forming longitudinal and transversal (T) tubules (red). SR sacs (yellow) form junctions with it. The space between the two membranes (dyadic/triadic cleft) is punctuated by electron-dense structures, presumably RyRs. Dyad and triad junctions form precise arrangements with myofilaments and mitochondria within the sarcomere, the repeating unit of muscle. (**b**) Dyad junction of canine cardiac muscle. (**c**) Triad junction of human skeletal muscle. Biopsied human skeletal muscle provided by Dr. Sheila Riazi, U. Toronto. Images obtained by thin-section electron microscopy stained with osmium tetroxide, uranyl acetate, and potassium ferrocyanide. (**d**) Schematics of the main proteins involved in EC coupling. EC coupling is initiated by an action potential propagated down the T-tubules, where the change in membrane potential initiates a conformational change in the DHPR. Skeletal muscle-type DHPR and RyR1 form an intricate quaternary complex mediated in part by Stac3 and junctophilin (JP2). Triadin (TRI) and junctin (JUN) interact with RyR1 in the SR membrane, while JP2 anchors the SR membrane to the sarcolemma. These molecular scaffolds allow for optimal propagation of the EC signal via conformational coupling, resulting in a transient opening of RyR1's ion channel and subsequent calcium release from the SR. Calcium binds to troponin, allowing interaction between actin and myosin; this results in shortening of the sarcomere and muscle contraction. SERCA restores the gradient of Ca^{2+} across the SR membrane. In cardiac muscle and other tissues, there is no conformational coupling between RyR2 and cardiac-type DHPR, and it is the calcium entry via the DHPR that prompts the RyR2 to open (calcium-induced calcium release mechanism, or CICR). (Panel **d** created in BioRender. Wallace, J. (2025) https://BioRender.com/m4izqsv)

levels: the transmembrane domain containing the channel, the core, and the peripheral shell. The complexity of the channel is epitomized by its 13 distinct domains per protomer. Some regions of the RyR remain unresolved; they likely correspond to intrinsically disordered regions, providing functionality beyond the classical view of structured domains (Holehouse and Kragelund 2024).

9.2.1 Comparison of RyR with the Other Intracellular Ca^{2+} Channel, the IP_3R

Several domains of RyR are homologous to those in the IP_3 receptor, the closest related protein from an evolutionary point of view, and the other main intracellular Ca^{2+} channel. Specifically, the two N′-terminal beta trefoil domains of IP_3R (βTF1 and βTF2) share homology with the N′ terminal domains (NTD) A and B of RyR, while the armadillo repeat 1 (ARM1) has homology with the NTDC of RyR. In the IP_3R, the βTF2 and ARM1 domains host the IP_3 binding site; the small molecule binds at their interface. Small portions of the helical domains (HD), ARM2, ARM3, and intervening lateral domain (ILD) of the IP_3R share homology with the handle, HD1, central domain (CD), and U-motif of RyR, respectively. The TMD of both ion channels is highly conserved. The TMD is followed by two homologous domains: the linker domain (LNK) in IP_3R and the C-terminal domain (CTD) in RyR. The C′-terminal portion of the IP_3R sequence (the CTD domain) consists of a 68-residue α-helix. The four CTDs form a central helical bundle that reaches the top of the structure, interacting with the βTF2 domains (see (Baker et al. 2023) for a review on the IP_3R structure). The additional domains of RyR consist of expansions of the helical repeats or insertions of distinct domains into the basic molecular plan of the IP_3R, resulting in a considerable expansion of the cytoplasmic shell and doubling of the overall molecular weight (see Fig. 9.2).

The next section describes the different pieces of the RyR machinery and sets the stage for their orchestrated movement in the following sections. Unless otherwise indicated, the residue numbers refer to the rabbit (r) RyR1 isoform.

9.2.2 The Peripheral Shell

The periphery of the cytoplasmic domain forms the cap of the mushroom, sometimes referred to as the top shell. It has a scaffold-like structure that facilitates diffusion of the Ca^{2+} ions while acting as a 12 nm spacer of the triadic/dyadic cleft. The large surface area offers multiple binding sites for molecular partners, ligands, small molecules, and drugs. It is formed by the NTD, the SPRY domains, the handle, the helical domains, and the P1 domains. As in the IP_3R, the NTD is constituted by two β-trefoil domains (NTDA and NTDB) per subunit clustered around the fourfold axis forming a vestibular ring, and a helical domain (NTDC). Downstream from the NTD are three rigid SPRY domains (SPRY1, SPRY2, and SPRY3), each with a core formed by two antiparallel beta sheets. The SPRY cluster is followed by the handle domain, a helical domain that forms a "flap" on the sides of the square, continued by a longer helical repeat known as HD1 and HD2, with HD2 forming a long section that loops out of the structure. In the RyR1 and RyR3 isoforms, the distal part of HD2 reaches the SPRY2 domain of the neighboring protomer, potentially enhancing coordination among subunits. In the RyR2 isoform, the distal portion of HD2 is less resolved, reflecting higher flexibility, which is likely to have functional implications. The P1 and P2 domains, each consisting of a twofold symmetric four-helical bundle, are inserted within the SPRY and the HD regions, respectively. The P1 domains form the four corners of the structure, while P2, a phosphorylation hotspot, protrudes toward the plasma membrane. The structure of the P2 domain has been solved by X-ray crystallography (Sharma et al. 2012; Yuchi et al. 2012), but is rarely resolved by cryo-EM due to its flexibil-

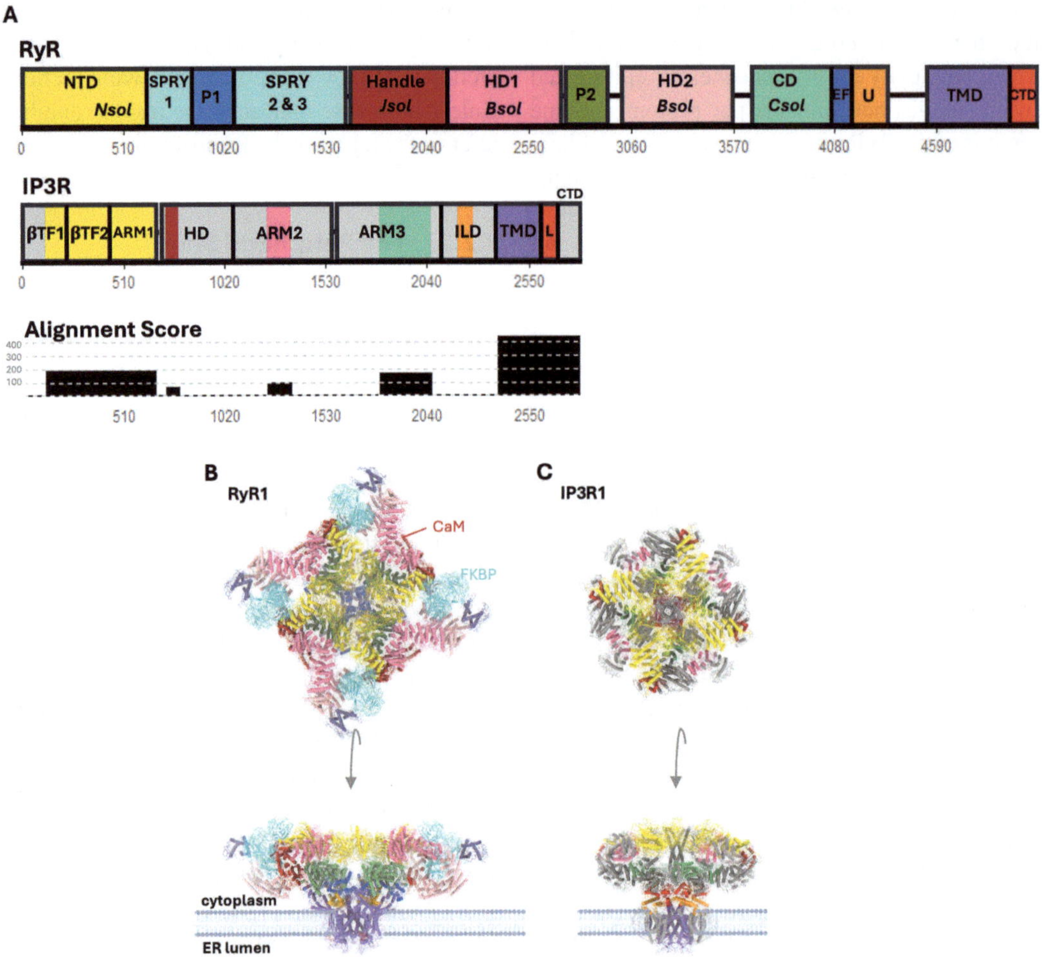

Fig. 9.2 Domains of the two intracellular calcium channels, RyR, and IP_3R. (**a**) Local sequence alignment of RyR1 and IP3R1; homologous domains are in matching color, and non-homologous domains are represented in gray. The terms in italic represent an alternative nomenclature used for RyR domains. (**b, c**) Comparison of the 3D structures of RyR1 and IP_3R1 with multiple segments having conserved sequence and structure. These are intervened by divergent regions (SPRY 1–3, P1, P2, handle) and extensions of the helical domains (HD), expanding out as a cytoplasmic shell. The transmembrane domain of both calcium channels is highly conserved. The CTD of RyR is homologous to the LNK domain of IP_3R, while the CTD of the IP_3R is a long, vertical α-helix with no homologous counterpart in RyR. Domains in (**b**) and (**c**) are color-coded according to domain boundaries in (**a**). The transparent cryo-EM density is overlaid on the atomic model. For RyR, the ligand FKBP is shown in cyan and the approximate position of the CaM binding site is also indicated

ity. In some reconstructions, it is discerned as a bridge of density with SPRY3, while in others it appears detached from it. Recognizable multi-domain features are the vestibular ring formed by the NTDA and NTDB domains from the four subunits, and four radial rhomboid structures, resembling four petals, formed by the NTD, SPRY1, SPRY3, and P1 of one subunit, and HD1 of the adjacent subunit.

9.2.3 The Core

The core of the structure is the only point of contact between the transmembrane domain and the peripheral domain, conveying hierarchy and an extra layer of fine control. It is dominated by the central domain (CD), which consists of 18 helices forming a concave-shaped helical repeat motif. At its TMD-facing surface, three negatively charged residues, described in more detail later, form part of the high-affinity Ca^{2+} binding site.

The EF hand domain sequence (residues 4071–4131) lies between the CD and the U-motif. It has two degenerated Ca^{2+}-binding EF loops, EF-1 and EF-2 (Fessenden et al. 2004; Xiong et al. 1998). This domain, named EF due to its canonic structure with one phenylalanine and a high content of acidic residues, forms a globular shape sticking out under the umbrella of the mushroom, near its stalk. It constitutes an important inhibitory Ca^{2+} and Mg^{2+} sensor.

The U-motif forms a clamp around the globular C-terminal domain (CTD) and also contacts the C-terminal end of S6, creating the primary binding site for ATP. ATP is essential for proper function of RyR1, mediating the transmission of long-range conformational changes by its multiple interactions with the CD, U-motif, and S6 (Chirasani et al. 2022).

The transmembrane domain starts after the U-motif. Between transmembrane helices S2 and S3, a 125-residue-long cytoplasmic extension forms a domain (23 L) that extends toward the EF hand of the neighboring subunit. As previously mentioned, reorientation of the EF hands, dictated by the Ca^{2+}/Mg^{2+} concentration, has important implications in Ca^{2+}-induced inactivation and Mg^{2+} inhibition.

The last transmembrane helix, S6, has a long cytoplasmic section that is continued by the CTD, a domain stabilized by a central zinc finger motif (Taweechat et al. 2024; des Georges et al. 2016; Yan et al. 2015). The carbonyl group of T5001, on its CD-facing surface, form the CTD moiety of the high-affinity site for Ca^{2+}. This interaction is crucial for channel gating, as described later.

9.2.4 The Transmembrane Domain

The transmembrane domain is composed of six alpha helices per subunit, S1-S6. S1-S4 of each subunit form a helical bundle structurally similar to voltage-sensing domains of voltage-sensitive cation channels, and the S5-S6 helices from each subunit interact at the center, with the four S6 helices forming a right-handed helical bundle that makes up the pore (Fig. 9.3a). Between S5 and S6 of the adjacent subunit there is a short, tilted pore helix known as the P-loop (residues 4879–4888), with its C-terminal end pointing toward a central cavity (Fig. 9.3b). The S4-S5 linker, a 12-residue amphipathic alpha helix that follows the cytoplasmic leaflet of the ER membrane, spaces S1-S4 from the central S5-S6 bundle such that the linker interacts with the S6 helices of the neighboring subunit. This leads to a domain-swapped inter-subunit arrangement (Fig. 9.3a, b) that increases the stability of the channel. All these features mirror the characteristic organization of 6-TM voltage-sensing cation channels (Samso et al. 2005; Yu et al. 2005). Considering the low sequence similarity of RyR to voltage-sensitive cation channels, this is an example of structural convergence. In the case of RyR, two more transmembrane helices per subunit are discerned in some cryo-EM density maps near S1 (des Georges et al. 2016), with one of them assigned to sequence 4318–4344 (Wei et al. 2025). Cryo-EM of RyR embedded in nanodiscs revealed two lipids at the interphase between S4 and S5 of neighboring subunits (Fig. 9.3c). The position of the two annular lipids is conserved among the RyR1 and RyR2 isoforms and are only resolved in the closed state (Nayak and Samso 2022; Nayak et al. 2024). Deletion of FF4923–4924 at this lipid interphase results in lethal fetal akinesia (Xu et al. 2020), implying that this interaction is crucial in stabilizing the closed state of the protein. The S6 α-helix is tilted 50–55° with respect to the membrane and protrudes 25 Å into the cytoplasm, where it interacts with the U-motif.

On the luminal side of the channel a long, highly negatively charged luminal loop downstream of S5 sticks into the lumen. As luminal Ca^{2+} is known to regulate the channel, it is tempting to attribute a Ca^{2+} sensing role for this loop. However, comparative analysis of multiple near-atomic 3D reconstructions reveals a relatively immobile structure, with movements of less than 2 Å and no significant impact on the attached α-helices. This suggests an indirect Ca^{2+}-sensing function, potentially by acting as an

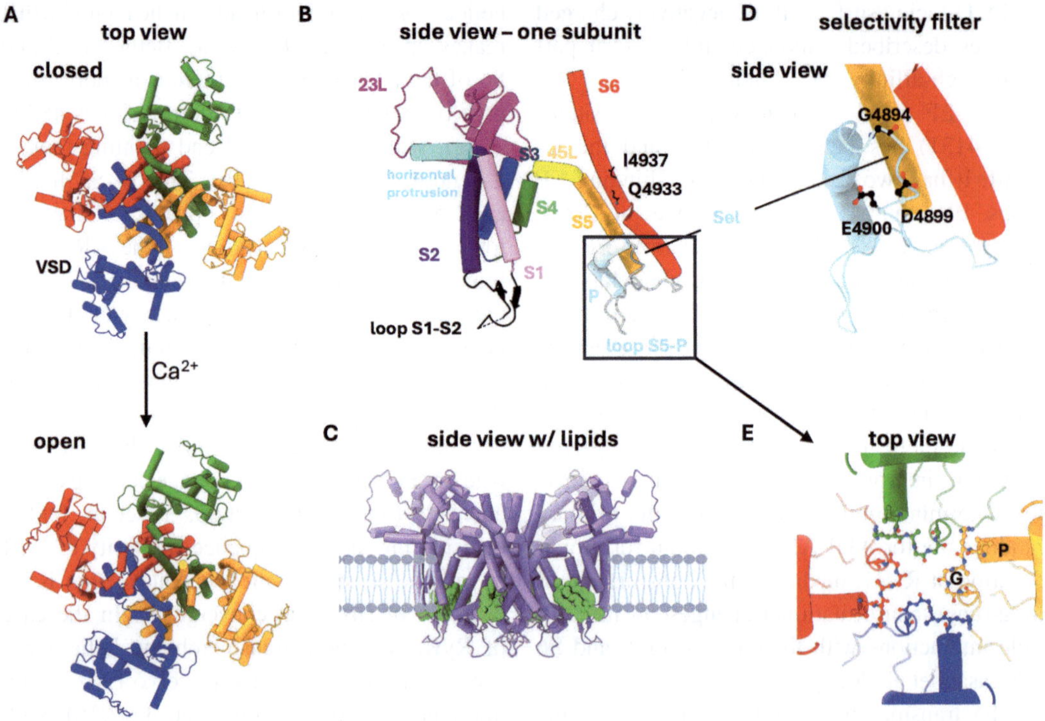

Fig. 9.3 Structural organization of the transmembrane domain of RyR. (**a**, **b**) Each of the four subunits has a peripheral voltage sensor-like domain (VSD) formed by transmembrane helices S1–S4, while helices S5-S6 form the central pore. The S4-S5 linker (45 L) interacts with S5-S6 of the adjacent subunit, forming the swapped-domain arrangement. The pore helix (P), between S5 and S6, protrudes toward a central cavity. (**c**) Two annular lipids (green) lodge between S4 and S5 of adjacent subunits. (**d**) Two negatively charged residues in the selectivity filter line the entrance to the channel. (**e**) Top view of the selectivity filter with its narrowest part defined by the carbonyl groups of G4894 (G)

anchor for another Ca^{2+} sensing protein such as calsequestrin.

9.2.5 The Ion Conduction Path

Starting from the luminal side, the ion-conducting path consists of the selectivity filter, the central cavity, the ion gate, and the cytoplasmic portion of the S6 bundle. In the open channel, the selectivity filter and the ion gate constitute two primary resistance points.

The selectivity filter, characterized by the GGGIGDE motif (residues 4894–4900 in rRyR1) (Gao et al. 2000), is stabilized by multiple close-knit interactions among S6, the pore helix, and proximal regions of the luminal loops (Nayak et al. 2024). Surrounding the luminal mouth of the channel, residues D4899 and E4900 create a region of high electronegative density, attracting cations locally and effectively excluding anions (Wang et al. 2005) (Fig. 9.3d). The narrowest segment of the ion conduction path, formed by the carbonyl groups of G4894 (Fig. 9.3e), has a radius of 3 Å—sufficient to pass a hydrated Ca^{2+} ion, which has a first solvation shell of radius 2.4 Å (Zhang et al. 2020; Gillespie et al. 2014). In the open state, the selectivity filter passage represents the channel's tightest point and forms a minimum energy well for Ca^{2+}. The cation permeates the channel through a knock-on mechanism (Zhang et al. 2020). Hydrated K^{+} and Na^{+} can also permeate the channel. The conductance is 800–1000 pS for K^{+}, compared to 100–200 pS for Ca^{2+} in RyR1, with a relative Ca^{2+} to K^{+} permeability of 7:1 (Wang et al. 2005;

Smith et al. 1988). In contrast, Ca_V (DHPR) channels are practically impermeable to K^+ in the presence of Ca^{2+} (Yang et al. 1993). The steep Ca^{2+} gradient across the SR and the high flow rate enable large cytoplasmic Ca^{2+} signals in excitable cells such as neurons and muscle.

The selectivity filter is followed by the central cavity, a region of predicted low binding affinity for cations (Zhang et al. 2020; Heinz et al. 2018). This suggests that, in the open state, the cavity maintains a degree of continuity with the cytosol, similar to models proposed for other Ca^{2+} channels (Jorgensen et al. 2016).

In the closed state the S6 helical bundle narrows and, at residues Q4933 and I4937, the radius reduces to 1 Å or less—an insurmountable barrier for ion passage. During channel opening, the C′ half of the S6 helix tilts out, and I4937 rotates away from the fourfold axis, acting as a gatekeeper residue. This structural change widens the passage to a radius of 3.5 Å, enabling Ca^{2+} flow while still presenting a slight barrier to ion passage.

The cytoplasmic portion of S6 features negatively charged residues, among which D4938 and D4945 form two rings of negative charges that facilitate ion flux (Xu et al. 2006) while exhibiting relatively low Ca^{2+} binding affinity (Heinz et al. 2018). Upon exiting the ion gate, Ca^{2+} may also escape through lateral fenestrations (Heinz et al. 2018), which could account for the subconductance states observed in the presence of ryanodine (Buck et al. 1992), toxins (Tripathy et al. 1998), or in the absence of the FKBP ligand (Ahern et al. 1997; Brillantes et al. 1994).

9.3 A Bird's Eye View of the Gating-Associated Conformational Changes

All three RyR isoforms are activated by Ca^{2+}, constituting the basis of the CICR mechanism. As noted earlier, an exception takes place in skeletal muscle, where initial RyR1 activation is triggered through conformational coupling with the DHPR. To capture RyR in the open state for structural determination, exogenous activators such as PCB95 (Samso et al. 2009; Bai et al. 2016) or caffeine (des Georges et al. 2016) have traditionally been used to potentiate the activating effect of Ca^{2+} and ATP, although the 3D structure of open RyR was achieved in a more physiological state with just Ca^{2+} and ATP (Nayak and Samso 2022).

Opening of the RyR channel involves a concerted movement encompassing most, if not all, of the protein's domains (Fig. 9.4). The most obvious motion occurs in the cytoplasmic shell. The four radial rhomboid features in the shell open as the petals of a flower, tilting and moving away from the fourfold axis. The vestibular domains (NTDA and NTDB) separate and shift upward, while the distal domains move downward by 10 Å (des Georges et al. 2016; Samso et al. 2009; Bai et al. 2016). The flexion angle, defined as the angle between the long axis of the rhomboid and the membrane plane, quantifies this movement (Steele and Samso 2019). The conformational change upon opening, summarized in Fig. 9.3a and 9.4a, also applies to the RyR2 (Peng et al. 2016) and RyR3 (Chen et al. 2024) isoforms.

Upon gating, the S6 helical bundle relaxes, and the C′-terminal half of the four helices tilts out. In addition, each helix undergoes a twist that moves the gating residue aside and permits Ca^{2+} flow. The gating residue is I4937 in rRyR1/I4938 in hRyR1, I4868 in mRyR2/I4869 in hRyR2, and I4759 in minkRyR3/I4770 in hRyR3. The canonical alpha helical structure near the center of the sequence of S6 in RyR1 is interrupted by a wider π helix that may facilitate movement of the C′ terminal halves of the S6 helices (Nayak and Samso 2022). As the CTDs separate and move away from the fourfold axis, their movement is communicated directly to the S6 helices. The S4-S5 linkers, which form a square frame around the S6 bundle, rotate out at their S5 end, releasing this frame and enabling expansion of the S6 bundle. How channel opening relates to the motion of the cytoplasmic shell remained unclear until high-resolution cryo-EM structures became available in 2015. The molecular mechanisms driving this motion are described in the next section.

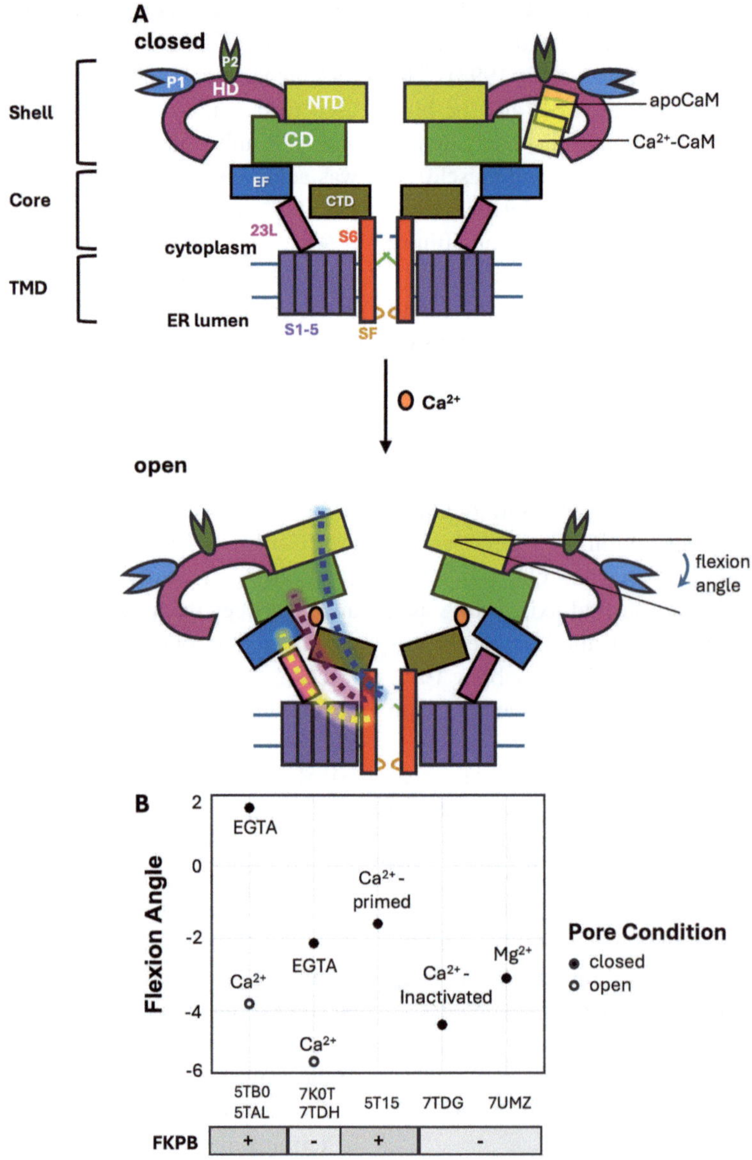

Fig. 9.4 Overall view of the allosteric pathways of RyR and the flexion angle. (**a**) In the activation pathway (magenta dashed line for one of the subunits), binding of Ca^{2+} to the interphase between the CD and CTD domain pulls the S6 helices open, enabling flow of Ca^{2+} down its gradient from the SR to the cytoplasm. The change in orientation of the CD and CTD results in a drooping of the entire cytoplasmic shell. In the inactivation pathway (yellow line), higher concentration of Ca^{2+} induces a pivot in the EF hands and their interaction with the 23 L, which closes the channel. In a third activation pathway exclusive of skeletal—type EC coupling (blue dashed line), inner rigid levers couple the motion of the DHPR (interacting at the top of the cytoplasmic shell of RyR1) to a region near the Ca^{2+} activating site. ApoCaM and Ca^{2+}-CaM modulate the channel; they are represented with yellow squares in one of the subunits of the top panel. The bottom panel illustrates the flexion angle, which measures the downward movement of the cytoplasmic shell. (**b**) Flexion angle of RyR1 measured for 3D reconstructions determined under the conditions indicated (pdb code shown below). The presence of FKBP coordinates and buttresses the four subunits, resulting in a more "upward" conformation and stabilizing the closed state, whereas its absence results in lower flexion angles, priming the channel, especially in the closed state. The presence of Ca^{2+} and Mg^{2+} lowers the flexion angle

9.4 Complex Mechanisms of Activation and Inactivation by Ca^{2+} and Mg^{2+}

Ca^{2+}, Mg^{2+}, and ATP are the primary soluble physiological regulators of the RyR. In skeletal muscle, in the absence of fatigue, ATP levels remain relatively stable. The same can be said about Mg^{2+} which is mostly complexed with ATP within the cell. However, Mg^{2+} and Ca^{2+} have a complex interplay, as both cations compete for binding sites on the RyR.

9.4.1 Ca^{2+}-Induced Activation

The "resting state" of skeletal muscle and the "diastole state" of the cardiac cell, in which the RyR is closed, are characterized by submicromolar cytoplasmic Ca^{2+} concentrations. Ca^{2+} induced activation, or CICR, is mediated by the binding of Ca^{2+} to the high-affinity site on the RyR, starting at low micromolar levels (Fig. 9.5a). Cryo-EM structures of RyR revealed the high-affinity Ca^{2+} binding site (CAS) at the interface between the CD and the CTD. In rRyR1, the binding site involved residues E3893, E3967, and Q3970 in the CD, and the carbonyl group of T5001 in the CTD (des Georges et al. 2016; Nayak and Samso 2022). The equivalent residues observed interacting with Ca^{2+} in the other two isoforms are E3847, E3921, Q3924, and T4930 in mRyR2 (Kobayashi et al. 2022)), and residues E3732, E3805, and T4823 in mink RyR3 (Chen et al. 2024). By coordinating to these residues, Ca^{2+} brings the two domains together, inducing a small tilt in the CD and a larger tilt in the CTD (red arrows in Fig. 9.5b illustrate this motion). As the CTD is continuous to the C′ terminal portion of the S6 helix, its movement directly influences S6 (Fig. 9.5c). The tilt in the CTD pulls outward and lifts the S6 helix, resulting in unwinding of the helical bundle and enabling Ca^{2+} flow (see transition from "EGTA closed" (i.e., absence of Ca^{2+}), to "Ca^{2+} open" in Fig. 9.5b). Investigation of the Ca^{2+} activation pathway in detail predicts additional participation of the U-motif, closely associated with the CTD (Chirasani et al. 2022).

9.4.2 The Elusive Luminal Ca^{2+} Sensor

RyR senses luminal Ca^{2+} concentration: increased SR Ca^{2+} loading triggers spontaneous Ca^{2+} release while depletion leads to channel closure (Laver 2007; Shannon et al. 2000), a mechanism that has been implicated in CICR termination in cardiac muscle (Zima et al. 2008; Gyorke et al. 2002; Terentyev et al. 2002). Nevertheless, the exact mechanism of luminal Ca^{2+} sensing remains unclear. One hypothesis is that RyR directly senses Ca^{2+} concentration at its luminal side (Ching et al. 2000; Gyorke and Gyorke 1998), but no specific residues have been identified yet. The negatively charged luminal loop after S5 is one candidate; however, its position—and that of surrounding residues—are practically unchanged under varying Ca^{2+} conditions. Alternatively, luminal Ca^{2+} sensing may involve molecular partners, with calsequestrin being a major candidate (Donoso et al. 1995; Szegedi et al. 1999; Royer et al. 2010). Calsequestrin could relay the signal directly to the RyR or act indirectly through a quaternary complex with triadin and junctin (Zhang et al. 1997; Rios and Gyorke 2009; Beard and Dulhunty 2015). Mutations in calsequestrin have been linked to pathological RyR leak, supporting this possibility (Lahat et al. 2001; Protasi et al. 2009). Junctin and/or triadin, RyR-interacting membrane proteins with luminal domains enriched in negatively charged residues, could also constitute the luminal sensor (Gyorke et al. 2002). The other main hypothesis is the "feed-through" mechanism (Xu and Meissner 1998), where Ca^{2+} permeates the channel and reaches the cytosolic RyR Ca^{2+} activation and inactivation sensors. Residue E4872 (mRyR2), initially proposed as a luminal sensor of Ca^{2+}, could fall into this category as it is located on the cytoplasmic side of S6 (Chen et al. 2014). A combination of both mechanisms, luminal

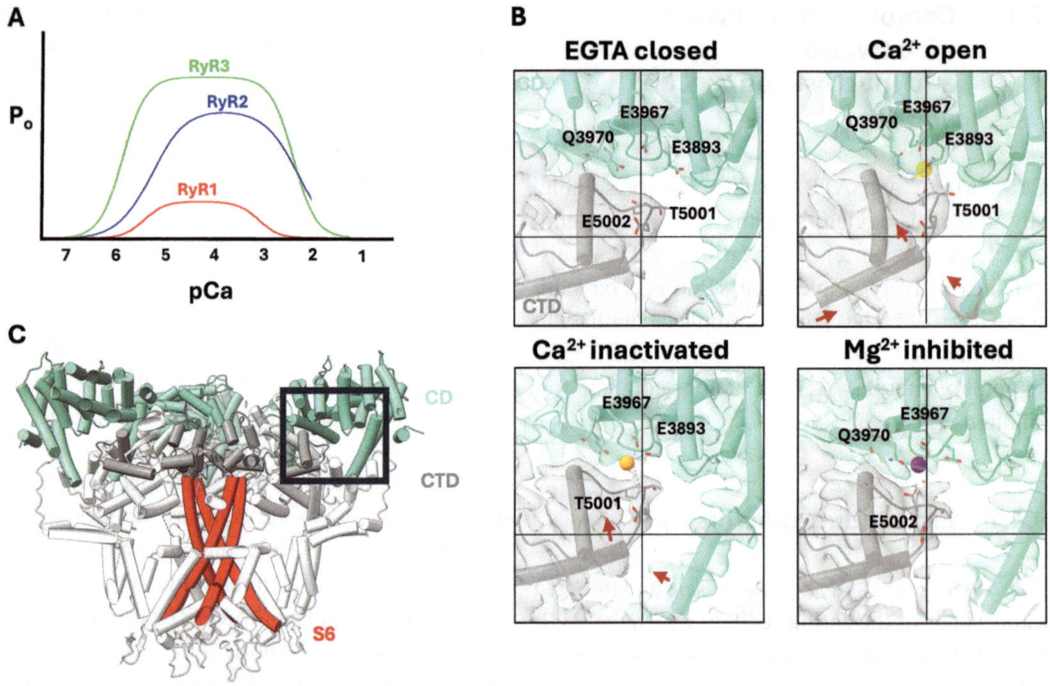

Fig. 9.5 High-affinity Ca^{2+} binding site of RyR1. (**a**) The probability of opening (Po) of RyR has a biphasic regulation by Ca^{2+}. The ascending and descending branches of the bell curve are controlled by the high and low-affinity binding sites, respectively. RyR2 and RyR3 inactivate at much higher concentration of Ca^{2+} (Meissner 2017). (**b**) Structure of the high-affinity Ca^{2+} activating site (CAS) at the interphase between the CD (green) and the CTD (dark gray). The relative position of the two domains changes according to the conditions indicated. Arrows highlight the main movements of select α-helices within the CD and CTD with respect to the conformation in the absence of divalent cation (EGTA closed). In the open state, Ca^{2+} binding results in outward movement of the CTD, pulling S6 away from the fourfold axis. However, in Ca^{2+}-induced inactivation, the CD-CTD block rotates further, moving inwards. The crossing lines in each frame are in a fixed position to emphasize the changes between conditions. The Ca^{2+} ion is in yellow and the Mg^{2+} ion is in purple. Mg^{2+} binds to the CAS site, but its larger sphere of hydration keeps the CD and CTD separated, inhibiting the channel. For the metal-liganded conditions, only chelating residues are shown. (**c**) Structure of the RyR1 core and TMD showing the location of the high-affinity Ca^{2+} activating site with respect to S6. The cytoplasmic shell has been left out for clarity, and the S6 helix is shown in red

sensing and feed-through, has also been proposed (Laver 2007).

9.4.3 Ca^{2+}-Induced Inactivation

RyR is regulated by Ca^{2+} concentration in a biphasic manner. Ca^{2+} at low concentrations activates, but at higher Ca^{2+} concentrations, the activity of RyR diminishes, with IC_{50} values between 0.20 and 0.70 mM for RyR1 (Meissner et al. 1997; Nayak and Samso, 2022; Laver et al. 1995). Ca^{2+}-induced inactivation is most relevant for RyR1 (Laver et al. 1995; Meissner et al. 1986), while RyR2 (Laver et al. 1995; Meissner 2002) and RyR3 (Chen et al. 1997) require a tenfold higher Ca^{2+} concentration to be inactivated (Fig. 9.5a). To account for the inactivation mechanism requires the RyR to have a low-affinity cytoplasmic Ca^{2+} binding site. The EF hand is an emerging candidate for such an inhibitory site. This domain is a classic Ca^{2+} binding domain; however, the sequence of the EF hands in the RyR diverges from canonical EF hands, resulting in low affinity for Ca^{2+}. It is precisely this characteristic that makes it ideal for inhibition, as more Ca^{2+} concentration is required to trigger an effect. The EF hands are not

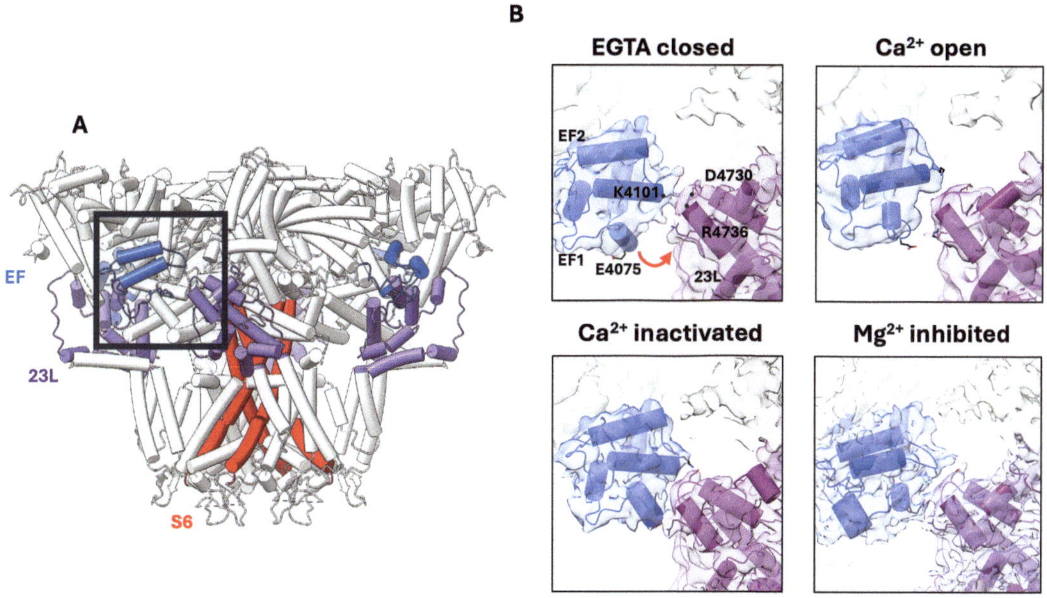

Fig. 9.6 Low-affinity binding site of RyR1 for Ca^{2+} and Mg^{2+}. (**a**) Structure of the RyR1 core and TMD showing location of the low-affinity Ca^{2+}/Mg^{2+} inhibitory site (CaMgIS) that responds to high concentrations of Ca^{2+} or Mg^{2+}. The cytoplasmic shell has been left out for clarity. The EF hand domain is shown in blue, the S2-S3 loop (23 L) is shown in purple, and the S6 is shown in red. (**b**) Relative position of the EF hand domain and the 23 L under the conditions indicated. At high concentration of either of the two divalent cations (panels "Ca^{2+}-inactivated" and "Mg^{2+}-inhibited"), the EF hand domain changes in orientation, pivoting around K4101-D4730 to form the E4075-R4736 salt bridge with the 23 L of the neighboring subunit. This adds an extra inter-subunit linkage that tightens the base of the stalk, inhibiting the channel. MH mutations at the positions indicated with a black circle result in inactivation-deficient RyR1 channels

well resolved in any of the existing cryo-EM 3D reconstructions so far, suggesting they are highly mobile domains. Nevertheless, in the presence of Ca^{2+}, the EF hands undergo a substantial Ca^{2+}-dependent conformational change that is visible despite the low resolution. In this dynamic rearrangement, the EF hand domain approaches the relatively immobile 23 L domain of the neighboring subunit (Fig. 9.6a). In the EGTA condition, the closest points between the two domains are $K4101^{EF}$-$D4730^{23L}$. Presence of Ca^{2+} induces a 10° pivot around this closest point, bringing $E4075^{EF}$-$R4736^{23L}$ progressively closer together. The distance between their Cα atoms changes from 20 Å for EGTA-closed channels to 11–12 Å for open channels (see Fig. 9.6b, transition from "EGTA closed" to "Ca^{2+}-open"). Under inactivating Ca^{2+} conditions, the distance between their Cα atoms is 10 Å and salt bridge formation occurs (see Fig. 9.6b, transition from "Ca^{2+}-open" to "Ca^{2+}-inactivated"). Residues K4101 and D4730 also approach by 1 Å throughout this transition. Thus, two salt bridges form between the two domains ($K4101^{EF}$-$D4730^{23L}$ and $E4075^{EF}$-$R4736^{23L}$ in rRyR1) in the high Ca^{2+}-inactivated state, but not in the open state (Nayak and Samso 2022). This movement has a significantly larger amplitude than its neighboring domains (Nayak and Samso 2022; Nayak et al. 2024; Iyer et al. 2022), including the adjoining CD and U-motif (see heatmap in Fig. 9.7a), which can only be explained by Ca^{2+} affecting the conformation of the EF hands directly. In the EF hand domain, it is pertinent to keep in mind that the two Ca^{2+}-binding EF loops are on the opposite side from the residues interacting with the 23 L (Fig. 9.6b). Ca^{2+} binding induces a change in the relative orientation of alpha helices within the helix-loop-helix motif, which explains the reorientation of the entire EF hand domain with

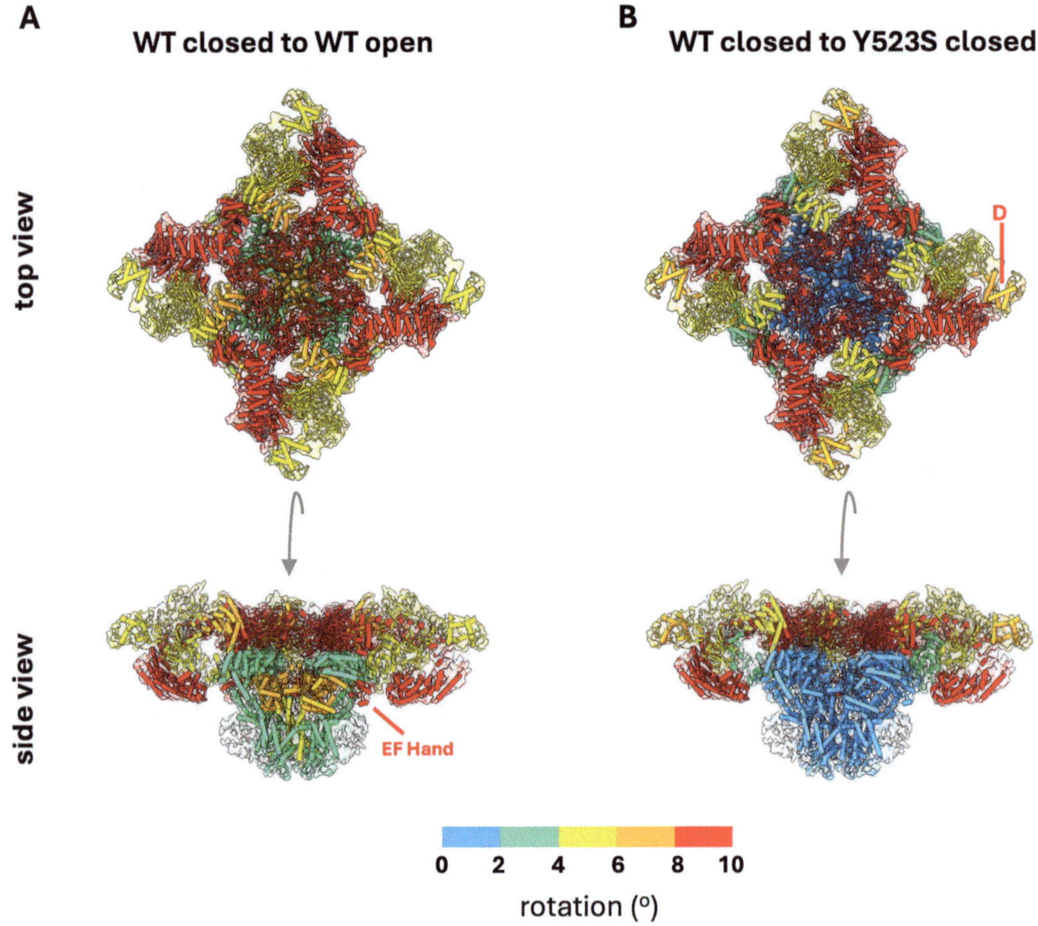

Fig. 9.7 Conformational changes of the channel in healthy and disease states. (**a**) Opening of the channel results in conformational changes throughout the protein, with distinct domains undergoing different extents of rotation and translation. The EF hand domain undergoes larger-magnitude movements than surrounding domains due to the direct influence of Ca^{2+}. (**b**) Disease mutations also cause conformational changes, increasing the probability of Ca^{2+} leak in closed channels. The heat map compares the cryo-EM structures of WT and mutant RyR1 (with a severe MH/CCD mutation; (Iyer et al. 2022)) in the closed state. Although the pore of the mutant RyR1 is closed (not visible here), the channel adopts a primed conformation, increasing the probability of transition to the open state and accounting for the leaky phenotype. The degree of rotation for each domain is indicated by the heat map. The binding site of the therapeutic drug dantrolene (D) is indicated in one of the subunits

respect to incoming and outgoing domains (i.e., the CD and the U-motif).

The two salt bridges between the EF hand domain and the 23 L add an extra linkage between the four subunits at the base of the protein's stalk (Nayak and Samso 2022; Nayak et al. 2024) (Fig. 9.7a) and tighten the U-motif, CTD, and the cytoplasmic region of S6 more closely together. This interaction is pivotal because it stabilizes the closed state. An important pathophysiological repercussion of this interaction is the impact of malignant hyperthermia (MH) and central core disease (CCD) mutations that encumber the formation of these salt bridges. In these cases, inactivation requires a one-order-of-magnitude higher Ca^{2+} concentration (Gomez et al. 2016; Gomez and Yamaguchi 2014). This inhibitory pathway, initiated by binding of Ca^{2+} (and Mg^{2+}, as will be seen below), acts via the interface between the EF hand and the 23 L of the adjacent subunit and is supported by in silico and mutagenesis studies (Chirasani et al. 2024). These studies also determined that, by comparing several cryo-EM structures, the EF hand and 23 L domains are more separated in RyR2 when compared to RyR1 under similar conditions, suggesting that this discrepancy may underlie the lower inhibitory Ca^{2+} affinity of the RyR2 isoform (Chirasani et al. 2024).

The Ca^{2+}-inactivated conformation of RyR1 displays a larger downward movement of the cytoplasmic shell and a characteristic lower flexion angle when compared to the open state (see Fig. 9.4b). These structural changes are associated with an inward rotation of the CD-CTD domain block. The further inward rotation of the CD-CTD block, together with the reorientation of the EF hand, has the important implication of pushing the S6 helices toward the fourfold axis, closing the channel (Nayak and Samso 2022).

In summary, when the high-affinity Ca^{2+} site is empty, the channel is closed, when it is occupied, the channel opens, and when both the high- and low-affinity sites are occupied at higher Ca^{2+} concentration, the channel inactivates, adopting a distinct conformation that differs from that in the "EGTA-closed" state (Nayak and Samso 2022).

Upon Ca^{2+} release, cytoplasmic Ca^{2+} concentration rises in the nanodomain between the T-tubule and the junctional SR (Stephenson 2024), where local Ca^{2+} concentrations can reach 100 μM (Langer and Peskoff 1996), sufficient to start inactivating RyR1. Although this could, in principle, constitute a negative feedback mechanism to terminate Ca^{2+} release from the SR in skeletal muscle, its significance may not be as prominent. Given that the EF hand domain is a binding site shared by both Ca^{2+} and Mg^{2+}, as detailed in the following section, and considering the high concentrations of Mg^{2+} in the cell, it is likely that under physiological conditions, the inhibitory effect of Mg^{2+} at the EF hand domain dominates over that of Ca^{2+}.

9.4.4 In Skeletal Muscle, the Strong Inhibition by Mg^{2+} Is Only Overcome by Conformational Coupling to the DHPR

A main effector of the RyR is Mg^{2+}, which is found at high concentrations in the cell. In the absence of fatigue, the concentration of this cation remains fairly constant, and there is no gradient across the ER. In skeletal muscle, the total concentration of Mg^{2+} is 8 mM, with 7 mM complexed to ATP and 1 mM free Mg^{2+} (Csernoch et al. 1998). Even at the optimally activating Ca^{2+} concentration of 50 μM, Mg^{2+} fully inhibits the channel at its physiological concentration (Nayak et al. 2024; Laver et al. 1997; Lamb and Stephenson 1992), an inhibition that is only overcome by the action of DHPR (Stephenson 2024; Lamb and Stephenson 1991). Several hypotheses have been postulated to explain the inhibitory effect of Mg^{2+}: (1) Mg^{2+} could compete for Ca^{2+} flow through the channel, (2) Mg^{2+} could bind to the low-affinity inactivation site and have the same effect as Ca^{2+}, and/or (3) Mg^{2+} could compete for the high-affinity Ca^{2+} binding site with an opposite effect (Meissner et al. 1986; Lamb and Stephenson 1991; Gillespie et al. 2012; Nagasaki and Kasai 1984). In RyR1, all three mechanisms appear to coexist.

Despite having a larger hydration diameter than Ca^{2+}, Mg^{2+} permeates the open ryanodine receptor (RyR) channel with similar conductance. Under physiological conditions, Mg^{2+} current accounts for 30% of the combined Ca^{2+}/Mg^{2+} flow through the channel (Gillespie et al. 2012; Nagasaki and Kasai 1984). The other two mechanisms are described in more detail below.

Cryo-EM showed that, in the absence of Ca^{2+}, the high-affinity Ca^{2+} activation site of RyR1 gets occupied by Mg^{2+} (Fig. 9.5, panel "Mg^{2+}-inhibited") (Nayak et al. 2024). Mg^{2+} coordinated with residues E3967 and E3970 from the CD, and E5002 in the CTD, sharing a similar network of interactions with Ca^{2+}, but with substantial structural differences. The relative orientation of CD with respect to the CTD in the Mg^{2+} closed condition was practically the same as in the presence of EGTA, suggesting that, unlike Ca^{2+}, Mg^{2+} does not bring the CD and CTD closer together. Rather, the larger size of the partially hydrated Mg^{2+} acts as a spacer between CD-CTD, preventing the transition to the open state (Nayak et al. 2024). Ca^{2+} has a much higher affinity for this site than Mg^{2+}, which is confirmed by the higher concentration of Mg^{2+} than Ca^{2+} needed to fill this site. This implies substitution of Mg^{2+} by Ca^{2+} once cytoplasmic Ca^{2+} surpasses the low μM activation threshold. Overall, Mg^{2+} binds to the high-affinity Ca^{2+} binding site but produces an antagonistic effect on the channel.

Mg^{2+} also binds to the low-affinity Ca^{2+} binding site and produces the same inhibitory effect as Ca^{2+}. Specifically, cryo-EM structures determined that an Mg^{2+} induced reorientation of the EF hand results in the formation of two salt bridges with the 23 L, as occurs in Ca^{2+}-induced inactivation (Fig. 9.5, compare "Ca^{2+}-inactivated" and "Mg^{2+}-inhibited"). This suggests that the EF hands are a low-affinity sensor for both Ca^{2+} and Mg^{2+}, in line with other studies (Chirasani et al. 2024); hence, we refer to this site as a Ca^{2+}/Mg^{2+} inhibition site (CaMgIS). The combined effect of Mg^{2+} binding at the high and low-affinity Ca^{2+} binding sites is reflected in the cytoplasmic shell as a reduced flexion angle, intermediate between the EGTA state and the open state (Fig. 9.4b).

In the presence of high concentrations of Mg^{2+}, a localized extra density appeared within the cytoplasmic region of the S6 bundle, "suspended" between the side chains of D4945 of the four subunits and attributed to Mg^{2+} (Nayak et al. 2024). Coordination of a partially hydrated Mg^{2+} within the ion pathway appears to afford an extra layer of inhibition by Mg^{2+}. Additionally, the presence of Mg^{2+} at this site results in tightening of the cytoplasmic section of the S6 helical bundle and creates a tripartite interaction network involving D4945'-R4944-D4938' (the apostrophe indicates S6 residues from the neighboring subunit), and additional interaction between E4942 from S6 and H4832 in the S4–S5 linker.

Given the strong inhibition provided by Mg^{2+} in the case of RyR1, how does the channel open in vivo? Without conformational coupling with the DHPR, opening of the RyR1 is impossible. This highlights the close physical and functional connection between the DHPR, RyR1, and ancillary proteins, which links the conformational change at the voltage sensor with the opening of the RyR gate. While little is known about the structure of this quaternary interaction, there is a feature within RyR1 that deserves attention. This is a 100 Å-long structure formed by four α-helices, two from the NTDC and two from the CD, connected in series ($461–480^{NTDC}$, $485–495^{NTDC}$, $3697–3710^{CD}$, and $3753–3769^{CD}$), that span from the T-tubule facing surface of the RyR1 all the way to the vicinity of the high-affinity Ca^{2+} binding site. This multi-helix structure forms four long "levers" (one per RyR protomer) that stay rigid and tilt outward when transitioning from the open to the closed state (Iyer et al. 2022). An exciting hypothesis is that these levers could form part of a long-range conformational pathway conveying the effect of DHPR activation to the RyR1. In the simplest form of this mechanism, activation of the DHPR could result in the separation of the four protomers at critical points via these levers. Specifically, their outward tilt would (a) cause loss of the EF hand-mediated interaction between adjacent protomers, overcoming inhibition at the CaMgIS site, and (b) induce separation of the S6 helical bundle, disrupting Mg^{2+} coordination at the D4945 site and allowing Ca^{2+} flow (Iyer et al. 2022). This proposed mechanism would explain how RyR1 can be activated despite the in vitro inhibitory Mg^{2+} concentrations (Nayak et al. 2024; Lamb and Stephenson 1992). A quantitative kinetic model that explores the interplay between Ca^{2+}, Mg^{2+}, and DHPR (Stephenson 2024) aligns with the structural concepts presented here.

Although absence of Mg^{2+} and nominal zero Ca^{2+} are conditions routinely used in structural and functional studies to attain the closed state of the channel, the metal-free condition is not physiological. It is important to note that in vivo, transitions likely occur between the Mg^{2+}-inhibited, the Ca^{2+}-inactivated, and the Ca^{2+}-open conformations.

9.5 Molecular Partners of RyR and Conformational Changes Induced by FKBP and CaM

An intricate quaternary structure of RyR1 with its molecular partners takes place in the triad junction of skeletal muscle. The minimum ensemble of triadic proteins that reproduces skeletal-type EC coupling in a heterologous system is formed by RyR1, $Ca_V1.1$, i.e., the channel-forming portion of the DHPR, the β1a subunit of the DHPR, Stac3, and either junctophilin 1 (JP1), JP2, or JP3 (Perni et al. 2017; Perni and Beam 2022). Additional protein partners include FKBP12, calmodulin (CaM), and S100A on the cytoplasmic shell, calsequestrin on the luminal side, and the single-pass transmembrane proteins triadin and junctin (Rebbeck et al. 2014). A distinctive feature of the cardiomyocyte when compared to skeletal muscle cells is that there is a dedicated set of isoforms for several EC coupling proteins, and that RyRs and DHPRs are not physically coupled. The position of some of these ligands is indicated in the schematics of Fig. 9.1b. RyR also interacts with neighboring RyRs in both cardiac and skeletal muscles, forming characteristic geometries in each tissue: corner-to corner in skeletal muscle, and oblique and side-to-side cardiac muscle (Franzini-Armstrong and Protasi 1997; Cabra et al. 2016; Asghari et al. 2014). This physical connection between neighboring RyRs underlies coupled gating, where RyRs open and close simultaneously (Marx et al. 1998).

The DHPR is a heteropentameric protein built around the α1 subunit which is the channel-forming portion; $Ca_V1.1$ in skeletal muscle and $Ca_V1.2$ in heart (Tanabe et al. 1988). Auxiliary subunits are the extracellular α2-δ subunit (products of cleavage of the same gene) (De Jongh et al. 1990), the intracellular β subunit (Beurg et al. 1999), and the transmembrane γ subunit (Yao et al. 2024). The intricate structural and functional interaction between RyR1s and DHPRs ensures the reliable transmission of the activation signal during activation, while in the resting state, DHPR appears to act as an inhibitor of RyR1 (Eltit et al. 2011; Zhou et al. 2006). Signaling in the retrograde direction also occurs, as evidenced by the enhanced channel activity of the DHPR promoted by RyR1 (Nakai et al. 1996). Groups of four DHPRs (tetrads) associate with every other RyR1, forming a distinctive and intriguing geometry (Block et al. 1988). This raises the question of how the non-interacting RyR1s get activated. One hypothesis suggests activation via coupled gating (Marx et al. 1998), while another proposed hypothesis, activation via CICR, was later discarded as CICR could not be observed in mammalian skeletal muscle under physiological conditions (Figueroa et al. 2012). Cryo-electron tomography and subtomogram averaging of skeletal muscle triads enabled easy recognition of the RyR1s in their typical checkerboard arrangement. Additionally, it revealed an electron-dense layer within the SR attributed to calsequestrin, and transmembrane densities that may correspond to triadin and/or junctin (Renken et al. 2009; Chen and Kudryashev 2020). Increased resolution enabled the first visualization of DHPR in situ (Xu et al. 2024).

Using single-particle cryo-EM, the binding site and effects of the small ligands FKBP, CaM, and S100 have been determined.

FKBP12 and FKBP12.6 (with molecular weight of 12 and 12.6 kDa, respectively) interact with the three isoforms of RyR in vitro, but only FKBP12 is expressed in skeletal muscle, while both FKBP12 and FKBP12.6 are expressed in heart. Absence of FKBP results in increased sensitivity of both RyR isoforms to Ca^{2+} activation. In addition, FKBP-depleted RyRs present long-lived subconductance states, which is rarely seen in native channels (Ahern et al. 1994; Ondrias et al. 1998; Sood et al. 2008).

3D reconstructions of RyR1-FKBP12 and RyR2-FKBP12.6 demonstrated that FKBP binds to the periphery of the cytoplasmic shell of RyR at four equivalent sites formed by the concave interface between the SPRY1 and handle domain, with additional interactions with SPRY3 (Steele and Samso 2019; Sharma et al. 2006; Wagenknecht et al. 1997). Multivariate statistical analysis and comparison of flexion angles between 3D reconstructions of RyR with and without FKBP established that, under closed-state conditions, RyR without FKBP displays flexion angles that resemble the "downward" conformation of the cytoplasmic shell seen in open RyR-FKBP. Without FKBP, the fully "upright" conformation was not observed (Steele and Samso 2019) (Fig. 9.4b). This suggests a "buttressing" effect of FKBP on RyR. Further, the "priming" observed in its absence could facilitate the transition to the open state, explaining the leaky behavior of the channel. FKBP binding to the concave SPRY1/handle site is also expected to limit the flexibility of the cytoplasmic shell and coordinate the four subunits, which could prevent subconductance states (Steele and Samso 2019). A similar conclusion was reached for RyR2-FKB12.6 (Miotto et al. 2024).

Calmodulin (CaM) is a 17 kDa Ca^{2+}-binding protein with two lobes, the N′ and the C′ terminal lobe, each containing a set of two EF hands. The binding of Ca^{2+} to the EF hands of CaM results in a structural rearrangement that exposes two hydrophobic surfaces, such that apoCaM and Ca^{2+}-CaM recognize different regions in the RyR and produce different effects. ApoCaM activates RyR1, Ca^{2+}-CaM inhibits RyR1 (Tripathy et al. 1995), and both forms of CaM inhibit RyR2 (Fruen et al. 2000; Xu and Meissner 2004). In fact, it has been proposed that Ca^{2+}-CaM plays a direct role in the inactivation of RyR1 (Yamaguchi et al. 2011) and RyR2 (Stern and Cheng 2004; Tian et al. 2013). At low resolution, apoCaM mapped to a region between the handle domain and HD1 of RyR1, while Ca^{2+}-CaM "slid" down and filled the space deeper between the handle domain and the distal part of HD2 (Samso and Wagenknecht 2002; Wagenknecht et al. 1994). The two distinct locations could explain their different effects on RyR1 function. In addition, the overlapping locations suggested a Ca^{2+}-dependent sliding mechanism, making CaM a mobile subunit of RyR (Samso and Wagenknecht 2002). The binding sites have since been confirmed at higher resolution, revealing detail at the molecular level. In RyR2, the N′ lobe of apoCaM binds to the upper region of a cleft in the HD1 domain, whereas the C′ lobe binds to the lower region of that cleft and is surrounded by the CD and handle domain (schematics in Fig. 9.4a) (Gong et al. 2019). Five surface patches mediate the interaction between both proteins, including the interaction between the C′ lobe of apoCaM and its target peptide on the RyR, helix $\alpha -1$ in the CD (residues 3593–3607 for porcine RyR2). Ca^{2+}-CaM binds further down the cleft, with the N′ lobe interacting extensively with the outer side of the CD and the C′ lobe mostly sticking out into the space between the handle and HD2, and both lobes interacting with helix $\alpha -1$ (Gong et al. 2019). The pore remained open, but a slight anticlockwise rotation of the CD domains resembles the open-to-closed transition of the channel, in line with the inhibitory modulation of CaM on RyR2 (Gong et al. 2019). A similar relocation of CaM can be seen for the RyR1 isoform. In RyR1, apoCaM bound in a very similar location and conformation as in RyR2-apoCaM (Woll et al. 2021). In the presence of Ca^{2+}, the C-lobe of Ca^{2+}-CaM descended closer to the tip of HD2 (Melville et al. 2022), but the N′ lobe remained in the upwards position, resulting in a more stretched out CaM. The conformational dynamics of CaM as it responds to changes in Ca^{2+} concentration occurs between 35 ms and 450 ms, as measured by stopped-flow fluorescence lifetime FRET. CaM relocates faster during the increase in Ca^{2+} than during the decrease, and at a faster rate for RyR1 than RyR2, which are on a time scale compatible with skeletal and cardiac muscle contraction processes (Rebbeck et al. 2024).

S100A1, a 21 kDa homodimer, is another Ca^{2+} binding protein with an important regulatory role in multiple tissues including skeletal muscle and heart (Kato and Kimura 1985; Schafer and Heizmann 1996). One of its targets is RyR, where it has an activating effect at low Ca^{2+} concentration (Treves et al. 1997; Kettlewell et al. 2005; Wright et al. 2008; Fulle et al. 2020). S100A1 forms a homodimer, where each monomer contains a pseudo-EF hand and a bona fide EF hand. Cryo-EM revealed that both apo- and Ca^{2+}-S100A1 occupy the space between HD1 and HD2 and the handle domains in RyR1, with Ca^{2+}-S100A1 moving lower into this space and inducing tighter inter-protomer interactions along with a 1 Å anticlockwise rotation of the cytoplasmic shell (Weninger et al. 2024). The same study revealed that Ca^{2+}-S100A1 interacts with residues 3625–3639 (mRyR1), a sequence that spans from the handle domain to HD2 and overlaps with helix α -1, the target peptide of CaM. Overall, the RyR binding sites for S100A1 and CaM overlap at high and low Ca^{2+} concentrations, thus simultaneous occupation is sterically impossible. When CaM and S100A1 were added simultaneously, CaM occupied the site (Weninger et al. 2024), indicating a higher affinity for CaM. Interestingly, while one monomer of S100A interacts with the RyR, the free monomer faces outwards, suggesting that it could act as an adaptor for other binding partners (Weninger et al. 2024).

9.6 Structural Changes Caused by Mutations in RyR

Numerous disease-causing point mutations have been described for RyR1 and RyR2 (Robinson et al. 2006). Mutations in RyR1 cause MH, a pharmacogenetic episode triggered by exposure to volatile anesthetics or succinylcholine muscle relaxants, or by excessive heat (McCarthy et al. 2000; MacLennan and Phillips 1992; Capacchione and Muldoon 2009; Ward et al. 1986). An MH episode can be lethal unless dantrolene, the only existing treatment, is administered promptly (Krause et al. 2004; Lopez et al. 1987). RyR1 mutations also cause debilitating myopathies such as CCD, multi-minicore disease, and nemaline myopathy; all of which manifest inactive regions of the muscle that lack contractile fibers (Treves et al. 2005; Rossi et al. 2007; Zvaritch et al. 2009). Mutations in RyR2 cause two forms of arrhythmia: catecholaminergic polymorphic ventricular tachycardia (CPVT) and arrhythmogenic right ventricular dysplasia type 2 (ARVD2) (Laitinen et al. 2001; Priori et al. 2001; Tiso et al. 2001; Kannankeril et al. 2006). CPVT can be triggered by exercise, more frequently observed in young people playing competitive sports, or under conditions of emotional stress, in the absence of structural heart defects. ARVD2 is characterized by partial degeneration of the right ventricle, electrical instability, and sudden death (Tiso et al. 2001). RyR2 mutations cause 30–50% cardiac death by the age of 30 (Postma et al. 2005; Betzenhauser and Marks 2010; Brini 2004). Most RyR disease mutations cause the channel to become leakier (Allen 2002; Lehnart et al. 2008; Lopez et al. 1988), although in some cases the channels have reduced permeability (Kobayashi et al. 2022; Avila et al. 2003).

With the current structural knowledge of RyR, a preliminary structure-based classification of all known mutations assigned them to four major categories (Iyer et al. 2023): (1) mutations affecting domain-domain interactions, (2) intradomain mutations changing the conformation of domains; (3) mutations at ligand-binding sites, and (4) mutations in S6, directly affecting the pore. Multiple cryo-EM 3D structures of RyR mutants have been solved, revealing surprisingly pronounced effects. A common pattern is a "priming" of the structure, where the cytoplasmic shell droops down to a lower flexion angle. Under experimental closed-state conditions (typically 2–5 mM EGTA) the pore remains closed, but the primed conformation results in a lower energy barrier toward the open state.

MH/CCD mutation RyR1 R163C, in the NTDA domain, is near the interface between the cytoplasmic shell and the core. This mutation alters the NTDA–CD interaction, inducing both a priming of the cytoplasmic shell and a shrinking of the CAS, akin to the closing of this cavity by Ca^{2+} (Iyer et al. 2025). This likely explains the increased sensitivity to Ca^{2+} of this mutant observed in functional studies (Yang et al. 2006). In contrast, the RyR2 CPVT mutation R176Q, in an equivalent position to that of R163C in RyR1, produced no noticeable effect on the structure. One explanation could be the less detrimental nature of the Q (versus C) mutation, which was demonstrated by the higher gain-of-function of the reciprocal mutation R176C (Iyer et al. 2025). On the other hand, RyR2 mutations may require additional sources of destabilization in order to manifest an effect. Along these lines, hyperphosphorylation was used in addition to mutations to simulate the effect of increased β-adrenergic signaling, which can trigger CPVT (Wehrens et al. 2006). Cryo-EM of hyperphosphorylated RyR2 CPVT mutants showed that R2474S, in the HD1 domain, caused the cytoplasmic shell to move downward and outward, inducing a priming halfway toward the open state. Mutations R420W/Q, in a residue participating in interactions between NTDB and NTDC, also caused priming but to a lesser extent (Miotto et al. 2022, 2024). Mutation R615C in pRyR1 dismantled stabilizing salt bridge interactions at the interface of domains NTDC and HD1, resulting in a downward shift of HD1 and loss of inter-subunit contacts at the NTDA/NTDB vestibular ring. Overall, this resulted in a priming of the channel (Woll et al. 2021). The severe RyR1 mutant Y522S, in the NTDC domain, not only induces Ca^{2+} leak but also affects the RyR1-DHPR communication via retrograde coupling, increasing its voltage sensitivity by 40 mV (Avila and Dirksen 2001). Cryo-EM of this mutant showed that, under closed-state conditions, the cytoplasmic shell was extremely primed, in a quasi-open conformation. Noticeable changes took place in the NTDC domain, a presumed region of interaction with the DHPR, whereby the NTDC rotated 5° and moved outward 2 Å, potentially explaining the altered communication with the DHPR phenotype of this mutation. Based on the cryo-EM structures published to date, it appears that although mutations causing leakiness induce priming of the cytoplasmic domain, the degree of priming, and the allosteric relay from the original local distortion is different for every mutation, which is expected to result in distinct biophysical properties of the channel. The structural details of loss-of-function mutations have also started to be elucidated, however not as much data exits (Kobayashi et al. 2022). From a therapeutic perspective, these structural insights could be instrumental to design therapies tailored to specific types of mutations, guiding personalized medicine for patients with RyR mutations.

9.7 Development of Drugs to Treat Hereditary RyR Disorders

Disease mutations in RyR typically lead to a gain-of-function phenotype (Allen 2002; Lehnart et al. 2008), thus the primary therapeutic strategies focus on identifying inhibitors of the channel. With regard to RyR1, dantrolene, a hydantoin derivative, is the only approved drug to arrest MH episodes (Kolb et al. 1982; Hopkins et al. 2021) and is also used to treat muscle spasticity (Ward et al. 1986). However, it is not approved for CCD due to its deleterious side effects when administered chronically. Cryo-EM studies showed that dantrolene binds to the P1 domains, which delineate the four corners of the cytoplasmic shell. Dantrolene was visible as an extra density within the cavity formed by the P1 4-helix bundle (Fig. 9.7b) and pulled in the outer helices by 10 Å (Iyer et al. 2025). In the P1 domain, mutating W882 of RyR1 completely abolished the inhibitory effect of dantrolene, while mutating W996 or R1000 had a partial effect. The drug strongly alleviated the priming induced by the RyR1 Y523S mutation, although it did not return the channel to the WT conformation. Comparing the closed and open structures of RyR1 Y523S in the presence of dantrolene showed that the channel was able to gate, but with less amplitude of movement than in wt RyR1 (Iyer et al. 2025). The conformational change of the P1 domain induced by dantrolene could also affect the quaternary arrangements between RyR1s, known to interact via the P1 domains in skeletal muscle (Yin et al. 2008). A 3D reconstruction of wt RyR1 with ARM210, a benzothiazepine derivative undergoing clinical trials, showed that the drug also bound to the P1 domain, inducing a mild closure of its 4-helix bundle (Melville et al. 2022). This suggests that the P1 domain could be a hotspot for drug binding. Several drugs are being considered to treat RyR2-associated arrhythmogenesis. These include flecainide (Watanabe et al. 2009; Hilliard et al. 2010), carvedilol (Zhou et al. 2011), S107 (Lehnart et al. 2008), and an unnatural verticilide enantiomer (Batiste et al. 2019).

The search for drugs targeting RyR by high throughput screening (HTS) is an ongoing challenge, which is complicated by the intracellular location of RyR limiting the available readout techniques. Additionally, in the case of RyR1, there is the difficulty to replicate conformational coupling in HTS platforms. Ongoing HTS efforts include the use of fluorescence lifetime FRET which tracks intramolecular movements of the receptor (Rebbeck et al. 2020) and time-lapse fluorescent measurement of ER Ca^{2+} content (Murayama et al. 2018; Yamazawa et al. 2021). HTS and optimization of lead compounds offer a promising path to develop effective and safe inhibitors of RyR with optimal pharmacological properties.

9.8 Conclusions and Outlook

Cryo-EM of RyR at near-atomic resolution is beginning to integrate a wealth of functional data into a structural framework, which is expected to solve long-standing questions in the field. Progress in cryo-EM and image processing will continue to shed light on how different ligands and ancillary proteins control gating, how their effect varies across isoforms, how disease-causing mutations destabilize specific conformations, and how drugs can restore wild-type regulation. This chapter highlighted the considerable conformational changes that drive gating, inactivation, and inhibition of the channel, under the effect of Ca^{2+}, Mg^{2+}, and CaM. The binding of these effectors makes for distinct interweaving between allosteric pathways. However, a more comprehensive effort will be required to understand the operation of this intricate macromolecular machine. Time-resolved cryo-EM and FRET, combined with large-scale molecular dynamics and artificial intelligence, will open a window into short-lived intermediates and processes on the milli- to nano-second timescale. In addition, high-resolution cryo-electron tomography of the entire EC coupling machinery in its native membrane environment will be crucial to better define the supramolecular complex and isolate the role of its different players.

The effect of mutations on RyR conformation is also starting to be understood. Although all gain-of-function channels studied so far have a primed conformation, the allosteric signal relay is contorted in different ways depending on the specific mutation. Dantrolene, while it did not fully restore the WT conformation of a specific RyR disease mutant, impinged on these allosteric pathways toward a more controlled gating. This suggests that different drugs could be tailored for specific mutation types, providing a pathway for HTS toward personalized medicine. At any rate, classical cryo-EM of isolated complexes will continue to be an essential tool as new findings and therapeutic drugs emerge.

References

Ahern GP, Junankar PR, Dulhunty AF (1994) Single channel activity of the ryanodine receptor calcium release channel is modulated by FK-506. FEBS Lett 352(3):369–374

Ahern GP, Junankar PR, Dulhunty AF (1997) Subconductance states in single-channel activity of skeletal muscle ryanodine receptors after removal of FKBP12. Biophys J 72(1):146–162

Allen PD (2002) Leaky "feet" and sudden death. Circ Res 91(3):181–182

Asghari P, Scriven DR, Sanatani S, Gandhi SK, Campbell AI, Moore ED (2014) Nonuniform and variable arrangements of ryanodine receptors within mammalian ventricular couplons. Circ Res 115(2):252–262

Ashley CC, Ridgway EB (1968) Simultaneous recording of membrane potential, calcium transient and tension in single muscle fibers. Nature 219(5159):1168–1169

Avila G, Dirksen RT (2001) Functional effects of central core disease mutations in the cytoplasmic region of the skeletal muscle ryanodine receptor. J Gen Physiol 118(3):277–290

Avila G, O'Connell KM, Dirksen RT (2003) The pore region of the skeletal muscle ryanodine receptor is a primary locus for excitation-contraction uncoupling in central core disease. J Gen Physiol 121(4):277–286

Bai XC, Yan Z, Wu J, Li Z, Yan N (2016) The central domain of RyR1 is the transducer for long-range allosteric gating of channel opening. Cell Res 26(9):995–1006

Baker MR, Fan G, Arige V, Yule DI, Serysheva II (2023) Understanding IP(3)R channels: from structural underpinnings to ligand-dependent conformational landscape. Cell Calcium 114:102770

Batiste SM, Blackwell DJ, Kim K, Kryshtal DO, Gomez-Hurtado N, Rebbeck RT et al (2019) Unnatural verticilide enantiomer inhibits type 2 ryanodine receptor-mediated calcium leak and is antiarrhythmic. Proc Natl Acad Sci USA 116(11):4810–4815

Beard NA, Dulhunty AF (2015) C-terminal residues of skeletal muscle calsequestrin are essential for calcium binding and for skeletal ryanodine receptor inhibition. Skelet Muscle 5:6

Betzenhauser MJ, Marks AR (2010) Ryanodine receptor channelopathies. Pflugers Arch 460(2):467–480

Beurg M, Sukhareva M, Ahern CA, Conklin MW, Perez-Reyes E, Powers PA et al (1999) Differential regulation of skeletal muscle L-type Ca2+ current and excitation-contraction coupling by the dihydropyridine receptor beta subunit. Biophys J 76(4):1744–1756

Block BA, Imagawa T, Campbell KP, Franzini-Armstrong C (1988) Structural evidence for direct interaction between the molecular components of the transverse tubule/sarcoplasmic reticulum junction in skeletal muscle. J Cell Biol 107(6 Pt 2):2587–2600

Brillantes AB, Ondrias K, Scott A, Kobrinsky E, Ondriasova E, Moschella MC et al (1994) Stabilization of calcium release channel (ryanodine receptor) function by FK506-binding protein. Cell 77(4):513–523

Brini M (2004) Ryanodine receptor defects in muscle genetic diseases. Biochem Biophys Res Commun 322(4):1245–1255

Buck E, Zimanyi I, Abramson JJ, Pessah IN (1992) Ryanodine stabilizes multiple conformational states of the skeletal muscle calcium release channel. J Biol Chem 267(33):23560–23567

Cabra V, Murayama T, Samso M (2016) Ultrastructural analysis of self-associated RyR2s. Biophys J 110(12):2651–2662

Capacchione JF, Muldoon SM (2009) The relationship between exertional heat illness, exertional rhabdomyolysis, and malignant hyperthermia. Anesth Analg 109(4):1065–1069

Chen W, Kudryashev M (2020) Structure of RyR1 in native membranes. EMBO Rep 21(5):e49891

Chen SR, Li X, Ebisawa K, Zhang L (1997) Functional characterization of the recombinant type 3 Ca2+ release channel (ryanodine receptor) expressed in HEK293 cells. J Biol Chem 272(39):24234–24246

Chen W, Wang R, Chen B, Zhong X, Kong H, Bai Y et al (2014) The ryanodine receptor store-sensing gate controls Ca2+ waves and Ca2+-triggered arrhythmias. Nat Med 20(2):184–192

Chen YS, Garcia-Castaneda M, Charalambous M, Rossi D, Sorrentino V, Van Petegem F (2024) Cryo-EM investigation of ryanodine receptor type 3. Nat Commun 15(1):8630

Ching LL, Williams AJ, Sitsapesan R (2000) Evidence for Ca(2+) activation and inactivation sites on the luminal side of the cardiac ryanodine receptor complex. Circ Res 87(3):201–206

Chirasani VR, Popov KI, Meissner G, Dokholyan NV (2022) Mapping co-regulatory interactions among ligand-binding sites in ryanodine receptor 1. Proteins 90(2):385–394

Chirasani VR, Elferdink M, Kral M, Carter JS, Heitmann S, Meissner G et al (2024) Structural and functional interactions between the EF hand domain and S2-S3 loop in the type-1 ryanodine receptor ion channel. J Biol Chem 300(2):105606

Csernoch L, Bernengo JC, Szentesi P, Jacquemond V (1998) Measurements of intracellular Mg2+ concentration in mouse skeletal muscle fibers with the fluorescent indicator mag-indo-1. Biophys J 75(2):957–967

Cully TR, Choi RH, Bjorksten AR, Stephenson DG, Murphy RM, Launikonis BS (2018) Junctional membrane Ca(2+) dynamics in human muscle fibers are altered by malignant hyperthermia causative RyR mutation. Proc Natl Acad Sci USA 115(32):8215–8220

Dayal A, Schrotter K, Pan Y, Fohr K, Melzer W, Grabner M (2017) The Ca(2+) influx through the mammalian skeletal muscle dihydropyridine receptor is irrelevant for muscle performance. Nat Commun 8(1):475

De Jongh KS, Warner C, Catterall WA (1990) Subunits of purified calcium channels. Alpha 2 and delta are encoded by the same gene. J Biol Chem 265(25):14738–14741

des Georges A, Clarke OB, Zalk R, Yuan Q, Condon KJ, Grassucci RA et al (2016) Structural basis for gating and activation of RyR1. Cell 167(1):145–57 e17

Donoso P, Prieto H, Hidalgo C (1995) Luminal calcium regulates calcium release in triads isolated from frog and rabbit skeletal muscle. Biophys J 68(2):507–515

Eltit JM, Li H, Ward CW, Molinski T, Pessah IN, Allen PD et al (2011) Orthograde dihydropyridine receptor signal regulates ryanodine receptor passive leak. Proc Natl Acad Sci USA 108(17):7046–7051

Fabiato A (1983) Calcium-induced release of calcium from the cardiac sarcoplasmic reticulum. Am J Phys 245(1):C1–C14

Fessenden JD, Feng W, Pessah IN, Allen PD (2004) Mutational analysis of putative calcium binding motifs within the skeletal ryanodine receptor isoform, RyR1. J Biol Chem 279(51):53028–53035

Figueroa L, Shkryl VM, Zhou J, Manno C, Momotake A, Brum G et al (2012) Synthetic localized calcium transients directly probe signalling mechanisms in skeletal muscle. J Physiol 590(Pt 6):1389–1411

Flucher BE, Franzini-Armstrong C (1996) Formation of junctions involved in excitation-contraction coupling in skeletal and cardiac muscle. Proc Natl Acad Sci USA 93(15):8101–8106

Flucher BE, Takekura H, Franzini-Armstrong C (1993) Development of the excitation-contraction coupling apparatus in skeletal muscle: association of sarcoplasmic reticulum and transverse tubules with myofibrils. Dev Biol 160(1):135–147

Franzini-Armstrong C, Nunzi G (1983) Junctional feet and particles in the triads of a fast-twitch muscle fibre. J Muscle Res Cell Motil 4(2):233–252

Franzini-Armstrong C, Porter KR (1964) Sarcolemmal invaginations constituting the T system in fish muscle fibers. J Cell Biol 22(3):675–696

Franzini-Armstrong C, Protasi F (1997) Ryanodine receptors of striated muscles: a complex channel capable of multiple interactions. Physiol Rev 77(3):699–729

Franzini-Armstrong C, Protasi F, Ramesh V (1999) Shape, size, and distribution of Ca(2+) release units and couplons in skeletal and cardiac muscles. Biophys J 77(3):1528–1539

Fruen BR, Bardy JM, Byrem TM, Strasburg GM, Louis CF (2000) Differential Ca(2+) sensitivity of skeletal and cardiac muscle ryanodine receptors in the presence of calmodulin. Am J Physiol Cell Physiol 279(3):C724–C733

Fulle S, Belia S, Fano IG (2020) The Arianna thread: the matching of S-100 family with the RyR's muscle receptor. Eur J Transl Myol 30(1):8839

Gao L, Balshaw D, Xu L, Tripathy A, Xin C, Meissner G (2000) Evidence for a role of the lumenal M3-M4 loop in skeletal muscle ca(2+) release channel (ryanodine receptor) activity and conductance. Biophys J 79(2):828–840

Giannini G, Conti A, Mammarella S, Scrobogna M, Sorrentino V (1995) The ryanodine receptor/calcium channel genes are widely and differentially expressed in murine brain and peripheral tissues. J Cell Biol 128(5):893–904

Gillespie D, Chen H, Fill M (2012) Is ryanodine receptor a calcium or magnesium channel? Roles of K+ and Mg2+ during Ca2+ release. Cell Calcium 51(6):427–433

Gillespie D, Xu L, Meissner G (2014) Selecting ions by size in a calcium channel: the ryanodine receptor case study. Biophys J 107(10):2263–2273

Gomez AC, Yamaguchi N (2014) Two regions of the ryanodine receptor calcium channel are involved in Ca(2+)-dependent inactivation. Biochemistry 53(8):1373–1379

Gomez AC, Holford TW, Yamaguchi N (2016) Malignant hyperthermia-associated mutations in the S2-S3 cytoplasmic loop of type 1 ryanodine receptor calcium channel impair calcium-dependent inactivation. Am J Physiol Cell Physiol 311(5):C749–C757

Gong D, Chi X, Wei J, Zhou G, Huang G, Zhang L et al (2019) Modulation of cardiac ryanodine receptor 2 by calmodulin. Nature 572(7769):347–351

Gyorke I, Gyorke S (1998) Regulation of the cardiac ryanodine receptor channel by luminal Ca2+ involves luminal Ca2+ sensing sites. Biophys J 75(6):2801–2810

Gyorke S, Gyorke I, Lukyanenko V, Terentyev D, Viatchenko-Karpinski S, Wiesner TF (2002) Regulation of sarcoplasmic reticulum calcium release by luminal calcium in cardiac muscle. Front Biosci 7:d1454–d1463

Harvey KE, LaVigne EK, Dar MS, Salyer AE, Pratt EPS, Sample PA et al (2022) RyR2/IRBIT regulates insulin gene transcript, insulin content, and secretion in the insulinoma cell line INS-1. Sci Rep 12(1):7713

Heinz LP, Kopec W, de Groot BL, Fink RHA (2018) In silico assessment of the conduction mechanism of the ryanodine receptor 1 reveals previously unknown exit pathways. Sci Rep 8(1):6886

Hilliard FA, Steele DS, Laver D, Yang Z, Le Marchand SJ, Chopra N et al (2010) Flecainide inhibits arrhythmogenic Ca2+ waves by open state block of ryanodine receptor Ca2+ release channels and reduction of Ca2+ spark mass. J Mol Cell Cardiol 48(2):293–301

Holehouse AS, Kragelund BB (2024) The molecular basis for cellular function of intrinsically disordered protein regions. Nat Rev Mol Cell Biol 25(3):187–211

Hopkins PM, Girard T, Dalay S, Jenkins B, Thacker A, Patteril M et al (2021) Malignant hyperthermia 2020: guideline from the Association of Anaesthetists. Anaesthesia 76(5):655–664

Hosoi E, Nishizaki C, Gallagher KL, Wyre HW, Matsuo Y, Sei Y (2001) Expression of the ryanodine receptor isoforms in immune cells. J Immunol 167(9):4887–4894

Hou Y, Jayasinghe I, Crossman DJ, Baddeley D, Soeller C (2015) Nanoscale analysis of ryanodine receptor clusters in dyadic couplings of rat cardiac myocytes. J Mol Cell Cardiol 80:45–55

Iyer KA, Hu Y, Klose T, Murayama T, Samso M (2022) Molecular mechanism of the severe MH/CCD mutation Y522S in skeletal ryanodine receptor (RyR1) by cryo-EM. Proc Natl Acad Sci USA 119(30):e2122140119

Iyer KA, Barnakov V, Samso M (2023) Three-dimensional perspective on ryanodine receptor mutations causing skeletal and cardiac muscle-related diseases. Curr Opin Pharmacol 68:102327

Iyer KA, Kobayashi T, Murayama T, Samso M (2025) Dantrolene inhibition of ryanodine receptor 1 carrying the severe malignant hyperthermia mutation Y522S visualized by cryo-EM. Structure 33(2):383–348.e4

Jorgensen C, Furini S, Domene C (2016) Energetics of ion permeation in an open-activated TRPV1 channel. Biophys J 111(6):1214–1222

Kannankeril PJ, Mitchell BM, Goonasekera SA, Chelu MG, Zhang W, Sood S et al (2006) Mice with the R176Q cardiac ryanodine receptor mutation exhibit catecholamine-induced ventricular tachycardia and cardiomyopathy. Proc Natl Acad Sci USA 103(32):12179–12184

Kato K, Kimura S (1985) S100ao (alpha alpha) protein is mainly located in the heart and striated muscles. Biochim Biophys Acta 842(2–3):146–150

Kettlewell S, Most P, Currie S, Koch WJ, Smith GL (2005) S100A1 increases the gain of excitation-contraction coupling in isolated rabbit ventricular cardiomyocytes. J Mol Cell Cardiol 39(6):900–910

Kobayashi T, Tsutsumi A, Kurebayashi N, Saito K, Kodama M, Sakurai T et al (2022) Molecular basis for gating of cardiac ryanodine receptor explains the mechanisms for gain- and loss-of function mutations. Nat Commun 13(1):2821

Kobayashi T, Yamazawa T, Kurebayashi N, Konishi M, Tanihata J, Sugihara M et al (2025) RyR1-mediated Ca(2+)-induced Ca(2+) release plays a negligible role in excitation-contraction coupling of normal skeletal muscle. Proc Natl Acad Sci U S A 122:e2500449122

Kolb ME, Horne ML, Martz R (1982) Dantrolene in human malignant hyperthermia. Anesthesiology 56(4):254–262

Krause T, Gerbershagen MU, Fiege M, Weisshorn R, Wappler F (2004) Dantrolene–a review of its pharmacology, therapeutic use and new developments. Anaesthesia 59(4):364–373

Lahat H, Pras E, Olender T, Avidan N, Ben-Asher E, Man O et al (2001) A missense mutation in a highly conserved region of CASQ2 is associated with autosomal recessive catecholamine-induced polymorphic ventricular tachycardia in Bedouin families from Israel. Am J Hum Genet 69(6):1378–1384

Laitinen PJ, Brown KM, Piippo K, Swan H, Devaney JM, Brahmbhatt B et al (2001) Mutations of the cardiac ryanodine receptor (RyR2) gene in familial polymorphic ventricular tachycardia. Circulation 103(4):485–490

Lamb GD, Stephenson DG (1991) Effect of Mg2+ on the control of Ca2+ release in skeletal muscle fibres of the toad. J Physiol 434:507–528

Lamb GD, Stephenson DG (1992) Importance of Mg2+ in excitation-contraction coupling in skeletal muscle. Physiology 7(6):270–274

Langer GA, Peskoff A (1996) Calcium concentration and movement in the diadic cleft space of the cardiac ventricular cell. Biophys J 70(3):1169–1182

Laver DR (2007) Ca2+ stores regulate ryanodine receptor Ca2+ release channels via luminal and cytosolic Ca2+ sites. Biophys J 92:3541–3555

Laver DR, Roden LD, Ahern GP, Eager KR, Junankar PR, Dulhunty AF (1995) Cytoplasmic Ca2+ inhibits the ryanodine receptor from cardiac muscle. J Membr Biol 147(1):7–22

Laver DR, Baynes TM, Dulhunty AF (1997) Magnesium inhibition of ryanodine-receptor calcium channels: evidence for two independent mechanisms. J Membr Biol 156(3):213–229

Ledbetter MW, Preiner JK, Louis CF, Mickelson JR (1994) Tissue distribution of ryanodine receptor isoforms and alleles determined by reverse transcription polymerase chain reaction. J Biol Chem 269(50):31544–31551

Lehnart SE, Mongillo M, Bellinger A, Lindegger N, Chen BX, Hsueh W et al (2008) Leaky Ca2+ release channel/ryanodine receptor 2 causes seizures and sudden cardiac death in mice. J Clin Invest 118(6):2230–2245

Lopez JR, Allen P, Alamo L, Ryan JF, Jones DE, Sreter F (1987) Dantrolene prevents the malignant hyperthermic syndrome by reducing free intracellular calcium concentration in skeletal muscle of susceptible swine. Cell Calcium 8(5):385–396

Lopez JR, Allen PD, Alamo L, Jones D, Sreter FA (1988) Myoplasmic free [Ca2+] during a malignant hyperthermia episode in swine. Muscle Nerve 11(1):82–88

MacLennan DH, Phillips MS (1992) Malignant hyperthermia. Science 256(5058):789–794

Marx SO, Ondrias K, Marks AR (1998) Coupled gating between individual skeletal muscle Ca2+ release channels (ryanodine receptors). Science 281(5378):818–821

McCarthy TV, Quane KA, Lynch PJ (2000) Ryanodine receptor mutations in malignant hyperthermia and central core disease. Hum Mutat 15(5):410–417

Meissner G (2002) Regulation of mammalian ryanodine receptors. Front Biosci 7:d2072–d2080

Meissner G (2017) The structural basis of ryanodine receptor ion channel function. J Gen Physiol 149(12):1065–1089

Meissner G, Darling E, Eveleth J (1986) Kinetics of rapid Ca2+ release by sarcoplasmic reticulum. Effects of Ca2+, Mg2+, and adenine nucleotides. Biochemistry 25(1):236–244

Meissner G, Rios E, Tripathy A, Pasek DA (1997) Regulation of skeletal muscle Ca2+ release channel (ryanodine receptor) by Ca2+ and monovalent cations and anions. J Biol Chem 272(3):1628–1638

Melville Z, Dridi H, Yuan Q, Reiken S, Wronska A, Liu Y et al (2022) A drug and ATP binding site in type 1 ryanodine receptor. Structure 30(7):1025–34 e4

Miotto MC, Weninger G, Dridi H, Yuan Q, Liu Y, Wronska A et al (2022) Structural analyses of human ryanodine receptor type 2 channels reveal the mechanisms for sudden cardiac death and treatment. Sci Adv 8(29):eabo1272

Miotto MC, Reiken S, Wronska A, Yuan Q, Dridi H, Liu Y et al (2024) Structural basis for ryanodine receptor type 2 leak in heart failure and arrhythmogenic disorders. Nat Commun 15(1):8080

Murayama T, Kurebayashi N, Ishigami-Yuasa M, Mori S, Suzuki Y, Akima R et al (2018) Efficient high-throughput screening by endoplasmic reticulum Ca(2+) measurement to identify inhibitors of ryanodine receptor Ca(2+)-release channels. Mol Pharmacol 94(1):722–730

Nagasaki K, Kasai M (1984) Channel selectivity and gating specificity of calcium-induced calcium release channel in isolated sarcoplasmic reticulum. J Biochem 96(6):1769–1775

Nakai J, Dirksen RT, Nguyen HT, Pessah IN, Beam KG, Allen PD (1996) Enhanced dihydropyridine receptor channel activity in the presence of ryanodine receptor. Nature 380(6569):72–75

Nayak AR, Samso M (2022) Ca(2+)-inactivation of the mammalian ryanodine receptor type 1 in a lipidic environment revealed by cryo-EM. Elife 11:e75568

Nayak AR, Rangubpit W, Will AH, Hu Y, Castro-Hartmann P, Lobo JJ et al (2024) Interplay between Mg(2+) and Ca(2+) at multiple sites of the ryanodine receptor. Nat Commun 15(1):4115

O'Connell PJ, Klyachko VA, Ahern GP (2002) Identification of functional type 1 ryanodine receptors in mouse dendritic cells. FEBS Lett 512(1–3):67–70

Ondrias K, Marx SO, Gaburjakova M, Marks AR. FKBP12 modulates gating of the ryanodine receptor/calcium release channel. Ann N Y Acad Sci 1998;853:149-56

Paolini C, Fessenden JD, Pessah IN, Franzini-Armstrong C (2004a) Evidence for conformational coupling between two calcium channels. Proc Natl Acad Sci USA 101(34):12748–12752

Paolini C, Protasi F, Franzini-Armstrong C (2004b) The relative position of RyR feet and DHPR tetrads in skeletal muscle. J Mol Biol 342(1):145–153

Peng W, Shen H, Wu J, Guo W, Pan X, Wang R et al (2016) Structural basis for the gating mechanism of the type 2 ryanodine receptor RyR2. Science 354(6310)

Perni S, Beam K (2022) Junctophilins 1, 2, and 3 all support voltage-induced Ca2+ release despite considerable divergence. J Gen Physiol 154(9)

Perni S, Lavorato M, Beam KG (2017) De novo reconstitution reveals the proteins required for skeletal muscle voltage-induced Ca(2+) release. Proc Natl Acad Sci USA 114(52):13822–13827

Postma A, Denjoy I, Kamblock J, Alders M, Lupoglazoff J, Vaksmann G et al (2005) Catecholaminergic polymorphic ventricular tachycardia: RYR2 mutations, bradycardia, and follow up of the patients. J Med Genet 42(11):863–870

Priori SG, Napolitano C, Tiso N, Memmi M, Vignati G, Bloise R et al (2001) Mutations in the cardiac ryanodine receptor gene (hRyR2) underlie catecholaminergic polymorphic ventricular tachycardia. Circulation 103(2):196–200

Protasi F, Paolini C, Dainese M (2009) Calsequestrin-1: a new candidate gene for malignant hyperthermia and exertional/environmental heat stroke. J Physiol 587(Pt 13):3095–3100

Rebbeck RT, Karunasekara Y, Board PG, Beard NA, Casarotto MG, Dulhunty AF (2014) Skeletal muscle excitation-contraction coupling: who are the dancing partners? Int J Biochem Cell Biol 48:28–38

Rebbeck RT, Singh DP, Janicek KA, Bers DM, Thomas DD, Launikonis BS et al (2020) RyR1-targeted drug discovery pipeline integrating FRET-based high-throughput screening and human myofiber dynamic Ca(2+) assays. Sci Rep 10(1):1791

Rebbeck RT, Svensson B, Zhang J, Samso M, Thomas DD, Bers DM et al (2024) Kinetics and mapping of ca-driven calmodulin conformations on skeletal and cardiac muscle ryanodine receptors. Nat Commun 15(1):5120

Renken C, Hsieh CE, Marko M, Rath B, Leith A, Wagenknecht T et al (2009) Structure of frozen-hydrated triad junctions: a case study in motif searching inside tomograms. J Struct Biol 165(2):53–63

Rios E, Gyorke S (2009) Calsequestrin, triadin and more: the molecules that modulate calcium release in cardiac and skeletal muscle. J Physiol 587(Pt 13):3069–3070

Rios E, Pizarro G (1991) Voltage sensor of excitation-contraction coupling in skeletal muscle. Physiol Rev 71(3):849–908

Robinson R, Carpenter D, Shaw MA, Halsall J, Hopkins P (2006) Mutations in RYR1 in malignant hyperthermia and central core disease. Hum Mutat 27(10):977–989

Rossi D, Sorrentino V (2002) Molecular genetics of ryanodine receptors Ca2+-release channels. Cell Calcium 32(5–6):307–319

Rossi D, De Smet P, Lyfenko A, Galli L, Lorenzini S, Franci D et al (2007) A truncation in the RYR1 gene associated with central core lesions in skeletal muscle fibres. J Med Genet 44(2):e67

Royer L, Sztretye M, Manno C, Pouvreau S, Zhou J, Knollmann BC et al (2010) Paradoxical buffering of calcium by calsequestrin demonstrated for the calcium store of skeletal muscle. J Gen Physiol 136(3):325–338

Samso M, Wagenknecht T (2002) Apocalmodulin and Ca2+-calmodulin bind to neighboring locations on the ryanodine receptor. J Biol Chem 277(2):1349–1353

Samso M, Wagenknecht T, Allen PD (2005) Internal structure and visualization of transmembrane domains of the RyR1 calcium release channel by cryo-EM. Nat Struct Mol Biol 12(6):539–544

Samso M, Feng W, Pessah IN, Allen PD (2009) Coordinated movement of cytoplasmic and transmembrane domains of RyR1 upon gating. PLoS Biol 7(4):980–995

Santulli G, Pagano G, Sardu C, Xie W, Reiken S, D'Ascia SL et al (2015) Calcium release channel RyR2 regulates insulin release and glucose homeostasis. J Clin Invest 125(11):4316

Schafer BW, Heizmann CW (1996) The S100 family of EF-hand calcium-binding proteins: functions and pathology. Trends Biochem Sci 21(4):134–140

Schneider MF, Chandler WK (1973) Voltage dependent charge movement of skeletal muscle: a possible step in excitation-contraction coupling. Nature 242(5395):244–246

Scriven DR, Asghari P, Schulson MN, Moore ED (2010) Analysis of Cav1.2 and ryanodine receptor clusters in rat ventricular myocytes. Biophys J 99(12):3923–3929

Scriven DR, Asghari P, Moore ED (2013) Microarchitecture of the dyad. Cardiovasc Res 98(2):169–176

Shannon TR, Ginsburg KS, Bers DM (2000) Potentiation of fractional sarcoplasmic reticulum calcium release by total and free intra-sarcoplasmic reticulum calcium concentration. Biophys J 78(1):334–343

Sharma MR, Jeyakumar LH, Fleischer S, Wagenknecht T (2006) Three-dimensional visualization of FKBP12.6 binding to an open conformation of cardiac ryanodine receptor. Biophys J 90(1):164–172

Sharma P, Ishiyama N, Nair U, Li W, Dong A, Miyake T et al (2012) Structural determination of the phosphorylation domain of the ryanodine receptor. FEBS J 279(20):3952–3964

Smith JS, Imagawa T, Ma J, Fill M, Campbell KP, Coronado R (1988) Purified ryanodine receptor from rabbit skeletal muscle is the calcium-release channel of sarcoplasmic reticulum. J Gen Physiol 92(1):1–26

Sood S, Chelu MG, van Oort RJ, Skapura D, Santonastasi M, Dobrev D et al (2008) Intracellular calcium leak due to FKBP12.6 deficiency in mice facilitates the inducibility of atrial fibrillation. Heart Rhythm 5(7):1047–1054

Steele TWE, Samso M (2019) The FKBP12 subunit modifies the long-range allosterism of the ryanodine receptor. J Struct Biol 205(2):180–188

Stephenson DG (2024) Modeling the mechanism of Ca2+ release in skeletal muscle by DHPRs easing inhibition at RyR I1-sites. J Gen Physiol 156(10)

Stern MD, Cheng H (2004) Putting out the fire: what terminates calcium-induced calcium release in cardiac muscle? Cell Calcium 35(6):591–601

Szegedi C, Sarkozi S, Herzog A, Jona I, Varsanyi M. Calsequestrin: more than 'only' a luminal Ca2+ buffer inside the sarcoplasmic reticulum. Biochem J 1999;337 (Pt 1)(Pt 1):19–22

Tanabe T, Beam KG, Powell JA, Numa S (1988) Restoration of excitation-contraction coupling and slow calcium current in dysgenic muscle by dihydropyridine receptor complementary DNA. Nature 336(6195):134–139

Taweechat P, Boonamnaj P, Samso M, Sompornpisut P (2024) Significance of Zn(2+) in RyR1 for structural integrity and ligand binding: insight from molecular dynamics. J Phys Chem B 128(19):4670–4684

Terentyev D, Viatchenko-Karpinski S, Valdivia HH, Escobar AL, Gyorke S (2002) Luminal Ca2+ controls termination and refractory behavior of Ca2+-induced Ca2+ release in cardiac myocytes. Circ Res 91(5):414–420

Tian X, Tang Y, Liu Y, Wang R, Chen SR (2013) Calmodulin modulates the termination threshold for cardiac ryanodine receptor-mediated Ca2+ release. Biochem J 455(3):367–375

Tiso N, Stephan DA, Nava A, Bagattin A, Devaney JM, Stanchi F et al (2001) Identification of mutations in the cardiac ryanodine receptor gene in families affected with arrhythmogenic right ventricular cardiomyopathy type 2 (ARVD2). Hum Mol Genet 10(3):189–194

Treves S, Scutari E, Robert M, Groh S, Ottolia M, Prestipino G et al (1997) Interaction of S100A1 with the Ca2+ release channel (ryanodine receptor) of skeletal muscle. Biochemistry 36(38):11496–11503

Treves S, Anderson AA, Ducreux S, Divet A, Bleunven C, Grasso C et al (2005) Ryanodine receptor 1 mutations, dysregulation of calcium homeostasis and neuromuscular disorders. Neuromuscul Disord 15(9–10):577–587

Tripathy A, Xu L, Mann G, Meissner G (1995) Calmodulin activation and inhibition of skeletal muscle Ca2+ release channel (ryanodine receptor). Biophys J 69(1):106–119

Tripathy A, Resch W, Xu L, Valdivia HH, Meissner G (1998) Imperatoxin a induces subconductance states in Ca2+ release channels (ryanodine receptors) of cardiac and skeletal muscle. J Gen Physiol 111(5):679–690

Vukcevic M, Spagnoli GC, Iezzi G, Zorzato F, Treves S (2008) Ryanodine receptor activation by Ca v 1.2 is involved in dendritic cell major histocompatibility complex class II surface expression. J Biol Chem 283(50):34913–34922

Wagenknecht T, Grassucci R, Frank J, Saito A, Inui M, Fleischer S (1989) Three-dimensional architecture of the calcium channel/foot structure of sarcoplasmic reticulum. Nature 338(6211):167–170

Wagenknecht T, Berkowitz J, Grassucci R, Timerman AP, Fleischer S (1994) Localization of calmodulin binding sites on the ryanodine receptor from skeletal muscle by electron microscopy. Biophys J 67(6):2286–2295

Wagenknecht T, Radermacher M, Grassucci R, Berkowitz J, Xin HB, Fleischer S (1997) Locations of calmodulin and FK506-binding protein on the three-dimensional architecture of the skeletal muscle ryanodine receptor. J Biol Chem 272(51):32463–32471

Walker SM, Schrodt GR (1965) Continuity of the T system with the sarcolemma in rat skeletal muscle fibers. J Cell Biol 27(3):671–677

Wang Y, Xu L, Pasek DA, Gillespie D, Meissner G (2005) Probing the role of negatively charged amino acid residues in ion permeation of skeletal muscle ryanodine receptor. Biophys J 89:256

Ward A, Chaffman MO, Sorkin EM (1986) Dantrolene. A review of its pharmacodynamic and pharmacokinetic properties and therapeutic use in malignant hyperthermia, the neuroleptic malignant syndrome and an update of its use in muscle spasticity. Drugs 32(2):130–168

Watanabe H, Chopra N, Laver D, Hwang HS, Davies SS, Roach DE et al (2009) Flecainide prevents catecholaminergic polymorphic ventricular tachycardia in mice and humans. Nat Med 15(4):380–383

Wehrens XH, Lehnart SE, Reiken S, Vest JA, Wronska A, Marks AR (2006) Ryanodine receptor/calcium release channel PKA phosphorylation: a critical mediator of heart failure progression. Proc Natl Acad Sci USA 103(3):511–518

Wei R, Chen Q, Zhang L, Liu C, Liu C, Yin CC et al (2025) Structural insights into transmembrane helix S0 facilitated RyR1 channel gating by Ca(2+)/ATP. Nat Commun 16(1):1936

Weninger G, Miotto MC, Tchagou C, Reiken S, Dridi H, Brandenburg S et al (2024) Structural insights into the regulation of RyR1 by S100A1. Proc Natl Acad Sci U S A 121(27):e2400497121

Wolf IM, Diercks BP, Gattkowski E, Czarniak F, Kempski J, Werner R et al (2015) Frontrunners of T cell activation: initial, localized Ca2+ signals mediated by NAADP and the type 1 ryanodine receptor. Sci Signal 8(398):ra102

Woll KA, Haji-Ghassemi O, Van Petegem F (2021) Pathological conformations of disease mutant ryanodine receptors revealed by cryo-EM. Nat Commun 12(1):807

Wright NT, Prosser BL, Varney KM, Zimmer DB, Schneider MF, Weber DJ (2008) S100A1 and calmodulin compete for the same binding site on ryanodine receptor. J Biol Chem 283(39):26676–26683

Xiong H, Feng X, Gao L, Xu L, Pasek DA, Seok J-H et al (1998) Identification of a two EF-hand Ca2+ binding domain in lobster skeletal muscle ryanodine receptor/Ca2+ release channel. Biochemistry 37(14):4804–4814

Xu L, Meissner G (1998) Regulation of cardiac muscle Ca2+ release channel by sarcoplasmic reticulum lumenal Ca2+. Biophys J 75(5):2302–2312

Xu L, Meissner G (2004) Mechanism of calmodulin inhibition of cardiac sarcoplasmic reticulum Ca2+ release channel (ryanodine receptor). Biophys J 86(2):797–804

Xu L, Wang Y, Gillespie D, Meissner G (2006) Two rings of negative charges in the cytosolic vestibule of type-1 ryanodine receptor modulate ion fluxes. Biophys J 90(2):443–453

Xu L, Harms FL, Chirasani VR, Pasek DA, Kortum F, Meinecke P et al (2020) Single-channel properties of skeletal muscle ryanodine receptor pore Delta(4923)FF(4924) in two brothers with a lethal form of fetal akinesia. Cell Calcium 87:102182

Xu J, Liao C, Yin CC, Li G, Zhu Y, Sun F (2024) In situ structural insights into the excitation-contraction coupling mechanism of skeletal muscle. Sci Adv 10:eadl1126

Yamaguchi N, Prosser BL, Ghassemi F, Xu L, Pasek DA, Eu JP et al (2011) Modulation of sarcoplasmic reticulum Ca2+ release in skeletal muscle expressing ryanodine receptor impaired in regulation by calmodulin and S100A1. Am J Physiol Cell Physiol 300(5):C998–C1012

Yamazawa T, Kobayashi T, Kurebayashi N, Konishi M, Noguchi S, Inoue T et al (2021) A novel RyR1-selective inhibitor prevents and rescues sudden death in mouse models of malignant hyperthermia and heat stroke. Nat Commun 12(1):4293

Yan Z, Bai XC, Yan C, Wu J, Li Z, Xie T et al (2015) Structure of the rabbit ryanodine receptor RyR1 at near-atomic resolution. Nature 517(7532):50–55

Yang J, Ellinor PT, Sather WA, Zhang JF, Tsien RW (1993) Molecular determinants of Ca2+ selectivity and ion permeation in L-type Ca2+ channels. Nature 366(6451):158–161

Yang T, Riehl J, Esteve E, Matthaei KI, Goth S, Allen PD et al (2006) Pharmacologic and functional characterization of malignant hyperthermia in the R163C RyR1 knock-in mouse. Anesthesiology 105(6):1164–1175

Yao X, Gao S, Yan N (2024) Structural biology of voltage-gated calcium channels. Channels (Austin) 18(1):2290807

Yin CC, D'Cruz LG, Lai FA (2008) Ryanodine receptor arrays: not just a pretty pattern? Trends Cell Biol 18(4):149–156

Yu FH, Yarov-Yarovoy V, Gutman GA, Catterall WA (2005) Overview of molecular relationships in the voltage-gated ion channel superfamily. Pharmacol Rev 57(4):387–395

Yuchi Z, Lau K, Van Petegem F (2012) Disease mutations in the ryanodine receptor central region: crystal structures of a phosphorylation hot spot domain. Structure 20(7):1201–1211

Zhang L, Kelley J, Schmeisser G, Kobayashi YM, Jones LR (1997) Complex formation between junctin, triadin, calsequestrin, and the ryanodine receptor. Proteins of the cardiac junctional sarcoplasmic reticulum membrane. J Biol Chem 272(37):23389–23397

Zhang A, Yu H, Liu C, Song C (2020) The Ca(2+) permeation mechanism of the ryanodine receptor revealed by a multi-site ion model. Nat Commun 11(1):922

Zhou J, Yi J, Royer L, Launikonis BS, Gonzalez A, Garcia J et al (2006) A probable role of dihydropyridine receptors in repression of Ca2+ sparks demonstrated in cultured mammalian muscle. Am J Physiol Cell Physiol 290(2):C539–C553

Zhou Q, Xiao J, Jiang D, Wang R, Vembaiyan K, Wang A et al (2011) Carvedilol and its new analogs suppress arrhythmogenic store overload-induced Ca2+ release. Nat Med 17(8):1003–1009

Zima AV, Picht E, Bers DM, Blatter LA (2008) Partial inhibition of sarcoplasmic reticulum ca release evokes long-lasting ca release events in ventricular myocytes: role of luminal ca in termination of ca release. Biophys J 94(5):1867–1879

Zvaritch E, Kraeva N, Bombardier E, McCloy RA, Depreux F, Holmyard D et al (2009) Ca2+ dysregulation in Ryr1(I4895T/wt) mice causes congenital myopathy with progressive formation of minicores, cores, and nemaline rods. Proc Natl Acad Sci USA 106(51):21813–21818

10 Perspectives on TRPV Channels in the Structural and Computational Revolution Era

Eric Catalina-Hernandez, Mario López-Martín, and Alex Peralvarez-Marin

Abstract

Transient Receptor Potential Vanilloid (TRPV) channels are a subfamily of six membrane proteins involved in various physiological processes. TRPVs share a pore channel between the fifth and sixth domains, which can be activated or inactivated by temperature, pain, and cellular homeostasis. Dysregulation of TRPVs can lead to pain, inflammation, and numerous pathological conditions. Understanding the molecular mechanisms that control TRPV channel activation/inhibition, gating, and interactions with ligands is crucial for targeted drug design. Molecular dynamics (MD) simulations have emerged as a powerful computational tool to study TRPVs' function and ligand interaction. In this chapter, we review the use of MD and other computational tools to investigate the function and gating mechanism of TRPVs, as well as the binding of lipids and small molecules in these channels. In short, we describe the changes induced by channel opening caused by temperature and upon ligand or lipid binding. Furthermore, we review the studies that delve into describing binding sites or discovering new TRPV channel modulators.

Keywords

TRPV channels · Ion channel · Molecular dynamics · Heat activation · Ligand binding · Lipid binding

E. Catalina-Hernandez · M. López-Martín · A. Peralvarez-Marin (✉)
Unit of Biophysics, Department of Biochemistry and Molecular Biology, Facultat de Medicina, Av. Can Domènech s/n, Universitat Autònoma de Barcelona, Catalonia, Spain

Institute of Neurosciences, Universitat Autònoma de Barcelona, Catalonia, Spain
e-mail: alex.peralvarez@uab.cat

10.1 TRPV Channels

Transient Receptor Potential Vanilloid (TRPV) channels are a subfamily of the larger TRP channel family, which are integral membrane proteins involved in various physiological processes (Nilius and Owsianik 2011). The TRPV subfamily consists of six members: TRPV1, TRPV2, TRPV3, TRPV4, TRPV5, and TRPV6 (Fig. 10.1) (Venkatachalam and Montell 2007). TRPV channels share several common features in terms of structure, topology, oligomerization, and function. Structurally, they consist of six transmembrane domains with a pore region between the fifth and sixth domains,

A. Rosenhouse-Dantsker (ed.), *Decoding Ion Channels Structure and Function*, Advances in Experimental Medicine and Biology 1497, https://doi.org/10.1007/978-3-032-07523-9_10

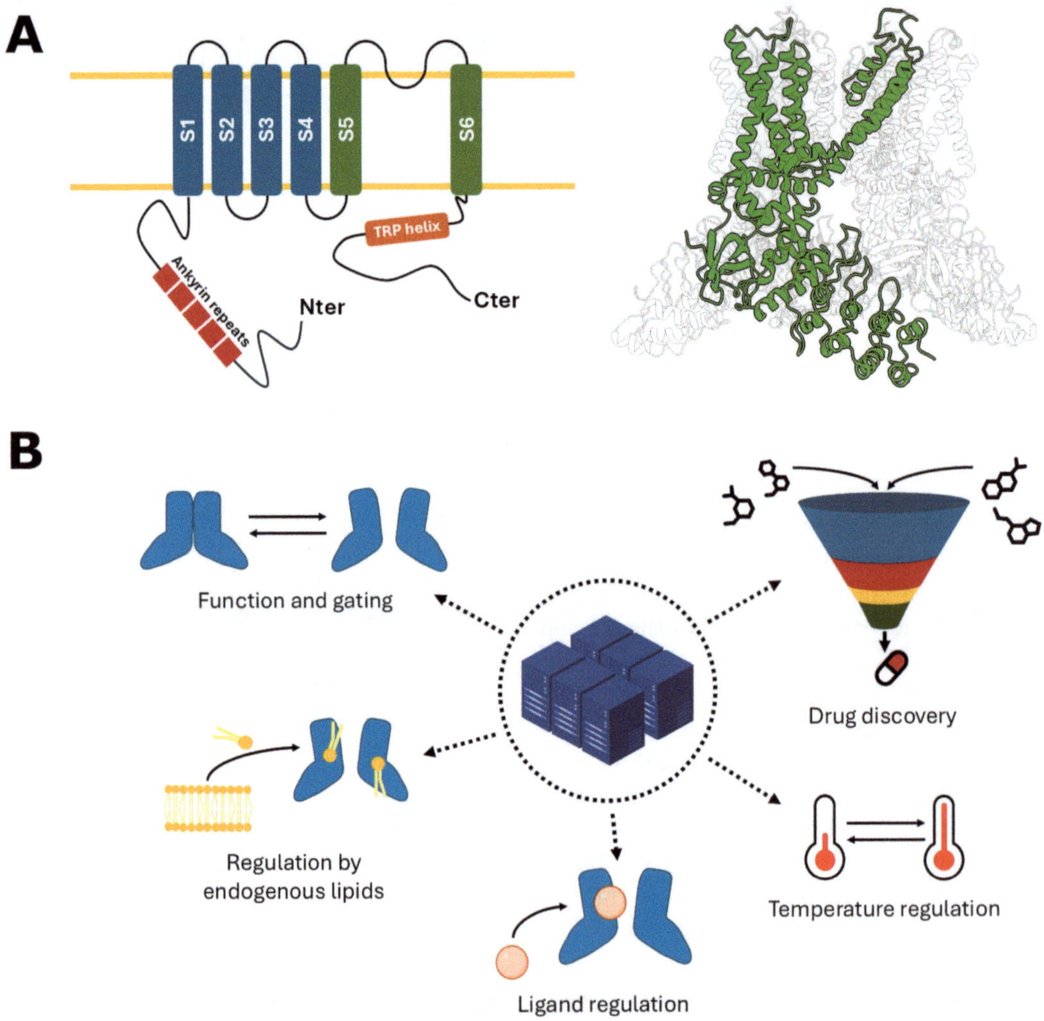

Fig. 10.1 (**a**) Schematic diagram of the monomeric TRPV topology (left) and representative homotetrameric TRPV structure (right) (PDB IDs: 6BO8, open state). (**b**) Graphical abstract of the main topic discussed in this chapter

forming a cation-permeable channel (Ramsey et al. 2006). Topologically, TRPV channels have intracellular N- and C-termini, which contain regulatory domains such as ankyrin repeats that are crucial for protein–protein interactions and channel regulation. Moreover, the six channels share intracellular domains, most notably the ankyrin repeat domains (ARD), which contribute to tetramerization and protein–protein interactions (Nilius and Owsianik 2011). Despite the high structural homology, TRPVs can be separated into two subgroups: TRPV1-TRPV4 and TRPV5-TRPV6, which share approximately 20% of sequence identity (Montell 2005). TRPV1, TRPV2, TRPV3, and TRPV4 share 30-50% sequence identity among them, with TRPV4 being the most dissimilar (Clapham et al. 2001; Samanta et al. 2018). In contrast, TRPV5 and TRPV6 share a much higher degree of amino acid sequence among them, approximately 75%, which supports their classification as adistinct, highly calcium-selective group (Montell 2005; Clapham et al. 2001; Samanta et al. 2018).

Oligomerization is a key feature, as TRPV channels typically form tetrameric complexes, which are essential for their functional properties. Functionally, these channels are involved in sensing and responding to various physical and chemical stimuli, including temperature, osmotic pressure, and mechanical stress (Clapham 2003). They play critical roles in processes such as pain perception, thermoregulation, and cation/calcium homeostasis. For specific knowledge on each channel, the reader is referred to the available literature (Rosenbaum and Islas 2024; Huynh et al. 2014; Zhao et al. 2021; Seebohm and Schreiber 2021; Pedersen et al. 2005; Owsianik et al. 2006). Here, we present a brief introductory overview of these channels before discussing the insights obtained from structural and computational studies.

10.1.1 TRPV1

TRPV1, commonly known as the capsaicin receptor, is activated by noxious heat, acidic conditions, and capsaicin, the pungent compound in chili peppers. It is predominantly expressed in sensory neurons and is crucial for pain perception and thermoregulation. Dysregulation of TRPV1 is implicated in chronic pain conditions and inflammatory diseases, making it a target for analgesic drug development.

10.1.2 TRPV2

TRPV2 is activated by high temperatures and mechanical stress (Perálvarez-Marín et al. 2013). It is expressed in various tissues, including the heart, immune cells, and the nervous system. TRPV2 plays a role in immune response, muscle function, and cardiac physiology. Abnormal TRPV2 activity has been associated with cardiac hypertrophy and neurodegenerative diseases, highlighting its importance in maintaining cellular integrity under stress.

10.1.3 TRPV3

TRPV3 is sensitive to warm temperatures and certain chemical stimuli. It is primarily found in the skin, where it contributes to thermoregulation and skin barrier function. Mutations in TRPV3 can lead to skin disorders such as Olmsted syndrome, characterized by severe itching and skin thickening. Understanding TRPV3's role in skin physiology is essential for developing treatments for such dermatological conditions.

10.1.4 TRPV4

TRPV4 responds to osmotic pressure, mechanical stress, and moderate heat. It is widely expressed in tissues such as the skin, lungs, and kidneys. TRPV4 is involved in maintaining cellular osmotic balance and mechanotransduction. Dysregulation of TRPV4 is linked to skeletal dysplasias, neuropathic pain, and pulmonary diseases, underscoring its diverse physiological roles.

10.1.5 TRPV5

TRPV5 is primarily expressed in the kidney and is vital for calcium reabsorption. It plays a significant role in maintaining systemic calcium homeostasis. Mutations in TRPV5 can disrupt calcium balance, leading to disorders such as hypercalciuria and nephrolithiasis. Research into TRPV5 function is crucial for understanding and managing calcium-related metabolic disorders.

10.1.6 TRPV6

TRPV6 is also involved in calcium absorption, particularly in the intestine. It is essential for maintaining calcium balance and bone health. Abnormal TRPV6 function is associated with conditions such as hypercalciuria and certain cancers. Investigating TRPV6's role in calcium metabolism provides insights into potential therapeutic targets for these diseases.

In summary, TRPV channels are integral to various physiological processes, from pain perception and thermoregulation to calcium homeostasis and mechanotransduction. Their dysregulation is linked to a range of pathological conditions, making them important targets for therapeutic intervention. For a comprehensive understanding of TRPV channels, readers are encouraged to refer to detailed reviews and studies, such as those by Rosenbaum and Islas (2024), which describes in detail the basal properties and physiological potential of these channels.

10.2 TRPV Channels in the Context of Structural and Computational Biology

The study of TRPV channels has been significantly advanced by several groundbreaking discoveries, many of which have been recognized with Nobel Prizes. These advancements have not only deepened our understanding of TRPV channels but also highlighted their importance in drug discovery and human pathophysiology.

The 2021 Nobel Prize in Physiology or Medicine, awarded to David Julius and Ardem Patapoutian for their discovery of the TRPV1 channel, was a landmark achievement. Their work elucidated how TRPV1 channels are involved in sensing temperature and pain, providing a molecular basis for these sensations (Dubochet et al. 1988; Julius and Basbaum 2001). This discovery has opened new avenues for developing pain management therapies, as understanding the structure and function of TRPV1 is crucial for targeting it effectively in drug design (Liao et al. 2013).

The development of cryo-electron microscopy (cryo-EM), which earned Jacques Dubochet, Joachim Frank, and Richard Henderson the 2017 Nobel Prize in Chemistry, has revolutionized structural biology (Karplus and Petsko 1990). Cryo-EM allows scientists to visualize biomolecules at near-atomic resolution, making it possible to study the detailed structures of TRPV channels (Jumper et al. 2021). This technique has provided critical insights into the conformational states and functional mechanisms of these channels, which are essential for understanding their role in various physiological processes (Julius 2021). The TRPV1 ion channel was among the first structures to be resolved at near-atomic resolution using cryo-EM, marking a significant milestone in the cryo-EM revolution (Dubochet et al. 2017). This breakthrough provided unprecedented insights into the structural dynamics of TRPV1, which is crucial for understanding its role in pain sensation and temperature regulation (Karplus 2013). The high-resolution cryo-EM structures of TRPV1, such as those determined by Liao et al. (2013) (Baker 2024) and subsequent studies, have paved the way for detailed mechanistic studies and drug design efforts targeting this important ion channel (Frank 2006; Cao et al. 2013).

Molecular dynamics (MD) simulations, recognized by the 2013 Nobel Prize in Chemistry awarded to Martin Karplus, Michael Levitt, and Arieh Warshel, in combination with high-resolution structural biology, are key tools for studying TRPV channels (Henderson 1995). MD simulation is a computational technique that enables the study of molecular behavior at the atomic level. Starting with the initial coordinates of the system—typically comprising a protein or molecule of interest along with its environment (e.g., water, ions, ligands, lipids)—Newton's equations of motion are numerically solved to predict the system's behavior over time. Accurate predictions rely on force fields, which are mathematical models that describe the bonded and nonbonded interactions between the various atoms in the system. Simulations are now extensively used to explore a wide range of macromolecular processes, including protein conformational changes, lipid bilayer dynamics, protein–lipid and protein–ligand interactions, and the formation and stability of protein complexes, among others.

MD simulations have emerged as a powerful computational tool in the understanding of intricate TRPV channel function and regulation mechanisms. MD simulations provide detailed insights into the dynamic behavior of these channels at an atomic level, allowing for the modeling of the physical movement of atoms and molecules over time, which complements structural data obtained by X-ray or cryo-EM techniques. By simulating the interactions within TRPV channels and their environments, MD simulations help elucidate conformational changes, the mechanisms of ion permeation, ligand binding pathways, and the effects of modulators in the channel structure. This section will explore the significant contribution of MD simulations to our understanding of TRPV channel molecular function and gating mechanism, and the binding of ligands or lipids into TRPV channels.

The 2024 Nobel Prize in Chemistry was awarded to Demis Hassabis, John Jumper, and David Baker for their work on AlphaFold and computational protein design has further accelerated the study of TRPV channels (Levitt 2001). AlphaFold's ability to predict protein structures with high accuracy has enabled researchers to model TRPV channels and their interactions with unprecedented precision (Warshel and Levitt 1976). This breakthrough has significant implications for drug discovery, as it allows for the rapid identification of potential binding sites and the design of novel therapeutics targeting TRPV channels (Henderson and Unwin 1975).

These Nobel-recognized advancements are tightly interwoven with the identification and study of TRPV channels, underscoring the relevance of structural and computational dynamics in this field. The integration of cryo-EM, molecular dynamics, and artificial intelligence (AI) driven protein structure prediction has provided a comprehensive toolkit for unraveling the complexities of TRPV channels (Dubochet and McDowall 1981). This multidisciplinary approach is crucial for understanding the role of TRPV channels in human pathophysiology, including their involvement in pain perception, inflammation, and various diseases (Julius 2013).

The relevance of TRPV channels in drug discovery cannot be overstated. These channels are implicated in a wide range of physiological processes and pathological conditions, making them attractive targets for therapeutic intervention (Pedersen and Nilius 2007). The detailed structural and dynamic information obtained through advanced computational techniques enables the rational design of drugs that can modulate TRPV channel activity with high specificity and efficacy. This has the potential to lead to the development of new treatments for chronic pain, inflammatory diseases, and other conditions associated with TRPV channel dysfunction.

In summary, the recent revolutions in structural and computational biology, as recognized by multiple Nobel Prizes, have profoundly impacted the study of TRPV channels. The synergy between cryo-EM, molecular dynamics, and AI-driven protein structure prediction has provided unprecedented insights into the structure and function of any protein. Computational and experimental integrated approaches are essential for advancing our understanding of TRPV channels and their role in human health and disease, paving the way for innovative drug discovery and therapeutic strategies. In this chapter, we aim at providing a detailed overview of what has been achieved at the structural computational level regarding TRPV channels. For a transversal perspective on what can be achieved at the computational level, we provide a general overview in Table 10.1. Furthermore, for clarity, the methodologies pertinent to each TRPV channel will be described in the subsequent sections.

10.3 TRPV1

10.3.1 TRPV1 Function and Gating

The increasing amount of structural data generated during the last decade has provided significant insights into key aspects of the TRPV gating mechanism. Studies have shown that the extracellular gate of TRPV1 acts as a selectivity filter, whereas the intracellular gate forms a hydrophobic barrier, opening in response to heat or ligands like capsaicin (Melnick and Kaviany 2018). Recently, Trofimov and collaborators identified three major states of TRPV1: α-closed, π-closed, and π-open (Fig. 10.2a), with the π-open being fully hydrated and ion-conductive, whereas the

Table 10.1 Classification of all the articles included in the chapter, organized by channel and topic as outlined in the text

Channel	Topic	Computational method	Further references
TRPV1	Function and gating	accelerated Molecular Dynamics (Chu et al. 2023) MD and ML (Zheng and Wen 2021) bias-exchange metadynamics (Jorgensen et al. 2016) Metadynamics to calculate free energy (Kasimova et al. 2018)	Melnick and Kaviany (2018), Trofimov et al. (2024), Zheng and Wen (2019), Wen et al. (2016), Wen and Zheng (2018), Chugunov et al. (2016), Zheng and Qin (2015), Ladrón-de-Guevara et al. (2020), Kasimova et al. (2018), Zamri et al. (2022), Darré et al. (2015), Chu et al. (2023), Trofimov et al. (2023), Jorgensen et al. (2016), Liu et al. (2023), Feng et al. (2016), Raymond et al. (2014), Nguyen et al. (2021), Ortega-Guerrero et al. (2016), Fan et al. (2024), Suárez-Suárez et al. (2024), and Steinberg et al. (2017)
	Ligand and lipid binding, and drug discovery	Metadynamics (Jorgensen and Domene 2018; Domene et al. 2022) Molecular docking and MD (Bustos et al. 2023) Virtual screening (Roshni et al. 2024) Pharmacophore construction, molecular docking, virtual screening, MD (Feng et al. 2015)	Roshni et al. (2024), Feng et al. (2015), Abdelilah et al. (2023), Fan et al. (2024), Muller et al. (2020), Jorgensen and Domene (2018), Darré and Domene (2015), Hanson et al. (2015), Domene et al. (2022), Bustos et al. (2023), and Poblete et al. (2015)
TRPV2	Function and gating	SMD, PMF (Yang et al. 2016) Rosetta docking (Alipour et al. 2023)	Ives et al. (2023), Alipour et al. (2023), and Yang et al. (2016)
	Ligand and lipid binding and drug discovery	LiGaMD (Catalina-Hernández et al. 2024)	Catalina-Hernández et al. (2024), Feng et al. (2023), Su et al. (2023), Doñate-Macián et al. (2019c)
TRPV3	Function and gating	Metadynamics (Wang et al. 2020) HREMD, PCA, ML (Burns et al. 2023)	Trofimov et al. (2024), Nguyen et al. (2021), Ives et al. (2023), Nadezhdin et al. (2021), Wang et al. (2020), and Burns et al. (2023)
	Ligand and lipid binding and drug discovery		Zhang et al. (2021b)
TRPV4	Function and gating	MD and US (Huang and Chen 2023)	Huang and Chen (2023), Zhen et al. (2023), Takahashi et al. (2014), Li et al. (2020), Inada et al. (2012), and Garcia-Elias et al. (2015)
	Ligand and lipid binding and drug discovery	3D-QSAR, molecular docking and MD (Abdelilah et al. 2024) MD and virtual screening (Doñate-Macian et al. 2022)	Doñate-Macian et al. (2022), Abdelilah et al. (2024), and Berna-Erro et al. (2017)
TRPV5	Function and gating	CGMD, SMD, US simulations, WHAM analysis and PMF calculations (Gu et al. 2023)	Ives et al. (2023), Flores-Aldama et al. (2023), Wang et al. (2016, 2017, 2019), Gu et al. (2023), Hughes et al. (2018a), and De Jesús-Pérez et al. (2024)
	Ligand and lipid binding and drug discovery		De Jesús-Pérez et al. (2024)
TRPV6	Function and gating	US (Liu et al. 2023)	Trofimov et al. (2024), Liu et al. (2023), Ives et al. (2023), Flores-Aldama et al. (2023), Sakipov et al. (2018), Neuberger et al. (2023b), and Wang et al. (2023)
	Ligand and lipid binding and drug discovery	Constant velocity steered MD and US (Neuberger et al. 2023b) LipIDens and CG (Ansell et al. 2023)	Bhardwaj et al. (2024), Neuberger et al. (2023b), Hughes et al. (2018b), and Ansell et al. (2023)

The "Further References" column includes, to date, all references of computational studies regarding TRPV channels. The "Computational method" column specifies any study using a computational method different than molecular dynamics (MD) simulations

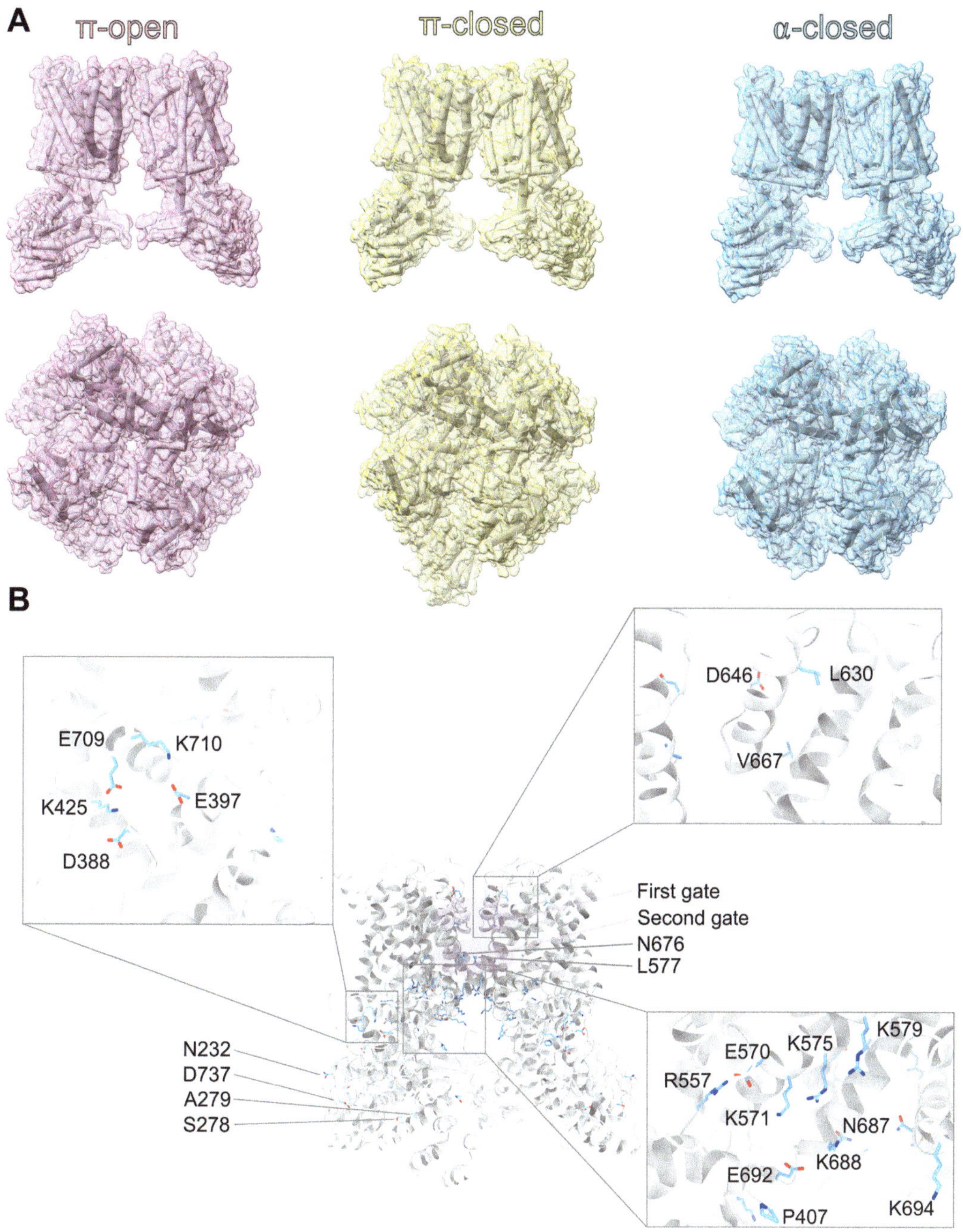

Fig. 10.2 (**a**) Surface and cylinder representations of TRPV1 in three distinct conformational spaces: π-open (left, light pink, PDB 7L2W), π-closed (middle, light yellow, PDB 7L2P), and α-closed (right, light blue, PDB 7MZD). Side (two chains) and top (all four chains) views are shown, highlighting the conformational differences between open and closed states. (**b**) Structural overview of TRPV1 in the open state (PDB: 7L2W) indicating key residues discussed throughout the chapter. Labels include residues directly from TRPV1 or, where other TRPV family members are referenced, the equivalent TRPV1 residue (as presented in Table 10.2) are shown. The first and second gates are indicated

other states are nonpermeable due to the presence of hydrophobic gates (Trofimov et al. 2024).

Heat-induced activation is one of the well-known landmarks of TRPV1-4 (thermoTRPVs) regulation (Benham et al. 2003). However, the different channels respond to a different range of temperatures, and the resulting activation can be potentiated or inhibited by the presence of other ligands and stimuli (Benham et al. 2003), highlighting the complexity of the structural and dynamic rearrangements involved in the process.

Investigations into the heat activation mechanism of TRPV1 have revealed key structural and dynamic features (Zheng and Wen 2019). An approach used consistently to study the conformational changes involved in heat-induced pore gating consists of using classical MD to simulate the system at different temperatures. For example, Wen and collaborators used classical MD simulations of TRPV1 at 30 °C and 60 °C to identify significant conformational changes involving the membrane proximal domain (MPD) linker, the outer pore, and the TRP helix, suggesting that these structural changes are key for heat-induced activation (Wen et al. 2016). A similar study revealed conformational changes in the S2-S3 and S4-S5 linkers (Fig. 10.1), along with an increase in protein–water electrostatic interaction, supporting the role of desolvation of polar/charged residues in temperature-sensitive gating (Wen and Zheng 2018). Melnick and Kaviany expanded this approach by modeling lipids into the vanilloid pocket and showing that, upon heating, the lipid is ejected from the pocket (Melnick and Kaviany 2018), a hypothesis initially based on structural data (Gao et al. 2016). Moreover, Chugunov et al. added to the previous knowledge, revealing the large temperature-dependent conformational transitions of the pore (Chugunov et al. 2016).

However, the large size of TRPV1 poses a significant challenge to observe conformational changes involved in the pore opening, as these usually occur on time scales over the microsecond range (Anandakrishnan et al. 2015). An alternative is using coarse-grained MD (CGMD), in which heavy atoms are modeled in groups, rather than individually, reducing the size and complexity of the system and allowing longer simulation scales (Souza et al. 2021). Zheng and Qin used both all-atom MD and CGMD simulations to predict key residues that control the temperature sensitivity of TRPV1, many of which are located in the MPD (Zheng and Qin 2015). Ladrón-de-Guevara and colleagues, on the other hand, suggested that the ankyrin repeat domain (ARD) had a crucial role in temperature-dependent gating, as ARD undergoes conformational changes at high temperatures (Ladrón-de-Guevara et al. 2020). Recently, our lab has used CGMD to characterize the lipid fingerprinting of the whole TRPV subfamily (Lopez-Martin et al. 2025). However, CGMD poses several limitations due to the absence of H-bonds, π-π interactions, and secondary structure changes, among others (Marrink et al. 2023).

Another way of accelerating the dynamics in MD simulations is by using enhanced sampling methods, such as metadynamics. In metadynamics, additional energy is provided to the system, enabling it to explore a broader conformational space, thereby allowing conformational changes that are not observable in conventional MD. In this regard, Kasimova et al. performed MD simulations and metadynamics to do free energy calculations, finding that the transition from closed to open states is influenced by the polar character of the molecular surface (Kasimova et al. 2018). Furthermore, IA-driven techniques, such as machine learning (ML) or deep learning (DL), can also be employed to study TRPV channels. For instance, Zheng and Wen conducted MD simulations in both open and closed states and used ML to identify key residues distinguishing these states (Zheng and Wen 2021).

Other stimuli can also affect TRPV1 channel gating. For instance, Zamri et al. explored how hydrostatic pressure influences the gating mechanism using classical MD simulations at different pressures (Zamri et al. 2022). They concluded that under elevated pressure, TRPV1 exhibits increased dynamism and movement, suggesting that hydrostatic pressure promotes channel opening.

Table 10.2 Residues represented in Fig. 10.2 (TRPV1) and the equivalent residues to the corresponding structure (TRPV3, TRPV4, TRPV5, or TRPV6)

TRPV3	TRPV1	TRPV4	TRPV1	TRPV5	TRPV1	TRPV6	TRPV1
D641	D646	K462	K425	E294	E397	Q483	E570
E682	N687	E745	E709	R606	K710	L490	L577
E689	K694	D425	D388	R302	P407	R302	P407
D727	D737	I715N	I679	E588	E692	K484	K571
G573	G563	R269H	N232	L530R	L630	R584	K688
R567	R557	R315W	S278	A563T	V667		
		R316H	A279				

MD simulations have also been used to study the channel's ion permeation. Darré et al. investigated the permeation and dynamics of the open channel in the presence of sodium, potassium, and calcium (Darré et al. 2015). In this study, they performed simulations of systems containing either sodium or calcium ions, which can induce structural changes in the selectivity filter, and potassium, an ion that does not induce conformational changes.

Chu and colleagues employed accelerated MD simulations, an enhanced sampling method that boosts the system energy potential to increase the sampling of the configurational space, to investigate the activation mechanism of oxidation-induced, showing that the formation of a disulfide bond between cysteines leads to conformational changes, partially activating the channel (Chu et al. 2023). Besides, in a previous study, Trofimov and colleagues proposed a novel mechanism for ion transport in TRPV1, in which the hydrophobic gate allows ions to pass through in a fully hydrated state, where N676 (Fig. 10.2b) plays a crucial role in interacting with water molecules (Trofimov et al. 2023). Jorgensen et al. investigated the energetics of ion permeation in the open TRPV1 channel using MD with bias-exchange metadynamics (where multiple replicas are run in parallel and periodically exchange their biases, to enhance conformational exploration (Piana and Laio 2007)). They identified key binding sites for monovalent ions, such as sodium, and provided insights into the ion permeation mechanism (Jorgensen et al. 2016). In 2023, simulations performed by Liu et al. revealed that TRPV1 has only one calcium binding site, resulting in less effective sodium blockage, in contrast with the two binding sites in TRPV6 (Liu et al. 2023).

Inhibitor studies have identified and evaluated diarylurea small molecule inhibitors of TRPV1, using molecular docking, MD simulations, and biological assays to explore interactions and identify key binding residues, and assessing the compounds' activities (Feng et al. 2016). The mechanism of action of RVD-hemopressin (RVD-Hp), another channel blocker, was demonstrated using MD simulations by Raymond and colleagues (Raymond et al. 2014). Their study revealed that RVD-HP interacts with the selectivity filter of TRPV1, preventing ion permeation.

Lipids also play a key role in regulating TRPV1 activity through both direct molecular interactions and indirect effects on the lipid environment. In this regard, cholesterol directly binds to specific recognition motifs on TRPV1, such as cholesterol recognition/interaction amino acid consensus (CRAC) and the cholesterol consensus motif (CCM). Molecular docking and mutational studies have identified residues such as R557 and R575 (Fig. 10.2b) that stabilize TRPV1 in its closed conformation, thus inhibiting channel activation by agonists like capsaicin (Morales-Lázaro and Rosenbaum 2019; Saha et al. 2017).

Studies on ligand and lipid binding effects have shown that monoterpenes (menthol and camphor) activate TRPV1, where conserved Gly and Arg residues in the S4-S5 linker are crucial for this interaction (Nguyen et al. 2021). Besides, Raymond and colleagues studied the structural characterization of the TRPV1 C-terminal domain, focusing on temperature sensing and phosphatidylinositol 4,5-bisphosphate (PIP2) regulation. They concluded that the C-terminal domain does undergo conformational changes at elevated temperatures, and also upon PIP2 binding (Raymond et al. 2014).

Finally, Ortega-Guerrero explored the potential use of TRPV1 as a target for cancer therapy through carbon nanotube (CNT)-based drug delivery systems (DDSs). They concluded that the channel must be dilated to facilitate drug diffusion, providing a basis for designing DDSs that exploit the overexpression of TRPV1 channels in cancer cells (Ortega-Guerrero et al. 2016).

10.3.2 TRPV1 Ligand Binding and Drug Discovery

TRPV1 is modulated by a myriad of different ligands, in addition to its role in temperature regulation. While structural studies have provided invaluable insights into the molecular mechanisms underlying protein–ligand interactions, they alone cannot fully elucidate the exact binding mechanisms and the accompanying protein conformational changes (Du et al. 2016). Computational methods, including MD simulations, have significantly advanced our understanding of TRPV1 channel interaction with various ligands and lipids, as well as finding novel regulatory compounds.

In this context, in 2024, Roshni et al. used virtual screening and MD to identify improved antagonists for TRPV1 by docking a library of anti-inflammatory compounds and studying their stability and reactivity in the binding pocket (Roshni et al. 2024). In a previous study, Feng and collaborators constructed a 3D homologue of hTRPV1 based on a cryo-EM structure from rats (Feng et al. 2015). They performed molecular docking and MD simulations to explore interactions with agonists like resiniferatoxin (RTX) and capsaicin, as well as antagonists like AMG98120 and BCTC. Their simulations revealed that agonists induce conformational changes to open the selectivity filter. Furthermore, they constructed a pharmacophore based on several antagonists and used it to perform virtual screening, leading to the identification of several hTRPV1 antagonists that were consequently experimentally validated. A similar approach was used by Abdelilah and colleagues (Abdelilah et al. 2023). In their study, they combined three-dimensional quantitative structure–activity relationship (3D-QSAR, used to understand the relationship between 3D structure and biological activity (Zhu et al. 2009)), molecular docking and MD simulations to design and identify novel TRPV1 antagonists, highlighting the importance of considering multiple computational approaches to predict and optimize the pharmacokinetic and pharmacodynamic properties of potential drug candidates. With 3D-QSAR, they identified steric, electrostatic, and hydrophobic fields as significant contributors to the biological activity of the compounds and performed molecular docking with these compounds. MD simulations were used to analyze the stability of the best-docked compounds in the TRPV1 binding pocket.

Structural knowledge about known modulators has also been expanded using MD. Fan et al. provided insights into the structural basis of TRPV1 inhibition by SAF312 (also known as Libvatrep), showing that SAF312 binds to the vanilloid binding pocket (VBP) and is stabilized by interacting with cholesterol (Fan et al. 2024). Muller and collaborators investigated the interaction of anandamide (AEA) with TRPV1, identifying that AEA could bind to the S1-S4 tunnel or the VBP, with a preference for the former binding site (Muller et al. 2020). Jorgensen and Domene used metadynamics to study the free energy of insertion of anesthetics like chloroform and isoflurane, suggesting that anesthetics may act as secondary agonists, sensitizing the channel to activation by other stimuli (Jorgensen and Domene 2018).

But perhaps the most extensive computational work in this regard has been focused on capsaicin, the best-known TRPV1 agonist. In early computational studies, Darré and Domene identified key binding modes of capsaicin to TRPV1 using MD and metadynamics (Darré and Domene 2015), while Hanson and collaborators suggested that capsaicin accesses its binding site on TRPV1 by flipping from the extracellular to the intracellular leaflet (Hanson et al. 2015). Moreover, Domene and colleagues used well-tempered metadynamic simulations (used to focus the computational effort on the physically relevant regions of the conformational space (Barducci et al. 2008)) to describe a membrane-mediated mechanism for capsaicin to reach its binding site in TRPV1, which entails a membrane-mediated mechanism (Domene et al. 2022). Furthermore, Bustos et al. presented a pipeline to design and select capsaicin analogues (Bustos et al. 2023). They introduced chemical modifications to create a library of capsaicin derivatives, performed molecular docking and MD simulations with the best-scored compounds, demonstrating improved binding affinity and pharmacokinetic properties.

Endogenous lipids also play a crucial role in the regulation of TRPV1, and higher-resolution cryo-EM structures show an increasing number of lipid-like densities occupying different pockets and regions on the proteins (Gao et al. 2016; Zhang et al. 2021a; Kwon et al. 2021; Neuberger et al. 2023a). However, these interactions tend to be transient and flexible, making it difficult to obtain precise structural details. In 2015, Poblete and colleagues used molecular docking and MD to identify the binding site of PIP2 between the TRP box and the S4-S5 linker, the key role of R575, R579, and K694 (Fig. 10.2b) residues, and unveil how the PIP2 binding induces conformational changes in the S6 and TRP domains, leading to channel opening (Poblete et al. 2015).

10.4 TRPV2

10.4.1 TRPV2 Function and Gating

Ives et al. conducted a comparative study of nonselective ion channels, such as TRPV2 and TRPV3, and Ca^{2+}-selective ion channels, such as TRPV5 and TRPV6 (Ives et al. 2023). By using MD with an external electric field and harmonic restraints to the α-carbon atoms of the lower gate residues, they maintained the pore open and drove ion permeation. This approach revealed a two-binding-site knock-on mechanism in TRPV2 primarily involving sites formed by carbonyl oxygen atoms and residues like Ile and Asp at the hydrophobic gate. In contrast to the three-binding site knock-on mechanism observed in Ca^{2+}-selective channels, this highlights the structural and functional diversity within the TRPV family.

In another study, Alipour et al. used steered molecular dynamics (SMD) to calculate the potential of mean force (PMF) for Ca^{2+} permeation through the TRPV2 channel (Alipour et al. 2023). SMD works by applying an external force to a specific atom or group of atoms along a predefined reaction coordinate. The system's response to this applied force can then be analyzed to extract the PMF (Park and Schulten 2004; Park et al. 2003). Using this technique, they identified two gates within the channel: the selectivity filter (first gate) and the hydrophobic region (second gate) (Fig. 10.2b). The second gate presented a higher energy barrier due to dehydration effects, acting as the limiting gate for ion permeation.

Not all computational studies require molecular dynamics. Yang et al. used Rosetta-based molecular docking to analyze the binding of resiniferatoxin and capsaicin to TRPV2 and its mutants (Yang et al. 2016). Their study revealed that resiniferatoxin acts as a molecular glue, bridging the S4-S5 linker to the S1-S4 domain, thereby opening the channel. Conversely, capsaicin failed to activate the mutant due to structural constraints that prevented the formation of this bridge. This research not only validated the current model for capsaicin activation of TRPV1 but also provided insights into the structural requirements for ligand-induced activation of TRPV channels.

10.4.2 TRPV2 Ligand Binding and Drug Discovery

In contrast to TRPV1, the pharmacological profile of TRPV2 is less understood. To date, no TRPV2-specific molecule has been described yet and even though some unspecific drugs, such as cannabidiol (CBD) or 2-APB, have been shown to modulate TRPV2 activity (Hu et al. 2004), their exact binding mechanisms remain unclear. In this context, computational techniques have emerged as valuable tools to explore and elucidate the complex pharmacology of this channel.

In a study performed by our group, we expanded the pharmacological toolbox for TRPV2 by identifying new potential agonists with higher efficacy than probenecid (Catalina-Hernández et al. 2024). We generated a library of 272 compounds using known TRPV2 ligands (e.g., probenecid and CBD) as scaffolds and performed virtual screening and ligand Gaussian accelerated MD (LiGaMD) simulations. LiGaMD is an enhanced sampling method similar to accelerated MD but offering more robust and accurate results. By adding a Gaussian-like potential specifically to the ligand, it is possible to accelerate the binding/unbinding process and even accurately calculate the binding free energies (Miao et al. 2020). Using this approach, we identified a promising compound, 4-(piperidine-1-sulfonyl)-benzoic acid (PSBA), which was validated in vivo, as well as proposing a binding site for both PSBA and probenecid. Additionally, our simulations highlighted the importance of studying membrane proteins within their native lipid environment, as the identified ligands showed stabilization owing to interactions with lipid molecules.

Feng and colleagues embedded TRPV2 in a lipid bilayer to examine the interactions between TRPV2 and plasma membrane lipids, highlighting the importance of lipid interactions in stabilizing TRPV2 and modulating its function. They found that PIP2 lipids play a significant role in binding to specific sites on TRPV2, while CBD binding leads to substantial conformational changes in the channel. Additionally, they observed low specificity for surface-bound lipids, as both POPC (1-palmitoyl-2-oleoyl-glycero-3-phosphocholine) and POPE (1-palmitoyl-2-oleoyl-sn-glycero-3-phosphoethanolamine) bound indistinctively. Finally, by modeling structural lipids into the VBP and the voltage sensor-like domain (VSLD) pocket, they revealed how their presence affects the channel stability and dynamics (Feng et al. 2023).

In another study, Su et al. combined cryo-EM experiments with MD simulations to investigate the binding of endogenous cholesterol and 2-aminoethoxydiphenyl borate (2-APB) in TRPV2 and their effects on channel function. Using the structural data as starting coordinates, their simulations revealed that cholesterol stably binds to the VBP of TRPV2, stabilizing the S4-S5 linker and preventing channel activation. Conversely, 2-APB activates TRPV2 by replacing cholesterol in the VBP, facilitating channel opening. Moreover, they identified key interacting residues in the pocket to validate them experimentally. Finally, by using CGMD, they approximated the energy barrier required for both ligands to unbind the pocket, showing that 2-APB faces a significantly larger barrier than cholesterol. This provides a thermodynamic explanation for the competitive binding of 2-APB and cholesterol to the VBP (Su et al. 2023).

10.5 TRPV3

10.5.1 TRPV3 Function and Gating

Similarly to TRPV1, structural data show three clearly differentiated major states for TRPV3: α-closed, π-closed, and π-open (Fig. 10.2a). Trofimov and collaborators employed classical MD to characterize these states and found out that, similar to TRPV1, the π-open state is fully hydrated during the whole 200-ns trajectory, while both π- and α-closed are nonconductive (Trofimov et al. 2024). To maintain the π-open state conductive, harmonic restraints were applied to the residues on the hydrophobic gate, indicating that other factors are required to observe the channel in a fully conductive state in unconstrained simulations. In parallel, Nadezhdin et al. used conventional MD to analyze the structural dynamics and conductance properties of TRPV3, and differentiated three states: closed, sensitized (with the pore slightly wider, but still nonconducting), and open state, conducting both water and ions. The sensitized state represents an intermediate step with significant conformational changes and the withdrawal of a lipid from the vanilloid binding site, emphasizing the fundamental key role of endogenous membrane lipids in the regulation of the conformational dynamics of the channel (Nadezhdin et al. 2021).

Wang and collaborators employed MD and metadynamics to study the dual effects of protons on TRPV3. They found that extracellular protons inhibit the channel through interaction with D641, while residues E682, E689, and D727 (Fig. 10.2b) are crucial for the sensitization of TRPV3 by intracellular protons (Wang et al. 2020). The metadynamics simulations revealed that charge neutralization of these residues stabilizes the sensitized state, providing insights into the molecular basis of proton regulation in TRPV3. In another study, Macikova et al. used MD at different temperatures to identify highly flexible regions in and around the β sheet-finger 3 intersubunit, which connected the rigid β-sheet, MPD, and TRP helix to the channel gate (Macikova et al. 2019). This allowed them to hypothesize that heat-induced changes in the flexible interface are transmitted to the channel gate through these rigid regions.

Finally, Burns and colleagues concluded that TRPV3's temperature sensitivity and gating are regulated by multiple structurally localized contact modes that respond to temperature changes. Using Hamiltonian Replica Exchange molecular dynamics (H-REMD) and principal component analysis (PCA), they identified temperature-sensitive contact modes at specific regions such as the ARD, helix-turn-helix (HTH), and transmembrane domain (Burns et al. 2023). In H-REMD, multiple MD dynamics simulations—*replicas*—are run simultaneously. While in typical REMD each replica is run at a different temperature, in H-REMD, each replica is run at the same temperature, but with different Hamiltonians (that is, the functions that describe the total energy of the system). At set intervals during the simulation, the configurations of the replicas can be swapped according to a probability based on the energy difference between the configurations being exchanged, ensuring the overall thermodynamic accuracy of the simulation. This method allows the system to overcome energy barriers and sample rare configurations that would be difficult to access in a classical molecular dynamics simulation (Hritz and Oostenbrink 2008). Additionally, network analysis and ML techniques were used to investigate allosteric pathways and predict gate expansion, highlighting the complex allosteric network influencing TRPV3 gating.

10.5.2 TRPV3 Ligand Binding and Drug Discovery

So far, very few studies have employed MD to study ligand–protein interactions on TRPV3. Nguyen and colleagues used classical MD to complement biochemical data and show that the monoterpenes menthol and camphor activate TRPV3 through interactions with conserved G573 and R567 residues in the S4–S5 linker of TRPV3 (Nguyen et al. 2021). Similarly, Nadezhdin et al. performed MD simulations to complement their structural cryo-EM data (Nadezhdin et al. 2021). By modeling different lipids into the vanilloid pocket and running 500-ns-long trajectories, they showed all tested lipids (cholesterol, POPC, POPI [1-palmitoyl-2-oleoyl-sn-glycero-3-phosphoinositol], POPE, POPG [1-palmitoyl-2-oleoyl-sn-glycero-3-phospho-(1'-rac-glycerol)], and POPS [1-palmitoyl-2-oleoyl-sn-glycero-3-phospho-L-serine]) remained stable in the VBP, suggesting that the site possesses low lipid specificity.

In an alternative fashion, Zhang and collaborators used structure-based virtual screening to identify potential TRPV3 inhibitors from a database of 7989 compounds (Zhang et al. 2021b). They employed *LibDock* and *CDOCKER* protocols, which involve high-temperature MD to generate different ligand conformations that are later optimized. They developed a derivative compound, named PC5, that showed promising activity and selectivity, making it a potential lead compound for further development.

10.6 TRPV4

10.6.1 TRPV4 Function and Gating

TRPV4 is a particularly interesting case, since it can be activated by many stimuli and it is involved in a widespread range of physiological activities (Darby et al. 2016). The stimuli can be both physical, such as osmotic pressure, mechanical stress, heat, and chemical, such as endogenous ligands or lipids (Garcia-Elias et al. 2014). TRPV4 also has an influence on the actin cytoskeleton, regulating the extension and retraction of cellular processes (Martin et al. 2012). Recently, the interaction between the cytoskeletal remodeling GTPase RhoA and TRPV4 was described (McCray et al. 2021), causing the inhibition of TRPV4 channel activity and Rho, ultimately affecting cytoskeletal dynamics.

Structurally, Garcia-Elias and collaborators highlighted the importance of interactions between the pre-S1, TRP, and ARD-S1 linker domains for the proper folding, assembly, and trafficking of TRPV channels. Disruptions in these interactions, due to mutations in residues such as K462, E745, or D425 in TRPV4 (Fig. 10.2b), lead to defective channel assembly and function. Huang and Chen examined the hydration properties of the TRPV4 pore and found that the I715N mutation in the lower pore region increased pore hydration and reduced the ion permeation free energy barrier by about 50%, employing conventional MD and Umbrella Sampling (US) (Huang and Chen 2023). Umbrella Sampling (US) is an enhanced sampling technique used to calculate the free energy profile of processes separated by high-energy barriers. It involves dividing the reaction coordinate (e.g., ion crossing a channel) into several windows and applying a harmonic bias potential in each, enhancing sampling within those regions. Methods like Weighted Histogram Analysis Method (WHAM) are then used to reconstruct the unbiased free energy profile (Torrie and Valleau 1977).

10.6.2 TRPV4 Ligand Binding and Drug Discovery

The role of TRPV4 as a regulator of osmotic pressure via mechanotransduction across multiple tissues makes it a very promising target to treat a myriad of different pathologies. Although, similarly to TRPV2 and TRPV3, not many modulators have been identified to date, computational methods are playing a fundamental role to find and characterize potent, specific new drugs for this channel, as well as characterizing binding site of endogenous modulators that can be exploited for drug discovery.

In another study, Doñate-Macian and colleagues combined different in silico techniques to identify and characterize a potential binding site of HC067047, a known TRPV4-specific inhibitor. They identified druggable cavities on the surface of TRPV4 by using the SiteMap algorithm (Halgren 2007), performed molecular docking of HC067047 on those predicted pockets, and used classical MD and computational mutagenesis to analyze the stability of the binding, estimate the binding energy, and identify key residues in the protein–ligand interaction. This data allowed them to generate mutants of TRPV4 on those specific residues to validate the results experimentally (Doñate-Macian et al. 2022). Moreover, they used the newly characterized binding site as the target site for a virtual screening assay that, combined with MD and experimental validation, allowed them to find three potential new TRPV4-specific inhibitors.

Abdelilah and collaborators also employed a combination of computational methods to identify new potent TRPV4 inhibitors (Abdelilah et al. 2024). They developed a 3D-QSAR model using a library of 2,4-dimethyl-[4,5-bithiazol]-2-yl amino derivatives with anti-TRPV4 activity. The model was then used to develop new TRPV4 antagonists, which were further characterized using molecular docking and classical MD.

In a recent study, Zhen and colleagues used MD to validate the structurally identified cytosol-facing cavity in the S1 to S4 bundle as a critical binding site for TRPV4 agonists like GSK101 and Agonist-1. Ligand binding to this site induced conformational changes that open the channel, while Ruthenium Red (RR) inhibited TRPV4 by binding to the extracellular entrance of the selectivity filter, blocking ion permeation (Zhen et al. 2023).

Regarding the regulation of TRPV4 by endogenous ligands, Berna-Erro and colleagues identified and characterized the binding site of 5′,6′-apoxyeicosatrienoic acid (EET) by molecular docking and classical MD, which involved residues from S2-S3 and S4-S5 linkers (Berna-Erro et al. 2017). The same approach was used by Takahashi et al. to study the effect of PIP2 in TRPV4 (Takahashi et al. 2014). They concluded that PIP2 binding to the TRPV4 ARD negatively regulates channel activity. By combining docking, MD, and experimental data, they showed that disease-associated mutations such as R269H, R315W, and R316H disrupt PIP2 binding, leading to increased channel activity and cellular toxicity.

The ARD is also important for the adenosine triphosphate (ATP) binding (Li et al. 2020). Li and colleagues used MD to demonstrate that well-known clinical mutations located in the ARD, such as K197R, L199F, and E183K, alter ATP binding abilities. The K197R mutation contributes to channel instability and significantly weakens ATP binding, while the L199F mutation slightly weakens ATP binding by reducing hydrogen bonds with ATP. Conversely, the E183K mutation enhances ATP binding by increasing the number of hydrogen bonds between ATP and TRPV4-ARD. Similarly, Inada and collaborators also studied the ARD in TRPV4, this time by using SMD to "stretch" both ends of the ARD region to test the folding stability of the domain. Their results suggested that the ATP binding increases the stability of the TRPV4-ARD, which may play a role in regulating channel activity (Inada et al. 2012).

10.7 TRPV5

10.7.1 TRPV5 Function and Gating

Contrary to thermoTRPVs, TRPV5 and TRPV6 are highly selective for Ca^{2+} ions and are involved, among others, in the reabsorption of this ion in the kidney and intestines, respectively (Van Abel et al. 2005). Therefore, understanding the differential structural and dynamical properties of these channels is crucial for the understanding of a wide number of human pathologies, including different types of cancer.

As mentioned in the TRPV1 section, Ives and collaborators used MD simulations to compare nonselective and Ca^{2+}-selective channels and described a three-binding site knock-on mechanism for ion permeation on the last ones: one formed by carboxylate oxygen atoms, another by carbonyl oxygen atoms, and the third formed by Ile and Asp residues at the hydrophobic gate (Ives et al. 2023), further discussed below for TRPV6.

Flores-Aldama and collaborators used classical MD simulations to find putative Ca^{2+}-binding sites in the ankyrin repeat domains of TRPV5 and TRPV6. Moreover, they analyzed how Ca^{2+}-bound structures decreased the distance between the helix-loop-helix (HLH) domain and the TRP domain helix (TDh)—residues E294 and R606 (Fig. 10.2b), respectively—, as well as between the ankyrin repeat domain 6 (ARD6) and the HLH domain (residues R302 and E588). Therefore, the simulations provided a dynamic explanation for the implication of both pairs of interactions in calcium-mediated channel inactivation (Flores-Aldama et al. 2023).

Wang and collaborators studied the implications of the clinical variants L530R and A563T. They performed homology modeling of the human TRPV5 using the rat TRPV6 structure as a template and then performed classical MD simulations. Trajectory analysis revealed that the A563T mutation causes increased secondary structure stability and induces dynamical and electrostatic changes in the pore, contributing to an increased Ca^{2+} transport ability (Wang et al. 2016, 2019). On the other hand, the L530R variation damages the secondary structure of the transmembrane helix 5 (TM5) and interrupts the Ca^{2+} transport activity of TRPV5, disrupting the hydrophobic interactions and leading to structural and functional impairments (Wang et al. 2017). Moreover, Gu and colleagues employed a combination of CGMD, SMD, and US to study the conformational transitions of TRPV5 induced by mechanical force: the S6 helix rotates and plays a crucial role in the gating mechanism, a mechanism that differs from TRPV6 (Gu et al. 2023). Moreover, WHAM analyses were conducted to calculate the PMF and demonstrate that TRPV5 can be activated by mechanical force, suggesting potential applications for therapeutic purposes.

10.7.2 TRPV5 Ligand Binding and Drug Discovery

Hughes and collaborators used MD to complement their cryo-EM data and confirm the stability of econazole binding to a hydrophobic pocket between the S3-S4 helices, the S4-S5 linker, and the S6 helix (a region equivalent to the VBP in TRPV1) (Hughes et al. 2018a). The simulations also revealed a different degree of flexibility of econazole that moderately correlated with the number of contacts with POPC molecules in the membrane, highlighting, as previously mentioned for TRPV2, the paramount importance of the lipid environment in the binding of ligands within the transmembrane regions of the channels.

Along the same lines as Hughes and collaborators, De Jesús-Pérez and colleagues use cryo-EM and MD simulations to study the binding of econazole and PIP2 (De Jesús-Pérez et al. 2024). They describe how endogenous PIP2 opens the channel, but econazole can outcompete PIP2 and inhibit the TRPV5 channel, leading to pore closure.

10.8 TRPV6

10.8.1 TRPV6 Function and Gating

Similar to TRPV5, the ion permeation mechanism in TRPV6 channels involves a cooperative knock-on mechanism, which is distinct from other TRPV channels (Liu et al. 2023; Sakipov et al. 2018). As mentioned above, the Ives et al. MD simulations have revealed that TRPV6 utilizes a three-binding site knock-on mechanism: one formed by carboxylate oxygen atoms, another by carbonyl oxygen atoms, and the third formed by Ile and Asp residues at the hydrophobic gate (Ives et al. 2023). Besides, US calculations allowed to estimate the PMF for Ca^{2+} permeation, which showed a much higher preference for this cation in TRPV6 when compared to TRPV1. The strong binding affinity of calcium to the selectivity filter lower binding site was reduced when an additional ion was bound to the upper binding site, suggesting a cooperative knock-on mechanism between the three pore binding sites that facilitates the permeation of Ca^{2+} selectively (Liu et al. 2023; Ives et al. 2023).

Structurally, TRPV6 channels exhibit the three major structural states previously described for other TRPVs: α-closed, π-closed, and π-open (Fig. 10.2a). The first two states present hydrophobic gates, whereas the latter is less hydrophobic and allows ion and water permeability. These states are critical for understanding the gating mechanisms of TRPV6 (Trofimov et al. 2024). As in TRPV5, the interaction with the HLH domain and the TRP helix is pivotal in the fast activation of TRPV6 (Flores-Aldama et al. 2023).

10.8.2 TRPV6 Ligand Binding and Drug Discovery

Several compounds have been identified as inhibitors of TRPV6. Most of them are nonspecific drugs that also target other channels, including other TRPVs. These include 2-APB, econazole, and some tetrahydrocannabinol derivatives. There are, in addition to those, the more specific 4-phenylcyclohexyl piperazine derivatives (PCHPDs) and the peptide SOR-C13, a specific TRPV6 inhibitor in clinical trials as treatment for ovarian and pancreatic cancer (Stewart 2020).

To characterize the mechanism of action of PCHPD inhibitors, Bhardwaj et al. combined structural data obtained by X-ray crystallography and cryo-EM with classical MD. Starting with the piperazine derivative cis-22a in either the vanilloid site or the channels' pore intracellular entrance as the initial configuration, the simulations revealed stable binding of the ligand to both these pockets, as well as identification of key residues involved in the binding (Bhardwaj et al. 2024). Interestingly, one of the identified residues, Q483, has previously been involved in endogenous lipid binding in the apo TRPV6 structures (Singh et al. 2018).

The same approach was employed by Neuberger and colleagues to characterize the exact orientation of the phytocannabinoid tetrahydrocannabivarin (THCV) in the structurally determined portal site of TRPV6, between the helices S5 and S6 that connect the pore with the membrane environment (Neuberger et al. 2023b). Due to the high flexibility of the THCV acyl chain, cryo-EM data are unable to determine the exact orientation. However, MD simulations revealed higher stability of THCV when the tricyclic ring is orientated toward the central pore and the acyl chain in contact with the membrane environment, thanks to stabilizing hydrogen between the hydroxyl group of THCV and the carbonyl oxygen of L490. In addition, they used constant velocity steered MD (cv-SMD) simulations to identify the pathways for THCV to leave its binding site, revealing two possible routes: a horizontal route along the S4-S5 linker and a vertical route along the pore helix. In cv-SMD (Zhang et al. 2012), the pulling velocity is kept constant, providing valuable insights into force-dependent processes, such as rupture forces. Lastly, they used umbrella sampling MD simulations to calculate the potential of mean force along the identified pathways, showing that the horizontal route is energetically more favorable.

In a combination of homology modeling, molecular docking, and classical MD, another study determined a potential binding site for PIP2, between the N-linker, the S4-S5 linker, and the S6 helix, in excellent agreement with what was observed in a cryo-EM structure of TRPV5 (Hughes et al. 2018b). An identical approach was used by Wang et al. to study the conformational changes induced by PIP2 binding to TRPV6. Their simulations on a TRPV5-based TRPV6 model identified a widening in the channel pore, possibly contributing to the opening of the channel. Positively charged residues (e.g., R302, K484, R584 (Fig. 10.2b)), also identified in the study of Hughes and Pumroy, were crucial for the stabilization of the PIP2 headgroup binding to TRPV6 (Wang et al. 2023).

Finally, Ansell and collaborators developed LipIDens, a pipeline that integrates multiscale molecular dynamics (coarse-grained and atomic) simulations with cryo-EM data to identify and refine lipid densities in cryo-EM maps, which often suffer from low resolution (Ansell et al. 2023). They tested their pipeline on several membrane proteins, including TRPV6, identifying several densities corresponding to PIP2, cholesterol, and POPC, among others.

10.9 Other Computational Approaches and Future Perspectives

The use of computational techniques has significantly enhanced our knowledge about TRPV channels. In particular, molecular dynamics has experienced substantial growth and development in recent years, driven by "structural revolution" brought about by the remarkable improvements in cryo-EM, which have provided more and higher-quality structural data. Combined with the improvement in computational power available—even in standard desktop computers—molecular dynamics is rapidly becoming a standard tool to refine, complement, and expand experimental data.

However, the use of MD to study TRPV channels still poses considerable challenges. First, TRPVs are tetrameric structures of considerable size (complete structures extending considerably beyond 2400 residues). Moreover, as membrane proteins, they need to be simulated embedded into a lipid bilayer, which enlarges the system even further. Overall, this significantly increases the computational cost of the simulation, limiting the simulation time. In turn, this can impede the observation of molecular phenomena that occur on larger time scales.

There are, of course, alternative approaches to overcome these limitations. In this review, we have discussed several enhanced sampling methods, including SMD, Umbrella Sampling, accelerated MD, and metadynamics. These methods help address time scale limitations by introducing a bias into the simulation. Another alternative is reducing the complexity of the system by using coarse-grained models, lowering the resolution and allowing for longer simulation time scales, but also limiting the accuracy.

The computational study of TRPV channels encompasses a variety of techniques beyond MD simulations. Other computational methodologies have been extensively used in TRPV studies. Molecular docking, for instance, used to predict how ligands interact with TRPV channels and evaluate the strength of these interactions (Feng et al. 2015; Abdelilah et al. 2023; Catalina-Hernández et al. 2024; Abdelilah et al. 2024; Caballero 2022; Wei et al. 2023; Yang et al. 2015). These methods comprise a vast range of different strategies and algorithms around a central basis: explore different conformations of ligands on different pockets of the protein to find the most likely binding site and pose, which is determined by a scoring function. With the rise of machine learning and deep learning, new docking methods with higher accuracy are being developed (McNutt et al. 2021; Corso et al. 2024).

Homology modeling has long been extensively used to obtain TRPV structures that were unavailable (Wang et al. 2019; Chik et al. 2012; Lee et al. 2011). The emergence of AlphaFold in 2020 (Jumper et al. 2021; Abramson et al. 2024) has revolutionized the study of TRPV channels. As mentioned earlier, TRPV channels are large, multi-domain proteins whose complete experimental structures are often difficult to resolve due to their dynamic nature and membrane localization, even with the advancements in cryo-EM. AlphaFold provides high-accuracy predictions of their 3D structures, offering critical insights into their complex architecture and functional mechanisms (Nguyen et al. 2024; Muazzam Ali Shah and Ou 2023). In addition, the last model of AlphaFold (Abramson et al. 2024) included prediction of complexes between proteins and of nonprotein molecules such as DNA, RNA, lipids, ions, and small molecules, which will surely aid in Computer-Aided Drug Design (CADD) for TRPVs.

Classic bioinformatics tools also play a significant role in the study of TRPV channels. Sequence alignment and phylogenetic analysis can reveal evolutionary relationships and conserved motifs (Doñate-Macián et al. 2019a; Flores-Aldama et al. 2020; Morini et al. 2022), shedding light on the functional importance of specific regions. Computational tools can be used to decipher the protein topology in the membrane and folding (Doñate-Macian et al. 2015; Huffer et al. 2020). Moreover, proteomics and interactomics studies can be performed to study protein–protein interactions (PPI) and reveal molecular mechanisms in physiology and pathology (Diehl 2022; Doñate-Macián et al. 2018a, b, 2019b; Doñate-Macián and Perálvarez-Marín 2014; Gallo et al. 2024; Goretzki 2021; Jiang et al. 2022; Lacerda et al. 2022).

It is impossible to overlook a critical factor in the context of structural biology: the rapid rise of AI—often referred to as the "AI revolution"—has had a transformative impact across a wide range of scientific disciplines. Structural biology is no exception, and several significant advancements are and will be driven by AI. AlphaFold and neuronal network-based docking algorithms are excellent examples, but other approaches that employ machine learning and deep learning have also shown very promising results.

Analysis of pore conformations also poses a significant challenge and an opportunity window in the TRPV study: pore snapshots in Cryo-EM structures are static, as opposed to MD simulations, where pore dynamics, ion coordination, and water flow can be inferred from structural dynamics analyses (Trofimov et al. 2024; Darré et al. 2015). Kacher and collaborators developed a deep learning pipeline to describe the activation path of a voltage sensor domain for an ion channel, by using PyTorch and Neural Networks (Kacher et al. 2024). Similar algorithms can be used to describe the states between the open and closed states by employing Cryo-EM structures. Besides, an interesting model using Support Vector Machines (SVM) developed by Rao and colleagues is capable of identifying hydrophobic gates in the pore structure, which are enough to block the channel (Rao et al. 2019). In another study, Rao and collaborators studied the hydrophobic gating mechanism in several ion channels, revealing insights into the dewetting transitions (removal or absence of water in a specific region), which play a critical role in gating by controlling water and ion permeability (Rao et al. 2021). Furthermore, they explored the relationship between pore hydration and ionic conductance, establishing a threshold to differentiate conductive from nonconductive states.

Perhaps the field where AI-based techniques have had the greatest impact is drug discovery. In recent years, rational drug design, as opposed to traditional high-throughput screening methods, has gained significant attention. In this context, ML and DL methods are increasingly employed to integrate vast amounts of data, enabling the identification of key structural insights that can lead to the design of drugs with improved specificity and efficacy.

One way of applying these AI-based methods is based on ligand-based approaches, such as Quantitative Structure–Activity Relationship (QSAR) models (Gianibbi et al. 2024; Taskin and Sevin 2009). QSAR establishes mathematical relationships between molecular structures, such as ligands, and their biological activities on ion channels. Starting from large datasets of known ion channel modulators and structural features, predictive models to estimate the potential effects of new compounds have been developed, combining ML methods of partial least squares regression. For instance, Abdelilah and colleagues performed a combination of a 3D-QSAR, molecular docking, and MD simulations to design novel potent TRPV1 antagonists with strong inhibitory activity (Abdelilah et al. 2023). Studies on other ion channels have also been performed, such as in voltage-gated sodium channel 1.7 (NaV1.7), involved in pain generation (Kong et al. 2020), or voltage-gated sodium ion channel 1.5 (NaV 1.5) (Khalifa et al. 2020), in which QSAR models were developed using a gradient boosting machine.

Ligand-based drug discovery plays a crucial role in the development of therapeutics targeting TRPVs and other ion channels. Ligand-based approaches allow for modeling interactions at atomic and molecular levels, enabling precise predictions of binding affinities and drug efficacy. In combination with MD, virtual screening, and structure-based design, ligand-based drug design aids in identifying and optimizing compounds that selectively target TRPV channels (Feng et al. 2015; Catalina-Hernández et al. 2024; Li et al. 2024). Besides, recent studies are adding the use of ML to predict the top ligand-binding sites (Zheng and Wen 2021; Wei et al. 2023).

Nonetheless, the limitations of ligand-based approaches lie in the inability to find binding sites without prior knowledge, compromising the accuracy and reliability of these methods (Nussinov et al. 2014; Amaya-Rodriguez et al. 2023). Besides, cryptic binding sites are hidden or inaccessible pockets that become accessible only under specific binding events or conditions (Kuzmanic et al. 2020). These pockets can play critical roles in TRPVs in modulating channel gating, ligand specificity, or allosteric regulation (Singh et al. 2018). Markov State Models (MSM), enhanced sampling techniques, or adaptive sampling approaches have been used to detect cryptic pockets (Konovalov et al. 2021; Zimmerman and Bowman 2015; Cruz et al. 2022; Comitani and Gervasio 2018).

Furthermore, AI methods also offer invaluable potential as complements for classical experimental techniques. Patch-clamp, for instance, allows researchers to measure ionic currents through the channels, offering direct assessment of channel activity, conductance, dysregulation, and gating mechanisms in response to stimuli (Amarouch et al. 2013; Studer and McNaughton 2010). However, the outputs are considerably noisy and need to be cleaned, usually done manually and by experienced researchers. Deep-Channel (Celik et al. 2020) was introduced to automate the patch clamp data cleaning process, by combining Convolutional Neural Networks (CNN), Long short term memory (LSTM), and Hidden Markov Models (HMM) to identify the number of open channels.

Finally, Cryo-EM techniques also suffer from inherent limitations, such as the dynamism and structural variability in biological complexes, the loss of contrast at high resolution, or the evaluation of the quality of the Cryo-EM structural model quality (Zhu et al. 2022; DiIorio and Kulczyk 2023). Luckily, AI revolution has also reached—and helped—Cryo-EM field. For instance, algorithms, such as sharpening, and DL techniques, such as DeepSharpen, have been applied to improve the loss of contrast (Zehni et al. 2020). cryoAI (Levy et al. 2022),

CryoPoseNet (Nashed et al. 2021), and 3DFlex (Punjani and Fleet 2023) are DL algorithms that can help determine particle poses or initial 3D structures to aid with structural variability. Moreover, DL-based de novo Cryo-EM structural modeling, such as DeepTracer (Pfab et al. 2021) or DeepMM (He and Huang 2021), does not require any initial structural models to build Cryo-EM structures.

In conclusion, TRPV channels are central to numerous physiological and pathological processes. Advances in experimental and computational approaches have significantly enhanced our understanding of their structure, function, and pharmacological modulation. Despite these efforts, challenges remain in discovering specific and efficient drugs, understanding dynamic behavior, and elucidating channel gating mechanisms. Emerging technologies, such as machine learning, deep learning, and AI-driven modeling, provide promising pathways to address these gaps. Continued interdisciplinary research and innovative methodologies will be crucial in overcoming these challenges and unlocking the full therapeutic potential of TRPV channels. This chapter has aimed to provide a comprehensive overview of TRPV channels, setting the stage for further exploration and discovery in this dynamic field.

References

Abdelilah T, Oussama C, El Aissouq A, Youssef EO, Bouachrine M, Ouammou A (2023) Prediction of novel TRPV1 antagonist: a combination of 3D-QSAR, molecular docking, MD simulations and ADMET prediction. Phys Chem Res 11:353–368. Available from: https://www.physchemres.org/article_153565.html

Abdelilah T, Oussama C, Abdellah EA, Youssef EO, Mohammed B, Kamal M et al (2024) Computational integration of TRPV4 antagonists: 3D QSAR, molecular docking, molecular dynamics simulations, ADME/Tox, and retrosynthesis studies. ChemistrySelect 9(42):e202404026

Abramson J, Adler J, Dunger J, Evans R, Green T, Pritzel A, et al. (2024) Accurate structure prediction of biomolecular interactions with AlphaFold 3. Nature [Internet]. [cited 2024 Nov 15] 630(8016):493–500. Available from: https://www.nature.com/articles/s41586-024-07487-w

Alipour M, Hajipour-Verdom B, Abdolmaleki P, Javan M (2023) Molecular properties of Ca2+ transport through TRPV2 channel: a molecular dynamics simulations study. J Biomol Struct Dyn 41:3892–3899. https://doi.org/10.1080/07391102.2022.2058615

Amarouch MY, Syam N, Abriel H (2013) Biochemical, single-channel, whole-cell patch clamp, and pharmacological analyses of endogenous TRPM4 channels in HEK293 cells. Neurosci Lett 541:105–110

Amaya-Rodriguez CA, Carvajal-Zamorano K, Bustos D, Alegría-Arcos M, Castillo K (2023) A journey from molecule to physiology and in silico tools for drug discovery targeting the transient receptor potential vanilloid type 1 (TRPV1) channel. Front Pharmacol 14:1251061

Anandakrishnan R, Drozdetski A, Walker RC, Onufriev AV (2015) Speed of conformational change: comparing explicit and implicit solvent molecular dynamics simulations. Biophys J [Internet]. [cited 2024 Dec 16] 108:1153–64. Available from: http://www.cell.com/article/S000634951500003X/fulltext

Ansell TB, Song W, Coupland CE, Carrique L, Corey RA, Duncan AL et al (2023) LipIDens: simulation assisted interpretation of lipid densities in cryo-EM structures of membrane proteins. Nat Commun 14:7774

Baker D (2024) Press release: The Nobel Prize in Chemistry 2024 - NobelPrize.org [Internet]. [cited 2024 Dec 16]. Available from: https://www.nobelprize.org/prizes/chemistry/2024/press-release/

Barducci A, Bussi G, Parrinello M (2008) Well-tempered metadynamics: a smoothly converging and tunable free-energy method. Phys Rev Lett [Internet]. [cited 2024 Dec 17] 100:020603. Available from: https://journals.aps.org/prl/abstract/10.1103/PhysRevLett.100.020603

Benham CD, Gunthorpe MJ, Davis JB (2003) TRPV channels as temperature sensors. Cell Calcium 33:479–487

Berna-Erro A, Izquierdo-Serra M, Sepúlveda RV, Rubio-Moscardo F, Doñate-Macián P, Serra SA et al (2017) Structural determinants of 5′,6′-epoxyeicosatrienoic acid binding to and activation of TRPV4 channel. Sci Rep 7(1):10522

Bhardwaj R, Lindinger S, Neuberger A, Nadezhdin KD, Singh AK, Cunha MR et al (2024) Inactivation-mimicking block of the epithelial calcium channel TRPV6. Sci Adv 6:eabe1508. https://doi.org/10.1126/sciadv.abe1508

Burns D, Venditti V, Potoyan DA (2023) Temperature sensitive contact modes allosterically gate TRPV3. PLoS Comput Biol 19:e1011545. https://doi.org/10.1371/journal.pcbi.1011545

Bustos D, Galarza C, Ordoñez W, Brauchi S, Benso B (2023) Cost-effective pipeline for a rational design and selection of capsaicin analogues targeting TRPV1 channels. ACS Omega 8:11736–11749. https://doi.org/10.1021/acsomega.2c05672

Caballero J (2022) A new era for the design of TRPV1 antagonists and agonists with the use of structural information and molecular docking of capsaicin-like compounds. J Enzyme Inhib Med Chem [Internet]. [cited 2024 Dec 4] 37:2169–78. Available from: https://www.tandfonline.com/doi/abs/10.1080/14756366.2022.2110089

Cao E, Liao M, Cheng Y, Julius D (2013) TRPV1 structures in distinct conformations reveal activation mechanisms. Nature 504:113–118. https://doi.org/10.1038/nature12823

Catalina-Hernández È, López-Martín M, Masnou-Sánchez D, Martins M, Lorenz-Fonfria VA, Jiménez-Altayó F et al (2024) Experimental and computational biophysics to identify vasodilator drugs targeted at TRPV2 using agonists based on the probenecid scaffold. Comput Struct Biotechnol J 23:473–482

Celik N, O'Brien F, Brennan S, Rainbow RD, Dart C, Zheng Y et al (2020) Deep-Channel uses deep neural networks to detect single-molecule events from patch-clamp data. Commun Biol 3(1):3

Chik WDW, Mohamed R, Majeed ABA, Jusoh SA (2012) Sequence analysis and homology modeling of TRPV5 and TRPV6 channels. 2012 IEEE Symposium on Business, Engineering and Industrial Applications. p. 342–6

Chu Y, Zhang H, Yang M, Yu R (2023) Molecular dynamic simulations reveal the activation mechanisms of oxidation-induced TRPV1. Int J Mol Sci 24(11):9553. Available from: https://www.mdpi.com/1422-0067/24/11/9553

Chugunov AO, Volynsky PE, Krylov NA, Nolde DE, Efremov RG (2016) Temperature-sensitive gating of TRPV1 channel as probed by atomistic simulations of its trans- and juxtamembrane domains. Sci Rep 6:33112. https://doi.org/10.1038/srep33112

Clapham DE (2003) TRP channels as cellular sensors. Nature [Internet]. [cited 2024 Dec 16] 426:517–24. Available from: https://pubmed.ncbi.nlm.nih.gov/14654832/

Clapham DE, Runnels LW, Strübing C (2001) The TRP ion channel family. Nat Rev Neurosci [Internet]. [cited 2025 May 11] 2:387–96. Available from: https://www.nature.com/articles/35077544

Comitani F, Gervasio FL (2018) Exploring cryptic pockets formation in targets of pharmaceutical interest with SWISH. J Chem Theory Comput [Internet]. [cited 2024 Nov 15] 14:3321–31. Available from: https://pubs.acs.org/doi/full/10.1021/acs.jctc.8b00263

Corso G, Deng A, Polizzi N, Barzilay R, Jaakkola T (2024) Deep confident steps to new pockets: strategies for docking generalization. 12th International Conference on Learning Representations, ICLR 2024 [Internet]. [cited 2024 Dec 17]. Available from: https://arxiv.org/abs/2402.18396v1

Cruz MA, Frederick TE, Mallimadugula UL, Singh S, Vithani N, Zimmerman MI, et al. (2022) A cryptic pocket in Ebola VP35 allosterically controls RNA binding. Nat Commun [Internet]. [cited 2024 Nov 15] 13:1–10. Available from: https://www.nature.com/articles/s41467-022-29927-9

Darby WG, Grace MS, Baratchi S, McIntyre P (2016) Modulation of TRPV4 by diverse mechanisms. Int J Biochem Cell Biol 78:217–228

Darré L, Domene C (2015) Binding of Capsaicin to the TRPV1 ion channel. Mol Pharm 12:4454–4465. https://doi.org/10.1021/acs.molpharmaceut.5b00641

Darré L, Furini S, Domene C (2015) Permeation and dynamics of an open-activated TRPV1 channel. J Mol Biol 427:537–549. Available from: https://www.sciencedirect.com/science/article/pii/S0022283614006123

De Jesús-Pérez JJ, Gabrielle M, Raheem S, Fluck EC, Rohacs T, Moiseenkova-Bell VY (2024) Structural mechanism of TRPV5 inhibition by econazole. Structure 32:148–156.e5. https://doi.org/10.1016/j.str.2023.11.012

Diehl E (2022) Elucidating the protein interactome of the human ion channel TRPV4: a mutli-pronged approach to reveal the secrets of the fascinating N-terminal TRPV4 protein interactome in vitro and in cellulo. [Mainz]

DiIorio MC, Kulczyk AW (2023) Novel artificial intelligence-based approaches for Ab initio structure destermination and atomic model building for cryo-electron microscopy. Micromachines [Internet]. [cited 2024 Dec 21] 14:1674. Available from: https://www.mdpi.com/2072-666X/14/9/1674/htm

Domene C, Darré L, Oakes V, Gonzalez-Resines S (2022) A potential route of Capsaicin to its binding site in the TRPV1 ion channel. J Chem Inf Model [Internet]. [cited 2024 Nov 21] 62:2481–9. Available from: https://pubs.acs.org/doi/full/10.1021/acs.jcim.1c01441

Doñate-Macián P, Perálvarez-Marín A (2014) Dissecting domain-specific evolutionary pressure profiles of transient receptor potential vanilloid subfamily members 1 to 4. PLoS One 9:e110715

Doñate-Macian P, Bañó-Polo M, Vazquez-Ibar JL, Mingarro I, Perálvarez-Marín A (2015) Molecular and topological membrane folding determinants of transient receptor potential vanilloid 2 channel. Biochem Biophys Res Commun 462:221–226

Doñate-Macián P, Jungfleisch J, Pérez-Vilaró G, Rubio-Moscardo F, Perálvarez-Marín A, Diez J, et al. (2018a) The TRPV4 channel links calcium influx to DDX3X activity and viral infectivity. Nat Commun [Internet]. 2018 [cited 2023 Dec 8] 9:1–13. Available from: https://www.nature.com/articles/s41467-018-04776-7

Doñate-Macián P, Gómez A, Dégano IR, Perálvarez-Marín A, Doñate-Macián P, Gómez A, et al. (2018b) A TRPV2 interactome-based signature for prognosis in glioblastoma patients. Oncotarget [Internet]. [cited 2023 Dec 8] 9:18400–9. Available from: https://www.oncotarget.com/article/24843/text/

Doñate-Macián P, Crespi-Boixader A, Perálvarez-Marín A (2019a) Molecular evolution bioinformatics toward structural biology of TRPV1-4 channels. In: Ferrer-Montiel A, Hucho T (eds) TRP channels: methods and protocols. Springer, New York, pp 1–21. https://doi.org/10.1007/978-1-4939-9446-5_1

Doñate-Macián P, Enrich-Bengoa J, Dégano IR, Quintana DG, Perálvarez-Marín A (2019b) Trafficking of stretch-regulated TRPV2 and TRPV4 channels inferred through interactomics. Biomolecules 9(12):791

Doñate-Macián P, Álvarez-Marimon E, Sepulcre F, Vázquez-Ibar JL, Perálvarez-Marín A (2019c) The membrane proximal domain of TRPV1 and TRPV2 channels mediates protein–protein interactions and lipid binding in vitro. Int J Mol Sci 20(3):682

Doñate-Macian P, Duarte Y, Rubio-Moscardo F, Pérez-Vilaró G, Canan J, Díez J et al (2022) Structural determinants of TRPV4 inhibition and identification of new antagonists with antiviral activity. Br J Pharmacol 179:3576–3591. https://doi.org/10.1111/bph.15267

Du X, Li Y, Xia YL, Ai SM, Liang J, Sang P, et al. (2016) Insights into protein–ligand interactions: mechanisms, models, and methods. Int J Mol Sci [Internet]. [cited 2024 Dec 17] 17:144. Available from: https://www.mdpi.com/1422-0067/17/2/144/htm

Dubochet J, McDowall AW (1981) Vitrification of pure water for electron microscopy. J Microsc [Internet]. [cited 2024 Dec 16] 124:3–4. Available from: https://onlinelibrary.wiley.com/doi/full/10.1111/j.1365-2818.1981.tb02483.x

Dubochet J, Adrian M, Chang J, Homo J, Lepault J, McDowall A, et al. (1988) Cryo-electron microscopy of vitrified specimens. Q Rev Biophys [Internet]. [cited 2024 Dec 16] 21:129–228. Available from: https://pubmed.ncbi.nlm.nih.gov/3043536/

Dubochet J, Frank J, Henderson R (2017) Press release: The 2017 Nobel Prize in Chemistry - NobelPrize.org [Internet]. [cited 2024 Dec 16]. Available from: https://www.nobelprize.org/prizes/chemistry/2017/press-release/

Fan J, Ke H, Lei J, Wang J, Tominaga M, Lei X (2024) Structural basis of TRPV1 inhibition by SAF312 and cholesterol. Nat Commun 15:6689. https://doi.org/10.1038/s41467-024-51085-3

Feng Z, Pearce LV, Xu X, Yang X, Yang P, Blumberg PM et al (2015) Structural insight into tetrameric hTRPV1 from homology modeling, molecular docking, molecular dynamics simulation, virtual screening, and bioassay validations. J Chem Inf Model 55:572–588. https://doi.org/10.1021/ci5007189

Feng Z, Pearce LV, Zhang Y, Xing C, Herold BKA, Ma S et al (2016) Multi-functional Diarylurea small molecule inhibitors of TRPV1 with therapeutic potential for neuroinflammation. AAPS J 18:898–913. https://doi.org/10.1208/s12248-016-9888-z

Feng S, Pumroy RA, Protopopova AD, Moiseenkova-Bell VY, Im W (2023) Modulation of TRPV2 by endogenous and exogenous ligands: a computational study. Protein Sci 32:e4490. https://doi.org/10.1002/pro.4490

Flores-Aldama L, Vandewege MW, Zavala K, Colenso CK, Gonzalez W, Brauchi SE et al (2020) Evolutionary analyses reveal independent origins of gene repertoires and structural motifs associated to fast inactivation in calcium-selective TRPV channels. Sci Rep 10:8684. https://doi.org/10.1038/s41598-020-65679-6

Flores-Aldama L, Bustos D, Cabezas-Bratesco D, Gonzalez W, Brauchi SE (2023) Intracellular helix-loop-helix domain modulates inactivation kinetics of mammalian TRPV5 and TRPV6 channels. Int J Mol Sci 24:4470. Available from: https://www.mdpi.com/1422-0067/24/5/4470

Frank J (2006) Three-dimensional electron microscopy of macromolecular assemblies: visualization of biological molecules in their native state. [Internet]. [cited 2024 Dec 16] 1–456. Available from: https://academic.oup.com/book/6747

Gallo PN, Mihelc E, Eisert R, Bradshaw GA, Dimek F, Leffler A et al (2024) The dynamic TRPV2 ion channel proximity proteome reveals functional links of calcium flux with cellular adhesion factors NCAM and L1CAM in neurite outgrowth. Cell Calcium 121:102894. Available from: https://www.sciencedirect.com/science/article/pii/S0143416024000526

Gao Y, Cao E, Julius D, Cheng Y (2016) TRPV1 structures in nanodiscs reveal mechanisms of ligand and lipid action. Nature 534:347–351

Garcia-Elias A, Mrkonjić S, Jung C, Pardo-Pastor C, Vicente R, Valverde MA (2014) The trpv4 channel. Handb Exp Pharmacol [Internet]. [cited 2024 Dec 20] 222:293–319. Available from: https://link.springer.com/chapter/10.1007/978-3-642-54215-2_12

Garcia-Elias A, Berna-Erro A, Rubio-Moscardo F, Pardo-Pastor C, Mrkonjić S, Sepúlveda RV et al (2015) Interaction between the linker, pre-S1, and TRP domains determines folding, assembly, and trafficking of TRPV channels. Structure 23:1404–1413. https://doi.org/10.1016/j.str.2015.05.018

Gianibbi B, Visibelli A, Spinsanti G, Spiga O (2024) Three-dimensional quantitative structure–activity relationship study of transient receptor potential Vanilloid 1 channel antagonists reveals potential for drug design purposes. Int J Mol Sci 25:1251061

Goretzki B (2021) Interactions of the ion channel TRPV4 with regulatory lipids and proteins - an (un)structural study. [Mainz]

Gu Y, Li Y, Ma B, Ren K, Cao C, Gu N (2023) Probing conformational transition of TRPV5 induced by mechanical force using coarse-grained molecular dynamics. J Chem Inf Model [Internet]. [cited 2024 Nov 21] 63:6768–77. Available from: https://pubs.acs.org/doi/full/10.1021/acs.jcim.3c00614

Halgren T (2007) New method for fast and accurate binding-site identification and analysis. Chem Biol Drug Des [Internet]. [cited 2024 Dec 16] 69:146–8. Available from: https://onlinelibrary.wiley.com/doi/full/10.1111/j.1747-0285.2007.00483.x

Hanson SM, Newstead S, Swartz KJ, Sansom MSP (2015) Capsaicin interaction with TRPV1 channels in a lipid bilayer: molecular dynamics simulation. Biophys J 108:1425–1434. Available from: https://www.sciencedirect.com/science/article/pii/S0006349515001757

He J, Huang SY (2021) Full-length de novo protein structure determination from cryo-EM maps using deep learning. Bioinformatics [Internet]. [cited 2024 Dec 21] 37:3480–90. Available from: https://pubmed.ncbi.nlm.nih.gov/33978686/

Henderson R (1995) The potential and limitations of neutrons, electrons and X-rays for atomic resolution microscopy of unstained biological molecules. Q Rev Biophys [Internet]. [cited 2024 Dec 16] 28:171–93. Available from: https://pubmed.ncbi.nlm.nih.gov/7568675/
Henderson R, Unwin PNT (1975) Three-dimensional model of purple membrane obtained by electron microscopy. Nature [Internet]. [cited 2024 Dec 16] 257:28–32. Available from: https://pubmed.ncbi.nlm.nih.gov/1161000/
Hritz J, Oostenbrink C (2008) Hamiltonian replica exchange molecular dynamics using soft-core interactions. J Chem Phy [Internet]. [cited 2024 Dec 17] 128. Available from: https://pubs.aip.org/aip/jcp/article-abstract/128/14/144121/848242/Hamiltonian-replica-exchange-molecular-dynamics
Hu HZ, Gu Q, Wang C, Colton CK, Tang J, Kinoshita-Kawada M, et al. (2004) 2-aminoethoxydiphenyl borate is a common activator of TRPV1, TRPV2, and TRPV3. J Biol Chem [Internet]. [cited 2023 Dec 10] 279:35741–8. Available from: https://pubmed.ncbi.nlm.nih.gov/15194687/
Huang J, Chen J (2023) Hydrophobic gating in bundle-crossing ion channels: a case study of TRPV4. Biophys J 122:109a. https://doi.org/10.1016/j.bpj.2022.11.770
Huffer KE, Aleksandrova AA, Jara-Oseguera A, Forrest LR, Swartz KJ (2020) Global alignment and assessment of TRP channel transmembrane domain structures to explore functional mechanisms. Csanády L, Aldrich RW, Csanády L, Du J, Zubcevic L, editors. Elife 9:e58660. https://doi.org/10.7554/eLife.58660
Hughes TET, Lodowski DT, Huynh KW, Yazici A, Del Rosario J, Kapoor A et al (2018a) Structural basis of TRPV5 channel inhibition by econazole revealed by cryo-EM. Nat Struct Mol Biol 25:53–60. https://doi.org/10.1038/s41594-017-0009-1
Hughes TET, Pumroy RA, Yazici AT, Kasimova MA, Fluck EC, Huynh KW et al (2018b) Structural insights on TRPV5 gating by endogenous modulators. Nat Commun 9:4198. https://doi.org/10.1038/s41467-018-06753-6
Huynh KW, Cohen MR, Chakrapani S, Holdaway HA, Stewart PL, Moiseenkova-Bell VY (2014) Structural insight into the assembly of TRPV channels. Structure [Internet]. [cited 2024 Nov 15] 22:260–8. Available from: http://www.cell.com/article/S0969212613004589/fulltext
Inada H, Procko E, Sotomayor M, Gaudet R (2012) Structural and biochemical consequences of disease-causing mutations in the ankyrin repeat domain of the human TRPV4 channel. Biochemistry [Internet]. [cited 2024 Nov 21] 51:6195–206. Available from: https://pubs.acs.org/doi/full/10.1021/bi300279b
Ives CM, Thomson NJ, Zachariae U (2023) A cooperative knock-on mechanism underpins Ca2+-selective cation permeation in TRPV channels. J Gen Physiol 155:e202213226. https://doi.org/10.1085/jgp.202213226
Jiang Y, Han D, Zhao Y, Zhang C, Shi X, Gu W (2022) Multi-Omics analysis of the prognosis and biological function for TRPV Channel family in clear cell renal cell carcinoma. Front Immunol 13:872170
Jorgensen C, Domene C (2018) Location and character of volatile general anesthetics binding sites in the transmembrane domain of TRPV1. Mol Pharm 15:3920–3930. https://doi.org/10.1021/acs.molpharmaceut.8b00381
Jorgensen C, Furini S, Domene C (2016) Energetics of ion permeation in an open-activated TRPV1 channel. Biophys J 111:1214–1222. Available from: https://www.sciencedirect.com/science/article/pii/S0006349516306968
Julius D (2013) TRP channels and pain. Annu Rev Cell Dev Biol [Internet]. [cited 2024 Dec 16] 29:355–84. Available from: https://pubmed.ncbi.nlm.nih.gov/24099085/
Julius D (2021) The Nobel Prize in Physiology or Medicine 2021 - Advanced information - NobelPrize.org [Internet]. [cited 2024 Dec 16]. Available from: https://www.nobelprize.org/prizes/medicine/2021/advanced-information/
Julius D, Basbaum AI (2001) Molecular mechanisms of nociception. Nature [Internet]. [cited 2024 Dec 16] 413(6852):203–10. Available from: https://www.nature.com/articles/35093019
Jumper J, Evans R, Pritzel A, Green T, surnov M, Ronneberger O et al (2021) Highly accurate protein structure prediction with AlphaFold. Nature 596:583–589. Available from: https://www.nature.com/articles/s41586-021-03819-2
Kacher J, Sokolova OS, Tarek M (2024) A deep learning approach to uncover voltage-gated ion channels' intermediate states. J Phys Chem B 128(36):8724–8736
Karplus M (2013) Press release: The Nobel Prize in Chemistry 2013 - NobelPrize.org [Internet]. [cited 2024 Dec 16]. Available from: https://www.nobelprize.org/prizes/chemistry/2013/press-release/
Karplus M, Petsko GA (1990) Molecular dynamics simulations in biology. Nature [Internet]. [cited 2024 Dec 16] 347(6294):631–9. Available from: https://www.nature.com/articles/347631a0
Kasimova MA, Yazici A, Yudin Y, Granata D, Klein ML, Rohacs T et al (2018) Ion channel sensing: are fluctuations the crux of the matter? J Phys Chem Lett 9:1260–1264. https://doi.org/10.1021/acs.jpclett.7b03396
Khalifa N, Kumar Konda LS, Kristam R (2020) Machine learning-based QSAR models to predict sodium ion channel (Nav 1.5) blockers. Future Med Chem [Internet]. [cited 2024 Nov 13] 12:1829–43. Available from: https://pubmed.ncbi.nlm.nih.gov/33034205/
Kong W, Tu X, Huang W, Yang Y, Xie Z, Huang Z (2020) Prediction and optimization of NaV1.7 sodium channel inhibitors based on machine learning and simulated annealing. J Chem Inf Model [Internet]. [cited 2024 Nov 13] 60:2739–53. Available from: https://pubmed.ncbi.nlm.nih.gov/32421318/
Konovalov KA, Unarta IC, Cao S, Goonetilleke EC, Huang X (2021) Markov state models to study the functional dynamics of proteins in the wake of machine learning. JACS Au [Internet]. [cited 2024 Nov

15] 1:1330–41. Available from: https://pubs.acs.org/doi/full/10.1021/jacsau.1c00254

Kuzmanic A, Bowman GR, Juarez-Jimenez J, Michel J, Gervasio FL (2020) Investigating cryptic binding sites by molecular dynamics simulations. Acc Chem Res 53:654–661. https://doi.org/10.1021/acs.accounts.9b00613

Kwon DH, Zhang F, Suo Y, Bouvette J, Borgnia MJ, Lee SY (2021) Heat-dependent opening of TRPV1 in the presence of capsaicin. Nat Struct Mol Biol [Internet]. [cited 2024 Dec 22] 28(7):554–63. Available from: https://www.nature.com/articles/s41594-021-00616-3

Lacerda JT, Gomes PRL, Zanetti G, Mezzalira N, Lima OG, de Assis LVM et al (2022) Lack of TRPV1 channel modulates mouse gene expression and liver proteome with glucose metabolism changes. Int J Mol Sci 23:7014. Available from: https://www.mdpi.com/1422-0067/23/13/7014

Ladrón-de-Guevara E, Dominguez L, Rangel-Yescas GE, Fernández-Velasco DA, Torres-Larios A, Rosenbaum T et al (2020) The contribution of the Ankyrin repeat domain of TRPV1 as a thermal module. Biophys J 118:836–845. https://doi.org/10.1016/j.bpj.2019.10.041

Lee JH, Lee Y, Ryu H, Kang DW, Lee J, Lazar J et al (2011) Structural insights into transient receptor potential vanilloid type 1 (TRPV1) from homology modeling, flexible docking, and mutational studies. J Comput Aided Mol Des 25:317–327. https://doi.org/10.1007/s10822-011-9421-5

Levitt M (2001) The birth of computational structural biology. Nat Struct Biol [Internet]. [cited 2024 Dec 16] 8:392–3. Available from: https://pubmed.ncbi.nlm.nih.gov/11323711/

Levy A, Poitevin F, Martel J, Nashed Y, Peck A, Miolane N, et al. (2022) CryoAI: amortized inference of poses for Ab initio reconstruction of 3D molecular volumes from real cryo-EM images. Lecture Notes in Computer Science (including subseries Lecture Notes in Artificial Intelligence and Lecture Notes in Bioinformatics) [Internet]. [cited 2024 Dec 21] 13681 LNCS:540–57. Available from: https://link.springer.com/chapter/10.1007/978-3-031-19803-8_32

Li D, Kao T-H, Chang S-W (2020) The structural changes of the mutated ankyrin repeat domain of the human TRPV4 channel alter its ATP binding ability. J Mech Behav Biomed Mater 101:103407. Available from: https://www.sciencedirect.com/science/article/pii/S1751616119308185

Li X, Zhou C, He L, Xu Z, Li Z, Cheng J (2024) Identification of potential novel insect TRPV channel modulators by homology modeling, binding mode analysis, virtual screening studies and chemical optimization. New J Chem [Internet]. [cited 2024 Nov 14] 48:12688–96. Available from: https://pubs.rsc.org/en/content/articlehtml/2024/nj/d4nj02266a

Liao M, Cao E, Julius D, Cheng Y (2013) Structure of the TRPV1 ion channel determined by electron cryo-microscopy. Nature [Internet]. [cited 2024 Dec 16] 504(7478):107–12. Available from: https://www.nature.com/articles/nature12822

Liu C, Xue L, Song C (2023) Calcium binding and permeation in TRPV channels: insights from molecular dynamics simulations. J Gen Physiol 155:e202213261. https://doi.org/10.1085/jgp.202213261

Lopez-Martin M, Eric Catalina-Hernandez E, Peralvarez-Marin A (2025) Computational insights into the lipid code of transient receptor potential vanilloid channels reveal overlap with drug binding sites. Int J Biol Macromol 330:148108. https://doi.org/10.1016/j.ijbiomac.2025.148108

Macikova L, Vyklicka L, Barvik I, Sobolevsky AI, Vlachova V (2019) Cytoplasmic inter-subunit interface controls use-dependence of thermal activation of TRPV3 channel. Int J Mol Sci [Internet]. [cited 2024 Dec 16] 20:3990. Available from: https://www.mdpi.com/1422-0067/20/16/3990/htm

Marrink SJ, Monticelli L, Melo MN, Alessandri R, Tieleman DP, Souza PCT (2023) Two decades of martini: better beads, broader scope. Wiley Interdiscip Rev Comput Mol Sci [Internet] [cited 2024 Feb 16] 13:e1620. Available from: https://onlinelibrary.wiley.com/doi/full/10.1002/wcms.1620

Martin E, Dahan D, Cardouat G, Gillibert-Duplantier J, Marthan R, Savineau JP, et al. (2012) Involvement of TRPV1 and TRPV4 channels in migration of rat pulmonary arterial smooth muscle cells. Pflugers Arch [Internet]. [cited 2024 Dec 20] 464:261–72. Available from: https://link.springer.com/article/10.1007/s00424-012-1136-5

McCray BA, Diehl E, Sullivan JM, Aisenberg WH, Zaccor NW, Lau AR, et al. (2021) Neuropathy-causing TRPV4 mutations disrupt TRPV4-RhoA interactions and impair neurite extension. Nat Commun [Internet]. [cited 2024 Dec 20] 12:1–17. Available from: https://www.nature.com/articles/s41467-021-21699-y

McNutt AT, Francoeur P, Aggarwal R, Masuda T, Meli R, Ragoza M et al (2021) GNINA 1.0: molecular docking with deep learning. J Cheminform 13:43. Available from: https://www.ncbi.nlm.nih.gov/pmc/articles/PMC8191141/

Melnick C, Kaviany M (2018) Thermal actuation in TRPV1: role of embedded lipids and intracellular domains. J Theor Biol 444:38–49. Available from: https://www.sciencedirect.com/science/article/pii/S0022519318300535

Miao Y, Bhattarai A, Wang J (2020) Ligand Gaussian Accelerated Molecular Dynamics (LiGaMD): characterization of ligand binding thermodynamics and kinetics. J Chem Theory Comput [Internet]. [cited 2023 Dec 10] 16:5526–47. Available from: https://pubs.acs.org/doi/abs/10.1021/acs.jctc.0c00395

Montell C (2005) The TRP superfamily of cation channels. Science's STKE [Internet]. [cited 2025 May 11] 2005. Available from: https://doi.org/10.1126/stke.2722005re3

Morales-Lázaro SL, Rosenbaum T (2019) Cholesterol as a key molecule that regulates TRPV1 channel function. Adv Exp Med Biol [Internet]. [cited

2025 Mar 6] 1135:105–17. Available from: https://link.springer.com/chapter/10.1007/978-3-030-14265-0_6
Morini M, Bergqvist CA, Asturiano JF, Larhammar D, Dufour S (2022) Dynamic evolution of transient receptor potential vanilloid (TRPV) ion channel family with numerous gene duplications and losses. Front Endocrinol (Lausanne) 13:1013868
Muazzam Ali Shah S, Ou YY (2023) Disto-TRP: an approach for identifying transient receptor potential (TRP) channels using structural information generated by AlphaFold. Gene 871:147435
Muller C, Lynch DL, Hurst DP, Reggio PH (2020) A closer look at anandamide interaction with TRPV1. Front Mol Biosci 7:548857
Nadezhdin KD, Neuberger A, Trofimov YA, Krylov NA, Sinica V, Kupko N et al (2021) Structural mechanism of heat-induced opening of a temperature-sensitive TRP channel. Nat Struct Mol Biol 28:564–572. https://doi.org/10.1038/s41594-021-00615-4
Nashed YSG, Poitevin F, Gupta H, Woollard G, Kagan M, Yoon CH, et al. (2021) CryoPoseNet: end-to-end simultaneous learning of single-particle orientation and 3D map reconstruction from cryo-electron microscopy data. 4066–76
Neuberger A, Oda M, Nikolaev YA, Nadezhdin KD, Gracheva EO, Bagriantsev SN, et al. (2023a) Human TRPV1 structure and inhibition by the analgesic SB-366791. Nat Commun [Internet]. [cited 2024 Dec 22] 14:1–10. Available from: https://www.nature.com/articles/s41467-023-38162-9
Neuberger A, Trofimov YA, Yelshanskaya MV, Khau J, Nadezhdin KD, Khosrof LS et al (2023b) Molecular pathway and structural mechanism of human oncochannel TRPV6 inhibition by the phytocannabinoid tetrahydrocannabivarin. Nat Commun 14:4630. https://doi.org/10.1038/s41467-023-40362-2
Nguyen THD, Itoh SG, Okumura H, Tominaga M (2021) Structural basis for promiscuous action of monoterpenes on TRP channels. Commun Biol 4:293. https://doi.org/10.1038/s42003-021-01776-0
Nguyen PT, Harris BJ, Mateos DL, González AH, Murray AM, Yarov-Yarovoy V (2024) Structural modeling of ion channels using AlphaFold2, RoseTTAFold2, and ESMFold. Channels 18:2325032. https://doi.org/10.1080/19336950.2024.2325032
Nilius B, Owsianik G (2011) The transient receptor potential family of ion channels. Genome Biol [Internet]. [cited 2024 Dec 16] 12:1–11. Available from: https://genomebiology.biomedcentral.com/articles/10.1186/gb-2011-12-3-218
Nussinov R, Tsai CJ, Liu J (2014) Principles of allosteric interactions in cell signaling. J Am Chem Soc [Internet]. [cited 2024 Nov 15] 136:17692–701. Available from: https://pubs.acs.org/doi/full/10.1021/ja510028c
Ortega-Guerrero A, Espinosa-Duran JM, Velasco-Medina J (2016) TRPV1 channel as a target for cancer therapy using CNT-based drug delivery systems. Eur Biophys J [Internet]. [cited 2024 Nov 21] 45:423–33. Available from: https://link.springer.com/article/10.1007/s00249-016-1111-8
Owsianik G, Talavera K, Voets T, Nilius B (2006) Permeation and selectivity of TRP channels. Annu Rev Physiol 68:685–717
Park S, Schulten K (2004) Calculating potentials of mean force from steered molecular dynamics simulations. J Chem Phys 120:5946–5961. Available from: http://www.ncbi.nlm.nih.gov/pubmed/15267476
Park S, Khalili-Araghi F, Tajkhorshid E, Schulten K (2003) Free energy calculation from steered molecular dynamics simulations using Jarzynski's equality. J Chem Phys 119:3559–3566
Pedersen SF, Nilius B (2007) Transient receptor potential channels in mechanosensing and cell volume regulation. Methods Enzymol [Internet]. [cited 2024 Dec 16] 428:183–207. Available from: https://pubmed.ncbi.nlm.nih.gov/17875418/
Pedersen SF, Owsianik G, Nilius B (2005) TRP channels: an overview. Cell Calcium 38:233–252
Perálvarez-Marín A, Doñate-Macian P, Gaudet R (2013) What do we know about the transient receptor potential Vanilloid 2 (TRPV2) ion channel? FEBS J 280:5471–5487
Pfab J, Phan NM, Si D (2021) DeepTracer for fast de novo cryo-EM protein structure modeling and special studies on cov-related complexes. Proc Natl Acad Sci U S A [Internet]. [cited 2024 Dec 21] 118. Available from: https://pubmed.ncbi.nlm.nih.gov/33361332/
Piana S, Laio A (2007) A bias-exchange approach to protein folding. J Phys Chem B [Internet]. [cited 2024 Dec 17] 111:4553–9. Available from: https://pubs.acs.org/doi/abs/10.1021/jp067873l
Poblete H, Oyarzún I, Olivero P, Comer J, Zuñiga M, Sepulveda RV et al (2015) Molecular determinants of phosphatidylinositol 4,5-bisphosphate (PI(4,5)P2) binding to transient receptor potential V1 (TRPV1) channels*. J Biol Chem 290:2086–2098. Available from: https://www.sciencedirect.com/science/article/pii/S002192582057748X
Punjani A, Fleet DJ (2023) 3DFlex: determining structure and motion of flexible proteins from cryo-EM. Nature Methods [Internet]. [cited 2024 Dec 21] 20:860–70. Available from: https://www.nature.com/articles/s41592-023-01853-8
Ramsey IS, Delling M, Clapham DE (2006) An introduction to TRP channels. Annu Rev Physiol [Internet]. [cited 2024 Dec 16] 68:619–47. Available from: https:/ /pubmed.ncbi.nlm.nih.gov/16460286/
Rao S, Klesse G, Stansfeld PJ, Tucker SJ, Sansom MSP (2019) A heuristic derived from analysis of the ion channel structural proteome permits the rapid identification of hydrophobic gates. Proc Natl Acad Sci U S A [Internet]. [cited 2024 Nov 13] 116:13989–95. Available from: https://pubmed.ncbi.nlm.nih.gov/31235590/
Rao S, Klesse G, Lynch CI, Tucker SJ, Sansom MSP (2021) Molecular simulations of hydrophobic gating of pentameric ligand gated ion channels: insights into water and ions. J Phys Chem B [Internet]. [cited 2024 Dec 19] 125:981–94. Available from: https://pubs.acs.org/doi/full/10.1021/acs.jpcb.0c09285

Raymond KA, Twomey EC, Wei Y (2014) Characterization of temperature-sensing and PIP2-regulation of TRPV1 ion channel at the C-terminal domain using NMR spectroscopy and molecular dynamics simulations. J Integr OMICS 4:79–86

Rosenbaum T, Islas LD (2024) Molecular physiology of TRPV channels: controversies and future challenges. Annu Rev Physiol 85:293–316. https://doi.org/10.1146/annurev-physiol-030222-012349

Roshni J, Sivakumar M, Alzahrani FM, Halawani IF, Alzahrani KJ, Patil S, et al. (2024) Virtual screening, molecular dynamics and density functional theory on pain inhibitors against TRPV1 associating inflammatory conditions. J Biomol Struct Dyn [Internet]. [cited 2024 Nov 21] 42:6788–98. Available from: https://www.tandfonline.com/doi/abs/10.1080/07391102.2023.2237595

Saha S, Ghosh A, Tiwari N, Kumar A, Kumar A, Goswami C (2017) Preferential selection of Arginine at the lipid-water-interface of TRPV1 during vertebrate evolution correlates with its snorkeling behaviour and cholesterol interaction. Sci Rep [Internet]. [cited 2025 May 6] 7:1–21. Available from: https://www.nature.com/articles/s41598-017-16780-w

Sakipov S, Sobolevsky AI, Kurnikova MG (2018) Ion permeation mechanism in epithelial calcium channel TRVP6. Sci Rep 8(1):5715

Samanta A, Hughes TET, Moiseenkova-Bell VY (2018) Transient Receptor Potential (TRP) channels. Subcell Biochem [Internet]. [cited 2025 May 11] 87:141–65. Available from: https://link.springer.com/chapter/10.1007/978-981-10-7757-9_6

Seebohm G, Schreiber JA (2021) Beyond hot and spicy: TRPV channels and their pharmacological modulation. Cell Physiol Biochem 55:108–130. Available from: https://www.cellphysiolbiochem.com/Articles/000358/

Singh AK, Saotome K, McGoldrick LL, Sobolevsky AI (2018) Structural bases of TRP channel TRPV6 allosteric modulation by 2-APB. Nat Commun [Internet]. [cited 2024 Nov 15] 9:1–11. Available from: https://www.nature.com/articles/s41467-018-04828-y

Souza PCT, Alessandri R, Barnoud J, Thallmair S, Faustino I, Grünewald F, et al. (2021) Martini 3: a general purpose force field for coarse-grained molecular dynamics. Nature Methods [Internet]. [cited 2024 Dec 16] 18(4):382–8. Available from: https://www.nature.com/articles/s41592-021-01098-3

Steinberg X, Kasimova MA, Cabezas-Bratesco D, Galpin JD, Ladron-de-Guevara E, Villa F et al (2017) Conformational dynamics in TRPV1 channels reported by an encoded coumarin amino acid. Csanády L, editor. Elife 6:e28626. https://doi.org/10.7554/eLife.28626

Stewart JM (2020) TRPV6 as a target for cancer therapy. J Cancer 11:374–387

Studer M, McNaughton PA (2010) Modulation of single-channel properties of TRPV1 by phosphorylation. J Physiol [Internet]. [cited 2024 Nov 13] 588:3743–56. Available from: https://onlinelibrary.wiley.com/doi/full/10.1113/jphysiol.2010.190611

Su N, Zhen W, Zhang H, Xu L, Jin Y, Chen X et al (2023) Structural mechanisms of TRPV2 modulation by endogenous and exogenous ligands. Nat Chem Biol 19:72–80

Suárez-Suárez C, González-Pérez S, Márquez-Miranda V, Araya-Duran I, Vidal-Beltrán I, Vergara S et al (2024) The endocannabinoid peptide RVD-Hemopressin is a TRPV1 channel blocker. Biomolecules 14(9):1134. Available from: https://www.mdpi.com/2218-273X/14/9/1134

Takahashi N, Hamada-Nakahara S, Itoh Y, Takemura K, Shimada A, Ueda Y et al (2014) TRPV4 channel activity is modulated by direct interaction of the ankyrin domain to PI(4,5)P2. Nat Commun 5:4994. https://doi.org/10.1038/ncomms5994

Taskin T, Sevin F (2009) QSAR approach to correlate TRPV1 antagonist activity for a series of heteroaromatic urea. QSAR Comb Sci [Internet]. [cited 2024 Nov 15];28:1098–111. Available from: https://onlinelibrary.wiley.com/doi/full/10.1002/qsar.200810157

Torrie GM, Valleau JP (1977) Nonphysical sampling distributions in Monte Carlo free-energy estimation: umbrella sampling. J Comput Phys 23:187–199

Trofimov YA, Minakov AS, Krylov NA, Efremov RG (2023) Structural mechanism of ionic conductivity of the TRPV1 channel. Dokl Biochem Biophys [Internet]. [cited 2024 Nov 21] 508:1–5. Available from: https://link.springer.com/article/10.1134/S1607672922600245

Trofimov YA, Krylov NA, Minakov AS, Nadezhdin KD, Neuberger A, Sobolevsky AI et al (2024) Dynamic molecular portraits of ion-conducting pores characterize functional states of TRPV channels. Commun Chem 7(1):119

Van Abel M, Hoenderop JGJ, Bindels RJM (2005) The epithelial calcium channels TRPV5 and TRPV6: Regulation and implications for disease. Naunyn Schmiedebergs Arch Pharmacol [Internet]. [cited 2024 Dec 17] 371:295–306. Available from: https://link.springer.com/article/10.1007/s00210-005-1021-2

Venkatachalam K, Montell C (2007) TRP channels. Annu Rev Biochem [Internet]. [cited 2024 Dec 16] 76:387–417. Available from: https://pubmed.ncbi.nlm.nih.gov/17579562/

Wang L, Holmes RP, Peng J-B (2016) Molecular modeling of the structural and dynamical changes in calcium channel TRPV5 induced by the African-specific A563T variation. Biochem Int 55:1254–1264. https://doi.org/10.1021/acs.biochem.5b00732

Wang L, Holmes RP, Peng J-B (2017) The L530R variation associated with recurrent kidney stones impairs the structure and function of TRPV5. Biochem Biophys Res Commun 492:362–367. Available from: https://www.sciencedirect.com/science/article/pii/S0006291X17316856

Wang L, Holmes RP, Peng J-B (2019) Modeling the structural and dynamical changes of the epithelial calcium channel TRPV5 caused by the A563T variation based on the structure of

TRPV6. J Biomol Struct Dyn 37:3506–3512. https://doi.org/10.1080/07391102.2018.1518790

Wang H, Yang P, Lu Y, Wang J, Jeon J, Wang Q et al (2020) Mechanisms of proton inhibition and sensitization of the cation channel TRPV3. J Gen Physiol 153:e202012663. https://doi.org/10.1085/jgp.202012663

Wang L, Cai R, Chen X-Z, Peng J-B (2023) Molecular insights into the structural and dynamical changes of calcium channel TRPV6 induced by its interaction with phosphatidylinositol 4,5-bisphosphate. J Biomol Struct Dyn 41:6559–6568. https://doi.org/10.1080/07391102.2022.2109752

Warshel A, Levitt M (1976) Theoretical studies of enzymic reactions: dielectric, electrostatic and steric stabilization of the carbonium ion in the reaction of lysozyme. J Mol Biol 103:227–249

Wei X, Yang Q, Yang Z, Huang T, Yang H, Wang L et al (2023) Discovery of novel TRPV1 modulators through machine learning-based molecular docking and molecular similarity searching. Chem Biol Drug Des 102:409–423. https://doi.org/10.1111/cbdd.14270

Wen H, Zheng W (2018) Decrypting the heat activation mechanism of TRPV1 channel by molecular dynamics simulation. Biophys J [Internet]. [cited 2024 Nov 20] 114:40–52. Available from: http://www.cell.com/article/S000634951731161X/fulltext

Wen H, Qin F, Zheng W (2016) Toward elucidating the heat activation mechanism of the TRPV1 channel gating by molecular dynamics simulation. Proteins Struct Funct Bioinf 84:1938–1949. https://doi.org/10.1002/prot.25177

Yang F, Xiao X, Cheng W, Yang W, Yu P, Song Z et al (2015) Structural mechanism underlying capsaicin binding and activation of the TRPV1 ion channel. Nat Chem Biol 11:518–524. https://doi.org/10.1038/nchembio.1835

Yang F, Vu S, Yarov-Yarovoy V, Zheng J (2016) Rational design and validation of a vanilloid-sensitive TRPV2 ion channel. Proc Nat Acad Sci 113:E3657–E3666. https://doi.org/10.1073/pnas.1604180113

Zamri MHB, Ujihara Y, Nakamura M, Mofrad MRK, Sugita S (2022) Decoding the effect of hydrostatic pressure on TRPV1 lower-gate conformation by molecular-dynamics simulation. Int J Mol Sci [Internet]. [cited 2024 Nov 21] 23:7366. Available from: https://www.mdpi.com/1422-0067/23/13/7366/htm

Zehni M, Do MN, Zhao Z (2020) Deepsharpen: deep-learning based sharpening of 3D reconstruction map from cryo-electron microscopy. ISBI Workshops 2020 - International Symposium on Biomedical Imaging Workshops, Proceedings.

Zhang JL, Zheng QC, Li ZQ, Zhang HX (2012) Molecular dynamics simulations suggest Ligand's binding to nicotinamidase/pyrazinamidase. PLoS One [Internet]. [cited 2024 Dec 17] 7:e39546. Available from: https://journals.plos.org/plosone/article?id=10.1371/journal.pone.0039546

Zhang K, Julius D, Cheng Y (2021a) Structural snapshots of TRPV1 reveal mechanism of polymodal functionality. Cell [Internet]. [cited 2024 Dec 22] 184:5138-5150.e12. Available from: http://www.cell.com/article/S009286742100982X/fulltext

Zhang F, Lin Y, Min W, Hou Y, Yuan K, Wang J et al (2021b) Computational discovery, structural optimization and biological evaluation of novel inhibitors targeting transient receptor potential vanilloid type 3 (TRPV3). Bioorg Chem 114:105093. Available from: https://www.sciencedirect.com/science/article/pii/S0045206821004703

Zhao Y, McVeigh BM, Moiseenkova-Bell VY (2021) Structural pharmacology of TRP channels. J Mol Biol 433(17):166914. Academic Press

Zhen W, Zhao Z, Chang S, Chen X, Wan Y, Yang F (2023) Structural basis of ligand activation and inhibition in a mammalian TRPV4 ion channel. Cell Discov 9:70. https://doi.org/10.1038/s41421-023-00579-3

Zheng W, Qin F (2015) A combined coarse-grained and all-atom simulation of TRPV1 channel gating and heat activation. J Gen Physiol 145:443–456. https://doi.org/10.1085/jgp.201411335

Zheng W, Wen H (2019) Heat activation mechanism of TRPV1: new insights from molecular dynamics simulation. Temperature 6:120–131. https://doi.org/10.1080/23328940.2019.1578634

Zheng W, Wen H (2021) Predicting lipid and ligand binding sites in TRPV1 channel by molecular dynamics simulation and machine learning. Proteins Struct Funct Bioinf 89:966–977. https://doi.org/10.1002/prot.26075

Zhu YQ, Lei M, Lu AJ, Zhao X, Yin XJ, Gao QZ (2009) 3D-QSAR studies of boron-containing dipeptides as proteasome inhibitors with CoMFA and CoMSIA methods. Eur J Med Chem 44:1486–1499

Zhu Z, Deng Z, Wang Q, Wang Y, Zhang D, Xu R, et al. (2022) Simulation and Machine Learning Methods for Ion-Channel Structure Determination, Mechanistic Studies and Drug Design. Front Pharmacol [Internet]. [cited 2024 Dec 21] 13:939555. Available from: https://pmc.ncbi.nlm.nih.gov/articles/PMC9275593/

Zimmerman MI, Bowman GR (2015) FAST conformational searches by balancing exploration/exploitation trade-offs. J Chem Theory Comput [Internet]. [cited 2024 Nov 15] 11:5747–57. Available from: https://pubs.acs.org/doi/full/10.1021/acs.jctc.5b00737

Structure–Function Studies of Pentameric Ligand-Gated Ion Channels: Combining Experimental and Computational Approaches

11

Wayland W. L. Cheng and Mark J. Arcario

Abstract

Pentameric ligand-gated ion channels (pLGICs) are a family of cation- and anion-selective channels that mediate fast synaptic transmission and regulate excitability in the nervous system. They include the GABA(A) receptor, nicotinic acetylcholine receptor (nAChR), glycine receptor (GlyR), and serotonin receptor (5-HT3$_A$R). pLGICs are gated by neurotransmitters, such as GABA and acetylcholine, producing rapid activation and slow desensitization. They are also modulated by lipids, including anionic phospholipids, cholesterol, and fatty acids. Here, we examine recent cryo-electron microscopy structures of pLGICs that reveal the molecular basis of ligand binding, activation, and desensitization. These structures have enabled the application of computational techniques, providing additional insight into structure–function relationships. We present how these approaches have informed the location of the activation and desensitization gates, the pathways of ion conduction, and the sites and energetics of ligand and lipid interactions.

Keywords

Ion channel · Ligand · Gating · Conductance · Lipids

11.1 Introduction to pLGICs

Pentameric ligand-gated ion channels (pLGICs) are neurotransmitter-gated ion channels that include the human GABA(A) receptor (GABA$_A$R), nicotinic acetylcholine receptor (nAChR), glycine receptor (GlyR), and serotonin receptor (5-HT3$_A$R). They are a diverse family of ion channels comprised of many subunit isoforms (19 for GABA$_A$R, 17 for nAChR, 5 for GlyR, and 5 for 5-HT3$_A$R) forming homo- and hetero-pentameric assemblies with distinct functional properties. pLGICs are the essential determinants of neuronal signaling, and, as such, mutations that alter the function of these ion channels underlie many diseases, including epilepsy and neuromuscular disorders. pLGICs are also

W. W. L. Cheng (✉) · M. J. Arcario
Department of Anesthesiology, Washington University School of Medicine, Saint Louis, MO, USA
e-mail: wayland.cheng@wustl.edu

A. Rosenhouse-Dantsker (ed.), *Decoding Ion Channels Structure and Function*, Advances in Experimental Medicine and Biology 1497, https://doi.org/10.1007/978-3-032-07523-9_11

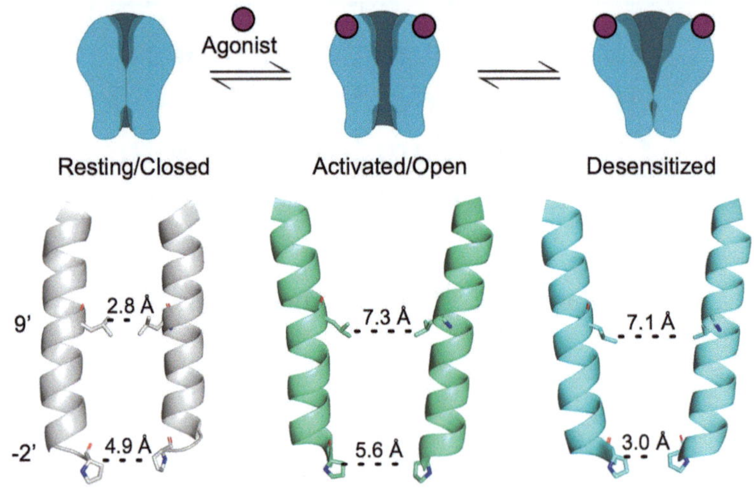

Fig. 11.1 pLGIC gating and conformational changes in the pore. Agonist binding promotes transition from a resting/closed state to an activated/open state. In the continued presence of an agonist, the channel transitions from an activated to a desensitized state. Shown below the schematic are the pore-forming M2 helices of two opposing subunits taken from structures of the zebrafish α1 GlyR in putative closed (PDB 6PXD), open (PDB 6 PM6), and desensitized conformations (PDB 6 PM5). The structures show that the narrowest region of the pore is at the 9′ leucine (L277) in the closed conformation, and the −2′ proline (P266) in the desensitized conformation

the targets of many drugs such as sedatives, neuromuscular blockers, and antiemetics. Therefore, there is great interest in understanding the structure–function relationships of these ion channels in the hope that such information could facilitate drug design.

In the absence of agonist (i.e., neurotransmitter), most pLGICs are in a closed, resting state (i.e., the open probability of an unliganded channel is very low) (Fig. 11.1). With the binding of agonist to the orthosteric site in the extracellular domain (ECD), the channel rapidly transitions to an activated state, opening the ion channel pore. When the agonist is removed, the channel will deactivate and close. Opening and closing of the ion channel is also referred to as gating. In the continued presence of agonist, many pLGICs enter a nonconducting, desensitized state (Fig. 11.1). While the activated state is kinetically favored, it is usually less stable than the desensitized state when agonist is bound. The transient and unstable activated state is, therefore, more challenging to study using structural and biochemical approaches. Activation and desensitization of pLGICs are also modulated by ligands through sites other than the orthosteric site (i.e., allosteric modulators), including lipids. For example, the endogenous neurosteroid, allopregnanolone, potentiates GABA activation of the $GABA_AR$ by binding to sites in the transmembrane domain (Sugasawa et al. 2020), while cholesterol is thought to promote desensitization of the nAChR also through direct binding (Zhuang et al. 2022).

There is considerable knowledge regarding structure–function relationships in many pLGICs, arising from decades of biophysical studies. Recently, cryo-electron microscopy (cryo-EM) structures have been determined for all the major receptor types in unliganded and liganded conformations, providing, in some cases, detailed structural snapshots of activation and desensitization. Here, we will present the major structural features of cation- and anion-selective pLGICs, focusing on the conserved conformational changes associated with activation and desensitization. Members of the pLGIC superfamily exist even in prokaryotes, and the well-characterized channels, ELIC (*Erwinia* ligand-gated ion channel) and GLIC (*Gloeobacter violaceus* ligand-gated ion channel), have also been the subject of many

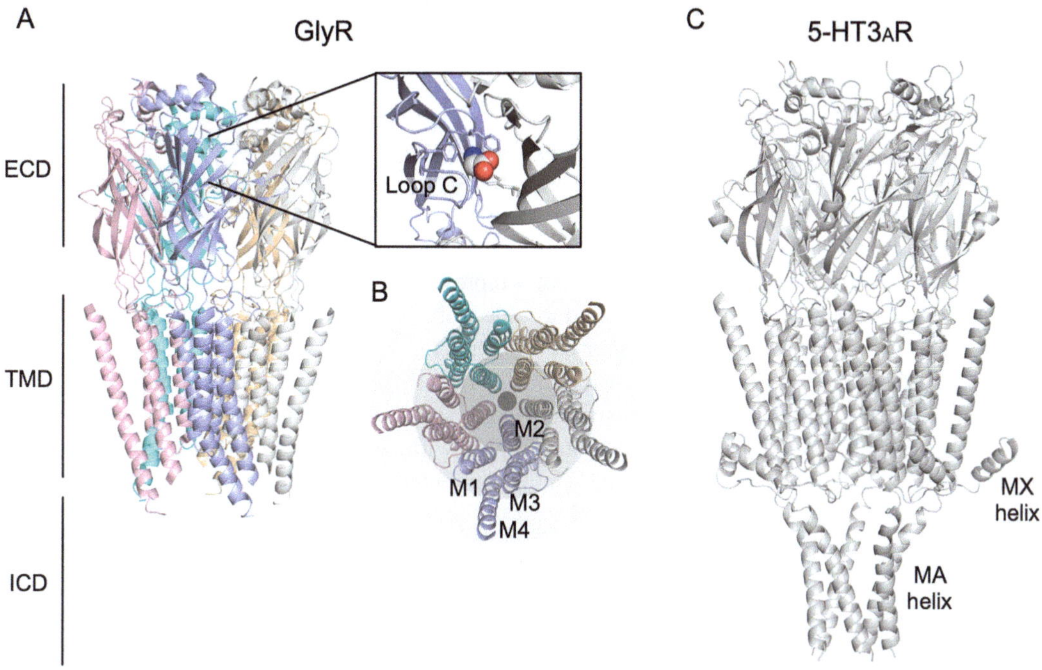

Fig. 11.2 Structural architecture of pentameric ligand-gated ion channels (pLGICs). (**a**) Cryo-EM structure of the pLGIC, zebrafish α1 GlyR (PDB 6PM5). The ion channel is a pentameric assembly with an extracellular domain that contains the agonist binding site, and a transmembrane domain with four transmembrane helices per subunit. In the top right is an image of the agonist binding site occupied by glycine. The agonist binding site is located between subunits, capped by loop C in the principal subunit, and consists of multiple aromatic residues that interact with the ligand. (**b**) View of the transmembrane helices from the extracellular side. The helices are roughly organized into three concentric layers: M2 lines the central ion conduction pore, M1 and M3 form the next layer, and M4 is in the periphery. (**c**) Cryo-EM structure of the mouse 5-$HT3_A$R (PDB 6DG8) with the MX and MA helices in the intracellular domain labeled

structural and biochemical studies. While our focus will be on the mammalian pLGICs, we will also highlight key findings from the prokaryotic pLGICs, which have been useful model systems, especially to study lipid regulation of these channels. With the many cryo-EM structures available for different pLGICs, computational approaches have played a key role in facilitating functional annotation of these structures and exploring the energetics and dynamics of processes such as ion conduction and ligand binding. We will cover a breadth of topics in this chapter, including ion conduction, gating, and ligand/lipid binding in pLGICs, but the unifying focus is how computational methods have enhanced our understanding of structure–function relationships.

11.2 Structural Snapshots of Activation and Desensitization

The structural architecture of all pLGICs, from prokaryotic to human channels, is overall quite similar (Fig. 11.2a). The channel is a pentameric assembly around a central ion pore. The large ECD consists of a β-sandwich with 10 β-strands connected by loops that form the conserved agonist binding site (orthosteric site) or interface with the transmembrane domain (β1–β2, β6–β7 also known as the Cys-loop, and β8–β9). The orthosteric site is capped by loop C and formed by the β-sheets of two adjacent subunits with multiple aromatic residues that interact with the agonist/neurotransmitter (Fig. 11.2a). In homomeric pLGICs such as the α1 GlyR, there are five identical binding sites, although it has been proposed that only two or three agonists are needed to gate the channel (Burzomato et al. 2004). In heteromeric pLGICs such as the synaptic $GABA_AR$ (α1β2γ2) or muscle-type nAChR ($\alpha_2\beta\delta\gamma$), there are two agonist binding sites formed by two different subunits (α1/β2 in the $GABA_AR$, and α/δ and α/γ in the nAChR) (Laverty et al. 2019; Kim et al. 2020; Rahman et al. 2020). Both agonists must bind to efficaciously activate the $GABA_AR$ or muscle-type nAChR (Auerbach 2013). In the transmembrane domain (TMD), each subunit consists of four helices (M1–M4) with M2 lining the central pore and M4 facing the membrane interface. The arrangement of these helices can be viewed as three concentric layers: the inner layer being the pore-facing M2, followed by the intervening M1 and M3 helices, and then M4 in the periphery (Fig. 11.2b). Much less is known about the structure of the intracellular domain (ICD) formed by the M3–M4 loop, which is significantly variable in length and sequence between pLGICs. This domain is absent in the prokaryotic pLGICs (e.g., ELIC, GLIC, sTeLIC, which is *Staphylococcus tevnia* ligand-gated ion channel). In eukaryotic pLGICs, the ICD is largely disordered and only some α-helices have been resolved in the structures of cation-selective pLGICs (e.g., MX and MA helices in nAChR and $5\text{-}HT3_AR$, Fig. 11.2c) (Rahman et al. 2020; Basak et al. 2018a, b; Noviello et al. 2021). The ICD plays an important role in the conductance, trafficking, and stability/clustering of these ion channels in the membrane, as it mediates interaction with other proteins, especially for those pLGICs localized to the postsynaptic membrane.

While there are multiple unliganded and liganded structures of all mammalian pLGICs, in only three cases have structures been captured in putative resting, activated, and desensitized states. These are the homomeric α1 GlyR (Yu et al. 2021; Kumar et al. 2020a), the α7 nAChR (Noviello et al. 2021), and the $5\text{-}HT3_AR$ (Basak et al. 2018a, b; Polovinkin et al. 2018). Structures of ELIC in all three states have also been captured by cryo-EM (Petroff 2nd et al. 2022; Dalal et al. 2025). We will discuss the key features of the anion-selective α1 GlyR and cation-selective α7 nAChR to summarize the conserved conformational changes associated with ligand binding, activation, and desensitization, and to highlight some differences. The activated structures of these channels have also been the subject of computational analysis of ion conduction, strengthening the conclusion that these structures represent open-channel conformations. We will describe the results of these analyses later in this chapter.

In both the α1 GlyR and α7 nAChR, the resting state structure exhibits a closed or nonconducting pore with one or more constrictions in the transmembrane domain. It is conventional to number the amino acids in M2 starting from the intracellular end to easily compare residue positions between pLGICs. In most pLGICs, the 9′ position is a conserved leucine that forms a narrow constriction; this hydrophobic residue produces a high energetic barrier to the passage of hydrated ions, and 9′ is the activation gate in all pLGICs for which a structure has been determined (Fig. 11.1). In the α7 nAChR, there are also hydrophobic constrictions at 13′ and 16′, which are found in other cation-selective pLGICs.

Agonist binding to the orthosteric site is associated with a characteristic capping or inward translation of loop C that contracts or reduces the volume of the binding pocket. This leads to a global contraction and counter-clockwise twisting of the ECD along with a tilting of each subunit. Consequently, the loops at the ECD–TMD interface move relative to the TMD. A key rearrangement at this interface involves the M2–M3 linker, which translates away from the pore axis and is associated with a tilting of M2 and M3, opening the ion channel pore. The activated, open-channel structure of the α7 nAChR was captured using the agonist epibatidine and the positive allosteric modulator, PNU-120596, that reduces channel desensitization (Noviello et al. 2021). A putative activated, open-channel structure of the α1 GlyR was also captured using picrotoxin to keep open the pore (which otherwise would be narrowed in the desensitized structure) (Kumar et al. 2020a), or by reconstituting the channel in SMA co-polymeric nanodiscs (Yu et al. 2021). In all cases, the 9′ leucine side chain is rotated out of the pore axis, yielding an expanded pore (Fig. 11.1).

In the presence of just agonist (epibatidine or glycine), the α1 GlyR and α7 nAChR are expected to be desensitized. Here, the pore conformations of these two channels are distinct. The α1 GlyR shows an open pore at 9′, but M2 has tilted such that the intracellular end of the pore (−2′ or P266) forms a constriction (Yu et al. 2021). Therefore, −2′ is thought to be the desensitization gate in α1 GlyR, and the same constriction at the intracellular end of the pore has been observed in desensitized structures of the $GABA_AR$ (Miller and Aricescu 2014; Masiulis et al. 2019) (Fig. 11.1). In contrast, the α7 nAChR desensitized structure shows constrictions at both −1′ and 9′, suggesting that the desensitized conformation is distinct in α7 nAChR (Noviello et al. 2021). Agonist-bound, putative desensitized structures of the prokaryotic pLGIC, ELIC, also show a 9′ constriction (Petroff 2nd et al. 2022), so there may be other cation-selective pLGICs that resemble α7 nAChR with regard to the desensitized conformation.

To summarize, structural snapshots of the α1 GlyR and α7 nAChR reveal a conserved mechanism of ligand activation. The binding of agonist to an orthosteric site produces similar conformational changes in the ECD, ECD–TMD interface, and TMD, yielding an open pore. However, the structural basis of desensitization may be different in these channels. It is also important to note that cryo-EM structures of pLGICs may be altered by reconstitution in detergent or lipid nanodiscs (Basak et al. 2018b; Yu et al. 2021; Dalal et al. 2024; Zhang et al. 2021). Recently, cryo-EM structures of ELIC were determined in liposomes, conditions in which the function of the ion channel was verified using a stopped-flow thallium flux assay (Dalal et al. 2025). This increases confidence that the structures represent snapshots of resting, activated, and desensitized states. Nevertheless, there are no high-resolution structures of other pLGICs in liposomes or cell vesicles. Therefore, there is some uncertainty about the functional relevance of pLGIC structures, and computational methods have been used to facilitate functional annotation.

11.3 Computational Studies of Ion Conduction

Molecular dynamics (MD) simulations of putative open-channel structures of pLGICs, including the α1 GlyR, GLIC, and the nematode pLGIC, GluCl, invariably lead to the rapid collapse or closure of the pore unless restraints are imposed on the protein backbone (Cheng and Coalson 2012; Cerdan et al. 2018; Calimet et al. 2013). It is unclear whether the rapid collapse of these structures represents deactivation or desensitization of the channel, although this would be occurring at a much faster time scale (tens to hundreds of nanoseconds) than is observed experimentally. This has led to some uncertainty regarding the functional relevance of the open-channel structures reported by cryo-EM or X-ray crystallography. Of note, the available open-channel structure of the α1 GlyR, at the time, had a pore diameter that was far wider than was previously reported for any pLGIC (8.8 Å- “wide-open” structure) (Du et al. 2015; Gonzalez-Gutierrez et al. 2017). To identify a stable open-channel structure of the α1 GlyR, several groups implemented simulation protocols where the “wide-open” structure of the α1 GlyR was restrained in different ways (e.g., symmetry restraints of the M2 helices, limiting the orientation of the 9′ leucine, a flat-bottomed harmonic restraint). These restraints were gradually released during equilibration of the system, and the resulting structures were used to perform unbiased MD simulations for several hundred nanoseconds. The simulations identified distinct stable open-channel conformations of the α1 GlyR (Cerdan et al. 2018; Damgen and Biggin 2020).

To validate these open structures, Cerdan and Cecchini conducted MD simulations to examine ion conduction, also called computational electrophysiology (Cerdan and Cecchini 2020). In these simulations, the driving force for ion conduction is either an applied electric field that produces a constant membrane potential like a voltage-clamp experiment, or a double membrane setup where an electrochemical gradient is produced by having two separate compartments with different ionic concentrations. The former is less computationally expensive due to the smaller system, while the latter can reproduce the physiologic ionic composition of extracellular and intracellular solutions. For the α1 GlyR, chloride conduction was assessed using a constant electric field. These measurements were performed using the TMD-only (i.e., the ECD was removed) and showed that, while the "wide-open" structure produced a conductance of >400 pS, the stable open structure obtained from MD simulations produced a single-channel conductance of 20 pS, which is more consistent with experimental measurements (Cerdan and Cecchini 2020). Further validating this approach, two cryo-EM structures of the α1 GlyR have since been reported in lipid nanodiscs with pore diameters (5–5.6 Å) more similar to what was identified in the stable MD conformation (Yu et al. 2021; Kumar et al. 2020a). Overall, these results illustrate how MD simulations can be used to annotate structures or to identify more stable conformations, if perhaps the crystal packing in X-ray crystallography, freezing in cryo-EM, or reconstitution conditions (e.g., detergents) lead to distortions of the structures. Despite the apparent success of work with the α1 GlyR, other open-channel structures for the 5-HT3$_A$R and ELIC have not been the subject of unbiased MD simulations to examine the stability of these structures.

Nevertheless, computational electrophysiology analyses have been performed for α1 GlyR, α7 nAChR, and 5-HT3$_A$R (in the case of the latter two, the cryo-EM structures were used, and the protein backbone structure was restrained throughout the simulation), yielding new insights into the ion conduction pathway. For the anion-selective α1 GlyR, the authors found that chloride conduction proceeded primarily through a lateral fenestration in the extracellular domain as opposed to the apical opening at the top of the ion channel (Cerdan et al. 2022). There are five lateral fenestrations in the pentameric assembly located between the β1–β2 and β8–β9 loops of adjacent subunits. This fenestration is lined by basic residues that provide an electrostatically favorable pathway for the passage of chloride ions (Fig. 11.3a), and mutations of these basic residues (e.g., K104E) produce an inward rectification (i.e., higher inward current than outward current, which is a higher chloride efflux than chloride influx). The lateral fenestration connects the bulk extracellular solution with the vestibule in the ECD above the transmembrane domain pore, and the authors propose a model where the concentration of chloride in this vestibule dictates the conductance of ions. For chloride efflux, the accumulation of chloride in the vestibule drives chloride out of the lateral fenestration, and this process is less impacted by the K104E mutation. For chloride influx, K104E reduces chloride translocation through the lateral fenestration into the vestibule, depleting chloride in the vestibule and reducing chloride influx. Therefore, a lateral fenestration is the primary pathway for chloride conductance in the α1 GlyR, and this pathway explains the inward rectification that is observed with mutations at the fenestration. It is interesting to note that a lateral fenestration has been suspected to exist in other anion-selective (e.g., GABA$_A$R) and cation-selective (e.g., 5-HT3$_A$R, nAChR, ELIC) pLGICs, although the contribution of this pathway to ion conduction remains to be determined. As expected, the electrostatic potential of the lateral fenestration is more electronegative in cation-selective channels, supporting the hypothesis that the lateral fenestration forms part of the conduction pathway for cations.

The open-channel structure of the α7 nAChR was characterized using the double membrane and applied electric field approaches (Zhuang et al. 2022). Both measurements showed sodium-selective current with an outward conductance of 223 pS, similar to experimental measurements. However, the inward conductance was markedly lower (44 pS); this outward rectification is not consistent with experimental measurements of α7 nAChR single-channel currents, which have a linear current–voltage relationship. Of note, measurements of sodium conduction in an α7

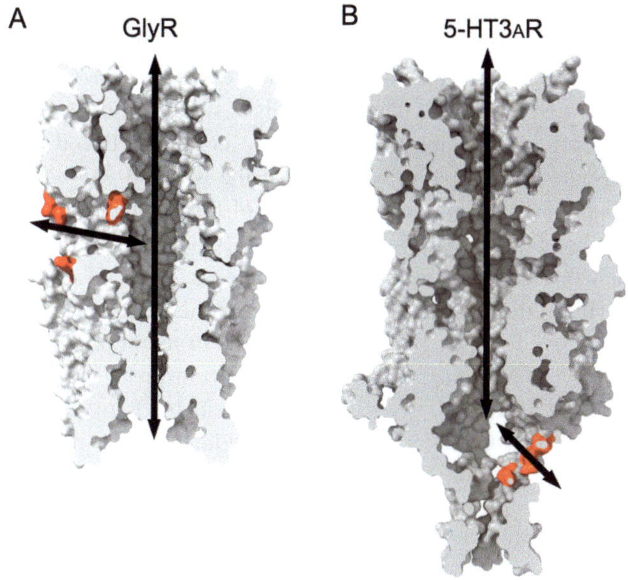

Fig. 11.3 The ion conduction pathway in the GlyR and 5-HT3$_A$R is formed by lateral fenestrations in the extracellular and intracellular domains. (**a**) Surface representation of the zebrafish α1 GlyR (PDB 6PM5) with part of the structure clipped out to reveal the central conduction pore and the lateral fenestration/pathway at one subunit in the ECD (central and lateral pathways indicated by arrows). The lateral fenestration is lined by several basic residues colored in red (K120, K159, R212, R229) in this anion-selective channel. (**b**) Surface representation of the 5-HT3$_A$R (PDB 6DG8) with part of the structure clipped out to reveal the central conduction pore and the lateral fenestration/pathway at one subunit in the ICD (central and lateral pathways indicated by arrows). The lateral fenestration is lined by several basic residues colored in red (R416, R420, R424) in this cation-selective channel

nAChR model consisting of just the TMD did not show any rectification, suggesting that the ECD is producing this phenomenon. The authors do not detail the passage of sodium ions through the apical or a possible lateral pathway in the ECD, but the results suggest that there is some limitation to the flow of sodium ions from the extracellular solution to the ECD vestibule. Therefore, while the open-channel structure of α7 nAChR does conduct sodium ions, some aspect of the structure or the MD simulation conditions is incorrect and does not recapitulate the inward conductance of the channel.

An open-channel structure has also been described for the cation-selective 5-HT3$_A$R (Basak et al. 2018b). Inward sodium permeation events were observed with an applied voltage of −200 mV and in the presence of 500 mM NaCl, but a conductance was not reported. Sodium conduction may be difficult to measure in 5-HT3$_A$R because the WT unitary conductance is very low (<1 pS). The simulations also reveal a barrier to water in the intracellular domain (ICD) along the central axis, where hydrophobic residues in the MA helix form a constriction. Here, sodium ions primarily pass through lateral portals in the ICD near a triad of arginine residues that are important in determining 5-HT3$_A$R conductance (Fig. 11.3b). Mutation of these arginines to remove the positive charge significantly increases 5-HT3$_A$R conductance. Interestingly, these lateral portals are occluded by the post-M3 loop in the apo structure of 5-HT3$_A$R, as well as 5-HT3$_A$R structures complexed with setron antagonists (Basak et al. 2018a, b, 2019). Therefore, MD simulations of the 5-HT3$_A$R open structure reveal additional complexity to the ion conduction pathway in the ICD that may also be true of the nAChR, which has an MA helix. Sodium conduction through a lateral fenestration in the ECD of the 5-HT3$_A$R was not described; however, such a pathway has been proposed for 5-HT3$_A$R using the agonist-bound crystal structure of 5-HT3$_A$R.

In summary, computational electrophysiology studies of pLGICs have provided molecular insights into the ion conduction pathway, including previously unappreciated lateral pathways in the ECD and ICD. Further work is needed to determine the prevalence of these pathways in other pLGICs and their possible role in the ion conduction and selectivity of these channels.

11.4 Computational Studies of the Ion Channel Gate

The conserved pore-lining 9′ leucine is the activation gate in all pLGICs (Fig. 11.1). In some cation-selective pLGICs, there are also aromatic or hydrophobic residues that form constrictions at 13′ and 16′ as well. The fact that these residues are hydrophobic is significant because they produce a dewetting of the pore (Seiferth et al. 2022). By eliminating water from this region, there is a high energetic barrier for the passage of hydrated cations or anions. The theory of hydrophobic dewetting predicts that a significant barrier exists even if the diameter of the pore exceeds that of sodium or chloride ions, although many apo/resting pLGIC structures have a pore diameter at 9′ that is small enough to sterically occlude these ions. To evaluate the energetic barrier imposed by the 9′ leucine, various computational methods have been used to determine the free energy profile of the permeation pathway for specific ions or molecules. The permeation free energy profile for sodium ions was determined for the closed/inhibited structure of the α7 nAChR, and the profile showed a single energetic barrier of 12 kJ/mol at the 9′ leucine (Zhuang et al. 2022). Similarly, the energetic barrier for the passage of water at the 9′ leucine in the resting/apo structure of 5-HT3$_A$R was estimated to be 10 kJ/mol (Basak et al. 2018b). In contrast, the activated, open-channel structures of these channels showed an absence of this energetic barrier. These analyses of α7 nAChR and 5-HT3$_A$R indicate that the 9′ leucine is the activation gate in these ion channels and likely all pLGICs.

It has been proposed that desensitization of pLGICs results from closure of a gate that is distinct from activation (Gielen and Corringer 2018). This is supported by a chimeric and mutagenesis study in the GABA$_A$R and α1 GlyR, showing that the M2 and M3 helices at the intracellular end of the channel dictate desensitization properties (Gielen et al. 2015). Furthermore, when examining block of these ion channels by picrotoxin, there is a large rebound current upon removal of picrotoxin, suggesting that picrotoxin prevents closure of the desensitization gate. Since picrotoxin binds to the ion channel pore at the intracellular end (2′ region), this was interpreted to mean that the desensitization gate is at the picrotoxin binding site, and that picrotoxin is stenting the channel open. Indeed, this is consistent with the cryo-EM structure of the agonist-bound α1 GlyR with picrotoxin, which shows an apparent open-channel conformation (Kumar et al. 2020a). The free energy of hydration along the permeation pathway for the desensitized structure of the α1 GlyR showed a modest energetic barrier (2.5 kJ/mol) at the −2′ position (proline), and MD simulations with a + 500 mV applied potential and with 500 mM NaCl showed no chloride permeation events over 200 ns in contrast to the picrotoxin-bound putative open-channel structure. Desensitized structures of the GABA$_A$R also show a narrow constriction at the −2′ position (Laverty et al. 2019; Miller and Aricescu 2014). Therefore, the current structural and computational evidence indicate that the desensitization gate in anion-selective channels is at the intracellular −2′ position.

The location of the desensitization gate does not appear to be the same for cation-selective pLGICs. The free energy profile for the permeation pathway in the putative desensitized structures of the α7 nAChR and 5-HT3$_A$R shows an energy barrier at 9′, but no other barrier at the intracellular end of the pore (Zhuang et al. 2022; Basak et al. 2018b). This is despite the fact that, in both α7 nAChR and 5-HT3$_A$R, the diameter of the pore at −1′ (E237 in α7 nAChR and E250 in 5-HT3$_A$R) is smaller than or similar to the diameter at the 9′ leucine. Similarly, the desensitized structure of the cation-selective prokaryotic pLGIC, ELIC, shows a single constriction at 9′, and the narrowest point at the intracellular end of M2 is a 2′ glutamine (Petroff 2nd et al. 2022; Dalal et al. 2024; Kumar et al. 2020b). The free energy calculations suggest that large hydrophilic residues like glutamate or glutamine at the intracellular end of M2 are unlikely to produce a significant barrier to sodium conduction. Based on the available cryo-EM structures, it appears that both the activation and desensitization gates in cation-selective pLGICs are formed by the 9′ leucine.

11.5 Structural Pharmacology of pLGICs

pLGICs are activated by neurotransmitters, but are also regulated by many other small-molecule agonists and antagonists that bind to the orthosteric site. Strong agonists such as neurotransmitters bind to the activated state with higher affinity and thereby stabilize the activated state relative to the resting state. The difference in affinity of the agonist between the activated and resting states is related to the energy that is conferred to gating. Partial agonists confer less stabilizing energy and are therefore less efficacious in activating/opening the ion channel. Neutral antagonists bind to the activated and resting states with equal affinity, and therefore prevent the channel from being activated by an agonist. However, it is difficult to differentiate neutral antagonists from inverse agonists (i.e., ligands that stabilize the resting state) because the activity of most pLGICs in the absence of an agonist is very low.

Cryo-EM structures have revealed the binding sites and structural consequences of many ligands and their cognate pLGICs. These include agonists, partial agonists, and inhibitors that bind to the orthosteric site, and allosteric modulators that bind to other sites in the ECD or TMD. Given the sheer breadth of information on this topic, we will limit this chapter to ligands that bind to the orthosteric site, highlighting cases in which insights have been gained from combining structural and computational approaches. As noted above, the orthosteric site or neurotransmitter binding site is located at the interface between the β-sheets of adjacent subunits and is also formed by loops B and C from one of the subunits (Fig. 11.4a). Multiple aromatic and polar residues form van der Waals and hydrogen bond interactions with the agonist, and as will be discussed in the next section, some of these interactions are mediated by water molecules in the orthosteric pocket. The binding of the agonist leads to a contraction of the orthosteric site, most notably by a "capping" of loop C (Fig. 11.4a).

A prevailing question in the field is: What is the structural basis of agonist efficacy? What are the molecular features that dictate whether a ligand is a strong agonist, a weak/partial agonist, or an inhibitor? Early work on the GlyR showed that the partial agonist, taurine, differs from the strong agonist, glycine, in its ability to stabilize a short-lived preactive state called a "flipped" state, and not in its ability to stabilize the open state relative to the flipped state (Lape et al. 2008). Therefore, at saturating agonist concentration, glycine drives most of the channels into the activated, open state, while for taurine, a large population of channels will remain in a closed, resting-like state. Molecular insights into agonist activity of the α1 GlyR were revealed from a recent cryo-EM study (Yu et al. 2021). Structures of α1 GlyR complexed with glycine showed a mixture of open and desensitized conformations. In contrast, α1 GlyR complexed with taurine or GABA, which are both partial agonists, showed a mixture of open, desensitized, and closed conformations, qualitatively consistent with the lower gating efficacy of these ligands. Examining the agonist binding site in the open structures, the authors note that the binding pocket is more contracted in the glycine (full agonist) structure than in the taurine or GABA (partial agonist) structures (Fig. 11.4b). This is primarily due to a contraction of loops C and B toward the agonist that is greater with glycine than with taurine and GABA. The authors propose that these structural differences are coupled to activation of the GlyR and underlie agonist stabilization of the open state. In further support of this idea, an α1 GlyR mutant that converts GABA to a full agonist was also examined. The structure of this mutant complexed with GABA

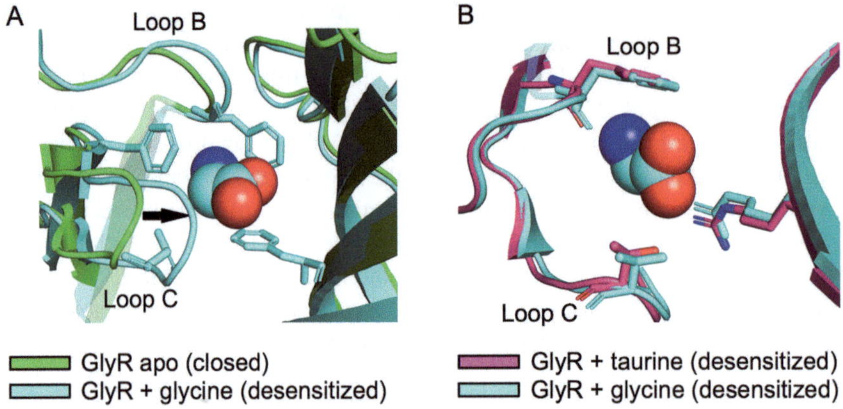

Fig. 11.4 Agonist-induced changes in the orthosteric site. (**a**) Structural comparison of the agonist binding site in the apo/resting (PDB 6PXD (Yu et al. 2021)) and glycine-bound desensitized (PDB 6 PM5 (Yu et al. 2021)) conformations of the zebrafish α1 GlyR. The binding of glycine is associated with a "capping" of loop C, contracting the binding site. The agonist binding pocket is formed by four aromatic residues: F79, F175, Y218, and F223. (**b**) Structural comparison of the agonist binding site in the zebrafish α1 GlyR, comparing glycine-bound desensitized (PDB 6 PM5 (Yu et al. 2021)) and taurine-bound desensitized (PDB 6PLS (Yu et al. 2021)) conformations. There is a slight contraction of the agonist binding site in the glycine-bound structure compared to the taurine-bound structure due to changes in the conformation of loops B and C. Residues displayed are F175 in loop B, T220 in loop C, and R81 in the β2-sheet

no longer showed a closed conformation, and the agonist binding pocket in the open structure was more compact, enabled by an additional hydrogen bond interaction. It is expected that the efficacy of an agonist depends on the difference in affinity of that agonist for the open and closed states of the ion channel (Auerbach 2013). This difference in affinity produces a net binding energy that contributes to gating. The proposed mechanism for partial agonist activity in the α1 GlyR seems plausible because the more compact binding pocket for glycine compared to taurine or GABA suggests a higher binding affinity of glycine for the open state compared to taurine or GABA. However, an analysis of the free energy of binding of these agonists to the GlyR has not been conducted.

Cryo-EM studies of the 5-HT3$_A$R have provided insights into the mechanism of competitive inhibitors (setrons) (Basak et al. 2019) and partial agonists (SMP-100 and ALB-148471) (Felt et al. 2024) in the 5-HT3$_A$R. Structures of 5-HT3$_A$R complexed with different setrons (e.g., ondansetron, granisetron, palonosetron, and alosetron) showed binding of the inhibitors in the orthosteric site. However, the structure of the 5-HT3$_A$R assumed an intermediate conformation between apo and serotonin-bound 5-HT3$_A$R with variable capping of loop C, counter-clockwise twisting of the ECD, and intermediate agonist-dependent changes in some of the ECD β-strands (e.g., β9 and β10, but not β1, β2, and β6). The pore shows a slight expansion of the pore-lining M2 helix but still a closed conformation with a constriction at 9′ (L260). The structures indicate that, while setrons bind competitively to the orthosteric site, they are not neutral antagonists but rather stabilize a closed intermediate conformation of the receptor. Structures of 5-HT3$_A$R complexed with the partial agonists, SMP-100 (SMP) and ALB-148471 (ALB), show ligand binding at the orthosteric site, and, as is the case with serotonin-bound 5-HT3$_A$R, two ligand-bound conformations, which were assigned as preactive (state 1) and open-like (state 2) conformations. Note that while we had described state 1 as a desensitized conformation in the previous discussion on the desensitization gate, there is uncertainty regarding the functional annotation of this structure. In support of state 1 being a preactive conformation, the authors see a qualitative trend where the partial agonist samples have more particles assigned to this state than the open-like state. The binding of SMP and ALB at the orthosteric site shows some differences compared to serotonin in the interactions with the protein and the orientation of the ligand within the pocket that may underlie the partial agonist activity. However, unlike the example of the GlyR, the size of the orthosteric pocket in 5-HT3$_A$R does not appear to vary between serotonin and the partial agonists.

11.6 Computational Studies of Ligand Binding

The cryo-EM structures of 5-HT3$_A$R complexed with inhibitors and partial agonists were further investigated using MD simulations of these ligand–receptor complexes (Basak et al. 2019; Felt et al. 2024). The simulations provided additional insight into the nature of the interactions and their dynamics, yielding a "fingerprint" of receptor–ligand interactions that was not appreciated from the static structures alone. For example, each setron makes interactions with a specific profile of residues, many of which are mediated by water molecules, that latter of which are not resolved in the structures. The simulations show that all the setrons occupy the orthosteric site with a much lower RMSD than serotonin, consistent with the much higher potency of these compounds relative to serotonin. Loop C is also more dynamic in the setron-bound models, fluctuating between a "closed" and "open" conformation due to the absence of a stabilizing salt bridge interaction. MD simulations of 5-HT3$_A$R complexed with the partial agonists SMP and ALB also showed interactions with the protein mediated by water molecules, and differences in ligand dynamics within the binding pocket. SMP sampled multiple dihedral angles between the bicyclic ring and fused aromatic rings, and such flexibility was even greater in simulations of the inhibitor, palonosetron. In contrast, ALB, which is a stronger partial agonist than SMP, exhibited less conformational flexibility. It was proposed that the increased efficacy of ALB is due to this reduced flexibility, allowing for stronger interactions between the bicyclic ring nitrogen and the receptor. In summary, MD simulations of agonist and inhibitor interactions in the orthosteric site of 5-HT3$_A$R revealed different patterns of binding and receptor–ligand dynamics, which may underlie the different activities of these ligands.

The efficacy of an agonist depends on the difference in affinity of that agonist for the open/high-affinity and closed/low-affinity states of the ligand-gated ion channel (Auerbach 2013). Thus, further validation of these structural models could be achieved by determining the difference in binding energy between agonist-bound structures and unliganded structures. Such an analysis was performed for ELIC using the agonist, propylamine, by performing a free energy perturbation (FEP) calculation. Snapshots of propylamine bound to the orthosteric site in the agonist-bound and unliganded structures were obtained in the final frames of 500 ns simulations, and these were used as the models for the FEP calculation, alchemically annihilating or creating propylamine within the binding site (Kumar et al. 2020b). The calculated binding energy difference was −9.3 kcal/mol, which is comparable to experimental values reported for the muscle-type nAChR, the only pLGIC in which these values have been determined (Nayak et al. 2019). In another study of agonist binding in the muscle-type nAChR (Singh et al. 2024), the authors examined the binding of four different agonists using docking followed by MD simulations. The simulations identified a sequence of ligand binding that involves a 180-degree flip of the agonist within the orthosteric site, a downward movement of loop C, and then an overall contraction and dehydration of the binding pocket. Assuming that the bound state after the 180-degree flip in the agonist represents binding to the closed/low-affinity state (m1 state), and that the bound state after the downward movement of loop C and contraction of the pocket represents binding to the activated/high-affinity state (m3 state), the authors determined binding affinities for each state using a Poisson-Boltzmann Surface Area (PBSA) calculation. For all four agonists, binding to the m3 state was more favorable than the m1 state, and the ratio of binding energies for m3/m1 was remarkably similar to experimental measurements of ligand gating efficiency. However, the absolute binding energies for two of the agonists deviated significantly from experimental values of efficacy. This is perhaps not unsurprising because PBSA is a computationally inexpensive approach that provides rough estimates of binding energies. Better estimates of binding energies could be achieved using an alchemical FEP simulation approach (Kumar et al. 2020b). Nevertheless, the MD simulations and binding energy calculations suggest models for agonist binding, and potential intermediates in the transition from a low- to high-affinity states in the nAChR. The goal is to relate agonist binding energies determined from structural models with their gating efficacies obtained from functional measurements. If the binding energy calculations are robust, it has been proposed that such an approach could be useful for computational drug discovery (Singh et al. 2024).

11.7 Lipid Modulation of pLGICs

Early studies reconstituting the muscle-type nAChR from the electric organ of sting rays (*Torpedo marmorata or californica*) in lipid vesicles showed that ion channel function (i.e., agonist-induced affinity change and ion translocation) required cholesterol, phosphatidylethanolamine (PE), and an anionic phospholipid such as phosphatidylserine (PS) or phosphatidic acid (PA) (Ochoa et al. 1983; Criado et al. 1982; Hamouda et al. 2006; Dalziel et al. 1980). The receptor was also sensitive to acyl chain composition, with function best promoted by long unsaturated chains, especially of PE (Criado et al. 1984; Fong and McNamee 1986). Of note, replacement of PE with other unsaturated phospholipids maintained agonist-induced affinity transitions, but ion translocation was reduced, indicating that there was some uncoupling between agonist binding and channel opening. nAChR and the $GABA_AR$ are also inhibited by polyunsaturated free fatty acids such as docosahexaenoic acid (DHA), which are thought to promote the nonconducting desensitized state of the receptor (Vijayaraghavan et al. 1995; Minota and Watanabe 1997; Nabekura et al. 1998; Hamano et al. 1996). The prokaryotic pLGICs, ELIC, and GLIC, also exhibit lipid sensitivity. ELIC requires PE and the anionic phospholipid, phosphatidylglycerol (PG), for maximal activation and ion translocation, while PG decreases channel desensitization (Petroff 2nd et al. 2022). Both ELIC and GLIC are also inhibited by DHA (Dietzen et al. 2022; Basak et al. 2017). The similarities in lipid sensitivity between ELIC and nAChR suggest ELIC is a useful model system to investigate structural mechanisms of lipid modulation in this family of ion channels (Petroff 2nd et al. 2022; Carswell et al. 2015a).

Alongside these early functional studies, the interaction of lipids with the nAChR was also assessed by electron spin resonance (ESR) spectroscopy. These measurements showed that a layer of lipids around the receptor has reduced mobility, suggestive of specific interactions with the protein (Marsh and Barrantes 1978; Marsh et al. 1981; Mantipragada et al. 2003). While phospholipids were thought to occupy annular lipid binding sites (i.e., sites at the periphery of the protein), cholesterol was proposed to occupy sites not accessible to phospholipids, but shared with free fatty acids, and these were called nonannular sites (i.e., sites in the protein) (Jones and McNamee 1988; Fernandez Nievas et al. 2008). However, the exact definition of annular and nonannular lipid binding sites is unclear, and except for the ESR data, there is no property of lipid interactions with pLGICs that necessitates categorizing binding sites in this way. Nevertheless, the many available cryo-EM structures of pLGICs, mostly in lipid nanodiscs, show lipid-like nonprotein densities interacting with intersubunit and intrasubunit grooves in the TMD. Of note, many of these lipid-binding sites are also binding sites for drugs, suggesting that lipids may allosterically modulate channel function through these interactions (Cheng et al. 2021). We will further discuss these binding sites below in the context of computational studies that examine lipid interactions.

Baenziger and colleagues have studied the mechanism of phospholipid modulation of the nAChR using Fourier transform infrared difference spectroscopy (FTIR). They showed that agonist-induced conformational changes detected by FTIR only occur in the presence of lipids that support channel activity (e.g., PA and cholesterol) (Baenziger et al. 2000), and propose that in the absence of these lipids, the nAChR adopts an uncoupled or desensitized state that is able to bind agonist but is unable to activate (daCosta and Baenziger 2009; daCosta et al. 2009). They also hypothesized that lipids promote a functional nAChR by altering interactions between M4 (a putative "lipid sensor") and the other transmembrane helices or interfacial (ECD-TMD) loops, such as the β6-β7 or Cys-loop (daCosta and Baenziger 2009). Consistent with this hypothesis, the introduction of aromatic residues predicted to strengthen interactions between M4 and M1/M3 in ELIC rendered this channel less sensitive to lipids (Carswell et al. 2015a; Carswell et al. 2015b). That is, the aromatic residues rescued ELIC function in liposomes composed of PC (phosphatidylcholine) only. Also consistent with this hypothesis, GLIC, which already has multiple aromatic residues that may stabilize M4 interactions with M1/M3, is less lipid sensitive than ELIC (i.e., channel function is maintained in a PC-only membrane, although there was a decrease in pH sensitivity in PC membranes) (Labriola et al. 2013). ELIC and GLIC may belong to two distinct archetypes of pLGICs: those with weak M4 interactions that are lipid sensitive (e.g., ELIC and nAChR), and those with strong M4 interactions that are less lipid sensitive (e.g., GLIC and GlyR) (Therien and Baenziger 2017). Because M4 faces and interacts extensively with the lipid membrane, it seems likely that M4 senses the lipid environment. Yet, the structural details of how this occurs remain unclear, and cryo-EM structures of pLGICs in lipid nanodiscs have not provided much insight into this problem. Since M4 is a dynamic structure, other approaches, such as pulsed EPR double electron-electron resonance (DEER) spectroscopy or molecular dynamics simulations, to characterize M4 dynamics in different lipid environments, may be useful to understand the role of this helix in lipid modulation (Basak et al. 2017).

11.8 Characterizing Lipid Interactions in pLGICs

Lipids may modulate the function of pLGICs by altering the physical properties of the lipid bilayer. Evidence for this includes the finding that amphiphiles such as DHA, Triton X-100, and capsaicin all inhibit $GABA_AR$ responses and increase the rate of desensitization, as does depletion of cholesterol from the membrane (Sogaard et al. 2006). It was proposed that these structurally diverse amphiphiles modulate $GABA_AR$ function by increasing membrane elasticity. In another study, agonist responses in the nAChR were found to be sensitive to membrane thickness; thick bilayers supported agonist-induced transitions in nAChR, even in a PC-only membrane (daCosta et al. 2013). Alternatively, lipids may modulate the function of pLGICs by binding to specific sites; in this case, the lipid is expected to exhibit conformationally selective binding. Due to the general low affinity of lipid interactions with membrane proteins, such a mechanism has not been definitively demonstrated. An exception to this is the endogenous lipophilic modulators, neurosteroids, for which there is substantial evidence from photoaffinity labeling and mutational analysis (Sugasawa et al. 2020; Chen et al. 2019; Cheng et al. 2018), as well as cryo-EM structures (Legesse et al. 2023), that modulation of the $GABA_AR$ by these compounds is mediated by binding to specific sites. For example, allopregnanolone potentiates the $GABA_AR$ by binding to intersubunit and intrasubunit sites in the TMD (Chen et al. 2019), while the sulfated neurosteroid, pregnenolone sulfate, inhibits the $GABA_AR$ by binding in the pore (Legesse et al. 2023). However, for phospholipids, fatty acids, or cholesterol, evidence that modulation is mediated by specific binding sites is less clear. Regarding the interactions of these lipids with pLGICs, including conformationally selective interactions, computational approaches have provided unique insights, which we will discuss for the remainder of this section.

Early cryo-EM structures from Nigel Unwin of the muscle-type nAChR in *Torpedo* ray electric membrane vesicles showed that the transmembrane (TM) helices are loosely packed, interspersed with nonprotein densities suggestive of lipids (Unwin 2005). To test this hypothesis, Brannigan and colleagues performed MD simulations of this structure with 15 cholesterol molecules embedded between the TM helices; unlike previous simulations without cholesterol, the overall architecture of the TMD was more stable throughout the simulation (Brannigan et al. 2008). Furthermore, the simulations showed preservation of intra-protein contacts at the ECD-TMD interface that were attributed to the presence of cholesterol interactions in the outer TMD. Subsequent higher-quality cryo-EM images of the nAChR by Unwin indeed showed evidence of cholesterol patches at specific regions between the transmembrane helices and bridging delta subunits of neighboring receptors (Unwin 2017; Unwin 2020). These results suggest that cholesterol occupancy at specific sites in the TMD of the muscle-type nAChR preserves its structure and may therefore also be important for its function.

A major challenge in examining lipid interactions with membrane proteins is the relatively short time scale of MD simulations when using explicit modeling of all atoms. The short time scale limits the extent of lipid diffusion and, consequently, observation of multiple binding and unbinding events. Coarse-grained MD (CGMD) simulations permit microsecond to millisecond-long simulations by reducing the atomic detail of the system, making it possible to observe these events, and this approach has been used to examine lipid interactions in multiple pLGICs. For example, the CGMD of the Unwin nAChR structure embedded in a complex neuronal-like membrane showed that phospholipids with polyunsaturated fatty acids and cholesterol have the highest affinity for the receptor, especially in the outer leaflet, with the former favoring intrasubunit sites around M4 and the latter favoring intersubunit sites (Sharp and Brannigan 2021). In a contrasting multiscale modeling study that used both CGMD and all-atom MD, cholesterol was found to bind most favorably to an inner leaflet M3/M4 intrasubunit site, interacting strongly with a tryptophan residue on the MX helix (Ananchenko et al. 2024). The difference in cholesterol interactions between these two studies may be attributed to the different structures used: the former used the Unwin nAChR structure, while the latter used higher-resolution muscle-type nAChR structures in lipid nanodiscs (Zarkadas et al. 2022). Indeed, the nanodisc-reconstituted nAChR structures have transmembrane helices that are 2 Å more compacted compared to the Unwin nAChR structure from native *Torpedo* membranes. Therefore, the results of these simulations are, not surprisingly, highly dependent on the structures used. CGMD was also applied to the glycine receptor, comparing lipid interactions with the closed and open/desensitized structures (Damgen and Biggin 2021; Jalalypour et al. 2024). Like the nAChR, cholesterol binds most favorably to an outer leaflet intersubunit site deeply buried between the M1 and M3 helices and contacting the pore-lining helix. State-dependent interactions were also observed: cholesterol bound more favorably to the activated or desensitized structures compared to the resting structure. A similar finding was observed for the α7 nAChR, where extensive all-atom simulations of the resting, open, and desensitized structures in the presence of cholesterol showed favorable binding of cholesterol at the outer leaflet intersubunit site (Zhuang et al. 2022). Markov state modeling to quantify the probabilities of transitioning between these states showed that binding of cholesterol to this site was associated with a stabilization of the desensitized state. Lastly, CGMD has been used to study interactions of phospholipids and the free fatty acid, DHA, with ELIC (Fig. 11.5) (Dietzen et al. 2022; Tong et al. 2019). The anionic phospholipid, PG, interacts more strongly with ELIC than the zwitterionic phospholipids, PE

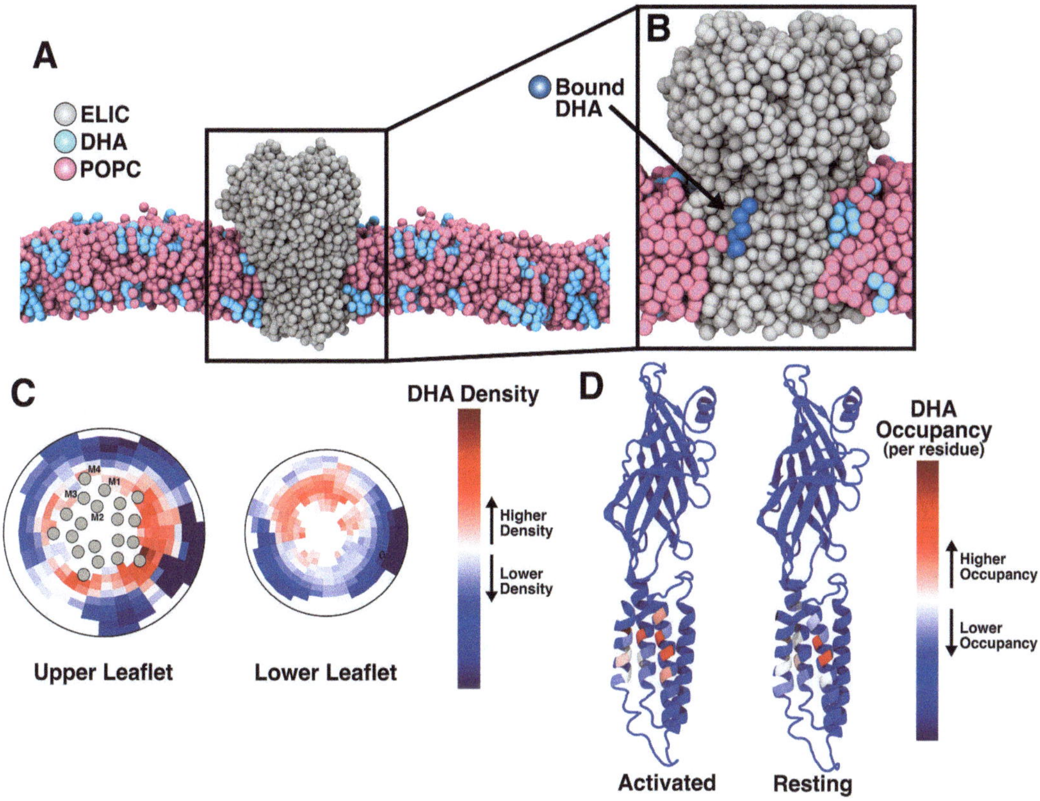

Fig. 11.5 CGMD analysis of DHA interactions with ELIC. (**a**) Coarse-grained representation of ELIC (gray spheres) in a putative desensitized conformation (PDB 8F34) embedded in a POPC (pink) membrane with free DHA (light blue). Water and ions have been omitted from the image for clarity, but are present in the simulation system. (**b**) Close-up image of (A) with a DHA fatty acid (dark blue) bound in a site between M3 and M4. Examples of the analysis of lipid interactions with ELIC from CGMD simulations include the density of lipid molecules in the boundary of the ion channel (**c**) and occupancy of lipids at specific ion channel residues (**d**). In (**c**), the time-averaged density of a lipid is measured in a radial grid surrounding the ion channel, with specific binding denoted by higher density. In (**d**), the time-averaged occupancy of a lipid-type is measured for each residue and plotted as a function of residue number mapped onto the backbone of the ion channel subunit. The data indicate that DHA is enriched at outer leaflet sites between M3/M4 and M1/M4

and PC, especially at an inner leaflet M1/M4 site. The strong interaction of PG with ELIC is consistent with measurements of lipid binding by native mass spectrometry. DHA interacts with ELIC at two outer leaflet intrasubunit sites between M1/M4 and M3/M4, and these findings by CGMD were consistent with photo-affinity labeling results of fatty acid binding sites in ELIC (Fig. 11.5). In summary, CGMD analysis suggests some specificity and state-dependence of lipid interactions with pLGICs. The most consistent observation in the nAChR and GlyR is that cholesterol binds to an outer leaflet intersubunit site with a preference for the activated or desensitized states of the channel.

The loss of atomic detail in a coarse-grained model may adversely impact the accuracy of modeling protein–lipid interactions. For example, the smaller headgroup of a phosphatidylethanolamine phospholipid is not well-represented in a coarse-grained model. A measurement of relative lipid binding affinity using an atomic model was achieved in ELIC using a free energy perturbation analysis called SAFEP (streamlined alchemical free energy perturbation) (Petroff 2nd et al. 2022). In this study, it was determined that the anionic phospholipid, PG, decreased ELIC desensitization, stabilizing the open/activated state of the channel relative to the desensitized state. An open structure of ELIC was determined using a gain-of-function mutant that does not desensitize, and that structure showed a phospholipid density at an outer leaflet intersubunit site (analogous to the cholesterol intersubunit site described above). Using SAFEP, it was shown that PG binds more favorably to this site compared to PE or PC (the other phospholipids used to assess ELIC activity or structure), indicating that PG is likely the phospholipid that occupies this site. Furthermore, PG binds to this site in the open structure with higher affinity than the desensitized structure, supporting the hypothesis that PG modulates ELIC desensitization by binding to this site.

11.9 Conclusion and Outlook

Cryo-EM structures of pLGICs have provided a template for understanding mechanisms of ion conduction and selectivity, ligand binding and gating, and lipid interactions. With the many available structures, the challenge remains to functionally annotate these structures and to understand the energetics and dynamics of processes such as ion conduction and ligand binding. Computational approaches have played an important role in this regard, helping to bridge the divide between structure and function. For example, MD simulations facilitated the identification of a plausible open-channel conformation of the α1 GlyR, and revealed ion conduction through a lateral fenestration in the ECD. Similarly, lateral fenestrations were identified in the ICD of the 5-HT3$_A$R. Based on the MD results, these fenestrations are likely to form the primary ion conduction pathway for these pLGICs. MD simulations have also begun to reveal how agonists, antagonists and partial agonists differentially interact with the orthosteric site to produce different effects on channel activity, and how phospholipids differentially interact with the TMD in different conformational states.

Looking ahead, a combination of experimental and computational approaches will be critical to further establish structure–function relationships in this family of ion channels. Importantly, open-channel structures are still lacking for some pLGICs, and except for the α1 GlyR, it remains uncertain whether other open-channel structures of the α7 nAChR, 5-HT3$_A$R, or ELIC are representative of the activated state without further analysis such as computational electrophysiology measurements. It is anticipated that more open-channel structures will become available using approaches such as time-resolved cryo-EM, and these should be validated by such computational studies.

pLGICs are regulated by many allosteric modulators or lipids that bind to sites in the TMD. These ligand/lipid interactions may be low-affinity and difficult to measure experimentally using conventional ligand-binding assays. Therefore, computational methods to examine ligand binding affinities in TMD sites may be useful. As described above, these include free energy perturbation analysis or CGMD of lipid interactions. There is a need to quantitatively relate measurements of ligand modulation and state-dependent ligand binding energies. It is also necessary to account for the partitioning of the ligand into the membrane in the case of membrane-accessible sites. Better models of ligand binding to the TMD of pLGICs will be necessary to predict small-molecule binding based on structural information.

Lastly, the ICD of eukaryotic pLGICs is largely unresolved in cryo-EM structures, and the next frontier of structure–function studies is likely to focus on the ICD. This domain is functionally important, determining properties of ion conduction and gating, allosteric modulation, stability within the plasma membrane, and intracellular signaling (Kracun et al. 2008; Lara et al. 2019; Kelley et al. 2003). The ICD is also the least conserved among pLGICs in terms of amino acid length and sequence, suggesting a divergence of regulatory mechanisms. However, the structural basis for regulation by the ICD is poorly understood for any pLGIC, and much of the ICD is predicted to be structurally disordered. Interestingly, a lipid crosslinking study suggested that the α1 GlyR ICD makes significant contact with the lipid membrane (Ferraro and Cascio 2018). MD simulations combined with other biochemical approaches will be useful to examine the structure and dynamics of this domain, as well as its interaction with the cell membrane.

References

Ananchenko A, Gao RY, Dehez F, Baenziger JE (2024) State-dependent binding of cholesterol and an anionic lipid to the muscle-type Torpedo nicotinic acetylcholine receptor. Commun Biol 7(1):437. https://doi.org/10.1038/s42003-024-06106-8

Auerbach A (2013) The energy and work of a ligand-gated ion channel. J Mol Biol 425(9):1461–1475. https://doi.org/10.1016/j.jmb.2013.01.027

Avstrikova M, Milan Rodriguez P, Burke SM, Hibbs RE, Changeux JP, Cecchini M (2025) Hidden complexity of alpha7 nicotinic acetylcholine receptor desensitization revealed by MD simulations and Markov state modeling. Proc Natl Acad Sci USA 122(7):e2420993122. https://doi.org/10.1073/pnas.2420993122

Baenziger JE, Morris ML, Darsaut TE, Ryan SE (2000) Effect of membrane lipid composition on the conformational equilibria of the nicotinic acetylcholine receptor. J Biol Chem 275(2):777–784

Basak S, Schmandt N, Gicheru Y, Chakrapani S (2017) Crystal structure and dynamics of a lipid-induced potential desensitized-state of a pentameric ligand-gated channel. elife 6:e23886. https://doi.org/10.7554/eLife.23886

Basak S, Gicheru Y, Samanta A, Molugu SK, Huang W, Fuente M et al (2018a) Cryo-EM structure of 5-HT3A receptor in its resting conformation. Nat Commun 9(1):514. https://doi.org/10.1038/s41467-018-02997-4

Basak S, Gicheru Y, Rao S, Sansom MSP, Chakrapani S (2018b) Cryo-EM reveals two distinct serotonin-bound conformations of full-length 5-HT3A receptor. Nature 563(7730):270–274. https://doi.org/10.1038/s41586-018-0660-7

Basak S, Gicheru Y, Kapoor A, Mayer ML, Filizola M, Chakrapani S (2019) Molecular mechanism of setron-mediated inhibition of full-length 5-HT3A receptor. Nat Commun 10(1):3225. https://doi.org/10.1038/s41467-019-11142-8

Brannigan G, Henin J, Law R, Eckenhoff R, Klein ML (2008) Embedded cholesterol in the nicotinic acetylcholine receptor. Proc Natl Acad Sci USA 105(38):14418–14423. https://doi.org/10.1073/pnas.0803029105

Burzomato V, Beato M, Groot-Kormelink PJ, Colquhoun D, Sivilotti LG (2004) Single-channel behavior of heteromeric alpha1beta glycine receptors: an attempt to detect a conformational change before the channel opens. J Neurosci 24(48):10924–10940. https://doi.org/10.1523/JNEUROSCI.3424-04.2004

Calimet N, Simoes M, Changeux JP, Karplus M, Taly A, Cecchini M (2013) A gating mechanism of pentameric ligand-gated ion channels. Proc Natl Acad Sci USA 110(42):E3987–E3996. https://doi.org/10.1073/pnas.1313785110

Carswell CL, Sun J, Baenziger JE (2015a) Intramembrane aromatic interactions influence the lipid sensitivities of pentameric ligand-gated ion channels. J Biol Chem 290(4):2496–2507. https://doi.org/10.1074/jbc.M114.624395

Carswell CL, Henault CM, Murlidaran S, Therien JPD, Juranka PF, Surujballi JA et al (2015b) Role of the fourth transmembrane alpha helix in the allosteric modulation of pentameric ligand-gated ion channels. Structure 23(9):1655–1664. https://doi.org/10.1016/j.str.2015.06.020

Cerdan AH, Cecchini M (2020) On the functional annotation of open-channel structures in the glycine receptor. Structure 28(6):690–3.e3. https://doi.org/10.1016/j.str.2020.05.003

Cerdan AH, Martin NE, Cecchini M (2018) An ion-permeable state of the glycine receptor captured by molecular dynamics. Structure 26(11):1555–62.e4. https://doi.org/10.1016/j.str.2018.07.019

Cerdan AH, Peverini L, Changeux JP, Corringer PJ, Cecchini M (2022) Lateral fenestrations in the extracellular domain of the glycine receptor contribute to the main chloride permeation pathway. Sci Adv 8(41):eadc9340. https://doi.org/10.1126/sciadv.adc9340

Chen ZW, Bracamontes JR, Budelier MM, Germann AL, Shin DJ, Kathiresan K et al (2019) Multiple functional neurosteroid binding sites on GABAA receptors. PLoS Biol 17(3):e3000157. https://doi.org/10.1371/journal.pbio.3000157

Cheng MH, Coalson RD (2012) Energetics and ion permeation characteristics in a glutamate-gated chloride (GluCl) receptor channel. J Phys Chem B 116(46):13637–13643. https://doi.org/10.1021/jp3074915

Cheng WWL, Chen ZW, Bracamontes JR, Budelier MM, Krishnan K, Shin DJ et al (2018) Mapping two neurosteroid modulatory sites in the prototypic pentameric ligand gated ion channel GLIC. J Biol Chem 293(8):3013–3027. https://doi.org/10.1074/jbc.RA117.000359

Cheng WWL, Arcario MJ, Petroff JT 2nd (2021) Druggable lipid binding sites in Pentameric ligand-gated ion channels and transient receptor potential channels. Front Physiol 12:798102. https://doi.org/10.3389/fphys.2021.798102

Criado M, Eibl H, Barrantes FJ (1982) Effects of lipids on acetylcholine receptor. Essential need of cholesterol for maintenance of agonist-induced state transitions in lipid vesicles. Biochemistry 21(15):3622–3629

Criado M, Eibl H, Barrantes FJ (1984) Functional properties of the acetylcholine receptor incorporated in model lipid membranes. Differential effects of chain length and head group of phospholipids on receptor affinity states and receptor-mediated ion translocation. J Biol Chem 259(14):9188–9198

daCosta CJ, Baenziger JE (2009) A lipid-dependent uncoupled conformation of the acetylcholine receptor. J Biol Chem 284(26):17819–17825. https://doi.org/10.1074/jbc.M900030200

daCosta CJ, Medaglia SA, Lavigne N, Wang S, Carswell CL, Baenziger JE (2009) Anionic lipids allosterically modulate multiple nicotinic acetylcholine receptor conformational equilibria. J Biol Chem 284(49):33841–33849. https://doi.org/10.1074/jbc.M109.048280

daCosta CJ, Dey L, Therien JP, Baenziger JE (2013) A distinct mechanism for activating uncoupled nicotinic acetylcholine receptors. Nat Chem Biol 9(11):701–707. https://doi.org/10.1038/nchembio.1338

Dalal V, Arcario MJ, Petroff JT 2nd, Tan BK, Dietzen NM, Rau MJ et al (2024) Lipid nanodisc scaffold and size alter the structure of a pentameric ligand-gated ion channel. Nat Commun 15(1):25. https://doi.org/10.1038/s41467-023-44366-w

Dalal V, Tan BT, Xu H, Cheng WWL (2025) Cryo-EM structures of a pentameric ligand-gated ion channel in liposomes. Elife 14:RP106728

Dalziel AW, Rollins ES, McNamee MG (1980) The effect of cholesterol on agonist-induced flux in reconstituted acetylcholine receptor vesicles. FEBS Lett 122(2):193–196. https://doi.org/10.1016/0014-5793(80)80435-2

Damgen MA, Biggin PC (2020) A refined open state of the glycine receptor obtained via molecular dynamics simulations. Structure 28(1):130–9.e2. https://doi.org/10.1016/j.str.2019.10.019

Damgen MA, Biggin PC (2021) State-dependent protein-lipid interactions of a pentameric ligand-gated ion channel in a neuronal membrane. PLoS Comput Biol 17(2):e1007856. https://doi.org/10.1371/journal.pcbi.1007856

Dietzen NM, Arcario MJ, Chen LJ, Petroff JT 2nd, Moreland KT, Krishnan K et al (2022) Polyunsaturated fatty acids inhibit a pentameric ligand-gated ion channel through one of two binding sites. Elife 11:e74306. https://doi.org/10.7554/eLife.74306

Du J, Lu W, Wu S, Cheng Y, Gouaux E (2015) Glycine receptor mechanism elucidated by electron cryo-microscopy. Nature 526(7572):224–229. https://doi.org/10.1038/nature14853

Felt K, Stauffer M, Salas-Estrada L, Guzzo PR, Xie D, Huang J et al (2024) Structural basis for partial agonism in 5-HT(3A) receptors. Nat Struct Mol Biol 31(4):598–609. https://doi.org/10.1038/s41594-023-01140-2

Fernandez Nievas GA, Barrantes FJ, Antollini SS (2008) Modulation of nicotinic acetylcholine receptor conformational state by free fatty acids and steroids. J Biol Chem 283(31):21478–21486. https://doi.org/10.1074/jbc.M800345200

Ferraro NA, Cascio M (2018) Cross-linking-mass spectrometry studies of cholesterol interactions with human alpha1 glycine receptor. Anal Chem 90(4):2508–2516. https://doi.org/10.1021/acs.analchem.7b03639

Fong TM, McNamee MG (1986) Correlation between acetylcholine receptor function and structural properties of membranes. Biochemistry 25(4):830–840

Gielen M, Corringer PJ (2018) The dual-gate model for pentameric ligand-gated ion channels activation and desensitization. J Physiol 596(10):1873–1902. https://doi.org/10.1113/JP275100

Gielen M, Thomas P, Smart TG (2015) The desensitization gate of inhibitory Cys-loop receptors. Nat Commun 6:6829. https://doi.org/10.1038/ncomms7829

Gonzalez-Gutierrez G, Wang Y, Cymes GD, Tajkhorshid E, Grosman C (2017) Chasing the open-state structure of pentameric ligand-gated ion channels. J Gen Physiol 149(12):1119–1138. https://doi.org/10.1085/jgp.201711803

Hamano H, Nabekura J, Nishikawa M, Ogawa T (1996) Docosahexaenoic acid reduces GABA response in substantia nigra neuron of rat. J Neurophysiol 75(3):1264–1270. https://doi.org/10.1152/jn.1996.75.3.1264

Hamouda AK, Sanghvi M, Sauls D, Machu TK, Blanton MP (2006) Assessing the lipid requirements of the Torpedo californica nicotinic acetylcholine receptor. Biochemistry 45(13):4327–4337. https://doi.org/10.1021/bi052281z

Jalalypour F, Howard RJ, Lindahl E (2024) Allosteric cholesterol site in glycine receptors characterized through molecular simulations. J Phys Chem B 128(20):4996–5007. https://doi.org/10.1021/acs.jpcb.4c01703

Jones OT, McNamee MG (1988) Annular and nonannular binding sites for cholesterol associated with the nicotinic acetylcholine receptor. Biochemistry 27(7):2364–2374. https://doi.org/10.1021/bi00407a018

Kelley SP, Dunlop JI, Kirkness EF, Lambert JJ, Peters JA (2003) A cytoplasmic region determines single-channel conductance in 5-HT3 receptors. Nature 424(6946):321–324. https://doi.org/10.1038/nature01788

Kim JJ, Gharpure A, Teng J, Zhuang Y, Howard RJ, Zhu S et al (2020) Shared structural mechanisms of general anaesthetics and benzodiazepines. Nature 585(7824):303–308. https://doi.org/10.1038/s41586-020-2654-5

Kracun S, Harkness PC, Gibb AJ, Millar NS (2008) Influence of the M3-M4 intracellular domain upon nicotinic acetylcholine receptor assembly, targeting and function. Br J Pharmacol 153(7):1474–1484. https://doi.org/10.1038/sj.bjp.0707676

Kumar A, Basak S, Rao S, Gicheru Y, Mayer ML, Sansom MSP et al (2020a) Mechanisms of activation and desensitization of full-length glycine receptor in lipid nanodiscs. Nat Commun 11(1):3752. https://doi.org/10.1038/s41467-020-17364-5

Kumar P, Wang Y, Zhang Z, Zhao Z, Cymes GD, Tajkhorshid E et al (2020b) Cryo-EM structures of a lipid-sensitive pentameric ligand-gated ion channel embedded in a phosphatidylcholine-only bilayer. Proc Natl Acad Sci USA 117(3):1788–1798. https://doi.org/10.1073/pnas.1906823117

Labriola JM, Pandhare A, Jansen M, Blanton MP, Corringer PJ, Baenziger JE (2013) Structural sensitivity of a prokaryotic pentameric ligand-gated ion channel to its membrane environment. J Biol Chem 288(16):11294–11303. https://doi.org/10.1074/jbc.M113.458133

Lape R, Colquhoun D, Sivilotti LG (2008) On the nature of partial agonism in the nicotinic receptor superfamily. Nature 454(7205):722–727. https://doi.org/10.1038/nature07139

Lara CO, Burgos CF, Silva-Grecchi T, Munoz-Montesino C, Aguayo LG, Fuentealba J et al (2019) Large intracellular domain-dependent effects of positive allosteric modulators on glycine receptors. ACS Chem Neurosci 10(5):2551–2559. https://doi.org/10.1021/acschemneuro.9b00050

Laverty D, Desai R, Uchanski T, Masiulis S, Stec WJ, Malinauskas T et al (2019) Cryo-EM structure of the human alpha1beta3gamma2 GABAA receptor in a lipid bilayer. Nature 565(7740):516–520. https://doi.org/10.1038/s41586-018-0833-4

Legesse DH, Fan C, Teng J, Zhuang Y, Howard RJ, Noviello CM et al (2023) Structural insights into opposing actions of neurosteroids on GABA(A) receptors. Nat Commun 14(1):5091. https://doi.org/10.1038/s41467-023-40800-1

Mantipragada SB, Horvath LI, Arias HR, Schwarzmann G, Sandhoff K, Barrantes FJ et al (2003) Lipid-protein interactions and effect of local anesthetics in acetylcholine receptor-rich membranes from Torpedo marmorata electric organ. Biochemistry 42(30):9167–9175. https://doi.org/10.1021/bi034485q

Marsh D, Barrantes FJ (1978) Immobilized lipid in acetylcholine receptor-rich membranes from Torpedo marmorata. Proc Natl Acad Sci USA 75(9):4329–4333. https://doi.org/10.1073/pnas.75.9.4329

Marsh D, Watts A, Barrantes FJ (1981) Phospholipid chain immobilization and steroid rotational immobilization in acetylcholine receptor-rich membranes from Torpedo marmorata. Biochim Biophys Acta 645(1):97–101. https://doi.org/10.1016/0005-2736(81)90516-2

Masiulis S, Desai R, Uchanski T, Serna Martin I, Laverty D, Karia D et al (2019) GABAA receptor signalling mechanisms revealed by structural pharmacology. Nature 565(7740):454–459. https://doi.org/10.1038/s41586-018-0832-5

Miller PS, Aricescu AR (2014) Crystal structure of a human GABAA receptor. Nature 512(7514):270–275. https://doi.org/10.1038/nature13293

Minota S, Watanabe S (1997) Inhibitory effects of arachidonic acid on nicotinic transmission in bullfrog sympathetic neurons. J Neurophysiol 78(5):2396–2401. https://doi.org/10.1152/jn.1997.78.5.2396

Nabekura J, Noguchi K, Witt MR, Nielsen M, Akaike N (1998) Functional modulation of human recombinant gamma-aminobutyric acid type A receptor by docosahexaenoic acid. J Biol Chem 273(18):11056–11061. https://doi.org/10.1074/jbc.273.18.11056

Nayak TK, Vij R, Bruhova I, Shandilya J, Auerbach A (2019) Efficiency measures the conversion of agonist binding energy into receptor conformational change. J Gen Physiol 151(4):465–477. https://doi.org/10.1085/jgp.201812215

Noviello CM, Gharpure A, Mukhtasimova N, Cabuco R, Baxter L, Borek D et al (2021) Structure and gating mechanism of the alpha7 nicotinic acetylcholine receptor. Cell 184(8):2121–34.e13. https://doi.org/10.1016/j.cell.2021.02.049

Ochoa EL, Dalziel AW, McNamee MG (1983) Reconstitution of acetylcholine receptor function in lipid vesicles of defined composition. Biochim Biophys Acta 727(1):151–162

Petroff JT 2nd, Dietzen NM, Santiago-McRae E, Deng B, Washington MS, Chen LJ et al (2022) Open-channel structure of a pentameric ligand-gated ion channel reveals a mechanism of leaflet-specific phospholipid modulation. Nat Commun 13(1):7017. https://doi.org/10.1038/s41467-022-34813-5

Polovinkin L, Hassaine G, Perot J, Neumann E, Jensen AA, Lefebvre SN et al (2018) Conformational transitions of the serotonin 5-HT3 receptor. Nature 563(7730):275–279. https://doi.org/10.1038/s41586-018-0672-3

Rahman MM, Teng J, Worrell BT, Noviello CM, Lee M, Karlin A et al (2020) Structure of the native muscle-type nicotinic receptor and inhibition by snake venom toxins. Neuron 106(6):952–62.e5. https://doi.org/10.1016/j.neuron.2020.03.012

Seiferth D, Biggin PC, Tucker SJ (2022) When is a hydrophobic gate not a hydrophobic gate? J Gen Physiol 154(11):e202213210. https://doi.org/10.1085/jgp.202213210

Sharp L, Brannigan G (2021) Spontaneous lipid binding to the nicotinic acetylcholine receptor in a native membrane. J Chem Phys 154(18):185102. https://doi.org/10.1063/5.0046333

Singh M, Indurthi DC, Mittal L, Auerbach A, Asthana S (2024) Conformational dynamics of a nicotinic receptor neurotransmitter site. eLife Sciences Publications, Ltd

Sogaard R, Werge TM, Bertelsen C, Lundbye C, Madsen KL, Nielsen CH et al (2006) GABA(A) receptor function is regulated by lipid bilayer elasticity. Biochemistry 45(43):13118–13129. https://doi.org/10.1021/bi060734+

Sugasawa Y, Cheng WW, Bracamontes JR, Chen ZW, Wang L, Germann AL et al (2020) Site-specific effects of neurosteroids on GABAA receptor activation and desensitization. Elife 9:e55331. https://doi.org/10.7554/eLife.55331

Therien JP, Baenziger JE (2017) Pentameric ligand-gated ion channels exhibit distinct transmembrane domain archetypes for folding/expression and function. Sci Rep 7(1):450. https://doi.org/10.1038/s41598-017-00573-2

Tong A, Petroff JT, Hsu FF, Schmidpeter PA, Nimigean CM, Sharp L, Brannigan G, Cheng WW (2019) Direct binding of phosphatidylglcyerol at specific sites modulates desensitization of a pentameric ligand-gated ion channel. bioRxiv 8:e50766. https://doi.org/10.1101/722512

Unwin N (2005) Refined structure of the nicotinic acetylcholine receptor at 4A resolution. J Mol Biol

346(4):967–989. https://doi.org/10.1016/j.jmb.2004.12.031

Unwin N (2017) Segregation of lipids near acetylcholine-receptor channels imaged by cryo-EM. IUCrJ 4(Pt 4):393–399. https://doi.org/10.1107/S2052252517005243

Unwin N (2020) Protein-lipid architecture of a cholinergic postsynaptic membrane. IUCrJ 7(Pt 5):852–859. https://doi.org/10.1107/S2052252520009446

Vijayaraghavan S, Huang B, Blumenthal EM, Berg DK (1995) Arachidonic acid as a possible negative feedback inhibitor of nicotinic acetylcholine receptors on neurons. J Neurosci 15(5 Pt 1):3679–3687

Yu J, Zhu H, Lape R, Greiner T, Du J, Lu W et al (2021) Mechanism of gating and partial agonist action in the glycine receptor. Cell 184(4):957–968. https://doi.org/10.1016/j.cell.2021.01.026

Zarkadas E, Pebay-Peyroula E, Thompson MJ, Schoehn G, Uchanski T, Steyaert J et al (2022) Conformational transitions and ligand-binding to a muscle-type nicotinic acetylcholine receptor. Neuron 110(8):1358–70.e5. https://doi.org/10.1016/j.neuron.2022.01.013

Zhang Y, Dijkman PM, Zou R, Zandl-Lang M, Sanchez RM, Eckhardt-Strelau L et al (2021) Asymmetric opening of the homopentameric 5-HT3A serotonin receptor in lipid bilayers. Nat Commun 12(1):1074. https://doi.org/10.1038/s41467-021-21016-7

Zhuang Y, Noviello CM, Hibbs RE, Howard RJ, Lindahl E (2022) Differential interactions of resting, activated, and desensitized states of the alpha7 nicotinic acetylcholine receptor with lipidic modulators. Proc Natl Acad Sci USA 119(43):e2208081119. https://doi.org/10.1073/pnas.2208081119

Structural Domains Associated with Permeation, Gating, and Selectivity of Aquaporin Ion Channels Found Across Phyla

12

Andrea J. Yool

Abstract

Aquaporins AQPs) are an ancient group of channels that arose early in phylogeny and radiated throughout the kingdoms of life. Classified as members of the broad Membrane Intrinsic Protein family, AQPs originally were envisioned as strictly water pores that were constitutively open. International research interest is driving new views of AQPs as multifunctional channels with complex control mechanisms that enable adaptive responses to physiological challenges. An expanding number of AQP classes are being discovered to allow permeation of diverse solutes in parallel to the archetypal substrate water, with channel activities that show subtype-specific regulation by intracellular and extracellular signals. To date, at least 15 classes of AQPs from mammals, plants, insects, and algae have been proposed to carry ions, enabling amino acid sequence comparisons that suggest ion channel functional domains arise from conserved transmembrane, loop, and terminal regions of the tetrameric AQP protein, serving as selectivity filters, gates, and sites for modulation. More AQP ion channel classes likely await discovery pending identification of relevant activators. Subtype-specific control of dual water-and-ion AQP classes orchestrates diverse roles for AQPs across the domains of life, promoting homeostasis, cell motility, nutrient acquisition, redox protection, sensory detection, and intercellular signaling in cells and tissues.

Keywords

Nonselective cation channel · Monovalent · MIP · Water channel · Cyclic GMP

12.1 Discovery of AQP Ion Channels

In groundbreaking work that launched the modern era of aquaporin (AQP) research, expression of cloned AQP1 in Xenopus oocytes was demonstrated to increase the membrane osmotic water permeability, thus identifying AQP1 as a water channel; however, it was reported to not allow transmembrane passage of ions (Agre et al. 1993a) based on two-electrode voltage clamp recordings of eight AQP1-expressing oocytes (Preston et al. 1992) in the absence of any activating stimulus (not yet

A. J. Yool (✉)
School of Biomedicine and The Institute for Photonics and Advanced Sensing, University of Adelaide, Adelaide, Australia
e-mail: andrea.yool@adelaide.edu.au

A. Rosenhouse-Dantsker (ed.), *Decoding Ion Channels Structure and Function*, Advances in Experimental Medicine and Biology 1497, https://doi.org/10.1007/978-3-032-07523-9_12

identified at that time). Against this doctrine, AQP ion channels faced a rocky start as a controversial idea. Embedding the concept that AQPs were not capable of ion channel function slowed progress in the field for years. Potential dual ion channel candidates remaining in the vast family of AQPs merit objective consideration; key approaches will be to find native stimuli as needed to activate ion channel gating, and to use an expression system such as yeast in which even small levels of ion conductance conferred by an introduced AQP ion channel can be readily detected by the rescue of growth in a transport-defective yeast line (Nourmohammadi et al. 2024).

Examples of diverse AQPs endowed with ion channel functionality are being discovered to exist across the domains of life (illustrated in Fig. 12.1). The first aquaporin identified as an ion channel was mammalian lens Major Intrinsic Protein (MIP) channel in 1985, followed by soybean Nodulin-26 in 1994, human AQP1 in 1996, mammalian AQP6 in 1999, Drosophila Big Brain in 2002, and subsequently an array of plant and algal PIP (Plasma membrane Intrinsic Protein) and TIP (Tonoplast Intrinsic Protein) channels (Fig. 12.1a). Mammalian AQP1 has been one of the most extensively studied AQP ion channels. In this rapidly emerging field, the available levels of detail for structure–function relationships in ion channel aquaporins vary by subtype; nonetheless, an accumulating set of examples provides a timely tool to assess subtype-specific specializations and evidence for a paradigm shift in our understanding of AQP channel properties.

Aquaporins initially were envisioned as transmembrane pathways just for water. In 1993, soon after the cloned aquaporin gene (*AQP1*) was demonstrated to encode a water channel (Preston et al. 1992), Peter Agre and colleagues suggested "(T)he term aquaporin should not be used to describe proteins permeated by ions ... or other molecules" (Agre et al. 1993b). In their Letter to the Editor, they recommended exclusion of MIP (now known as AQP0) because of its permeability to ions (Zampighi et al. 1985), and the bacterial GlpF (Glycerol Facilitator) based on its permeability to glycerol (Heller et al. 1980). Nonetheless, the name "aquaporin" was widely adopted for most members of the extensive family, justified by high amino acid sequence homologies and the hallmark presence of pairs of conserved signature motifs (each consisting of asparagine, proline, and typically alanine (NPA)) in all four subunits of the tetrameric channels (Reizer et al. 1993; Zardoya 2005). Two-electrode voltage clamp recordings were carried out on AQP0-expressing oocytes, with AQP1-expressing oocytes presumably included for comparison as pure water channels. Results of that study showed ion currents were associated with both AQP0 and AQP1 (induced by undefined stretch-activated signaling pathways) that exceeded control oocytes in current amplitude; however, these results were interpreted as artifacts that confirmed neither AQP0 nor AQP1 had ion channel functionality (Mulders et al. 1995). Unexpectedly, 1 year later, the archetypal water channel itself, human AQP1, turned out to be a gated ion channel (Yool et al. 1996), though the direct triggering stimulus remained to be defined as cyclic GMP in follow-on studies (Anthony et al. 2000).

Transmembrane topology of AQP1 was first evaluated using a chymotrypsin-sensitive epitope introduced by site-directed mutagenesis into a series of positions in AQP1; changes in water channel activity for mutant constructs expressed in oocytes showed that Loop C (linking the third and fourth transmembrane domains, M3 and M4, respectively) was extracellular, and the N- and C-termini and Loop D (linking M4 and M5) were intracellular (Preston et al. 1994). Crystal structures have since provided striking three-dimensional views of the AQP channel architecture that supported the original findings and added substantial detail. Loop B (linking M2 to M3) and Loop E (linking M5 to M6) contain the NPA motifs that line the intrasubunit water pores. Loop D between M4 and M5 is a gating domain (Yu et al. 2006). The central passage in AQP1 is lined by transmembrane domains M2 and M5, capped on the outer side by a fourfold symmetry of hydrophobic residues (Val50 in human AQP1) that impose a limiting diameter of less than 3 Å (Murata et al. 2000) in the closed state.

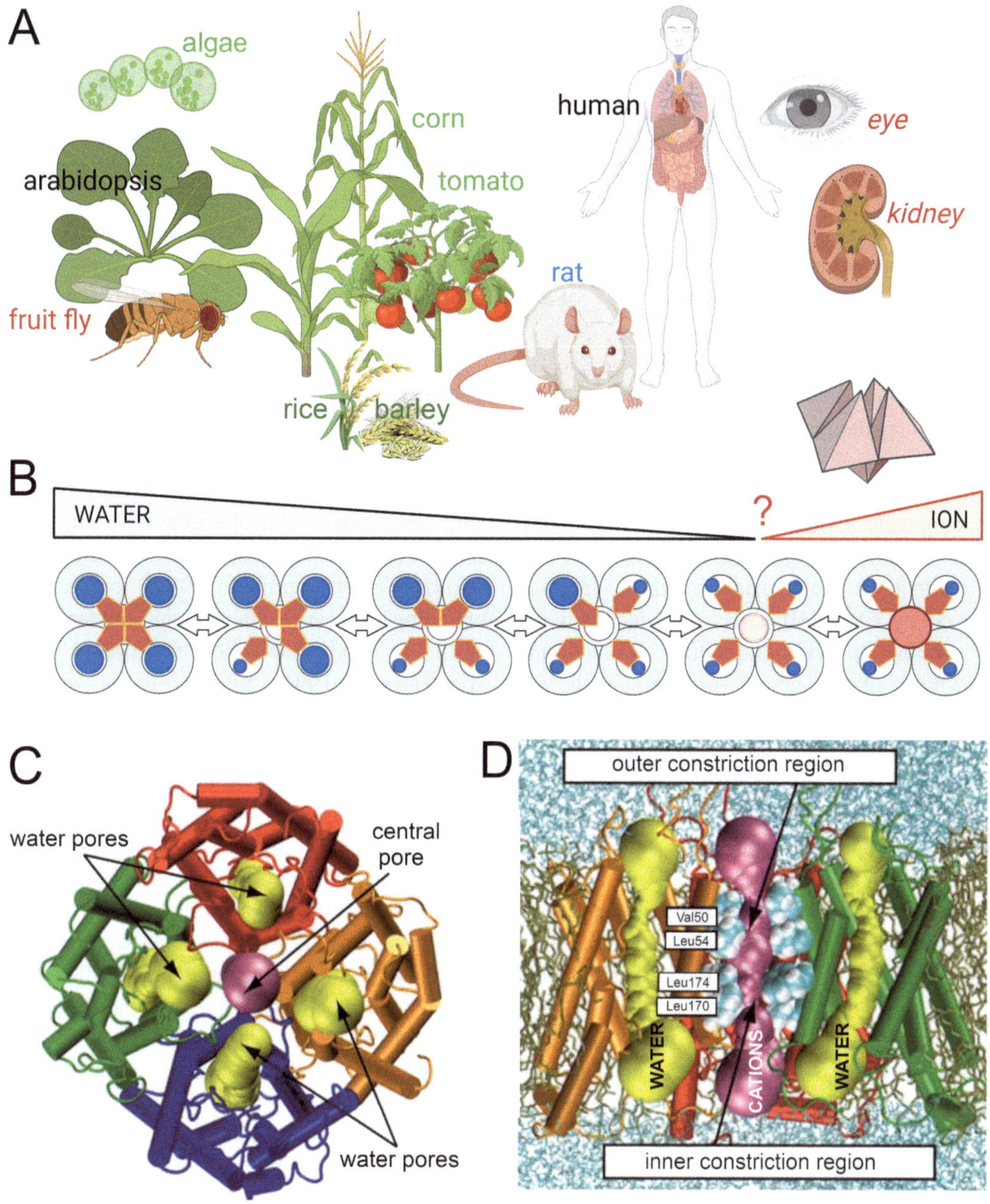

Fig. 12.1 Overview of aquaporin ion channels. (**A**) Illustration of the variety of organisms in which aquaporins with ion channel function have been discovered thus far. (**B**) Schematic diagram of a simplified state diagram illustrating a hypothetical bimodal system with two stable states, carrying water (left) or conducting ions (right), linked by substates reflecting conformation changes in hydrophobic central pore barrier residues at the subunit level (in random order). (**C**) Images from structural dynamics modeling (Yu et al. 2006) of the AQP1 tetramer (left) viewed from above, showing water and ion pore locations, and (**D**) viewed from the side, showing the positions of barrier residues in the human AQP1 central pore. (Panels A and B were created in BioRender. Yool, A. (2025) https://BioRender.com/j14n077. Images in C and D were used with permission from Elsevier, Copyright Clearance Center #501959597, relabeled to match human AQP1)

The predicted structure of the first cloned member of the MIP family, lens MIP (now AQP0), with six transmembrane domains and intracellular N- and C-termini, was viewed as consistent with likely functionality as a membrane channel (Gorin et al. 1984). Basic protein properties of AQPs, including tetrameric subunit organization, have been noted as a theme shared by many classes of ion channels, including potassium channels (Jan and Jan 2012; Yool and Weinstein 2002) and others. AQP0 is natively expressed in the lens fibers of the eye, where inherited autosomal dominant mutations result in cataracts, thought to be caused by trapping of wild-type AQP0 by mutant subunits via oligomerization in the endoplasmic reticulum, inducing cell toxicity (Varadaraj et al. 2008). AQP0 protein purified from the eye lens was shown to have water channel activity that was blocked by Ca^{2+} (10 μM) when reconstituted in combination with calmodulin (Girsch and Peracchia 1985), which interacts with the carboxyl-terminal domain (Peracchia and Girsch 1985). In the oocyte expression system, AQP0-mediated water permeability was confirmed, though unitary water flux was determined to be approximately 40-fold less than that of AQP1 channels (Chandy et al. 1997). Further evidence for AQP0 water channel functionality was provided by the loss of water flux after deletion of the C-terminal domain (Chandy et al. 1997).

Ion permeation through purified lens AQP0 was discovered in 1985 by James Hall and colleagues; their work showed reconstituted AQP0 ion channels in bilayers had four conductance states, with a main open state at 200 pS in 100 mM saline that was not blocked by mM Ca^{2+} (Zampighi et al. 1985). Additional work showed that bovine AQP0 ion channels in bilayers had two main conductance states of 380 and 160 pS, plus subconductance states in 100 mM KCl, were more permeable to anions than cations (P_{Cl}/P_K 1.8), and were not sensitive to Ca^{2+} (Ehring et al. 1990). Interestingly, Ehring and colleagues noted AQP0 channels occasionally closed directly from the maximally open single channel state to a nonconducting state equivalent to bilayer alone (in <1% of all closing events), though intermediate steps were more common. Chicken lens AQP0 (MIP28) in lipid bilayers showed similarly high unitary conductance values (290 pS and 60 pS) in 150 mM KCl saline, and a preference for anions over cations was seen by a 5-mV shift in reversal potential when KCl concentration was raised from 100 mM to 200 mM, indicating a permeability ratio (P_{Cl}/P_K) of 1.87 (Modesto et al. 1996).

Most classes of AQP ion channels which have been analyzed at the single-channel level, whether reconstituted in bilayers or phospholipid vesicles or recorded in cell-excised patches, have typically shown high unitary conductances, long open times, and multiple subconductance states. As seen in AQP0, subconductance states also have been described for AQP ion channels, including soybean Nodulin26 and human AQP1 (Anthony et al. 2000; Henderson et al. 2023; Weaver et al. 1994). Nitrogen-fixing soybeans express the AQP Nodulin-26 in the membranes of symbiosomes, prompting the prediction that Nodulin-26 had a transport function; this was confirmed when the purified protein was reconstituted into liposomes and assessed in planar lipid bilayers. Single-channel events showed a high conductance (3.1 nS in 1 M KCl saline) and multiple subconductance states. Both anions and cations were permeable, with a weak preference for anions, giving a P_{Cl}/P_K ratio of 1.2 based on the reversal potential shift seen when KCl concentration was changed from 1 M to 0.1 M (Weaver et al. 1994).

In contrast to the anionic selectivity of AQP0 and Nodulin-26, AQP1 ion channels were found to carry cations (Yool et al. 1996). Human AQP1 ion channels were first characterized in voltage clamp recordings in the oocyte expression system, demonstrating cation selectivity and activation by intracellular second messenger signaling (Yool et al. 1996). Additional lines of evidence compiled over almost three decades have confirmed that AQP1 channels are directly activated by binding of intracellular cyclic GMP and that ions permeate through the gated central pore, shown using electrophysiology, site-directed mutagenesis, optical probes for cation flux, pharmacology, and molecular dynamics modeling (Anthony et al. 2000; Yu et al. 2006; Boassa and Yool 2003; Campbell et al. 2012).

AQP1 in excised patches from oocytes showed a unitary conductance of 150 pS and subconductance states in symmetrical 100 mM K^+ salines with Cl^- or gluconate (Anthony et al. 2000). When reconstituted in proteoliposomes, human AQP1 in symmetrical 112 mM KCl saline showed a single-channel conductance of approximately 75 pS, with subconductance states and patterns of activation that implied possible cooperativity between multiple channels in a patch (Henderson et al. 2023). Control of extracellular and intracellular gates in tetrameric AQP0 channels has also been proposed to show cooperativity between subunits, modulated by CaM binding (Freites et al. 2019). Independent work using recombinant human AQP1 channels confirmed that they were activated by cGMP, blocked by a cGMP-related antagonist, and permeable to Na^+ and K^+ (Saparov et al. 2001)—all properties in agreement with published work (Anthony et al. 2000); however, the reconstituted AQP1 showed much smaller unitary conductances (<10 pS) and a lower proportion of active channels than reported for oocyte assays. Differences in unitary conductances for AQPs between preparations could be due to differences in membrane lipid composition (Henderson et al. 2022), post-translational modifications, including phosphorylation state at the time of harvest (Campbell et al. 2012; Qiu et al. 2020), interactions with other proteins (Adams et al. 2001; Amiry-Moghaddam et al. 2004; Cowan et al. 2000; Edamana et al. 2024; Ribeiro et al. 2022), or additional environmental factors.

12.2 Proposed Model for Ion Versus Water Permeable States

The structural correlates of the subconductance states observed in AQP ion channels have not been determined, but as a starting point, a working hypothesis can be suggested (Fig. 12.1b), building on a diagram sketched previously (Tyerman et al. 2021). The scheme shows a simple model comprising two opposing stable states—one favoring water flux via intrasubunit pores (left), and one favoring ion passage through the central pore (right), linked by a series of intermediate states depicting stepwise conformational changes in subunits that might precede central pore opening. A visual metaphor for the outcome is a paper fortune teller (inset top right), though real changes in channel conformation likely involve subtle shifts in barrier residues, not massive reorganization of whole transmembrane domains. The fifth position, with central barriers removed but the central pore not fully open, could encompass a number of ion current subconductance states, not illustrated. The ability of a channel to conduct ions without all four subunits being in an ion-permissive orientation seems unlikely, given that hydration across the full pore length appears necessary for ion conduction, based on simulation studies (Yu et al. 2006). Support for the bimodal idea comes from work showing ion and water conducting functions appear to be reciprocally gated in one of the plant AQP classes (AtPIP2;1), in which ion and water permeabilities were shown to be inversely governed by phosphorylation (Qiu et al. 2020). Apparently, reciprocal gating of water and ion channel activity for AQP1 was also observed in response to phosphorylation of consensus sites by protein kinase C (Zhang et al. 2007). Validity and details of the hypothetical state model in Fig. 12.1b remain to be determined, but could frame testable hypotheses for further work.

A reasonable prediction of the reciprocal gating idea illustrated in Fig. 12.1b is that a decrease in water channel activity should occur when AQP ion channels are opened. However, it is important to note that this outcome will depend on the proportion of AQP channels in a population that are active as ion channels, which appears to differ between subtypes. While findings for the plant AQP channel AtPIP2:1 support the model (Qiu et al. 2020), unpublished data for AQP1 (DN Birdsell; University of Arizona dissertation research, 1995) showed no appreciable reduction in oocyte swelling rates after AQP1 channel activation by cGMP. Similarly, a mutant of the barley AQP HvPIP2;8 carrying a phosphomimetic substitution in the C-terminal domain did show reciprocal regulation of water and ion fluxes, but the wild-type HvPIP2;8 did not (Tran et al. 2020). How can this puzzle be reconciled? Taking AQP1 as an example, simple calculations show that the typical microamp currents measured in AQP1-expressing oocytes actually are generated by a tiny proportion (approximately 0.002%) of the total population of AQP1 channels present in the oocyte plasma membrane. The proportion of active ion channels can be calculated directly as the total membrane conductance (typically μS) divided by the unitary conductance (150 pS), as a simplified estimate that does not account for subconductance states. Similarly, the estimated water channel number is determined from the whole oocyte swelling rate divided by the calculated unitary water flux rate (which is the whole cell swelling rate divided by the number of membrane protein channels estimated from freeze-fracture scanning electron microscopic images). In the final analysis, cGMP activation of AQP1-expressing oocytes rendered only 1/56,000 of the channels conductive; nonetheless, this 0.002% level of AQP1 ion channel contribution, when assessed in a quantitative model of kidney proximal tubule function, translated into cation flux that was comparable to the level seen in native peritubular membrane (Yool and Weinstein 2002).

Weinstein's model-based calculations further predicted that cGMP stimulation of just 0.002% of AQP1 channels in the proximal tubule would increase overall epithelial Na^+ reabsorption by 18%, and slightly depolarize the membrane potential by several mV, positing reasonable magnitudes that would be consistent with a realistic response (Yool and Weinstein 2002). The availability of human AQP1 to be activated as ion channels by cGMP was shown to be controlled by phosphorylation at Tyr 253 in the C-terminal domain (Campbell et al. 2012). Tyrosine dephosphorylation by pharmacological treatment or mutation of Tyr 253 to Cys removed the Tyr phosphorylation signal (confirmed by Western blot) and prevented ion channel activation by cGMP; in contrast, blocking serine and threonine phosphorylation consensus sites by pharmacology or mutagenesis did not alter cGMP-dependent activation responses (Campbell et al. 2012). This example illustrates the principle that when channels are highly abundant in a membrane, only a tiny fraction needs to be active as ion channels to confer physiologically meaningful outcomes. As aptly summarized by Bertil Hille, a respected leader in the ion channel field, "Signaling requires only small ion fluxes" (Hille 2001). The logical corollary is that tissues that depend on high expression levels of an AQP dual water-and-ion channel for water transport must tightly control the parallel ion channel capability in order to prevent collapse of membrane potential and ion gradients.

When coupled with molecular dynamics simulations, crystal structures have enabled impressive insights into the molecular mechanisms of AQP function, elucidating the five-pore organization of AQP1 with a central pore for cations lined by M2 and M5 (Yu et al. 2006), and four individual water pores in each of the subunits (Sui et al. 2001; Tajkhorshid et al. 2002), as shown in Fig. 12.1c. Molecular dynamics modeling identified four hydrophobic residues (Val50, Leu54, Leu170, and Leu174 in human AQP1) positioned as barriers to ion permeation in the closed central pore, with positions 50 and 54 located in four-fold symmetry on the outer half of M2, and positions 170 and 174 in the inner half of M5. Dynamics simulations showed that ligand binding to the Loop D gating domain induced widening of the central passage, allowing hydration followed by cation permeation (Yu et al. 2006), as depicted in Fig. 12.1d. Space-filling flow pathways are illustrated for both the ion and water conducting states. Biological assays confirmed that site-directed mutations in the predicted arginine-rich ligand binding site of Loop D dramatically impaired the rate of activation of the ion conductance (Yu et al. 2006; Kourghi et al. 2018).

Since the first introduction of the paradigm-shifting idea of dual water-and-ion channel AQPs, multiple converging lines of evidence have shown that many classes of AQPs can have ion channel activity, which has opened the field for the discovery of new classes of multifunctional AQPs.

12.3 Mechanisms of AQP Ion Channel Gating and Modulation

Precisely tuned filtering mechanisms, which enable high rates of throughput and selective permeability of intrasubunit pores to water, glycerol, and other neutral solutes, have been covered capably by other research groups, including those of Eric Beitz (Beitz et al. 2009), Roslyn Bill (Kitchen et al. 2019), Graça Soveral (Gotfryd et al. 2018) and many more; this topic is not revisited in this review. In contrast, ion channel gating in AQPs remains an under-researched area of work. Fortunately, accruing examples provide evidence for intracellular ligand binding, gating domains, phosphorylation sites, extracellular pH, and salinity as factors controlling AQP channel activities, here outlined for a cross-section of AQP classes.

Identification of Loop D as a gating domain for AQP1 ion channels was supported by molecular modeling and biological assays (Fig. 12.2). Site-directed mutations in Loop D did not prevent AQP1 water channel activity (confirming that the mutant constructs were expressed and targeted to the plasma membrane) but did greatly impact ion channel activation (Yu et al. 2006; Kourghi et al. 2018). Mutants in which conserved Arg residues at 159 and 160 were changed to Ala in human AQP1 (yellow highlight, Fig. 12.2a) showed substantially impaired ion channel activation responses to cGMP as compared with wild type (Yu et al. 2006; Kourghi et al. 2018). AQP4 (set in brackets because it has not yet been identified as an ion channel) has high sequence homology with AQP1, and when phosphorylated at Ser 180 (yellow highlight; a position equivalent to Arg 159 in AQP1) by Protein Kinase C, it showed decreased water permeability (McCoy et al. 2010; Zelenina et al. 2002), indicating Loop D exerts regulatory effects across AQP subtypes. Substitutions of Loop D residues with prolines, which introduce kinks in peptide structures, had differential effects on ion conductance responses that depended on position. Proline substitutions at Thr 157, Asp 158, or Asp 160 impaired ion channel activation; mutation of Asp 159 to Pro was tolerated without appreciable consequence, but Gly 165 to Pro had the opposite effect, potentiating the conductance response to cGMP (Kourghi et al. 2018). These results provide further evidence that Loop D conformation influences AQP1 ion channel gating.

Interestingly, amino acid sequence alignment shows that Loop D sequences are not conserved across the various types of ion channel AQPs discovered thus far; these classes appear to carry their own conserved motifs and might be predicted to be activated by different subtype-specific types of stimuli (Fig. 12.2a). Loop D, in contrast, is very highly conserved across AQP1 orthologs (Kourghi et al. 2018) in mammals, reptiles, and birds (Fig. 12.2b), and thus can be deduced to serve a function that offers a selective advantage; i.e., natural mutations in this domain during evolution have not been well tolerated. Sequence conservation for AQP1 Loop D argues that it enables an essential channel function in native tissues.

The catalyst in 2006 for discovering that Loop D in AQP1 was the gating domain came from the observation of differences between Loop D orientations in two crystal structures (Yu et al. 2006). In bovine AQP1 (prepared as thallium-derivatized crystals, imaged using multiwavelength anomalous diffraction) (Sui et al. 2001), the four Loop D segments (one from each subunit) were folded inward, covering the intracellular face of the central pore, as illustrated in Fig. 12.2c. In contrast, in human AQP1 (prepared as two-dimensional crystals, imaged by helium-cooled electron microscopy) (Murata et al. 2000), the arginine-rich Loop D domains were rotated outward, leaving the cytoplasmic vestibule of the central pore accessible. This bimodal conformation was confirmed in molecular dynamics simulations. Dynamic modeling indicated that binding of cyclic GMP to arginine residues in Loop D (Fig. 12.2d) triggered an outward shift of the gating domains (Fig. 12.2e), which in turn reoriented hydrophobic side chains of barrier residues lining the cen-

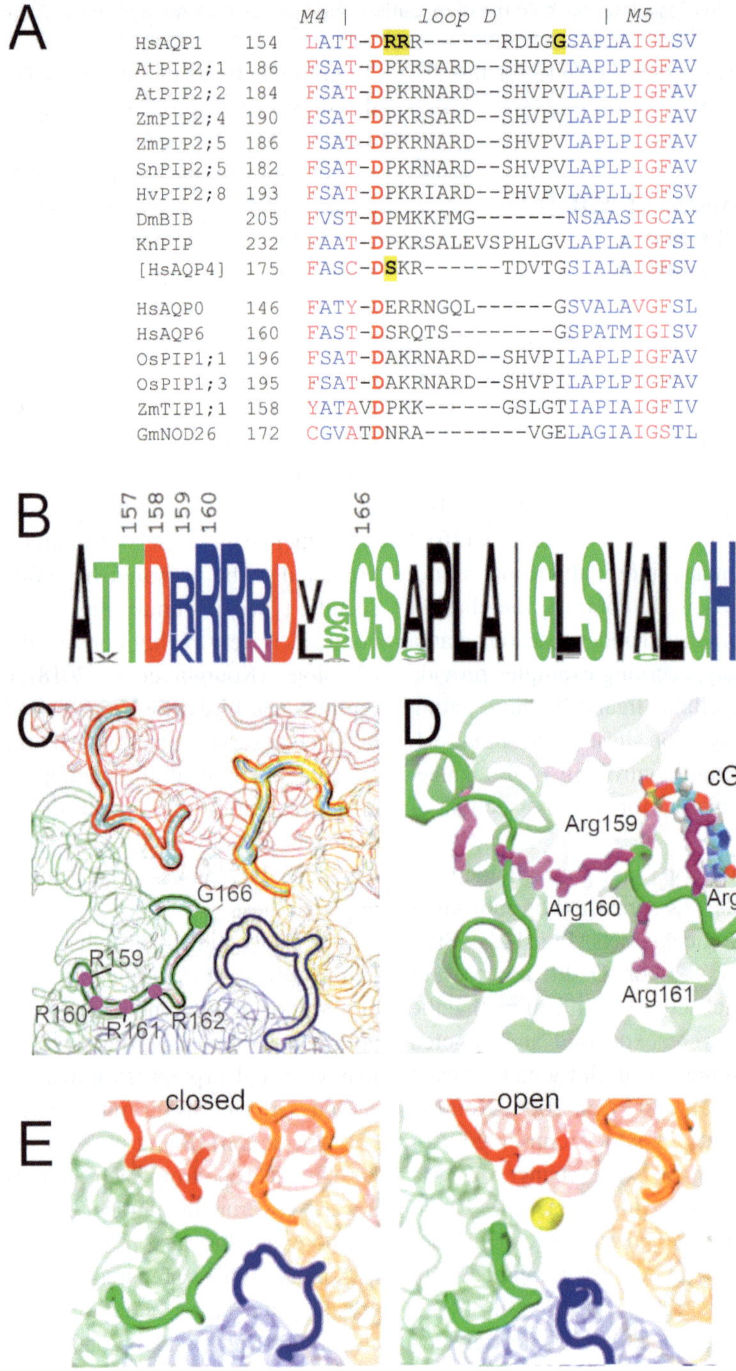

Fig. 12.2 Sequence comparisons and structural features of the Loop D gating domains. Loop D is located between the fourth and fifth transmembrane domains (M4, M5). (**A**) Multiple amino acid sequence alignment was done with NIH Cobalt (www.ncbi.nlm.nih.gov/tools/cobalt; conservation setting 3 bits), showing highly conserved (red), moderately conserved (blue), and variable regions (gray) that illustrate differences between AQP classes. Yellow highlights in AQP1 show Arg residues needed for cGMP binding, and the Gly residue which potentiates activation after mutation to Pro; yellow highlight in AQP4 shows Ser that, when phosphorylated by PKC, decreases water permeability. (**B**) Weblogo illustration of the frequency of occurrence of Loop D amino acids as a function of position, compiled for AQP1 channels from mammals, reptiles, and birds, and showing high sequence conservation (Kourghi et al. 2018). (**C**) Cartoon diagram of labeled Loop D residues in the closed configuration. (**D**) Molecular dynamics model-generated image capturing cGMP binding to the arginine-rich region in AQP1 Loop D. (**E**) Model-generated images showing Loop D of AQP1 viewed from the intracellular side in the ion channel closed state (left) and open state (right). (Panels B–E are from Yu et al. (2006), used with permission, Copyright Clearance Center #501959597, edited to create the cartoon in (C), and relabeled to match human AQP1)

tral pore, particularly at the internal constriction in M5 (Fig. 12.1d). Subtle changes in barrier residues were then sufficient to induce hydration of the pore, removing a major obstacle to ion permeation and modeled as enabling transition to the open state (Yu et al. 2006).

Single-channel recordings showed that AQP1 channels were relatively slow to activate (on the scale of minutes) after addition of cGMP to the intracellular face of excised patches (Anthony et al. 2000), unlike the fast response times measured in seconds for classic cyclic nucleotide-gated channels involved in sensory perception (Strassmaier and Karpen 2007). The long latency in AQP1 may indicate that multiple rate-limiting factors in a cascade of conformational changes ultimately lead to full pore hydration, prolonging the time until ion permeation. Since AQP1 in vivo serves in long-term functions such as fluid homeostasis, volume control, and cell motility, a slower onset of the ion conductance appears to be satisfactory. Quadruple mutation of the central pore barrier residues (Val 50, Leu 54, Leu 170, Leu 174) to alanine resulted in an inwardly rectifying current response to cGMP, with a shift in reversal potential of approximately +10 mV consistent with an increased permeability to extracellular tetraethylammonium (TEA^+) ion (Campbell et al. 2012).

A consensus on the structural basis for ion channel function in AQP6 is taking shape after a curious history of divergent viewpoints. Cloned in 1996 and characterized in oocytes, human AQP6 was originally identified as a water channel that was blocked by the classic AQP inhibitor mercuric ion, and not permeable to urea or glycerol (Ma et al. 1996). Holm and colleagues in 2004, working with rat AQP6 in oocytes, found that ^{14}C-glycerol and ^{14}C-urea did cross the membrane (Holm et al. 2004). Rat AQP6 differed in showing little baseline water channel activity, being activated rather than blocked by Hg^{2+}, and displaying an anion conductance (Yasui 2009).

Normally targeted to intracellular vesicles in the kidney collecting duct, rat AQP6, when expressed in oocytes, was routed to the plasma membrane, a feature that allowed detection of the anionic conductance after exposure to 100 μM Hg^{2+} or acidic pH (<5.5) (Yasui et al. 1999). Anion substitution shifted the reversal potential for rat AQP6, indicating a moderately anionic selectivity. Cations also flowed through the channel, though less well than anions, yet interestingly, Na^+ and $NMDG^+$ (N-methyl-D-glucamine) were equally permeable (Yasui et al. 1999), which suggested that the AQP6 ion pore, when open, is large. Na^+ entry through rat AQP6 was confirmed by $^{22}Na^+$ uptake (Hazama et al. 2002). Mutation of rat AQP6 Lys72 to Glu did not prevent the ion current, but blunted the reversal potential shift seen in wild type upon saline substitution of gluconate for Cl^-. It was initially surmised from the location of Lys72 that anions must flow through the intrasubunit pores (Yasui et al. 1999); however, a closer look at sequence alignments suggests that this residue is positioned at the internal end of M2 adjacent to Loop B, and thus sits between the central pore and intrasubunit pore domains, not diagnostic for either permeation pathway. Patch clamp recordings were published in 2002 by Hazama and colleagues for rat AQP6 channels in patches from Xenopus oocytes; the flickery, noisy events showed an estimated conductance of 49 pS in 100 mM NaCl, and were activated by 10 μM $HgCl_2$ (via interaction with Cys155 in M4 and Cys190 in Loop E, as identified by site-directed mutagenesis). Ion substitution of Cl^- by gluconate shifted the reversal potential approximately +10 mV; however, the authors concluded anions and cations were equally permeable. Prior work with voltage clamp recordings had supported the idea that Cl^- was more permeable than Na^+ (P_{Cl}/P_{Na} 3.6) in rat AQP6 (Yasui et al. 1999). The current model for AQP6 as an anion-conducting channel shows an intriguing resemblance to AQP1 in that the central pore, rather than the intrasubunit pore, is viewed as the pathway for ion conduction (Yamamoto et al. 2024), suggesting a coherent picture of general AQP structure–function features is emerging (see Sect. 12.4 below).

A third example of different mechanisms involved in AQP ion channel activation comes from *Drosophila* big brain (BIB), an insect AQP that was first discovered as a channel-like protein of unknown function encoded by a neurogenic gene

(Rao et al. 1990). BIB functions as a monovalent cation channel but interestingly showed no measurable water channel activity in oocytes expressing wild-type or terminal-deleted BIB constructs (Yanochko and Yool 2002). The monovalent conductance showed a permeability sequence of $K^+ > Na^+ \gg TEA^+$, and was blocked by divalent cations Ba^{2+} and Ca^{2+} but not Mg^{2+} (Yanochko and Yool 2004). BIB is regulated by patterns of tyrosine phosphorylation in the C-terminal domain. BIB ion currents were activated by tyrosine kinase inhibition, and conversely inactivated by tyrosine phosphorylation, consistent with a role for BIB as a signaling protein involved in the regulation of early nervous system development in the fly (Rao et al. 1990; Yanochko and Yool 2004).

AQP1 channel functions are modulated by phosphorylation. Tyrosine phosphorylation in the C-terminal domain increased the proportion of channels available to be gated as ion channels by cGMP (Campbell et al. 2012). Consensus sites for other kinases enabled opposite patterns of activation of AQP1 water versus ion channel permeabilities, mediated by protein kinase C (PKC) (Zhang et al. 2007). In that study, phorbol ester activation of PKC caused phosphorylation of AQP1 at consensus sites Thr 157 (located at the proximal end of Loop D) and Thr 239 (in the C-terminal domain), which increased the water permeability but decreased the amplitude of AQP1 cationic currents; neither response was seen in mutants lacking these Thr sites. In contrast, the AQP1 ion current response to cyclic nucleotides was not affected by deletion of PKC phosphorylation sites, consistent with a separate mechanism of action (Zhang et al. 2007), now known to involve direct cGMP binding at the Loop D domain. Some batches of AQP1-expressing oocytes displayed spontaneous activation of ionic conductance after being pricked by insertion of voltage clamp electrodes; this activation response (similar to the originally reported response to forskolin (Yool et al. 1996)) was blocked by pretreatment with the nonspecific kinase inhibitor H7, whereas cGMP-mediated activation was not altered by H7 treatment (Anthony et al. 2000). There are a number of possible mechanisms of action for the effects of phosphorylation state in controlling AQP1 function, which include stabilizing the ion or the water open state conformations, shifting affinity for cGMP, or adjusting the flexibility of Loop D.

12.4 Ionic Selectivity of AQP Conductances

A possible structure–functional basis for AQP ion channel selectivity is outlined in Fig. 12.3. AQP1 and BIB are cation-conducting AQPs that allow passage of a broad size range of monovalent cations, including K^+, Na^+, Cs^+, Li^+, and TEA^+ (Yool et al. 1996; Yanochko and Yool 2002; Pei et al. 2019). Ions moving through the AQP central pore are expected to transit with their associated hydration shells in place (Yu et al. 2006). The AQP ion pore lacks the pattern of amino acids imposing stacked carbonyl groups, which allow ion dehydration, a prerequisite for ion passage through narrow pores such as those in highly selective K^+ channels (Doyle et al. 1998). The filters in AQPs are needed only to create a generic preference for cations or anions, and thus are expected to be comparatively simple in design.

Insight into possible selectivity mechanisms comes from considering other nonselective ion channels, such as the excitatory nicotinic acetylcholine receptors and serotonin receptors, which conduct cations, and the inhibitory $GABA_A$ and glycine receptors, which conduct anions (Keramidas et al. 2004). In their review, Keramidas and colleagues summarized work showing that in anion-selective ligand-gated channels, the residues located at the inner boundary of the M2 transmembrane pore domains have a net positive charge; conversely, negatively charged residues are present at equivalent positions in the cation-selective ligand-gated channels. These rings of charge arrayed in the pentameric channel are accessible to permeating ions. An interesting detail for the nAChR and serotonin cation channels is that a positively charged Lys or Arg is located adjacent to the selectivity filter Glu residue, and viewed as being oriented away from the pore axis (Keramidas et

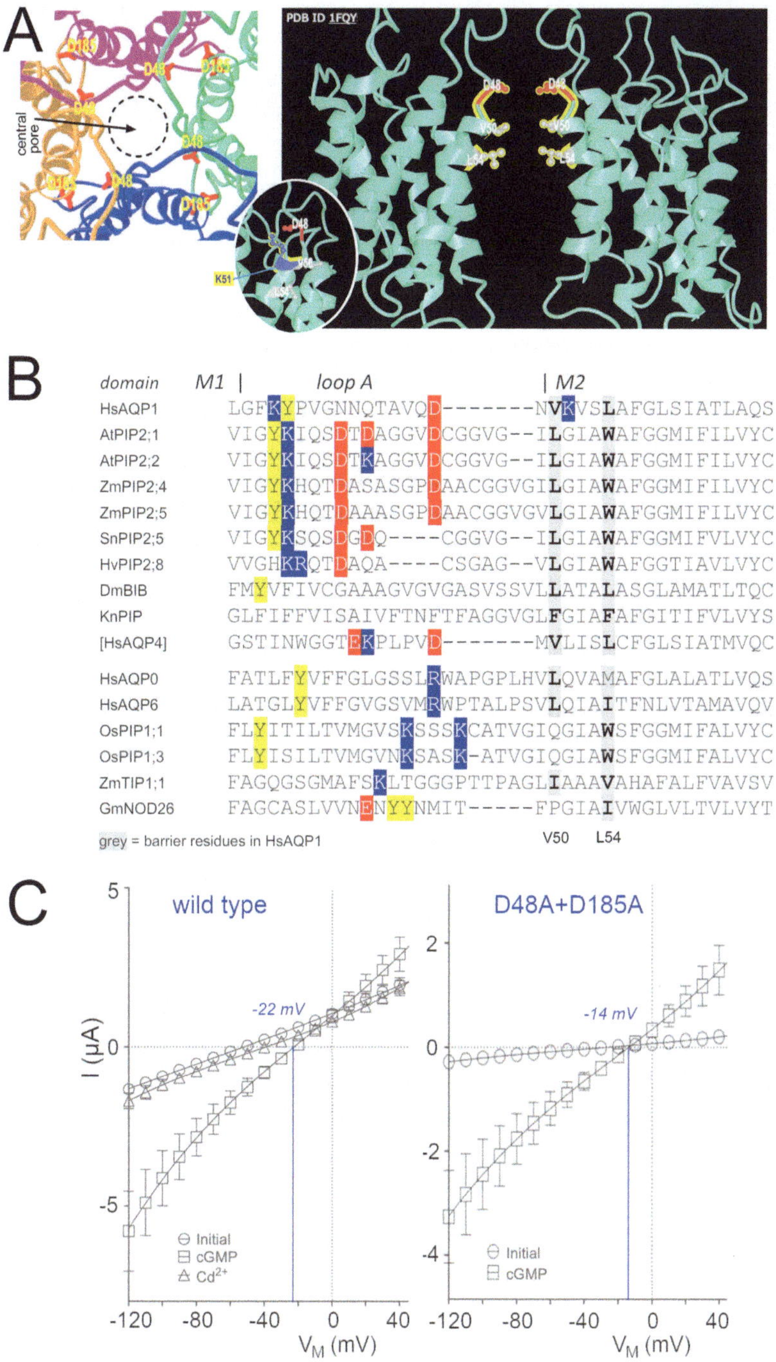

Fig. 12.3 Proposed model for the structural basis of ionic selectivity filters in aquaporin ion channels. (**A**) Images for an AQP1 crystal structure (PDB ID 1FQY) thought to be in an ion channel open-state-like conformation. Left: top view showing positions of two aspartate residues (D48 and D185) surrounding the central pore (dotted circle). Right: side view of M2 barrier residues Leu 54 and Val 50, with Asp 48 directly above in the permeation pathway. Oval inset: orientation showing the side chain of Lys 51 extending away from the central pore axis. (**B**) Amino acid sequence alignment of extracellular Loop A and the adjoining pore domain M2, showing positions of amino acids with negative (red), positive (blue) charges, and Tyr residues (yellow), for cation-selective AQPs (rows HsAQP1 to KnPIP), AQP4 (not shown to be an ion channel), and anion-selective AQPs (rows HsAQP0 to GmNOD26). (**C**) Current–voltage plots of ion currents recorded in oocytes expressing wild-type AQP1 (left) or a double mutant with Asp 48 and Asp 185 replaced by Ala (right) from Henderson et al (Henderson et al. 2023). (Panel A (left side) and Panel C graphs were used with permission from Elsevier, Copyright Clearance Center # 501959601, edited to mark x-axis positions of reversal potentials)

al. 2004). By analogy with ligand-gated channels, ion selectivity filters for ion channel AQPs would be predicted to involve clusters or rings of charged residues located near the internal or external vestibules of the central pore, creating a net charge opposite to that of the permeating ions.

The extracellular vestibule of human AQP1 has two negatively charged Asp residues in each of the subunits framing the central pore (Fig. 12.3a), and creating an eight-point ring of negative charges at the outer face (Henderson et al. 2023), with Asp 48 located near the external edge of the M2 pore lining domain, and Asp 185 near the external edge of the M5 pore lining domain. Considering a cation selectivity role for AQP1 D185, it is interesting to note that the equivalent position in human AQP6 channel is His 189, which is one of the pH-sensing residues for this channel, as noted below. His residues acquire a positive charge at acidic pH, in keeping with expectations for an appropriate residue in an anion-preferring pH-dependent channel. In the human AQP1 crystal structure, PDB ID 1FQY (Murata et al. 2000), chosen here because it was fortuitously captured with Loop D in an open-state-like conformation (Fig. 12.3a, right panel), the position of Asp 48 directly above the two central pore barrier residues Val 50 and Leu 54 suggests a position poised to influence ion permeation. Interestingly, the adjacent positively charged Lys at position 51 in human AQP1 appears to be oriented away from the pore axis (Fig. 12.3a, oval inset), which is reminiscent of a similar design in excitatory ligand-gated receptors (Keramidas et al. 2004), though perhaps just a coincidence. Justification of these interpretations awaits further investigation.

Amino acid sequence alignment of the M2 pore and adjacent loop regions of known AQP ion channels offers an inaugural opportunity to scan for differences in patterns of net charges in cation- versus anion-preferring subtypes identified thus far (Fig. 12.3b). Though speculative, ideas observed here could merit further testing. The cation-preferring channels (top rows, HsAQP1 to KnPIP) show a predominant pattern of negative Asp residues in the Loop A domain immediately upstream of the M2 pore region (fitting observed cation selectivity), apart from Drosophila Big Brain and algal KnPIP. Presumably, if the theory holds, other extracellular sites for BIB and KnPIP (such as Tyr in BIB and Asp in KnPIP, located near the outer edge of M5) or intracellular vestibule regions might provide negatively charged positions, but these remain to be identified. Human AQP4 has not been shown to be an ion channel; however, if an activating stimulus were identified, the pattern here predicts it would be cation-selective. The anion-preferring AQP channels (lower rows; HsAQP0 to GmNOD26) show a predominant pattern of positively charged Arg or Lys in Loop A close to M2, similarly in keeping with an anionic channel. The exception is GmNOD26, which has a negatively charged Glu residue, as well as tyrosine residues, which via their net negative charge of π electrons in aromatic rings can coordinate cation interactions (Gallivan and Dougherty 1999); no other candidates are evident immediately outside M5. The lack of obvious candidates for charge selectivity sites might be consistent with a low calculated ionic selectivity value of Nodulin-26 as compared to other anion AQPs; for Nodulin-26, the P_{Cl}/P_{K} ratio is 1.2, in contrast to P_{Cl}/P_{K} 1.8–1.9 in AQP0, and P_{Cl}/P_{Na} 3.6 in AQP6. Further studies will be required to see if the predictions suggested here by sequence alignments are met by experimental results.

Post hoc re-evaluation of prior work on human AQP1 that was modified by double mutagenesis of Asp 48 and Asp 185 to alanines (Henderson et al. 2023) might offer serendipitous support for the charged ring model of AQP ion selectivity, shown in Fig. 12.3. Though the study by Henderson and colleagues concluded that there were no significant differences in ionic conductance amplitudes between the wild type and D48A + D185A constructs expressed in oocytes, current–voltage plots in the same paper appear to show a positive shift in reversal potential from −22 mV in wild type to −14 mV in the Asp-deleted mutant (Fig. 12.3c), suggesting a reduction in ionic selectivity. Retesting these channel constructs and other charge site mutations with appropriate controls will be needed to confirm this proposal.

The anion-conducting channel AQP6 has now been analyzed with molecular dynamics modeling, providing fascinating results that have opened a new view of AQP6 structure–function properties, most notably the discovery that the anion-conducting pathway is routed through the central pore (Yamamoto et al. 2024), not the individual subunits. The proposed pH sensor is a pair of His residues located at the outer side of the M5 domain at 184 and 189 in human AQP6. Protonation of the His residues is proposed to cause opening of the central pore filter, formed by Leu 56, Ile 60, and Asn 63, located in the outer half of M2, allowing hydration of the central pore. Positions L56 and I60 in AQP6 are directly equivalent to the AQP1 central barrier residues V50 and L54; in contrast, the third position N63 in AQP6 is a glycine in AQP1 (Fig. 12.3b). The Asn at position 63 in AQP6 is suggested to be the first site of interaction for Cl^- ions as they transit through the pore (Yamamoto et al. 2024). The overall design of AQP6 is strikingly similar to that of AQP1, with an interesting difference; in AQP6, the activating stimulus (acidic pH) comes from the extracellular side and opens the outer M2 barrier residues first, whereas in AQP1, the activating stimulus (cGMP) comes from the cytoplasmic side and opens the inner M5 barrier residues first.

It appears that the field is coalescing on a unified view of AQP ion channels that could launch an exciting new era of discoveries.

12.5 Physiological Roles of AQPs as Multisolute Transporters

AQP ion channels in plants and animals have essential roles in enabling growth, survival, and adaptation to physiological challenges. Structural features and mechanisms of regulation of AQP ion channel activities are being found to have important consequences for maintaining health and viability, but as a dramatic counterpoint, they also raise concerns when AQP multifunctional channels are co-opted in disease progression.

Roles of AQPs in promoting cell motility and influencing neural cell fate are beneficial for the natural processes of nervous system development. Colorfully described as creating "bulldozers," AQP1 channels expressed in the leading edges of chick embryo neural crest cells were shown to be essential for facilitating the timely migration of the patterned waves of cells from their point of origin above the neural tube to their target peripheral destinations (McLennan et al. 2020). Whether the migrating neural crest cells rely on AQP1 water channel activity alone or utilize additional cation channel functionality has not been explored. As reviewed in 2022, cGMP is involved in controlling the development of afferent sensory pathways, for example, flipping responses of dorsal root ganglion growth cones to guidance cues such as semaphorin (Schmidt et al. 2022), indicating contributions of the AQP1 ion channels to neural crest outgrowth are conceivable as part of an array of cGMP-sensitive pathways. Expression of the Big Brain channel in developing fruit flies is essential for neural cell fate determination by a signaling process of lateral inhibition, in which expression of the neurogenic gene product BIB directs cells into epithelial as opposed to neuronal fates, and *Dmbib* gene deficiency results in lethal overgrowth of the nervous system (Rao et al. 1990). BIB channels regulated by C-terminal Tyr phosphorylation state showed no water channel activity, but the observed cation channel function (Yanochko and Yool 2002) is consistent with records of 20 mV depolarization of resting membrane potentials measured in grasshopper embryos in cells destined for non-neuronal rather than neuroblast cell fates (Goodman and Spitzer 1979). AQP1 expression in other highly motile cells, such as endothelial cells in angiogenesis and leukocyte immune cells needed for protection and surveillance, suggests that a broad potential range of AQP ion channel roles in cell motility remain to be explored (Smith and Stroka 2023).

While AQP1 ion channel involvement in facilitating cell migration benefits normal physiology, it becomes a problem in the context of cancer invasiveness and metastasis. Better understanding the AQP ion channel roles in cancer pathologies could help identify novel therapeutic approaches. Using a Li^+-selective photoswitchable probe SHL (Sabrina Heng Lithium) preloaded in living HT29 human colon cancer cells, AQP1 ion channel activity was monitored in real time by confocal imaging, with activity detected as lithium hot spots (Pei et al. 2019). The hot spots transitioned slowly between on and off states, and were clustered in the leading edges of the motile cancer cells, in agreement with prior experimental work that linked AQP1 expression to accelerating the rates of cell migration in a variety of aggressive cancer cell lines (Chow et al. 2020; De Ieso and Yool 2018; Khan et al. 2023), and implied relevance to mechanisms of metastasis. Both Li^{+} hot spots and cell migration were blocked by treatment with an AQP1 ion channel antagonist, AqB011 (Kourghi et al. 2016) and by siRNA knockdown of AQP1 (Pei et al. 2019). The brain cancer glioblastoma is a devastating disease associated with a high risk of death, typically within 2 years postdiagnosis, despite the best treatments available. The medical challenge results from the highly invasive nature of glioblastoma cells, a process that has been linked in experimental work in vitro to the activity of AQP1 ion channels (Varricchio et al. 2023; Varricchio and Yool 2023) located in the leading edges (McCoy and Sontheimer 2007). Agents such as arylsulfonamide AqB011, which block the AQP1 ion conductance by interaction with the Loop D gating domain (Kourghi et al. 2016), show promise in slowing glioblastoma cell migration and could be lead compounds for adjunct treatments aimed at holding the cancer cells in place, while prolonging the window of therapeutic opportunity for primary eradication treatments (Yool and Ramesh 2020).

The first example demonstrating that an aquaporin ionic conductance can have a physiologically relevant function was in the brain choroid plexus, in which natively expressed AQP1 channels generated a cGMP-dependent cation conductance that was capable of adjusting net rates of cerebral spinal fluid production (Boassa et al. 2006). Whole-cell patch electrophysiology of the choroid plexus cells confirmed permeability to Na^{+}, K^{+}, TEA^{+}, and Cs^{+}, voltage insensitivity, and cGMP dependence, with background K^{+} and Cl^{-} currents minimized by using Cs-methane sulfonate recording salines. Cation current responses to cGMP in primary cultures of rat choroid plexus were decimated after siRNA knockdown of AQP1 expression, which had no effect on native K^{+} or Cl^{-} currents. The AQP1 cation current was activated by stimulation with atrial natriuretic peptide (ANP), which works via endogenous receptor guanylate cyclase receptors; ANP caused a decrease in the rate of basal-to-apical fluid transport, an effect blocked by the AQP1 ion channel antagonist Cd^{2+}, which restored normal fluid transport rates (Boassa et al. 2006). Evidence for AQP1 ion channel function in choroid plexus provides a mechanism consistent with the known effects of ANP in slowing cerebral spinal fluid production in vivo (Johanson et al. 2006). It is possible that cation entry through AQP1 channels results in local accumulation of Na^{+} in the internal vestibule region that could reduce osmotically driven fluid efflux through the water pores, essentially putting the brakes on fluid excretion.

AQP1 ion channels in the retinal pigmented epithelium (RPE) of the eye provide a mechanism for fluid absorption from the apical-to-basolateral side (Baetz et al. 2012) in a process that helps protect against fluid accumulation in the subretinal space, a factor in retinal detachment diseases (Stamer et al. 2003). In monolayer cultures of human RPE, apical-to-basolateral fluid transport measured in a filter chamber was stimulated by treatment with a membrane-permeable cGMP agonist, and by receptor-mediated activation of cGMP by ANP; these responses were blocked by the AQP1 antagonist AqB013 but not the kinase inhibitor H-8, consistent with direct binding of cGMP to the target of action (Baetz et al. 2012) in the facilitation of fluid uptake by AQP1. It is possible that AQP1-mediated cation accumulation in the internal vestibule region could augment osmotically driven fluid influx through the water pores, essentially promoting fluid absorption.

AQP1 ion channels in red blood cells appear to be involved in the dynamic regulation of membrane compliance, a property that is essential as these 8 μm diameter cells course through the blood stream and must compress repeatedly, on average once a minute over their 4-month life-spans, to fit through the narrow diameters of capillaries (5 μm) in the microvasculature. The AQP1 ion conductance in human red blood cells was discovered to be an unexpected cotarget for the therapeutic actions of the furan compound 5-hydroxymethyl furfural and related derivatives in human sickle cell erythrocytes (Chow et al. 2021). Protection from the morphological stiffening (sickling) of red blood cells resulted from a reduction in a mysterious cation leak current first described in 1993 as "P_{sickle}"" (Joiner 1993) and now identified as AQP1, which, when blocked, protects cells from sickling (Chow et al. 2021), an effect augmented by the parallel complementary action of furan drugs in reducing mutant hemoglobin aggregation.

Plants use AQP ion channels to adapt to drought stress conditions. A nonselective cation current described as a Ca^{2+}- and pH-sensitive pathway for Na^+ entry into Arabidopsis roots was found in 2002 (Demidchik and Tester 2002), but the molecular identity of this channel remained unknown for years, until 2017, when AtPIP2;1 was identified as a dual water-and-ion-channel aquaporin with the same properties of localization, regulation, and function (Byrt et al. 2017), solving the mystery. AtPIP2;1 is regulated by phosphorylation at Ser residues in the C-terminal domain; mutations at the consensus sites to Asp (mimicking phosphorylation) were associated with a high ion conductance and low water permeability, whereas replacement by Ala (mimicking dephosphorylation) favored high water permeation and a low ion conductance (Qiu et al. 2020). Other plant AQPs, including AtPIP2;2, also carry nonselective cation conductances (Qiu et al. 2020; Byrt et al. 2017; Kourghi et al. 2017). Maize ZmPIP2;5 is a cation channel permeable to Li^+, illustrated using the SHL photoswitchable probe for optical monitoring of Li^+ entry across root cell membranes (Ahmed et al. 2024). Expression levels of the anion-conducting AQP channel OsPIP1;3, when boosted in roots of rice plants in response to drought stress, enhance the capacity for transporting nitrate anions in addition to water (Liu et al. 2020).

In summary, there is an impressive range of functional roles for AQP ion channels being uncovered, including the identification of new roles in health and disease, as well as the resolution of old mysteries around leak currents that have eluded molecular identification, even though they are recognized as key components of physiological responses.

12.6 Conclusion and Outlook

Aquaporins, once thought to be permeable to water alone, are being acknowledged as having complex repertoires of permeable substrates and diverse arrays of specializations and mechanisms of regulation. Physiological relevance of the AQP ion channels is highlighted by research advances that are defining the ion channel functions and the structural mechanisms that mediate their multi-functional roles. An adaptive benefit of AQP ion channels through evolution appears evident in the fact that amino acid sequences of ion channel pore and gating domains are highly conserved within given classes of aquaporins across species. AQP ion channels are key contributors to diverse processes well beyond fluid homeostasis alone. More classes of aquaporin ion channels are likely to be discovered pending identification of the signaling pathways that govern activation. A unified view of structure–function properties of AQP ion channels is emerging. Further work is anticipated to continue to challenge assumptions, expand the breadth of documented roles for these surprisingly sophisticated proteins, and spark new interest in AQP research across all the domains of life that could open a wealth of unexpected discoveries.

References

Adams ME, Mueller HA, Froehner SC (2001) In vivo requirement of the alpha-syntrophin PDZ domain for the sarcolemmal localization of nNOS and aquaporin-4. J Cell Biol 155(1):113–122. https://doi.org/10.1083/jcb.200106158

Agre P, Preston GM, Smith BL, Jung JS, Raina S, Moon C et al (1993a) Aquaporin CHIP: the archetypal molecular water channel. Am J Phys 265(4 Pt 2):F463–F476. https://doi.org/10.1152/ajprenal.1993.265.4.F463

Agre P, Sasaki S, Chrispeels MJ (1993b) Aquaporins: a family of water channel proteins. Am J Phys 265(3 Pt 2):F461. https://doi.org/10.1152/ajprenal.1993.265.3.F461

Ahmed J, Ismail A, Ding L, Yool AJ, Chaumont F (2024) A new method to measure aquaporin-facilitated membrane diffusion of hydrogen peroxide and cations in plant suspension cells. Plant Cell Environ 47(2):527–539. https://doi.org/10.1111/pce.14763

Amiry-Moghaddam M, Frydenlund DS, Ottersen OP (2004) Anchoring of aquaporin-4 in brain: molecular mechanisms and implications for the physiology and pathophysiology of water transport. Neuroscience 129(4):999–1010. https://doi.org/10.1016/j.neuroscience.2004.08.049

Anthony TL, Brooks HL, Boassa D, Leonov S, Yanochko GM, Regan JW et al (2000) Cloned human aquaporin-1 is a cyclic GMP-gated ion channel. Mol Pharmacol 57(3):576–588. https://doi.org/10.1124/mol.57.3.576

Baetz NW, Stamer WD, Yool AJ (2012) Stimulation of aquaporin-mediated fluid transport by cyclic GMP in human retinal pigment epithelium in vitro. Invest Ophthalmol Vis Sci 53(4):2127–2132. https://doi.org/10.1167/iovs.11-8471

Beitz E, Becker D, von Bulow J, Conrad C, Fricke N, Geadkaew A et al (2009) In vitro analysis and modification of aquaporin pore selectivity. Handb Exp Pharmacol 190:77–92. https://doi.org/10.1007/978-3-540-79885-9_4

Boassa D, Yool AJ (2003) Single amino acids in the carboxyl terminal domain of aquaporin-1 contribute to cGMP-dependent ion channel activation. BMC Physiol 3:12. https://doi.org/10.1186/1472-6793-3-12

Boassa D, Stamer WD, Yool AJ (2006) Ion channel function of aquaporin-1 natively expressed in choroid plexus. J Neurosci 26(30):7811–7819. https://doi.org/10.1523/jneurosci.0525-06.2006

Byrt CS, Zhao M, Kourghi M, Bose J, Henderson SW, Qiu J et al (2017) Non-selective cation channel activity of aquaporin AtPIP2;1 regulated by Ca^{2+} and pH. Plant Cell Environ 40(6):802–815. https://doi.org/10.1111/pce.12832

Campbell EM, Birdsell DN, Yool AJ (2012) The activity of human aquaporin 1 as a cGMP-gated cation channel is regulated by tyrosine phosphorylation in the carboxyl-terminal domain. Mol Pharmacol 81(1):97–105. https://doi.org/10.1124/mol.111.073692

Chandy G, Zampighi GA, Kreman M, Hall JE (1997) Comparison of the water transporting properties of MIP and AQP1. J Membr Biol 159(1):29–39. https://doi.org/10.1007/s002329900266

Chow PH, Kourghi M, Pei JV, Nourmohammadi S, Yool AJ (2020) 5-hydroxymethyl-furfural and structurally related compounds block the ion conductance in human Aquaporin-1 channels and slow cancer cell migration and invasion. Mol Pharmacol 98(1):38–48. https://doi.org/10.1124/mol.119.119172

Chow PH, Cox CD, Pei JV, Anabaraonye N, Nourmohammadi S, Henderson SW et al (2021) Inhibition of the Aquaporin-1 cation conductance by selected furan compounds reduces red blood cell sickling. Front Pharmacol 12:794791. https://doi.org/10.3389/fphar.2021.794791

Cowan CA, Yokoyama N, Bianchi LM, Henkemeyer M, Fritzsch B (2000) EphB2 guides axons at the midline and is necessary for normal vestibular function. Neuron 26(2):417–430. https://doi.org/10.1016/s0896-6273(00)81174-5

De Ieso ML, Yool AJ (2018) Mechanisms of aquaporin-facilitated cancer invasion and metastasis. Front Chem 6:135. https://doi.org/10.3389/fchem.2018.00135

Demidchik V, Tester M (2002) Sodium fluxes through nonselective cation channels in the plasma membrane of protoplasts from Arabidopsis roots. Plant Physiol 128(2):379–387. https://doi.org/10.1104/pp.010524

Doyle DA, Morais Cabral J, Pfuetzner RA, Kuo A, Gulbis JM, Cohen SL et al (1998) The structure of the potassium channel: molecular basis of K^+ conduction and selectivity. Science 280(5360):69–77. https://doi.org/10.1126/science.280.5360.69

Edamana S, Login FH, Riishede A, Dam VS, Kirkegaard T, Nejsum LN (2024) The water channels aquaporin-1 and aquaporin-3 interact with and affect the cell polarity protein Scribble in 3D in vitro models of breast cancer. Am J Physiol Cell Physiol 327(5):C1323–C1c34. https://doi.org/10.1152/ajpcell.00094.2024

Ehring GR, Zampighi G, Horwitz J, Bok D, Hall JE (1990) Properties of channels reconstituted from the major intrinsic protein of lens fiber membranes. J Gen Physiol 96(3):631–664. https://doi.org/10.1085/jgp.96.3.631

Freites JA, Németh-Cahalan KL, Hall JE, Tobias DJ (2019) Cooperativity and allostery in aquaporin 0 regulation by Ca^{2+}. Biochim Biophys Acta Biomembr 1861(5):988–996. https://doi.org/10.1016/j.bbamem.2019.02.007

Gallivan JP, Dougherty DA (1999) Cation-pi interactions in structural biology. Proc Natl Acad Sci USA 96(17):9459–9464. https://doi.org/10.1073/pnas.96.17.9459

Girsch SJ, Peracchia C (1985) Lens cell-to-cell channel protein: I. Self-assembly into liposomes and permeability regulation by calmodulin. J Membr Biol 83(3):217–225. https://doi.org/10.1007/bf01868696

Goodman CS, Spitzer NC (1979) Embryonic development of identified neurones: differentiation from neuroblast to neurone. Nature 280(5719):208–214. https://doi.org/10.1038/280208a0

Gorin MB, Yancey SB, Cline J, Revel JP, Horwitz J (1984) The major intrinsic protein (MIP) of the bovine lens fiber membrane: characterization and structure based on cDNA cloning. Cell 39(1):49–59. https://doi.org/10.1016/0092-8674(84)90190-9

Gotfryd K, Mosca AF, Missel JW, Truelsen SF, Wang K, Spulber M et al (2018) Human adipose glycerol flux is regulated by a pH gate in AQP10. Nat Commun 9(1):4749. https://doi.org/10.1038/s41467-018-07176-z

Hazama A, Kozono D, Guggino WB, Agre P, Yasui M (2002) Ion permeation of AQP6 water channel protein. Single channel recordings after Hg^{2+} activation. J Biol

Chem 277(32):29224–29230. https://doi.org/10.1074/jbc.M204258200
Heller KB, Lin EC, Wilson TH (1980) Substrate specificity and transport properties of the glycerol facilitator of Escherichia coli. J Bacteriol 144(1):274–278. https://doi.org/10.1128/jb.144.1.274-278.1980
Henderson SW, Nourmohammadi S, Ramesh SA, Yool AJ (2022) Aquaporin ion conductance properties defined by membrane environment, protein structure, and cell physiology. Biophys Rev 14(1):181–198. https://doi.org/10.1007/s12551-021-00925-3
Henderson SW, Nakayama Y, Whitelaw ML, Bruning JB, Anderson PA, Tyerman SD et al (2023) Proteoliposomes reconstituted with human aquaporin-1 reveal novel single-ion-channel properties. Biophys Rep (N Y) 3(1):100100. https://doi.org/10.1016/j.bpr.2023.100100
Hille B (2001) Ion channels of excitable membranes. Sinauer Press, Massachusettes USA
Holm LM, Klaerke DA, Zeuthen T (2004) Aquaporin 6 is permeable to glycerol and urea. Pflugers Arch 448(2):181–186. https://doi.org/10.1007/s00424-004-1245-x
Jan LY, Jan YN (2012) Voltage-gated potassium channels and the diversity of electrical signalling. J Physiol 590(11):2591–2599. https://doi.org/10.1113/jphysiol.2011.224212
Johanson CE, Donahue JE, Spangenberger A, Stopa EG, Duncan JA, Sharma HS (2006) Atrial natriuretic peptide: its putative role in modulating the choroid plexus-CSF system for intracranial pressure regulation. Acta Neurochir Suppl 96:451–456. https://doi.org/10.1007/3-211-30714-1_92
Joiner CH (1993) Cation transport and volume regulation in sickle red blood cells. Am J Phys 264(2 Pt 1):C251–C270. https://doi.org/10.1152/ajpcell.1993.264.2.C251
Keramidas A, Moorhouse AJ, Schofield PR, Barry PH (2004) Ligand-gated ion channels: mechanisms underlying ion selectivity. Prog Biophys Mol Biol 86(2):161–204. https://doi.org/10.1016/j.pbiomolbio.2003.09.002
Khan S, Lokman NA, Oehler MK, Ricciardelli C, Yool AJ (2023) Reducing the invasiveness of low- and high-grade endometrial cancers in both primary human cancer biopsies and cell lines by the inhibition of Aquaporin-1 channels. Cancers (Basel) 15(18):4507. https://doi.org/10.3390/cancers15184507
Kitchen P, Salman MM, Pickel SU, Jennings J, Tornroth-Horsefield S, Conner MT et al (2019) Water channel pore size determines exclusion properties but not solute selectivity. Sci Rep 9(1):20369. https://doi.org/10.1038/s41598-019-56814-z
Kourghi M, Pei JV, De Ieso ML, Flynn G, Yool AJ (2016) Bumetanide derivatives AqB007 and AqB011 selectively block the Aquaporin-1 ion channel conductance and slow cancer cell migration. Mol Pharmacol 89(1):133–140. https://doi.org/10.1124/mol.115.101618
Kourghi M, Nourmohammadi S, Pei JV, Qiu J, McGaughey S, Tyerman SD et al (2017) Divalent cations regulate the ion conductance properties of diverse classes of aquaporins. Int J Mol Sci 18(11):2323. https://doi.org/10.3390/ijms18112323
Kourghi M, De Ieso ML, Nourmohammadi S, Pei JV, Yool AJ (2018) Identification of loop D domain amino acids in the human Aquaporin-1 channel involved in activation of the ionic conductance and inhibition by AqB011. Front Chem 6:142. https://doi.org/10.3389/fchem.2018.00142
Liu S, Fukumoto T, Gena P, Feng P, Sun Q, Li Q, Matsumoto T, Kaneko T, Zhang H, Zhang Y, Zhong S, Zeng W, Katsuhara M, Kitagawa Y, Wang A, Calamita G, Ding X (2020) Ectopic expression of a rice plasma membrane intrinsic protein (OsPIP1;3) promotes plant growth and water uptake. Plant J. 102(4): 779–796. https://doi.org/10.1111/tpj.14662. Epub 2020 Feb 13
Ma T, Yang B, Kuo WL, Verkman AS (1996) cDNA cloning and gene structure of a novel water channel expressed exclusively in human kidney: evidence for a gene cluster of aquaporins at chromosome locus 12q13. Genomics 35(3):543–550. https://doi.org/10.1006/geno.1996.0396
McCoy E, Sontheimer H (2007) Expression and function of water channels (aquaporins) in migrating malignant astrocytes. Glia 55(10):1034–1043. https://doi.org/10.1002/glia.20524
McCoy ES, Haas BR, Sontheimer H (2010) Water permeability through aquaporin-4 is regulated by protein kinase C and becomes rate-limiting for glioma invasion. Neuroscience 168(4):971–981. https://doi.org/10.1016/j.neuroscience.2009.09.020
McLennan R, McKinney MC, Teddy JM, Morrison JA, Kasemeier-Kulesa JC, Ridenour DA et al (2020) Neural crest cells bulldoze through the microenvironment using Aquaporin 1 to stabilize filopodia. Development 147(1):dev185231. https://doi.org/10.1242/dev.185231
Modesto E, Lampe PD, Ribeiro MC, Spray DC, Campos de Carvalho AC (1996) Properties of chicken lens MIP channels reconstituted into planar lipid bilayers. J Membr Biol 154(3):239–249. https://doi.org/10.1007/s002329900148
Mulders SM, Preston GM, Deen PM, Guggino WB, van Os CH, Agre P (1995) Water channel properties of major intrinsic protein of lens. J Biol Chem 270(15):9010–9016. https://doi.org/10.1074/jbc.270.15.9010
Murata K, Mitsuoka K, Hirai T, Walz T, Agre P, Heymann JB et al (2000) Structural determinants of water permeation through aquaporin-1. Nature 407(6804):599–605. https://doi.org/10.1038/35036519
Nourmohammadi S, Henderson SW, Ramesh SA, Yool AJ (2024) Characterization of human aquaporin ion channels in a yeast expression system as a tool for novel ion channel discovery. Biosci Rep 44(8):BSR20240542. https://doi.org/10.1042/BSR20240542
Pei JV, Heng S, De Ieso ML, Sylvia G, Kourghi M, Nourmohammadi S et al (2019) Development of a photoswitchable lithium-sensitive probe to analyze nonselective cation channel activity in migrating cancer cells. Mol Pharmacol 95(5):573–583. https://doi.org/10.1124/mol.118.115428

Peracchia C, Girsch SJ (1985) Permeability and gating of lens gap junction channels incorporated into liposomes. Curr Eye Res 4(4):431–439. https://doi.org/10.3109/02713688509025157

Preston GM, Carroll TP, Guggino WB, Agre P (1992) Appearance of water channels in Xenopus oocytes expressing red cell CHIP28 protein. Science 256(5055):385–387. https://doi.org/10.1126/science.256.5055.385

Preston GM, Jung JS, Guggino WB, Agre P (1994) Membrane topology of aquaporin CHIP. Analysis of functional epitope-scanning mutants by vectorial proteolysis. J Biol Chem 269(3):1668–1673

Qiu J, McGaughey SA, Groszmann M, Tyerman SD, Byrt CS (2020) Phosphorylation influences water and ion channel function of AtPIP2;1. Plant Cell Environ 43(10):2428–2442. https://doi.org/10.1111/pce.13851

Rao Y, Jan LY, Jan YN (1990) Similarity of the product of the Drosophila neurogenic gene big brain to transmembrane channel proteins. Nature 345(6271):163–167. https://doi.org/10.1038/345163a0

Reizer J, Reizer A, Saier MH Jr (1993) The MIP family of integral membrane channel proteins: sequence comparisons, evolutionary relationships, reconstructed pathway of evolution, and proposed functional differentiation of the two repeated halves of the proteins. Crit Rev Biochem Mol Biol 28(3):235–257. https://doi.org/10.3109/10409239309086796

Ribeiro JC, Bernardino RL, Carrageta DF, Soveral G, Calamita G, Alves MG et al (2022) CFTR modulates aquaporin-mediated glycerol permeability in mouse Sertoli cells. Cell Mol Life Sci 79(12):592. https://doi.org/10.1007/s00018-022-04619-1

Saparov SM, Kozono D, Rothe U, Agre P, Pohl P (2001) Water and ion permeation of aquaporin-1 in planar lipid bilayers. Major differences in structural determinants and stoichiometry. J Biol Chem 276(34):31515–31520. https://doi.org/10.1074/jbc.M104267200

Schmidt H, Bottcher A, Gross T, Schmidtko A (2022) cGMP signalling in dorsal root ganglia and the spinal cord: various functions in development and adulthood. Br J Pharmacol 179(11):2361–2377. https://doi.org/10.1111/bph.15514

Smith IM, Stroka KM (2023) The multifaceted role of aquaporins in physiological cell migration. Am J Physiol Cell Physiol 325(1):C208–CC23. https://doi.org/10.1152/ajpcell.00502.2022

Stamer WD, Bok D, Hu J, Jaffe GJ, McKay BS (2003) Aquaporin-1 channels in human retinal pigment epithelium: role in transepithelial water movement. Invest Ophthalmol Vis Sci 44(6):2803–2808. https://doi.org/10.1167/iovs.03-0001

Strassmaier T, Karpen JW (2007) Novel N7- and N1-substituted cGMP derivatives are potent activators of cyclic nucleotide-gated channels. J Med Chem 50(17):4186–4194. https://doi.org/10.1021/jm0702581

Sui H, Han BG, Lee JK, Walian P, Jap BK (2001) Structural basis of water-specific transport through the AQP1 water channel. Nature 414(6866):872–878. https://doi.org/10.1038/414872a

Tajkhorshid E, Nollert P, Jensen M, Miercke LJ, O'Connell J, Stroud RM et al (2002) Control of the selectivity of the aquaporin water channel family by global orientational tuning. Science 296(5567):525–530. https://doi.org/10.1126/science.1067778

Tran STH, Horie T, Imran S, Qiu J, McGaughey S, Byrt CS et al (2020) A survey of barley PIP aquaporin ionic conductance reveals Ca^{2+}-sensitive HvPIP2;8 Na^{+} and K+ conductance. Int J Mol Sci 21(19):7135. https://doi.org/10.3390/ijms21197135

Tyerman SD, McGaughey SA, Qiu J, Yool AJ, Byrt CS (2021) Adaptable and multifunctional ion-conducting Aquaporins. Annu Rev Plant Biol 72:703–736. https://doi.org/10.1146/annurev-arplant-081720-013608

Varadaraj K, Kumari SS, Patil R, Wax MB, Mathias RT (2008) Functional characterization of a human aquaporin 0 mutation that leads to a congenital dominant lens cataract. Exp Eye Res 87(1):9–21. https://doi.org/10.1016/j.exer.2008.04.001

Varricchio A, Yool AJ (2023) Aquaporins and ion channels as dual targets in the design of novel glioblastoma therapeutics to limit invasiveness. Cancers (Basel) 15(3):849. https://doi.org/10.3390/cancers15030849

Varricchio A, Khan S, Price ZK, Davis RA, Ramesh SA, Yool AJ (2023) Pharmacological inhibition of membrane signaling mechanisms reduces the invasiveness of U87-MG and U251-MG glioblastoma cells in vitro. Cancers (Basel) 15(4):1027. https://doi.org/10.3390/cancers15041027

Weaver CD, Shomer NH, Louis CF, Roberts DM (1994) Nodulin 26, a nodule-specific symbiosome membrane protein from soybean, is an ion channel. J Biol Chem 269(27):17858–17862

Yamamoto E, Joo K, Lee J, Sansom MSP, Yasui M (2024) Molecular mechanism of anion permeation through aquaporin 6. Biophys J 123(16):2496–2505. https://doi.org/10.1016/j.bpj.2024.06.013

Yanochko GM, Yool AJ (2002) Regulated cationic channel function in Xenopus oocytes expressing Drosophila Big Brain. J Neurosci 22(7):2530–2540. https://doi.org/10.1523/jneurosci.22-07-02530.2002

Yanochko GM, Yool AJ (2004) Block by extracellular divalent cations of Drosophila Big Brain channels expressed in Xenopus oocytes. Biophys J 86(3):1470–1478. https://doi.org/10.1016/S0006-3495(04)74215-0

Yasui M (2009) pH regulated anion permeability of aquaporin-6. Handb Exp Pharmacol 190:299–308. https://doi.org/10.1007/978-3-540-79885-9_15

Yasui M, Hazama A, Kwon TH, Nielsen S, Guggino WB, Agre P (1999) Rapid gating and anion permeability of an intracellular aquaporin. Nature 402(6758):184–187. https://doi.org/10.1038/46045

Yool AJ, Ramesh S (2020) Molecular targets for combined therapeutic strategies to limit glioblastoma cell migration and invasion. Front Pharmacol 11:358. https://doi.org/10.3389/fphar.2020.00358

Yool AJ, Weinstein AM (2002) New roles for old holes: ion channel function in Aquaporin-1. News Physiol Sci 17:68–72. https://doi.org/10.1152/nips.01372.2001

Yool AJ, Stamer WD, Regan JW (1996) Forskolin stimulation of water and cation permeability in Aquaporin 1 water channels. Science 273(5279):1216–1218. https://doi.org/10.1126/science.273.5279.1216

Yu J, Yool AJ, Schulten K, Tajkhorshid E (2006) Mechanism of gating and ion conductivity of a possible tetrameric pore in Aquaporin-1. Structure 14(9):1411–1423. https://doi.org/10.1016/j.str.2006.07.006

Zampighi GA, Hall JE, Kreman M (1985) Purified lens junctional protein forms channels in planar lipid films. Proc Natl Acad Sci USA 82(24):8468–8472. https://doi.org/10.1073/pnas.82.24.8468

Zardoya R (2005) Phylogeny and evolution of the major intrinsic protein family. Biol Cell 97(6):397–414. https://doi.org/10.1042/bc20040134

Zelenina M, Zelenin S, Bondar AA, Brismar H, Aperia A (2002) Water permeability of Aquaporin-4 is decreased by protein kinase C and dopamine. Am J Physiol Renal Physiol 283(2):F309–F318. https://doi.org/10.1152/ajprenal.00260.2001

Zhang W, Zitron E, Hömme M, Kihm L, Morath C, Scherer D et al (2007) Aquaporin-1 channel function is positively regulated by protein kinase C. J Biol Chem 282(29):20933–20940. https://doi.org/10.1074/jbc.M703858200

Index

A. Rosenhouse-Dantsker (ed.), *Decoding Ion Channels Structure and Function*, Advances in Experimental Medicine and Biology 1497, https://doi.org/10.1007/978-3-032-07523-9

MIX
Papier aus verantwortungsvollen Quellen
Paper from responsible sources
FSC® C105338

If you have any concerns about our products,
you can contact us on
ProductSafety@springernature.com

In case Publisher is established outside the EU,
the EU authorized representative is:
Springer Nature Customer Service Center GmbH
Europaplatz 3, 69115 Heidelberg, Germany

Printed by Libri Plureos GmbH
in Hamburg, Germany